T0189909

LECTURES ON QUANTUM MECHANICS

LECTURE NOTES AND SUPPLEMENTS IN PHYSICS

John David Jackson and David Pines, *Editors*

Gabriel Barton	Introduction to Dispersion Techniques in Field Theory, 1965
Gordon Baym	Lectures on Quantum Mechanics, 1969 (3rd printing, with corrections, 1974)
Hans A. Bethe and Roman W. Jackiw	Intermediate Quantum Mechanics, 1968 (2nd printing, with corrections, 1973)
David Bohm	The Special Theory of Relativity, 1965
B. H. Bransden	Atomic Collision Theory, 1970
Willem Brouwer	Matrix Methods in Optical Instrument Design, 1964
R. Hagedorn	Relativistic Kinematics: A Guide to the Kinematic Problems of High-Energy Physics, 1964 (3rd printing, with corrections, Spring 1973)
John David Jackson	Mathematics for Quantum Mechanics: An Introductory Survey of Operators, Eigenvalues, and Linear Vector Spaces, 1962
Robert S. Knox and Albert Gold	Symmetry in the Solid State, 1964
K. Nishijima	Fields and Particles: Field Theory and Dispersion Relations, 1969
David Park	Introduction to Strong Interactions: A Lecture-Note Volume, 1966
David Pines	Elementary Excitations in Solids: Lectures on Phonons, Electrons, and Plasmons, 1964
R. D. Sard	Relativistic Mechanics: Special Relativity and Classical Particle Dynamics, 1970

LECTURES ON QUANTUM MECHANICS

GORDON BAYM
University of Illinois

CRC Press is an imprint of the
Taylor & Francis Group, an **informa** business

First published 1969 by Westview Press

Published 2018 by CRC Press
Taylor & Francis Group
6000 Broken Sound Parkway NW, Suite 300
Boca Raton, FL 33487-2742

CRC Press is an imprint of the Taylor & Francis Group, an informa business

Copyright © 1969, 1973, 1990 Taylor & Francis Group LLC

No claim to original U.S. Government works

This book contains information obtained from authentic and highly regarded sources. Reasonable efforts have been made to publish reliable data and information, but the author and publisher cannot assume responsibility for the validity of all materials or the consequences of their use. The authors and publishers have attempted to trace the copyright holders of all material reproduced in this publication and apologize to copyright holders if permission to publish in this form has not been obtained. If any copyright material has not been acknowledged please write and let us know so we may rectify in any future reprint.

Except as permitted under U.S. Copyright Law, no part of this book may be reprinted, reproduced, transmitted, or utilized in any form by any electronic, mechanical, or other means, now known or hereafter invented, including photocopying, microfilming, and recording, or in any information storage or retrieval system, without written permission from the publishers.

For permission to photocopy or use material electronically from this work, please access www. copyright.com (http://www.copyright.com/) or contact the Copyright Clearance Center, Inc. (CCC), 222 Rosewood Drive, Danvers, MA 01923, 978-750-8400. CCC is a not-for-profit organization that provides licenses and registration for a variety of users. For organizations that have been granted a photocopy license by the CCC, a separate system of payment has been arranged.

Trademark Notice: Product or corporate names may be trademarks or registered trademarks, and are used only for identification and explanation without intent to infringe.

Visit the Taylor & Francis Web site at
http://www.taylorandfrancis.com

and the CRC Press Web site at
http://www.crcpress.com

Library of Congress Catalog Card Number: 68-5611

ISBN 13: 978-0-8053-0667-5 (pbk)

PREFACE

These lecture notes comprise a three-semester graduate course in quantum mechanics given at the University of Illinois. There are a number of texts which present the basic topics very well; but since a fair quantity of the material discussed in my course was not available to the students in elementary quantum mechanics books, I was asked to prepare written notes. In retrospect these lecture notes seemed sufficiently interesting to warrant their publication in this format. The notes, presented here in slightly revised form, constitute a self-contained course in quantum mechanics from first principles to elementary relativistic one-particle mechanics. The student may want to look as well at one or more of the standard texts, such as K. Gottfried, *Quantum Mechanics* (W.A. Benjamin, New York, 1966); A. Messiah, *Quantum Mechanics* (North-Holland Publishing Company, Amsterdam, 1961), in two volumes; E. Merzbacher, *Quantum Mechanics* (John Wiley and Sons, New York, 1961); and for relativistic quantum mechanics, J. D. Bjorken and S. D. Drell, *Relativistic Quantum Mechanics* (McGraw-Hill Book Company, New York, 1964) and J. Sakurai, *Advanced Quantum Mechanics* (Addison-Wesley Publishing Company, Reading, Massachusetts, 1967). I occasionally refer in the notes to these books by author only. References to the Feynman Lectures are to R. P. Feynman, R. B. Leighton, and M. Sands, *The Feynman Lectures on Physics* (Addison-Wesley Publishing Company, Reading, Massachusetts, 1965), in three volumes.

Prerequisite to reading these notes is some familiarity with elementary quantum mechanics, at least at the undergraduate level; preferably the reader should already have met the uncertainty principle and the concept of a wave function. Mathematical prerequisites include sufficient acquaintance with complex variables to be able to do simple contour integrals and to understand words such as "poles" and "branch cuts." An elementary knowledge of Fourier transforms and series is necessary. I also assume an awareness of classical electrodynamics on the level of J. D. Jackson, *Classical Electrodynamics* (John Wiley and Sons, New York, 1962). Finally, I should mention that the figures in the notes are all sketches, not accurate graphs.

Among past and present colleagues I am particularly indebted, for their patient criticisms, discussions, and comments, to R.W. Hellwarth, L.P. Kadanoff, and D.G. Ravenhall; and to J.D. Jackson, who read the notes through and made many helpful suggestions about revising for publication. I am grateful to Mrs. Marie Brosig who typed the original notes, to W. F. Saam for proofreading this volume, and to Nina Baym, for substantial editorial assistance.

GORDON BAYM

Urbana, Illinois
January, 1968

CONTENTS

Chapter 1
PHOTON POLARIZATION

In order to become more familiar with the concepts and techniques of quantum mechanics, let us concentrate on one of the simplest quantum mechanical systems — one with which we are all very familiar — the polarization of light.

A classical light wave propagating in the z direction is described by the electric field vector

$$\mathbf{E}(\mathbf{r}, t) = \begin{pmatrix} E_x(\mathbf{r}, t) \\ E_y(\mathbf{r}, t) \\ 0 \end{pmatrix}, \tag{1-1}$$

since the electric (and magnetic) field vector is perpendicular to the direction of propagation of the light. In Gaussian units, $|\mathbf{H}| = |\mathbf{E}|$, $H_x = -E_y$, and $H_y = E_x$:

$$\mathbf{H} = \hat{\mathbf{z}} \times \mathbf{E}$$

Thus knowing E we know H.

Since electric fields are real we may write

$$E_x(\mathbf{r}, t) = E_x{}^0 \cos(kz - \omega t + \alpha_x)$$

$$E_y(\mathbf{r}, t) = E_y{}^0 \cos(kz - \omega t + \alpha_y), \tag{1-2}$$

where k is the wavenumber of the light,

$$k = \frac{2\pi}{\lambda}, \tag{1-3}$$

and ω is the angular frequency. α_x and α_y are the phases and E_x^0 and E_y^0 the (real) amplitudes of the electric field components.

It is often more convenient to use a complex notation for the electric field. We define the complex **E** vector by $\mathbf{E} = (E_x, E_y, 0)$ where

$$E_x = E_x^0 e^{i\alpha_x}, \qquad E_y = E_y^0 e^{i\alpha_y}, \tag{1-4}$$

and write

$$E_x(\mathbf{r}, t) = E_x e^{ikz - i\omega t}, \qquad E_y(\mathbf{r}, t) = E_y e^{ikz - i\omega t}, \tag{1-5}$$

remembering that in the end we should take the real part of the complex field.

The polarization state of the light is directly related to the **E** vectors. For example:

(i) if $E_y = 0$, the wave is plane polarized in the x direction.
(ii) if $E_x = 0$, the wave is plane polarized in the y direction.
(iii) if $E_x = E_y$, the wave is polarized at 45°, as in Fig. 1-1;
(iv) if $E_y = e^{i\pi/2} E_x = iE_x$, then the y component lags the x component by 90° and the wave is right circularly polarized:

$$\mathrm{Re}\, E_y(rt) \sim \cos(kz - \omega t + \tfrac{\pi}{2}), \qquad \mathrm{Re}\, E_x(rt) \sim \cos(kz - \omega t).$$

(v) similarly, if $E_y = -iE_x$, the wave is left circularly polarized.

Let us calculate the energy of the wave in terms of **E**. The energy density of an electromagnetic field is given, in Gaussian units, by

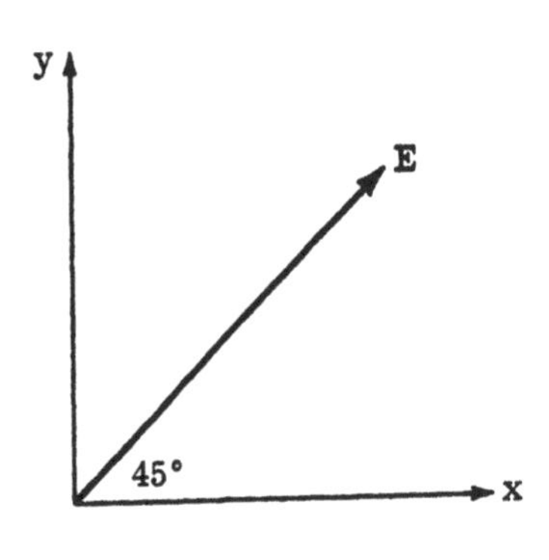

Fig. 1-1

The electric field for 45° polarization.

$$\varepsilon(\mathbf{rt}) = \frac{1}{8\pi}[(\mathrm{Re}\,E_x(\mathbf{rt}))^2 + (\mathrm{Re}\,E_y(\mathbf{rt}))^2 \qquad + (\mathrm{Re}\,H_x(\mathbf{rt}))^2 + (\mathrm{Re}\,H_y(\mathbf{rt}))^2]. \tag{1-6}$$

But for a plane light wave:

$$\mathrm{Re}\,H_x(\mathbf{rt}) = -\mathrm{Re}\,E_y(\mathbf{rt}), \qquad \mathrm{Re}\,H_y(\mathbf{rt}) = \mathrm{Re}\,E_x(\mathbf{rt}). \tag{1-7}$$

Thus

$$\varepsilon(\mathbf{rt}) = \frac{1}{4\pi}[(\mathrm{Re}\,E_x(\mathbf{rt}))^2 + (\mathrm{Re}\,E_y(\mathbf{rt}))^2]$$

$$= \frac{1}{4\pi}[|E_x|^2\cos^2(kz-\omega t+\alpha_x) + |E_y|^2\cos^2(kz-\omega t+\alpha_y)]. \tag{1-8}$$

Now let us assume that the wave occupies a volume V which is very many wavelengths long in the z direction. The total energy of the wave is thus

$$\varepsilon_{\text{total}} = \int_V d^3r\,\varepsilon(\mathbf{rt}) = \frac{1}{4\pi}\cdot\frac{V}{2}(|E_x|^2 + |E_y|^2) = \frac{|\mathbf{E}|^2}{8\pi}V, \tag{1-9}$$

and $|\mathbf{E}|^2/8\pi$ is the average energy per unit volume of the wave.

Consider what happens if we take a wave polarized at a 45° angle to the x axis and pass it through a *polaroid filter* that passes x polarized light, but not y polarized light. Then before the light passes through the polaroid

$$E_x = E_y = E \tag{1-10}$$

and after it passes through,

$$E_x = E, \qquad E_y = 0; \tag{1-11}$$

the beam comes out polarized in the x direction and its total energy is halved. The emerging electric field must be along $\hat{\mathbf{x}}$ since we know that a second such polaroid, whose axis is parallel to the first, would have no further effect on the beam.

Now let us reconsider the effect of the polaroid from a quantum mechanical point of view. First of all we know that in quantum mechanics the total energy of a wave of frequency ω cannot be arbitrary but must be an integral multiple of $\hbar\omega$:

$$\varepsilon_{\text{total}} = N\hbar\omega \tag{1-12}$$

where N is the number of photons in the wave. Thus when the energy of the wave is halved by the polaroid what must happen is that half the photons pass through and half don't. This is very weird — classically we would say that if any photon gets through, then since all photons are identical and all see identical conditions at the polaroid they should all pass through. The only way we can interpret what happens at the polaroid is to say that each photon has a *probability* one-half of passing through. We are really forced into a probabilistic point of view by the fact that the energy of electromagnetic radiation is quantized!

An immediate consequence of the fact that the passage of photons through the polaroid is governed by the laws of probability is that only rarely will *exactly* half the photons pass through. The mean number of photons passing through will be half the incident number — this must be the case if in the classical limit, i.e., when the beam consists of many, many photons, we are to recover from the probability laws the "deterministic" classical law for the passage of light through the polaroid. However, there will always be fluctuations about this mean number due to the finite size of $\hbar$. (The requirement that quantum mechanics yields the correct classical limit is called the *correspondence principle.)*

In general, to calculate the probability of the photon passing through the polaroid, we just have to ask for the fraction of the energy of a similar classical beam that is passed by the polaroid. This fraction is given by the ratio

$$\frac{|E_x|^2}{|E_x|^2 + |E_y|^2} = \frac{|E_x|^2}{|\mathbf{E}|^2} \tag{1-13}$$

For example, if the beam is polarized at an angle θ to the x axis, then $|E_x| = |\mathbf{E}| \cos\theta$. Hence a fraction $\cos^2\theta$ of the total energy passes through, and we would conclude that a single photon whose polarization vector was at an angle θ to the x axis would have a probability $\cos^2\theta$ of passing through an x-polaroid. Note that the photon, *if* it passes through, emerges, according to (1-11) polarized in the x direction!

Similarly, if we had a prism that passed only right circularly polarized light, then to calculate the fraction of the energy passed, we would write the beam as a coherent superposition of right circularly polarized light plus left circularly polarized light

$$\mathbf{E} = \mathbf{E}_{RCP} + \mathbf{E}_{LCP}; \tag{1-14}$$

Then the effect of the prism would be to throw away the $\mathbf{E}_{LCP}$ component and pass $\mathbf{E}_{RCP}$. The fraction of energy passed is thus

$$\frac{|\mathbf{E}_{RCP}|^2}{|\mathbf{E}_{RCP}|^2 + |\mathbf{E}_{LCP}|^2}, \tag{1-15}$$

and quantum mechanically we would interpret this fraction as the probability that one photon would pass through the prism, and emerge with right circular polarization.

Since all beams of light are superpositions of many beams consisting of one photon each, we shall turn our attention to the polarization properties of single photons. As we have already seen, it will be easy to discover the probability rules for one photon from our knowledge of the behavior of classical beams. The general laws of quantum mechanics are just generalizations of these rules.

For one photon, we have, from (1-9) and (1-12), that

$$|\mathbf{E}|^2 V = 8\pi\hbar\omega. \tag{1-16}$$

We shall define the *state vector* of the photon polarization

$$|\Psi\rangle = \begin{pmatrix} \psi_x \\ \psi_y \end{pmatrix}, \tag{1-17}$$

by writing

$$\psi_x = \sqrt{\frac{V}{8\pi\hbar\omega}}\, E_x, \qquad \psi_y = \sqrt{\frac{V}{8\pi\hbar\omega}}\, E_y. \tag{1-18}$$

The $|\Psi\rangle$ vectors are vectors in a *complex* two-dimensional space, since their components are complex numbers. From (1-16) it follows at once that $|\Psi\rangle$ has unit length:

$$|\psi_x|^2 + |\psi_y|^2 = 1. \tag{1-19}$$

In fact, the state vectors are independent of the volume V and depend only on the state of polarization of the photon. For example, if

$$|\Psi\rangle = \frac{1}{\sqrt{2}}\begin{pmatrix} e^{i\alpha} \\ e^{i\alpha} \end{pmatrix},$$

then the photon is polarized at 45° to the x axis. A knowledge of the $|\Psi\rangle$ vector gives us all the information we can have about the state of polarization of the photon.

Some special examples of these vectors are

$$|x\rangle = \begin{pmatrix} 1 \\ 0 \end{pmatrix} : \text{ x polarization}$$

$$|y\rangle = \begin{pmatrix} 0 \\ 1 \end{pmatrix} : \text{ y polarization}$$

$$|R\rangle = \frac{1}{\sqrt{2}}\begin{pmatrix} 1 \\ i \end{pmatrix} : \text{ right circular polarization}$$

$$|L\rangle = \frac{1}{\sqrt{2}}\begin{pmatrix} 1 \\ -i \end{pmatrix} : \text{ left circular polarization.}$$

Let us associate with each column vector $|\Psi\rangle$ a row vector $\langle\Psi|$ which we define by

$$\langle\Psi| = (\psi_x{}^* \quad \psi_y{}^*), \tag{1-20}$$

where * stands for complex conjugate. Also we shall define the *scalar product* of a row vector $\langle\Phi|$ and a column vector $|\Psi\rangle$ to be

$$\langle\Phi|\Psi\rangle = \phi_x{}^*\psi_x + \phi_y{}^*\psi_y = \langle\Psi|\Phi\rangle^*. \tag{1-21}$$

The normalization condition (1-19) on the $|\Psi\rangle$ vectors can thus be written as

$$\langle\Psi|\Psi\rangle = 1. \tag{1-22}$$

Clearly

$$\langle x|x\rangle = 1 = \langle y|y\rangle, \qquad \langle R|R\rangle = 1 = \langle L|L\rangle. \tag{1-23}$$

The vectors $|x\rangle$ and $|y\rangle$ are *orthogonal*, i.e., perpendicular, in the sense that

$$\langle x|y\rangle = 0. \tag{1-24}$$

They are said to form a *basis*, since any $|\Psi\rangle$ vector can be written as a linear superposition of them

$$|\Psi\rangle = \begin{pmatrix} \psi_x \\ \psi_y \end{pmatrix} = \psi_x|x\rangle + \psi_y|y\rangle. \tag{1-25}$$

Because they are orthogonal and satisfy the normalization condition (1-22), the basis they form is called *orthonormal.* Similarly, the set $|R\rangle$ and $|L\rangle$ form an orthonormal basis, since we can always write

$$|\Psi\rangle = \begin{pmatrix} \psi_x \\ \psi_y \end{pmatrix} = \frac{\psi_x - i\psi_y}{\sqrt{2}}|R\rangle + \frac{\psi_x + i\psi_y}{\sqrt{2}}|L\rangle. \tag{1-26}$$

If we take the scalar product of both sides of Eq. (1-25) with $\langle x|$ we see that

$$\langle x|\Psi\rangle = \psi_x\langle x|x\rangle + \psi_y\langle x|y\rangle = \psi_x. \tag{1-27}$$

Thus we can write (1-25) as

$$|\Psi\rangle = |x\rangle\langle x|\Psi\rangle + |y\rangle\langle y|\Psi\rangle, \tag{1-28}$$

and (1-26) as

$$|\Psi\rangle = |R\rangle\langle R|\Psi\rangle + |L\rangle\langle L|\Psi\rangle. \tag{1-29}$$

These equations are examples of the *superposition principle;* we can regard any arbitrary polarization as a coherent superposition of, e.g., x and y polarization states, or equivalently as a coherent superposition of right and left circularly polarized states.

Now let us return to the problem of passing a beam through an x polaroid. The classical rules tell us to regard the beam as a superposition of an x polarized beam and a y polarized beam, and that the effect of the polaroid is to throw away the y polarized component and pass only the x polarized component. The absolute value squared of the amplitude of the beam gives us its energy before it passes through, and the absolute square of its x component gives us its energy after it passes through. The fraction of the beam that passes through is given by (1-13). As we have seen, quantum mechanically this fraction gives us the probability of one photon with the initial polarization passing through the polaroid. Written in terms of $|\Psi\rangle$, (1-13) is:

$$\text{Probability} = \frac{|\psi_x|^2}{|\psi_x|^2 + |\psi_y|^2} = |\psi_x|^2 = |\langle x|\Psi\rangle|^2. \tag{1-30}$$

Thus $\langle x|\Psi\rangle$ is the amplitude of the x polarized component of $|\Psi\rangle$, and its absolute value squared is the probability that the photon in the state $|\Psi\rangle$ passes through the x polaroid. We call $\langle x|\Psi\rangle$ the *probability amplitude* for the photon to pass through the x polarizer.

Next consider passing light through a prism that passes right circular but rejects left circular. As we have said, to calculate the probability we write the beam as a coherent sum of right circularly polarized and left circularly polarized light, as in Eq. (1-14) or in terms of the state vectors, as in Eq. (1-29). Then $\langle R|\Psi\rangle$ is the amplitude of the component passed by the prism and, in accord with Eq. (1-15), $|\langle R|\Psi\rangle|^2$ is the probability that a photon in the state $|\Psi\rangle$ will pass through the prism.

The general rule is that if we have a prism that passes only light in the state $|\Phi\rangle$, rejecting light in states orthogonal to $|\Phi\rangle$, then the probability amplitude that a photon in the state $|\Psi\rangle$ will pass through the prism is

$$\langle\Phi|\Psi\rangle, \tag{1-31}$$

and the probability that the photon passes through is

$$|\langle\Phi|\Psi\rangle|^2. \tag{1-32}$$

Notice that this probability is independent of the phase of $|\Psi\rangle$ or $|\Phi\rangle$, though the probability amplitude depends on this phase.

TRANSFORMATION OF BASES

So far, in writing out $|\Psi\rangle$ vectors, as in Eq. (1-17), we have been using the x, y basis. We can equally well use any other basis in writing out the vectors. For example, if we use a basis x', y' related by an angle θ with respect to the old basis, as in Fig. 1-2, then we would write

$$|\Psi\rangle = \begin{pmatrix} \psi_{x'} \\ \psi_{y'} \end{pmatrix} = \begin{pmatrix} \langle x'|\Psi\rangle \\ \langle y'|\Psi\rangle \end{pmatrix}. \tag{1-33}$$

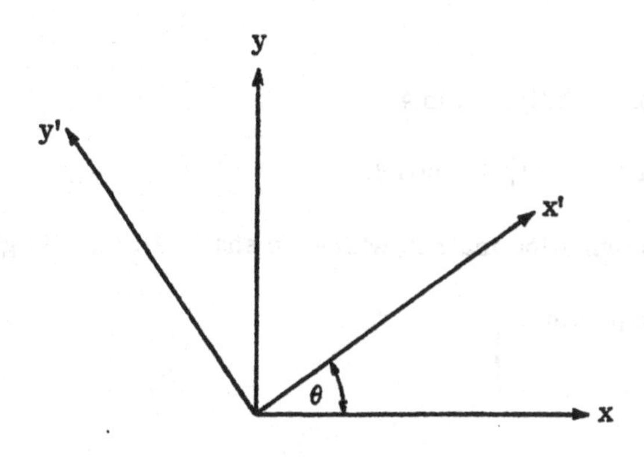

Fig. 1-2
Rotation of plane polarized basis by angle θ.

The problem we want to solve now is how are the components of the vector $|\Psi\rangle$ in the x', y' basis related to the components in the x, y basis. From Eq. (1-25) we can write

$$\langle x'|\Psi\rangle = \langle x'|x\rangle\langle x|\Psi\rangle + \langle x'|y\rangle\langle y|\Psi\rangle \tag{1-34}$$

and

$$\langle y'|\Psi\rangle = \langle y'|x\rangle\langle x|\Psi\rangle + \langle y'|y\rangle\langle y|\Psi\rangle. \tag{1-35}$$

Written in matrix language:

$$\begin{pmatrix} \langle x'|\Psi\rangle \\ \langle y'|\Psi\rangle \end{pmatrix} = \begin{pmatrix} \langle x'|x\rangle & \langle x'|y\rangle \\ \langle y'|x\rangle & \langle y'|y\rangle \end{pmatrix} \begin{pmatrix} \langle x|\Psi\rangle \\ \langle y|\Psi\rangle \end{pmatrix}. \tag{1-36}$$

Thus we know how to transform from the x, y basis to the x', y' basis as soon as we know the matrix in Eq. (1-36); we call this matrix the *transformation matrix* from the x, y basis to the x', y' basis. Notice that Eq. (1-36) is valid for any two bases, not just for the two linear polarized bases we have been considering.

For the particular example that x', y' is the plane polarized basis rotated through an angle θ from the x, y basis, we can write

$$|x\rangle = \cos\theta|x'\rangle - \sin\theta|y'\rangle$$

$$|y\rangle = \sin\theta|x'\rangle + \cos\theta|y'\rangle \tag{1-37}$$

so that

$$\langle x'|x\rangle = \cos\theta, \qquad \langle x'|y\rangle = \sin\theta,$$

$$\langle y'|x\rangle = -\sin\theta, \qquad \langle y'|y\rangle = \cos\theta, \tag{1-38}$$

and the transformation matrix, which we shall call $\mathcal{R}(\theta)$, is given by

$$\mathcal{R}(\theta) = \begin{pmatrix} \cos\theta & \sin\theta \\ -\sin\theta & \cos\theta \end{pmatrix}. \tag{1-39}$$

Thus

$$\begin{pmatrix} \langle x'|\Psi\rangle \\ \langle y'|\Psi\rangle \end{pmatrix} = \begin{pmatrix} \cos\theta & \sin\theta \\ -\sin\theta & \cos\theta \end{pmatrix} \begin{pmatrix} \langle x|\Psi\rangle \\ \langle y|\Psi\rangle \end{pmatrix}. \tag{1-40}$$

We can interpret this equation in either of two ways. First, it tells us the components of the $|\Psi\rangle$ vector in the rotated basis. On the other hand, it is completely equivalent to keep the vector fixed and rotate the basis, or to keep the basis fixed and rotate the vector in the opposite direction. Thus we can regard the vector on the left as a new vector $|\Psi'\rangle$ whose components in the x, y basis are the same as the components of $|\Psi\rangle$ in the x', y' basis, i.e.,

$$\langle x'|\Psi\rangle = \langle x|\Psi'\rangle,$$

$$\langle y'|\Psi\rangle = \langle y|\Psi'\rangle \tag{1-41}$$

When ψ_x and ψ_y are real, the vector $|\Psi'\rangle$ is the vector $|\Psi\rangle$ rotated clockwise by θ, as in Fig. 1-3. Written in terms of $\mathcal{R}(\theta)$,

$$|\Psi'\rangle = \mathcal{R}(\theta)|\Psi\rangle. \tag{1-42}$$

In general, if we transform a vector $|\Psi\rangle$ by $\mathcal{R}(\theta)$, the new components of $|\Psi\rangle$ will bear little resemblance to the old components. Let us ask if there are any state vectors that when operated upon by $\mathcal{R}(\theta)$ are at most multiplied by a constant

$$\mathcal{R}(\theta)|\Psi\rangle = c|\Psi\rangle. \tag{1-43}$$

If a vector satisfies a relation like this it is called an *eigenvector* of $\mathcal{R}(\theta)$, and the number c is called the *eigenvalue* of $\mathcal{R}(\theta)$ belonging to

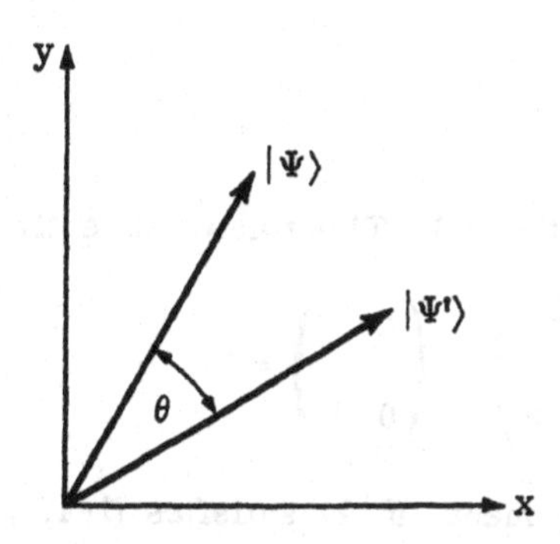

Fig. 1-3
Rotating the basis by angle θ is equivalent to rotating the vectors by angle θ in the opposite direction.

the eigenvector $|\Psi\rangle$. We shall see that the concept of eigenvectors and eigenvalues plays a very important role in quantum mechanics.

To find the eigenvectors and eigenvalues of $\mathcal{R}(\theta)$, let us write it as

$$\mathcal{R}(\theta) = \cos\theta \begin{pmatrix} 1 & 0 \\ 0 & 1 \end{pmatrix} + i \sin\theta \begin{pmatrix} 0 & -i \\ i & 0 \end{pmatrix}. \tag{1-44}$$

Let

$$S = \begin{pmatrix} 0 & -i \\ i & 0 \end{pmatrix}, \tag{1-45}$$

in the x, y basis. Then we can write the rotation matrix $\mathcal{R}(\theta)$ as

$$\mathcal{R}(\theta) = \cos\theta + iS \sin\theta, \tag{1-46}$$

where the unit matrix $\begin{pmatrix} 1 & 0 \\ 0 & 1 \end{pmatrix}$ is understood to multiply the $\cos\theta$.

We first find the eigenvectors and eigenvalues of S and then use Eq. (1-46).

Now if

$$S|\Psi\rangle = \lambda|\Psi\rangle, \tag{1-47}$$

it is easy to see that $\lambda^2 = 1$. This follows directly from the fact that

$$S^2 = \begin{pmatrix} 0 & -i \\ i & 0 \end{pmatrix}\begin{pmatrix} 0 & -i \\ i & 0 \end{pmatrix} = \begin{pmatrix} 1 & 0 \\ 0 & 1 \end{pmatrix} = 1, \tag{1-48}$$

by the following argument: If $|\Psi\rangle$ satisfies (1-47), then

$$|\Psi\rangle = S^2|\Psi\rangle = S(S|\Psi\rangle) = S(\lambda|\Psi\rangle) = \lambda S|\Psi\rangle = \lambda^2|\Psi\rangle, \tag{1-49}$$

and therefore $\lambda^2 = 1$ and

$$\lambda = \pm 1. \tag{1-50}$$

These are the eigenvalues of S. Now let us find the eigenvectors. The eigenvector corresponding to the eigenvalue +1 satisfies $S|\Psi\rangle = |\Psi\rangle$. We can find the eigenvector by inspection; it is just the vector for right circularly polarized light

$$|R\rangle = \frac{1}{\sqrt{2}}\begin{pmatrix} 1 \\ i \end{pmatrix},$$

as may be verified by a trivial calculation. Thus

$$S|R\rangle = |R\rangle \tag{1-51}$$

and similarly we find

$$S|L\rangle = -|L\rangle, \tag{1-52}$$

so that the eigenvector corresponding to the eigenvalue -1 is $|L\rangle$, the vector for left circularly polarized light.

It follows at once that $|R\rangle$ and $|L\rangle$ are also the eigenvectors of $\mathcal{R}(\theta)$ since

$$\begin{aligned}\mathcal{R}(\theta)|R\rangle &= (\cos\theta + i\sin\theta\, S)|R\rangle = (\cos\theta + i\sin\theta)|R\rangle \\ &= e^{i\theta}|R\rangle\end{aligned} \tag{1-53}$$

and

$$\mathcal{R}(\theta)|L\rangle = (\cos\theta + i\sin\theta S)|L\rangle = (\cos\theta - i\sin\theta)|L\rangle$$
$$= e^{-i\theta}|L\rangle. \quad (1\text{-}54)$$

Thus the components of the vectors for right and left circularly polarized light are changed only by a phase factor under a rotation of the basis. Are $|R\rangle$ and $|L\rangle$ the only eigenvectors of $\mathcal{R}(\theta)$? The answer is yes, as we can see from the fact that we can expand any arbitrary $|\Psi\rangle$ in terms of $|R\rangle$ and $|L\rangle$

$$|\Psi\rangle = |R\rangle\langle R|\Psi\rangle + |L\rangle\langle L|\Psi\rangle.$$

Under a rotation then

$$|\Psi\rangle \to \mathcal{R}(\theta)|\Psi\rangle = \mathcal{R}(\theta)|R\rangle\langle R|\Psi\rangle + \mathcal{R}(\theta)|L\rangle\langle L|\Psi\rangle$$
$$= e^{i\theta}|R\rangle\langle R|\Psi\rangle + e^{-i\theta}|L\rangle\langle L|\Psi\rangle. \quad (1\text{-}55)$$

Thus we can describe the effect on any $|\Psi\rangle$ of a rotation through angle θ, by saying that the right circularly polarized component is multiplied by a phase factor $e^{i\theta}$ and the left circularly polarized component is multiplied by a phase factor $e^{-i\theta}$. Only if $|\Psi\rangle$ is $|R\rangle$ or $|L\rangle$ is $\mathcal{R}|\Psi\rangle$ a constant times $|\Psi\rangle$.

We shall call S the *spin operator for the photon.* If a photon is right circularly polarized, then its state is an *eigenstate* (which means the same as "eigenvector") of S with eigenvalue +1. We shall say here that such a photon has spin 1. Similarly, left circularly polarized photons have spin −1. Any other state of the photon is a linear superposition of a photon with spin 1 (right circularly polarized) and a photon with spin −1 (left circularly polarized). As seen head on, the electric field vector of right circularly polarized light rotates counterclockwise, while the electric field vector of left circularly polarized light rotates clockwise.

ANGULAR MOMENTUM

The concept of the spin of the photon, which we have introduced in terms of the behavior under rotation has a direct physical meaning; $\hbar$ times the spin of a photon is the component of its *angular momentum* in the z direction. We can see this connection, if we start from the expression for the angular momentum, $\mathbf{L}$, of the classical electromagnetic field[1]:

$$\mathbf{L} = \frac{1}{4\pi c} \int d^3r\, \mathbf{r} \times [\mathbf{E}(\mathbf{r},t) \times \mathbf{H}(\mathbf{r},t)]. \tag{1-56}$$

Let us consider a plane wave traveling in the z direction. Then $\mathbf{E}$ and $\mathbf{H}$ are in the x, y plane and $\mathbf{L}$ therefore has no component in the z direction. This result is only true for the mythical case of a wave filling all space; for a beam of finite diameter in the x, y plane, $\mathbf{E} \times \mathbf{H}$ must have a component in the x, y plane, near the surface of the beam, in order that the wave satisfy Maxwell's equations.[2] There is actually a z component of the integrand coming from the surface of the beam, and due to the $\mathbf{r}$ in the integrand the total contribution of this term is proportional to the radius of the beam, times the surface area, i.e., proportional to the volume of the beam. We isolate this term simply as follows. When $\nabla \cdot \mathbf{E}(\mathbf{r},t) = 0$, as it is for light, we may introduce the (transverse) vector potential $\mathbf{A}(\mathbf{r},t)$, in terms of which

$$\mathbf{H}(\mathbf{r},t) = \nabla \times \mathbf{A}(\mathbf{r},t), \qquad \mathbf{E}(\mathbf{r},t) = -\frac{1}{c}\frac{\partial \mathbf{A}(\mathbf{r},t)}{\partial t}. \tag{1-57}$$

Then substituting for $\mathbf{H}$ in (1-56), integrating by parts, and assuming the wave to be of finite extent, so that the surface terms at infinity are zero, we find:

$$\mathbf{L} = \frac{1}{4\pi c} \int d^3r\, (\mathbf{E}(\mathbf{r},t) \times \mathbf{A}(\mathbf{r},t)) + \frac{1}{4\pi c}\int d^3r \sum_i E_i(\mathbf{r},t)(\mathbf{r} \times \nabla) A_i(\mathbf{r},t). \tag{1-58}$$

For a plane wave beam traveling in the z direction

$$A_x(\mathbf{r},t) = \frac{c}{\omega} E_x{}^0 \sin(kz - \omega t + \alpha_x),$$

[1] See J.D. Jackson, *Classical Electrodynamics* (John Wiley and Sons, New York, 1962).

[2] This is a famous problem; see J.M. Jauch and F. Rohrlich, *The Theory of Photons and Electrons* (Addison-Wesley, Cambridge, 1955), p. 34, for a detailed bibliography on the problem.

$$A_y(\mathbf{r}, t) = \frac{c}{\omega} E_y{}^0 \sin(kz - \omega t + \alpha_y) \tag{1-59}$$

except near the edges of the beam. Then the z component of the second term in (1-58) can be neglected. The reason is that $(\mathbf{r} \times \nabla)_z = \partial/\partial\varphi$ [where φ is the polar angle in cylindrical coordinates with axis along $\hat{\mathbf{z}}$] and thus only the surface of the beam contributes; the contribution is proportional to the surface area, not the volume. The z component of the angular momentum of the beam is thus

$$L_z = \frac{1}{4\pi c} \int d^3r \, [E_x(\mathbf{r}, t) A_y(\mathbf{r}, t) - E_y(\mathbf{r}, t) A_x(\mathbf{r}, t)],$$

and from (1-59) we find:

$$\begin{aligned} L_z &= \frac{1}{4\pi\omega} E_x{}^0 E_y{}^0 \int d^3r \, \sin(\alpha_y - \alpha_x) \\ &= \frac{V}{4\pi\omega} E_x{}^0 E_y{}^0 \sin(\alpha_y - \alpha_x). \end{aligned} \tag{1-60}$$

We have been using a real notation in these last few equations. Now let us write (1-60) in terms of the complex $\mathbf{E}$ vector that we have previously been using. Then

$$\begin{aligned} L_z &= \frac{V}{8\pi i\omega} E_x{}^0 E_y{}^0 (e^{i\alpha_y} e^{-i\alpha_x} - e^{-i\alpha_y} e^{i\alpha_x}) \\ &= \frac{V}{8\pi i\omega} (E_x{}^* E_y - E_x E_y{}^*) \\ &= \frac{V}{8\pi\omega} \left(\left| \frac{E_x - iE_y}{\sqrt{2}} \right|^2 - \left| \frac{E_x + iE_y}{\sqrt{2}} \right|^2 \right). \end{aligned} \tag{1-61}$$

If we think of the beam as a superposition of a right and a left circularly polarized beam, then

$$\frac{E_x - iE_y}{\sqrt{2}} = E_{RCP}$$

is the right circularly polarized component, and

$$\frac{E_x + iE_y}{\sqrt{2}} = E_{LCP}$$

is the left circularly polarized component. Thus

$$L_z = \frac{V}{8\pi\omega}\left(|E_{RCP}|^2 - |E_{LCP}|^2\right). \tag{1-62}$$

Up to here we have been completely classical. Now to make contact with quantum mechanics, we consider a beam consisting of only one photon. Then expressing **E** in terms of $|\Psi\rangle$, [Eq. (1-18)], we find

$$\begin{aligned} L_z &= \hbar\left(\left|\frac{\psi_x - i\psi_y}{\sqrt{2}}\right|^2 - \left|\frac{\psi_x + i\psi_y}{\sqrt{2}}\right|^2\right) \\ &= \hbar\,|\langle R|\Psi\rangle|^2 - \hbar|\langle L|\Psi\rangle|^2. \end{aligned} \tag{1-63}$$

How do we interpret this formula? First of all, it is an experimental fact that if a photon traveling in the z direction is absorbed by matter, then the z component of the angular momentum of the matter either increases by $\hbar$ or decreases by $\hbar$. It never remains the same, nor does it ever change by a value other than $\pm\hbar$. Furthermore, one cannot predict with certainty whether the transfer will be plus or minus $\hbar$. At best we can only predict the probabilities of it being plus or minus $\hbar$. We must interpret Eq. (1-63), therefore, as giving us the *average* value of the angular momentum transfer, averaged over many experiments in which we use a photon in the same polarization state $|\Psi\rangle$ each time. This interpretation is completely analogous to the probabilistic interpretation we gave to the classical formula for the energy that passes through a polaroid filter, for the case that the beam consists of only one photon.

The average value of the angular momentum transfer is $\hbar$ times the probability that the photon transfers angular momentum $\hbar$, plus $-\hbar$ times the probability that the photon transfers angular momentum $-\hbar$. If we then look at Eq. (1-63) for the average value of the angular momentum transfer, we see that $|\langle R|\Psi\rangle|^2$ is the probability that the z component of the angular momentum of the photon is $+\hbar$, and $|\langle L|\Psi\rangle|^2$ is the probability that it is $-\hbar$. In the special case that the photon is right circularly polarized, then $L_z = \hbar$ and we can say that the photon *definitely* has angular momentum $\hbar$. Similarly if the photon is left circularly polarized, then $L_z = -\hbar$, and we can say that the photon *definitely* has angular momentum $-\hbar$. In either case, the z component of the angular momentum is $\hbar$ times the spin value of the photon.

In general, if the photon is neither pure right nor pure left circularly polarized, we can't assign a definite value to its angular momentum; we can only speak of the probability of the angular momentum of the photon being observed to have value $\hbar$, and the probability of its being observed to have value $-\hbar$. As with the energy, the discreteness of the possible values of the angular momentum leads us into a probabilistic description, if we are to recover the classical description in the limit of a large number of photons.

Let us summarize our description of the z component of the angular momentum of photon. Corresponding to this physical quantity we have an *operator,* i.e., a matrix, $\hbar S$. Photons that are in an *eigenstate,* $|R\rangle$ or $|L\rangle$, of this operator can be assigned a definite value of the z component of angular momentum. This value is just the *eigenvalue* corresponding to the eigenstate that the photon is in. The eigenvalues of $\hbar S$ give the possible values of the z component of angular momentum of the photon. Any other photon state $|\Psi\rangle$ cannot be assigned a definite value of angular momentum, only probabilities for having the possible values. The probability of having a given value is calculated by taking the scalar product of $|\Psi\rangle$ with the eigenstate corresponding to this eigenvalue and then taking the absolute values squared of this product, e.g., $|\langle R|\Psi\rangle|^2$.

This connection between physical quantities and operators, the eigenvalues and the possible values of the physical quantities, and the eigenstates and the probabilities of having one of the possible values, is true in all cases in quantum mechanics, not merely for the particular case we have considered. It is the basis for calculating the expected results of measurements made on quantum mechanical systems.

We notice that the average value of L_z for a given state $|\Psi\rangle$, which is usually called the *expectation value* in that state and is written $\langle L_z\rangle$, is simply given by

$$\langle L_z\rangle = \langle\Psi|\hbar S|\Psi\rangle.$$

To prove this formula we note that from Eq. (1-29)

$$S|\Psi\rangle = S(|R\rangle\langle R|\Psi\rangle + |L\rangle\langle L|\Psi\rangle)$$

$$= |R\rangle\langle R|\Psi\rangle - |L\rangle\langle L|\Psi\rangle,$$

using Eqs. (1-51) and (1-52). Thus

$$\langle\Psi|S|\Psi\rangle = \langle\Psi|R\rangle\langle R|\Psi\rangle - \langle\Psi|L\rangle\langle L|\Psi\rangle, \tag{1-64}$$

so that

$$\hbar\,|\langle R|\Psi\rangle|^2 - \hbar\,|\langle L|\Psi\rangle|^2 = \langle\Psi|\,\hbar S\,|\Psi\rangle. \tag{1-65}$$

We can make a great advance in notation if we define the *outer product*, $|\Psi\rangle\langle\Phi|$, of two vectors $|\Psi\rangle$ and $|\Phi\rangle$, which is a matrix, as follows

$$|\Psi\rangle\langle\Phi| \equiv \begin{pmatrix} \psi_x \\ \psi_y \end{pmatrix} (\phi_x^* \quad \phi_y^*) = \begin{pmatrix} \psi_x\phi_x^* & \psi_x\phi_y^* \\ \psi_y\phi_x^* & \psi_y\phi_y^* \end{pmatrix}. \tag{1-66}$$

For example,

$$\begin{aligned} |x\rangle\langle x| &= \begin{pmatrix} 1 \\ 0 \end{pmatrix} (1 \quad 0) = \begin{pmatrix} 1 & 0 \\ 0 & 0 \end{pmatrix} \\ |y\rangle\langle y| &= \begin{pmatrix} 0 \\ 1 \end{pmatrix} (0 \quad 1) = \begin{pmatrix} 0 & 0 \\ 0 & 1 \end{pmatrix} \\ |x\rangle\langle y| &= \begin{pmatrix} 1 \\ 0 \end{pmatrix} (0 \quad 1) = \begin{pmatrix} 0 & 1 \\ 0 & 0 \end{pmatrix} \\ |y\rangle\langle x| &= \begin{pmatrix} 0 \\ 1 \end{pmatrix} (1 \quad 0) = \begin{pmatrix} 0 & 0 \\ 1 & 0 \end{pmatrix}, \end{aligned} \tag{1-67}$$

etc. It is important to notice that if we operate with $|\Psi\rangle\langle\Phi|$ on a vector $|\Theta\rangle$, then the result is the vector $|\Psi\rangle$ times $\langle\Phi|\Theta\rangle$, that is,

$$(|\Psi\rangle\langle\Phi|)|\Theta\rangle = |\Psi\rangle\,(\langle\Phi|\Theta\rangle). \tag{1-68}$$

[One should verify this in detail using Eqs. (1-66) and (1-21).] Thus we needn't write the parentheses in Eq. (1-68), and we can interpret an expression like $|\Psi\rangle\langle\Phi|\Theta\rangle$ in either of the two senses in Eq. (1-68).

Using this outer product notation we can write

$$|x\rangle\langle x| + |y\rangle\langle y| = 1, \tag{1-69}$$

the unit matrix. This equation, which is true for any orthonormal basis, is called the *completeness relation.* Notice that if we operate with both sides of (1-69) on a vector $|\Psi\rangle$ we find

$$|x\rangle\langle x|\Psi\rangle + |y\rangle\langle y|\Psi\rangle = |\Psi\rangle$$

which we recognize as being simply the expansion, Eq. (1-28), of $|\Psi\rangle$ in terms of the basis vectors $|x\rangle$ and $|y\rangle$.

Now if we multiply the completeness relation in the $|R\rangle$, $|L\rangle$ basis

$$1 = |R\rangle\langle R| + |L\rangle\langle L| \tag{1-70}$$

on the left by S, we have

$$S = S|R\rangle\langle R| + S|L\rangle\langle L|$$

which, by Eqs. (1-51) and (1-52), becomes

$$\begin{aligned} S &= (+1)|R\rangle\langle R| + (-1)|L\rangle\langle L| \\ &= |R\rangle\langle R| - |L\rangle\langle L|. \end{aligned} \tag{1-71}$$

[This expansion of a matrix in terms of its eigenvalues and eigenvectors will be generally valid for the matrices that represent physical quantities.] Had we known this expression for S before, we could have written down (1-65) immediately.

To end this mathematical interlude, let us note that if we have an arbitrary 2×2 matrix M, which in the x, y basis is explicitly

$$M = \begin{pmatrix} m_{xx} & m_{xy} \\ m_{yx} & m_{yy} \end{pmatrix},$$

then the matrix elements are given by

$$\begin{aligned} m_{xx} &= \langle x|M|x\rangle, \quad & m_{xy} &= \langle x|M|y\rangle \\ m_{yx} &= \langle y|M|x\rangle, \quad & m_{yy} &= \langle y|M|y\rangle. \end{aligned} \tag{1-72}$$

To see this, we note that, for example,

$$M|x\rangle = \begin{pmatrix} m_{xx} & m_{xy} \\ m_{yx} & m_{yy} \end{pmatrix} \begin{pmatrix} 1 \\ 0 \end{pmatrix} = \begin{pmatrix} m_{xx} \\ m_{yx} \end{pmatrix},$$

so that

$$\langle y|M|x\rangle = (0 \quad 1)\begin{pmatrix} m_{xx} \\ m_{yx} \end{pmatrix} = m_{yx},$$

etc. Equation (1-72) can be used for writing down the matrix elements in any basis; for example, in the R, L basis

$$m_{LR} = \langle L|M|R\rangle.$$

AMPLITUDE MECHANICS

In discussing the angular momentum of the photon, we saw that $|\langle R|\Psi\rangle|^2$ was the probability that a photon in state $|\Psi\rangle$ would transfer angular momentum $+\hbar$ in an experiment. Recall that $|\langle R|\Psi\rangle|^2$ also gave us the probability that a photon in state $|\Psi\rangle$ would pass through a prism that passed only right circularly polarized light. In both these cases, $|\langle R|\Psi\rangle|^2$ gives us the probability that the photon in state $|\Psi\rangle$ will behave as if it were in the state $|R\rangle$. The origin of this interpretation lies, of course, in the superposition principle which allowed us to write

$$|\Psi\rangle = |R\rangle\langle R|\Psi\rangle + |L\rangle\langle L|\Psi\rangle.$$

This is one of the strange features of quantum mechanics. Classically a system in a certain state, e.g., a particle at position $\mathbf{x}$, moving with velocity $\mathbf{v}$, never acts as if it were in a different state, say at $\mathbf{x}'$ with velocity $\mathbf{v}'$. But quantum mechanically, because of the superposition principle, a particle in one state, $|\Psi\rangle$, always has a probability of behaving as if it were in state $|\Phi\rangle$. This probability is $|\langle\Phi|\Psi\rangle|^2$, and it vanishes only when $|\Phi\rangle$ and $|\Psi\rangle$ are orthogonal. Only then will a photon in state $|\Psi\rangle$ not have a finite probability of exhibiting all the properties of a photon in state $|\Phi\rangle$.

We shall call $\langle\Phi|\Psi\rangle$ the *probability amplitude* for a photon in the state $|\Psi\rangle$ to be in the state $|\Phi\rangle$. Often one says that $|\langle\Phi|\Psi\rangle|^2$ is the probability for a photon in state $|\Psi\rangle$ *being* in state $|\Phi\rangle$. But, what one really means by this statement is, as we have seen, that $|\langle\Phi|\Psi\rangle|^2$ is the probability of a photon in state $|\Psi\rangle$ behaving in an experiment as if it were in state $|\Phi\rangle$.

We might be tempted to conclude, from measuring the angular momentum transferred to matter, that because the angular momentum transfer is always $+\hbar$ or $-\hbar$, photons were always either in the state $|R\rangle$, with a certain probability, α, or in the state $|L\rangle$ with probability $1-\alpha$. This is not a correct conclusion, for were it true we could never explain why an x polarized photon *never* passes through a polaroid whose axis lies in the y direction. After all, if

(a) an x polarized photon has probability $|\langle R|x\rangle|^2 = 1/2$ of being right circularly polarized, and a right circularly polarized photon has a probability $|\langle y|R\rangle|^2 = 1/2$ of passing through a y-polaroid, and

(b) an x polarized photon has probability $|\langle L|x\rangle|^2 = 1/2$ of being left circularly polarized, and a left circularly polarized photon has a probability $|\langle y|L\rangle|^2 = 1/2$ of passing through a y-polaroid, then the total probability of an x polarized photon passing through a y-polaroid would be

$$|\langle R|x\rangle|^2|\langle y|R\rangle|^2 + |\langle L|x\rangle|^2|\langle y|L\rangle|^2 = \frac{1}{2}. \tag{1-73}$$

Since x polarized photons *never* pass through y-polaroids, this must be a wrong way of doing the calculation. What we are missing, in thinking of x polarized photons as being either right circularly polarized or left circularly polarized, with equal probability, is the possibility of *interference* between the right and left circularly polarized amplitudes.

The correct calculation of the probability of an x polarized photon passing through a y-polaroid, would take $\langle y|x\rangle$, the probability amplitude for the x polarized photon being in the state $|y\rangle$, and square it to get the probability. Since $\langle y|x\rangle = 0$, this probability is zero. If we regard an x polarized photon as a *superposition* of right and left circularly polarized photons, then the calculation would be the following: The initial state is $|x\rangle$ which can be written as a superposition of $|R\rangle$ and $|L\rangle$:

$$|x\rangle = |R\rangle\langle R|x\rangle + |L\rangle\langle L|x\rangle.$$

Then the probability amplitude for passing through the y-polaroid is

$$\langle y|\Big[|R\rangle\langle R|x\rangle + |L\rangle\langle L|x\rangle\Big] = \langle y|R\rangle\langle R|x\rangle + \langle y|L\rangle\langle L|x\rangle. \tag{1-74}$$

The probability for passing through the polaroid is the absolute square of this amplitude:

$$\begin{aligned}\text{Probability} = |\langle y|R\rangle\langle R|x\rangle + \langle y|L\rangle\langle L|x\rangle|^2 &= |\langle y|R\rangle|^2|\langle R|x\rangle|^2 \\ &+ |\langle y|L\rangle|^2|\langle L|x\rangle|^2 + \langle y|R\rangle\langle R|x\rangle\langle y|L\rangle^*\langle L|x\rangle^* \\ &+ \langle y|R\rangle^*\langle R|x\rangle^*\langle y|L\rangle\langle L|x\rangle.\end{aligned} \tag{1-75}$$

The first two terms on the right give the probability we calculated in (1-73). The true probability is the sum of these two terms plus the last two terms, which represent the *interference* between the right and left circularly polarized photon. It is a trivial calculation to see that the interference terms *exactly* cancel the first two terms, giving a probability zero for an x polarized photon to pass through a y-polaroid; this is, of course, the correct answer.

Let us look more closely at expression (1-74) for the total probability amplitude. $\langle L|x\rangle$ is the probability amplitude for an x polarized photon to be left circularly polarized, and $\langle y|L\rangle$ is the probability amplitude for a left circularly polarized photon to pass through a y-polaroid. Similarly $\langle R|x\rangle$ is the probability amplitude for an x polarized photon to be right circularly polarized, and $\langle y|R\rangle$ is the probability amplitude for a right circularly polarized photon to pass through a y-polaroid. Expression (1-74) then tells us that the total probability amplitude for an x polarized photon to pass through a y-polaroid is the probability *amplitude,* $\langle y|R\rangle\langle R|x\rangle$, for it to pass through as a right circularly polarized photon, plus the probability *amplitude* $\langle y|L\rangle\langle L|x\rangle$ for it to pass through as a left circularly polarized photon. The amplitude for the photon to pass through as a right circularly polarized photon is $\langle L|x\rangle$, the probability amplitude for the x polarized photon to be L polarized, times $\langle y|L\rangle$, the amplitude for an L polarized photon to pass through a y-polaroid. To find the total probability of the photon passing through, we calculate the absolute value squared of the *total* probability amplitude. We do *not* calculate the separate probabilities and then add, as in (1-73).

Notice that the rules for the composition of *probability amplitudes* in quantum mechanics look very much like the classical rules for composition of probabilities:

1. The probability amplitude for two successive possibilities is the product of the amplitudes for the individual possibilities, e.g., the amplitude for the x polarized photon to be right circularly polarized *and* for the right circularly polarized photon to pass through the y-polaroid is $\langle R|x\rangle\langle y|R\rangle$, the product of the individual amplitudes.

2. The amplitude for a process that can take place in one of several *indistinguishable* ways is the sum of the amplitudes for each of the individual ways. For example, the total amplitude for the x polarized photon to pass through the y-polaroid is the sum of the amplitude for it to pass through as a right circularly polarized photon, $\langle y|R\rangle\langle R|x\rangle$, plus the amplitude for it to pass through as a left circularly polarized photon, $\langle y|L\rangle\langle L|x\rangle$, as in Eq. (1-74).

3. The total probability for the process to occur is the absolute value squared of the total amplitude calculated by 1 and 2.

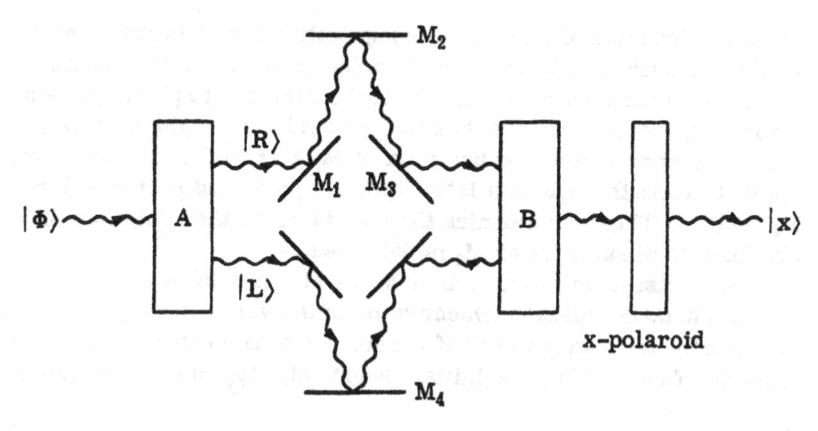

Fig. 1-4

Device for observing whether a photon passes through an x-polaroid as a right or left circularly polarized photon. Prism A splits the beam into right and left circularly polarized components, which are then reflected from mirrors [M_1, M_2, and M_3 for the right circularly polarized component] before being recombined by prism B.

In writing down rule 2 we referred to "indistinguishable ways." This means, for example, that if a photon initially in state $|\Phi\rangle$ passed through a y-polaroid, then just from a knowledge that it passed through, we could not tell if it went through as a right circular or as a left circular photon; these two ways are indistinguishable. If we attempt to distinguish the two possibilities by using some additional measuring device, then the presence of this device so changes the phase relation between the right and left components as to destroy the interference effect. If we had such a device present, and only if it was present, then, in fact, we would observe photons whose initial polarization was in the x direction, passing through the y-polaroid. Such a device is illustrated in Fig. 1-4. A prism A [composed of a birefringent material combined with a quarter wave plate] first splits the light into its right and left circularly polarized components. Then each of these components is passed through a separate sequence of three mirrors, and recombined again by prism B before passing through the polaroid filter. The purpose of the mirrors is to detect the path taken by the photon. A photon of wavelength λ has momentum

$$p = \frac{2\pi\hbar}{\lambda} \tag{1-76}$$

along its direction of motion. Consequently when it is reflected by a mirror, such as M_2, at normal incidence say, it reverses its direction and transfers momentum 2p to the mirror. If a photon passes through the device, and reaches the polaroid, as a right circularly polarized photon, it will cause mirror M_2 to recoil with momentum 2p; if it passes through as a left circularly polarized photon, mirror M_4 recoils. Thus to determine the path taken by the photon we simply observe whether M_2 or M_4 is deflected.

Now to use a mirror, such as M_2 or M_4, as a measuring device we must know its initial momentum p_y in the vertical, or y direction, sufficiently accurately to tell if it receives a momentum transfer 2p from the photon. Thus the initial uncertainty Δp_y in p_y must satisfy

$$\Delta p_y < 2p. \tag{1-77}$$

But from the Heisenberg uncertainty principle,

$$\Delta p_y \Delta y \gtrsim 2\pi\hbar, \tag{1-78}$$

the *position* of the mirror in the y direction must be uncertain by an amount Δy obeying

$$\Delta y \gtrsim \frac{2\pi\hbar}{\Delta p_y} \gtrsim \frac{2\pi\hbar}{2p} = \frac{\lambda}{2}. \tag{1-79}$$

Thus if the y coordinate of the mirror M_2 is uncertain by half a wavelength, the length of the path traveled by the photon in reflecting from M_2 is uncertain by λ [$\lambda/2$ coming and $\lambda/2$ going] – a full wavelength. Consequently the phase of the photon by the time it reaches prism B is *completely* uncertain, and the phase relation between the right and left circularly polarized components is lost.

Mathematically the initial photon state is

$$|\Phi\rangle = |R\rangle\langle R|\Phi\rangle + |L\rangle\langle L|\Phi\rangle; \tag{1-80}$$

let us say that the total phase change of the right circularly polarized component in passing through the device (up to the polaroid) is α_R, and α_L is the total phase change of the left circularly polarized component. Then the photon emerges from prism B in the state

$$|\Phi'\rangle = |R\rangle\langle R|\Phi\rangle\, e^{i\alpha_R} + |L\rangle\langle L|\Phi\rangle\, e^{i\alpha_L}. \tag{1-81}$$

The amplitude that the photon passes through the x-polaroid is

$$\langle x|R\rangle\langle R|\Phi\rangle e^{i\alpha_R} + \langle x|L\rangle\langle L|\Phi\rangle e^{i\alpha_L},$$

and the probability is

$$|\langle x|R\rangle|^2|\langle R|\Phi\rangle|^2 + |\langle x|L\rangle|^2|\langle L|\Phi\rangle|^2 \\ +2\text{Re}[\langle x|R\rangle\langle R|\Phi\rangle e^{i\alpha_R}\langle x|L\rangle^*\langle L|\Phi\rangle^* e^{-i\alpha_L} \tag{1-82}$$

The first two terms are the probabilities that the photon passes through as a right *or* left circularly polarized photon; the last term is the interference. When the mirrors M_2 and M_4 are set up with $\Delta p_y < 2p$ then α_L and α_R are uncertain to within 2π. The observed probability, a result of many measurements, is thus the average of (1-82) over all values of the phases. Since

$$\int_0^{2\pi} \frac{d(\alpha_R - \alpha_L)}{2\pi} e^{i(\alpha_R - \alpha_L)} = 0$$

the interference term averages to zero.

Notice that this loss of interference does not depend on our looking at the mirrors; it is due solely to our preparing the mirrors with $\Delta p_y < 2p$ so that they are capable of detecting the photon. If we wished to preserve the interference we must have the positions of all the mirrors fixed to an accuracy $\Delta y < \lambda/2$; this precludes using the mirrors to detect the photons.

UNPOLARIZED LIGHT

All along we have been speaking of photons in definite states of polarization. It is not obvious that we have room in our elaborate machinery for describing *unpolarized* light. The problem is that we have always assumed that we knew the state of polarization of the photon. But consider the following situation. Let us assume that we have a beam of monochromatic light which is composed of photons from two sources, one which always sends out photons in the state $|\Psi_1\rangle$, and one which always sends out photons in the state $|\Psi_2\rangle$. If the sources emit randomly, and are independent, we cannot say whether any particular photon in the beam came from one source or the other, but let us suppose that we know the relative strengths of the two sources. Then we can assign a probability, say p_1, for any photon in the beam to have come from source one, i.e., that it is in state $|\Psi_1\rangle$, and a probability, p_2, for the photon to have come from source two, and thus be in the state $|\Psi_2\rangle$. Of course, $p_1 + p_2 = 1$.

Let us ask how we would write the expectation value, for the photons in the beam, of any physical quantity, say the angular momentum in the direction of propagation. Recall that the average value of L_z was given by $\hbar$ times the probability that a photon transfers angular momentum $+\hbar$, plus $-\hbar$ times the probability that a photon transfers angular momentum $-\hbar$. Now the probability that the photon transfers angular momentum $+\hbar$ is

$$p_1|\langle R|\Psi_1\rangle|^2 + p_2|\langle R|\Psi_2\rangle|^2, \tag{1-83}$$

that is, the probability, p_1, that the photon was in state $|\Psi_1\rangle$, times $|\langle R|\Psi_1\rangle|^2$, the probability that a photon in the state $|\Psi_1\rangle$ will transfer angular momentum $+\hbar$, plus the probability, p_2, that the photon was in state $|\Psi_2\rangle$, times the probability that a photon in state $|\Psi_2\rangle$ will transfer angular momentum $+\hbar$. Similarly, the probability that the photons transfer angular momentum $-\hbar$ is

$$p_1|\langle L|\Psi_1\rangle|^2 + p_2|\langle L|\Psi_2\rangle|^2, \tag{1-84}$$

so that the average value of the angular momentum transfer is

$$\begin{aligned}
&\hbar(p_1|\langle R|\Psi_1\rangle|^2 + p_2|\langle R|\Psi_2\rangle|^2) - \hbar(p_1|\langle L|\Psi_1\rangle|^2 + p_2|\langle L|\Psi_2\rangle|^2) \\
&\quad = p_1(\hbar|\langle R|\Psi_1\rangle|^2 - \hbar|\langle L|\Psi_1\rangle|^2) + p_2(\hbar|\langle R|\Psi_2\rangle|^2 - \hbar|\langle L|\Psi_2\rangle|^2) \qquad (1\text{-}85) \\
&\quad = p_1\langle L_z\rangle_1 + p_2\langle L_z\rangle_2 = p_1\langle\Psi_1|\hbar S|\Psi_1\rangle + p_2\langle\Psi_2|\hbar S|\Psi_2\rangle.
\end{aligned}$$

Thus, the average value of L_z for the beam consisting of two photon states is the average of the expectation value of L_z for state $|\Psi_1\rangle$ and the expectation value of L_z for state $|\Psi_2\rangle$, weighted with the probabilities for finding these states in the beam.

It is very important to realize that the statement "the photon is either in state $|\Psi_1\rangle$ or $|\Psi_2\rangle$, but we don't know which," is *not* the same statement as "the photon is in a state which is a superposition of $|\Psi_1\rangle$ and $|\Psi_2\rangle$." This is the distinction we were emphasizing on p. 21. The point is that when we say that a photon is in a state $|\Phi\rangle$ that is a linear combination of states $|\Psi_1\rangle$ and $|\Psi_2\rangle$,

$$|\Phi\rangle = \alpha|\Psi_1\rangle + \beta|\Psi_2\rangle, \tag{1-86}$$

then we are implying that the relative phase of the coefficients α and β is certain. For example, if $|\Psi_1\rangle = |x\rangle$ and $|\Psi_2\rangle = |y\rangle$, then if $\alpha = 1/\sqrt{2}$ and $\beta = i/\sqrt{2}$, clearly $|\Phi\rangle = |R\rangle$, but if $\alpha = 1/\sqrt{2}$ and $\beta = -i/\sqrt{2}$, then $|\Phi\rangle = |L\rangle$, which is a state completely different from

$|R\rangle$. The relative phase must be fixed to specify the linear combination uniquely. On the other hand, when we say that a photon is either in state $|\Psi_1\rangle$ or state $|\Psi_2\rangle$, but we don't know which, then we are implying that there is no connection between the phases of these states, and hence no possibility of interference effects, which depend critically on delicate phase relations, taking place.

There are really two levels of probability in quantum mechanics. The first, which is called the *pure case*, is when we have a system that is in a definite state (often called a pure state). Then the behavior of this system in a given experiment is governed by the probability amplitude rules we have been discussing. The second case is when the system can be in any of several states, with certain probabilities. This case is called the *mixed case,* and we say that the photon is in a *mixed state.* Then one must calculate the results one expects in a given experiment for each of the separate states that can be present, and take the weighted average of the result over these states, as we have done in Eq. (1-85). In averaging over the various states that can be present, one uses the ordinary classical probability rules; there is no possibility of interference occurring between these states.

If there is complete uncertainty between the phases of α and β in (1-86), then we get the same results as if we had a mixed case in which the photon could be in state $|\Psi_1\rangle$ with probability $p_1 = |\alpha|^2$, and state $|\Psi_2\rangle$ with probability $p_2 = |\beta|^2$. Consider for example, the expectation value of L_z. For the state $|\Phi\rangle$, (1-86),

$$\begin{aligned}\langle L_z\rangle = \langle\Phi|\hbar S|\Phi\rangle &= (\alpha^*\langle\Psi_1| + \beta^*\langle\Psi_2|)\hbar S(\alpha|\Psi_1\rangle + \beta|\Psi_2\rangle)\\ &= |\alpha|^2\langle\Psi_1|\hbar S|\Psi_1\rangle + |\beta|^2\langle\Psi_2|\hbar S|\Psi_2\rangle \\ &\quad + \alpha^*\beta\langle\Psi_1|\hbar S|\Psi_2\rangle + \beta^*\alpha\langle\Psi_2|\hbar S|\Psi_1\rangle .\end{aligned} \tag{1-87}$$

The first two terms on the right are the classical average of the result for state $|\Psi_1\rangle$ and the result for state $|\Psi_2\rangle$, as in Eq. (1-85). The last two terms represent quantum mechanical interference effects, and depend critically on the relative phase of α and β. If this phase is completely uncertain, and can take on random values for the different photons in the beam, then to get the average value of L_z, we must average over all possible values of this relative phase. To calculate this average, let us write $\alpha = |\alpha| e^{i\varphi_1}$, $\beta = |\beta| e^{i\varphi_2}$. Clearly neither $|\alpha|^2$ nor $|\beta|^2$ in (1-87) depend on the phases φ_1, φ_2. But

$\alpha^*\beta = |\alpha\beta| e^{-i(\varphi_1 - \varphi_2)}$, and when we average over the relative phase, $\varphi_1 - \varphi_2$, then $\alpha^*\beta$ and $\beta^*\alpha$ will average to zero, the last two terms in (1-87) vanish, and we are left with the mixed case result (1-85) for the expectation value of L_z.

We are now ready to answer the question, "what is unpolarized light?" It is light that has equal probability of being in *any* polarization state; the state of unpolarized light is a mixed state.

It turns out that this description of unpolarized light is equivalent to a much simpler one: If $|\Psi_1\rangle$ and $|\Psi_2\rangle$ are *any* two states that form an orthonormal basis, then light that is in a mixed state with equal probability of being in state $|\Psi_1\rangle$ or $|\Psi_2\rangle$ has all the properties of unpolarized light. That is, we could not distinguish experimentally whether unpolarized light was in any state with equal probability, or just in states $|\Psi_1\rangle$ and $|\Psi_2\rangle$ with equal probability.

We can see this as follows. Our first description of unpolarized light calls for us, in calculating any physical quantity, such as $\langle L_z\rangle$, to average over all polarization states. Now because $|\Psi_1\rangle$ and $|\Psi_2\rangle$ form a basis, we can write any polarization state as a linear combination of them, as in Eq. (1-86). Averaging over all polarization states is equivalent to averaging over all values of α and β, subject only to $|\alpha|^2 + |\beta|^2 = 1$, since $\langle\Phi|\Phi\rangle = 1$.

Let us again consider, as an example, calculating $\langle L_z\rangle$; but I must emphasize that what we are doing is valid for *any* physical quantity and its corresponding operator. The expectation value of L_z is given in terms of α, β, $|\Psi_1\rangle$, and $|\Psi_2\rangle$ by Eq. (1-87). In averaging over all possible values of α and β, let us first average over their phases. Then, as we have seen, the last two terms in (1-87) average to zero. Next, we must average over the magnitudes of α and β. Clearly the average of $|\alpha|^2$ will equal the average of $|\beta|^2$ since α and β are completely equivalent. Also, since $|\alpha|^2 + |\beta|^2 = 1$, we see that the average of $|\alpha|^2$ equals the average of $|\beta|^2$ equals $^1/_2$. Thus in carrying out the average for an unpolarized beam we find

$$\langle L_z\rangle = \frac{1}{2}\langle\Psi_1|\hbar S|\Psi_1\rangle + \frac{1}{2}\langle\Psi_2|\hbar S|\Psi_2\rangle . \tag{1-88}$$

This is just the average we would have found had we assumed the photons to be in either state $|\Psi_1\rangle$ or $|\Psi_2\rangle$ with equal weight.

The average (1-88) for unpolarized light is, of course, independent of the basis $|\Psi_1\rangle$, $|\Psi_2\rangle$, since this was an arbitrary basis. We can see this in another way, if we make use of the completeness relation, e.g. (1-69), for a basis. Suppose $\{|\Phi_1\rangle, |\Phi_2\rangle\}$ is another orthonormal basis. Then we have

$$\langle L_z \rangle = \frac{1}{2}\langle \Psi_1|\hbar S|(|\Phi_1\rangle\langle\Phi_1| + |\Phi_2\rangle\langle\Phi_2|)|\Psi_1\rangle$$

$$+\frac{1}{2}\langle \Psi_2|\hbar S|(|\Phi_1\rangle\langle\Phi_1| + |\Phi_2\rangle\langle\Phi_2|)|\Psi_2\rangle$$

$$= \frac{1}{2}(\langle \Phi_1|\Psi_1\rangle\langle\Psi_1|\hbar S|\Phi_1\rangle + \langle\Phi_1|\Psi_2\rangle\langle\Psi_2|\hbar S|\Phi_1\rangle)$$

$$+\frac{1}{2}(\langle \Phi_2|\Psi_1\rangle\langle\Psi_1|\hbar S|\Phi_2\rangle + \langle\Phi_2|\Psi_2\rangle\langle\Psi_2|\hbar S|\Phi_2\rangle)$$

$$= \frac{1}{2}\langle \Phi_1|\hbar S|\Phi_1\rangle + \frac{1}{2}\langle\Phi_2|\hbar S|\Phi_2\rangle,$$

using the completeness relation for the Ψ basis. Thus unpolarized light is equally well a mixture of equal parts of photons in state $|\Phi_1\rangle$ and photons in state $|\Phi_2\rangle$. For example, the beam that emerges from prism B in Fig. 1-4 is essentially unpolarized when $\Delta y > \lambda/2$ for the mirrors M_4 or M_2, and we do not observe which path the photons took.

BEHAVIOR OF PHOTON POLARIZATION IN MATTER

Up to now we have been considering the effect on photons of prisms and polaroids that pass light in certain polarization states, but reject light in states orthogonal to these. Generally, if we shoot a photon at such a prism it has a certain probability for not emerging from the other side of the prism. Let us consider the passage of photons through crystals that to a first approximation allow all the photons to pass through, but change the state of polarization of the photons.

As an example, consider a "birefringent" crystal, such as calcite. This type of crystal has an axis, called the *optic axis,* with the property that the crystal has a different index of refraction for light polarized parallel to the optic axis than it has for light polarized perpendicular to the optic axis. Let us take a block of calcite with the optic axis in the x, y plane, and pass through photons in the z direction. The photons polarized perpendicular to the optic axis are called the *ordinary* ray, and the photons polarized parallel to the optic axis are called the *extraordinary* ray.

In calcite, for red light, the index of refraction, n_o, for the ordinary ray is about 10% greater than the index, n_e, for the extraordinary ray. This means that the extraordinary ray travels faster through the crystal than the ordinary ray. Thus for an arbitrarily polarized photon traveling in the z direction, the relative phase of the ordinary and extraordinary components will vary as the photon propagates through the crystal. Consequently the polarization state of the photon when it emerges from the calcite will, in general, be different than its state of polarization on entering the calcite. Let us try to describe this process mathematically.

Let $|o\rangle$ denote the state of polarization of the ordinary ray and let $|e\rangle$ denote the state of polarization of the extraordinary ray; $|e\rangle$ and $|o\rangle$ form an orthonormal basis. Recall that the electric field vector changes by a phase $e^{ikz-i\omega t}$ from point to point in space. Now for a given frequency, ω, the wavenumber k will be different for e and o polarizations:

$$k_e = \frac{\omega n_e}{c}$$

$$k_o = \frac{\omega n_o}{c}. \tag{1-89}$$

If the calcite is of length l in the z direction, then the difference in phase of an e wave between its point of entry in the calcite and its point of departure will be $k_e l$, and similarly an o wave changes phase by $k_o l$. [The time factor $e^{-i\omega t}$ is the same for both polarizations and doesn't effect their relative phase.] The effect of the calcite is to change an incident state $|e\rangle$ into $e^{ik_e l}|e\rangle$, and an incident $|o\rangle$ state into $e^{ik_o l}|o\rangle$ as the light passes through the crystal.

Thus, we can describe the effect of the calcite on the polarization state of a photon as follows. Let $|\Psi_{in}\rangle$ denote the polarization state of the photon as it enters the crystal. We can write $|\Psi_{in}\rangle$ as a linear superposition of $|e\rangle$ and $|o\rangle$:

$$|\Psi_{in}\rangle = |e\rangle\langle e|\Psi_{in}\rangle + |o\rangle\langle o|\Psi_{in}\rangle. \tag{1-90}$$

Then the effect of the calcite can be described by multiplying the $|e\rangle$ component by $e^{ik_e l}$ and the $|o\rangle$ component by $e^{ik_o l}$; the polarization state $|\Psi_{out}\rangle$ of the photon as it leaves the calcite is therefore

$$|\Psi_{out}\rangle = e^{ik_e l}|e\rangle\langle e|\Psi_{in}\rangle + e^{ik_o l}|o\rangle\langle o|\Psi_{in}\rangle = U_l|\Psi_{in}\rangle, \tag{1-91}$$

where

$$U_z = e^{ik_e z}|e\rangle\langle e| + e^{ik_o z}|o\rangle\langle o|. \tag{1-92}$$

The probability amplitude that if the photon enters in state $|\Psi\rangle$ it emerges in state $|\Phi\rangle$ (i.e., behaves as if in the state $|\Phi\rangle$) is called the *transition amplitude* from $|\Psi\rangle$ to $|\Phi\rangle$. To calculate this transition amplitude we use the fact that if the photon enters in state $|\Psi\rangle$ it emerges in state $U_l|\Psi\rangle$, and thus the probability amplitude for the emergent photon being in state $|\Phi\rangle$ is $\langle\Phi|U_l|\Psi\rangle$. The square of this amplitude, $|\langle\Phi|U_l|\Psi\rangle|^2$, is usually called the *transition probability* from $|\Psi\rangle$ to $|\Phi\rangle$. It is the probability that if a photon enters the crystal in state $|\Psi\rangle$ it will behave as if it were in the state $|\Phi\rangle$ when it emerges from the crystal. We shall meet the concept of transition probability again and again in quantum mechanics.

Clearly, the state of polarization, $|\Psi_z\rangle$, of the photon after traveling through a length z of the calcite is

$$|\Psi_z\rangle = U_z|\Psi_{in}\rangle, \tag{1-93}$$

It is instructive to consider how $|\Psi_z\rangle$ changes as we move in the z direction. To do this, let us notice that U_z obeys the simple property

$$U_{z+a} = U_a U_z. \tag{1-94}$$

Showing this is left as an exercise. Thus we can determine the polarization vector at the point z + a in terms of the vector at z, by writing

$$|\Psi_{z+a}\rangle = U_{z+a}|\Psi_{in}\rangle = U_a U_z|\Psi_{in}\rangle = U_a|\Psi_z\rangle. \tag{1-95}$$

Now let us suppose that the point z + a is infinitesimally close to z, that is, $k_o a \ll 1$ and $k_e a \ll 1$. Then we can expand the exponentials in the maxtrix U_a, (1-92), and write

$$U_a = (1 + ik_o a)|o\rangle\langle o| + (1 + ik_e a)|e\rangle\langle e|. \tag{1-96}$$

Note that because $|o\rangle$ and $|e\rangle$ form an orthonormal basis

$$|o\rangle\langle o| + |e\rangle\langle e| = 1.$$

Let us define the wavenumber matrix, K, by

$$K = k_e|e\rangle\langle e| + k_o|o\rangle\langle o|. \tag{1-97}$$

The eigenvalues of K are the possible wavenumbers of photons with frequency ω in the calcite. The eigenvectors are the polarizations that have these wavenumbers. A photon in any other polarization state can't be assigned a definite wavenumber, but must be regarded as a superposition of the two states $|e\rangle$ and $|o\rangle$ which have definite wavenumbers.

In terms of K we can write U_a, for infinitesimal a, as

$$U_a = 1 + iaK. \tag{1-98}$$

Hence $|\Psi_{z+a}\rangle = (1 + iaK)|\Psi_z\rangle$, or

$$|\Psi_{z+a}\rangle - |\Psi_z\rangle = iaK|\Psi_z\rangle. \tag{1-99}$$

If we divide both sides by a, and let a tend to zero, then the left side becomes the derivative of the vector $|\Psi_z\rangle$ with respect to z, and we have

$$\frac{d}{dz}|\Psi_z\rangle = iK|\Psi_z\rangle. \tag{1-100}$$

This equation, or equivalently Eq. (1-99), tells us how the polarization state changes as we move an infinitesimal distance along the wave. Let us write out Eq. (1-99) in components in the x, y basis:

$$\langle x|\Psi_{z+a}\rangle - \langle x|\Psi_z\rangle = ia\langle x|K|x\rangle\langle x|\Psi_z\rangle + ia\langle x|K|y\rangle\langle y|\Psi_z\rangle. \tag{1-101}$$

This equation says that the *change* in the x component of $|\Psi_z\rangle$ as we move down the beam an infinitesimal amount a, is made up of two pieces; the first proportional to the x component of $|\Psi_z\rangle$, with a constant of proportionality $ia\langle x|K|x\rangle$, and the second proportional to the y component of $|\Psi_z\rangle$ with a constant of proportionality $ia\langle x|K|y\rangle$. Similarly

$$\langle y|\Psi_{z+a}\rangle - \langle y|\Psi_z\rangle = ia\langle y|K|x\rangle\langle x|\Psi_z\rangle + ia\langle y|K|y\rangle\langle y|\Psi_z\rangle. \tag{1-102}$$

The vector $|\Psi_{z+a}\rangle$ is of unit length when $|\Psi_z\rangle$ is; this means that no photons are absorbed between z and z+a. This places a restriction on the matrix K, which we can see by direct calculation. In components

$$\langle\Psi_{z+a}|\Psi_{z+a}\rangle = |\langle x|\Psi_{z+a}\rangle|^2 + |\langle y|\Psi_{z+a}\rangle|^2.$$

Using Eqs. (1-101) and (1-102) and neglecting terms of order a^2, we find

$$\langle \Psi_{z+a}|\Psi_{z+a}\rangle = 1 = |\langle x|\Psi_z\rangle|^2 + |\langle y|\Psi_z\rangle|^2$$
$$+ ia(\langle x|K|x\rangle - \langle x|K|x\rangle^*)|\langle x|\Psi_z\rangle|^2$$
$$+ ia(\langle y|K|y\rangle - \langle y|K|y\rangle^*)|\langle y|\Psi_z\rangle|^2$$
$$+ ia(\langle x|K|y\rangle - \langle y|K|x\rangle^*)\langle y|\Psi_z\rangle\langle x|\Psi_z\rangle^*$$
$$+ ia(\langle y|K|x\rangle - \langle x|K|y\rangle^*)\langle x|\Psi_z\rangle\langle y|\Psi_z\rangle^*.$$

Since the first two terms on the right are $\langle \Psi_z|\Psi_z\rangle = 1$, the last four terms must vanish. Since $|\Psi_z\rangle$ is an arbitrary vector, these four terms will vanish only if the coefficients of the components of $|\Psi_z\rangle$ vanish. Thus

$$\langle x|K|x\rangle = \langle x|K|x\rangle^*, \qquad \langle y|K|y\rangle = \langle y|K|y\rangle^*$$

$$\langle x|K|y\rangle = \langle y|K|x\rangle^*, \qquad \langle y|K|x\rangle = \langle x|K|y\rangle^*. \tag{1-103}$$

That K satisfies these three conditions in fact, can be verified directly from its definition, Eq. (1-97).

If we define the Hermitian adjoint, $K^\dagger$, of K to be the matrix

$$K^\dagger = \begin{pmatrix} \langle x|K|x\rangle^* & \langle y|K|x\rangle^* \\ \langle x|K|y\rangle^* & \langle y|K|y\rangle^* \end{pmatrix}, \tag{1-104}$$

Then (1-103) says that

$$K^\dagger = K; \tag{1-105}$$

K is then said to be Hermitian.

Thus the condition that no photons be absorbed as the beam travels through the crystal implies that the matrix, K, which tells us the infinitesimal change in $|\Psi_z\rangle$ as we move an infinitesimal distance along the beam, must be Hermitian. We shall meet a similar argument again when we come to discuss the Schrödinger equation, which says how state vectors change in time. It is also easy to verify that

$$U_z^\dagger U_z = 1; \tag{1-106}$$

a matrix satisfying this relation is called *unitary*. The fact that U_z is unitary is equivalent to K being Hermitian.

PROBLEMS

1. A man sends a beam of red light along the z axis through a polaroid filter that passes only x polarized light. The beam is initially

polarized at 30° to the x axis, and the total energy content of the beam is quite accurately 10 joules. Estimate the fluctuations in the energy of the beam, i.e., the range of likely energy values, after it passes through the polaroid. How do the fluctuations depend on $\hbar$?

2. (a) Write down a basis corresponding to 45°, 135° polarizations.
   (b) Write down a basis that is neither plane nor circularly polarized.
3. (a) Calculate the transformation matrix from the x, y basis to the R, L basis.
   (b) Calculate the transformation matrix from the R, L basis to the basis devised in Problem 2(b).
   (c) Calculate the transformation matrix from the x, y basis to the basis in Problem 2(b), and show that it is the product of the matrix calculated in Problem 3(b) and the matrix calculated in Problem 3(a). In which order must you multiply these matrices?
   (d) Show in general that the product of
      (i) the transformation matrix from a basis, 1, to a basis, 2, with
      (ii) the transformation matrix from basis 2 to a third basis, 3,
   is the transformation matrix from basis 1 to basis 3. Hint: make extensive use of the completeness relation, e.g., (1-69), (1-70), for a basis.
4. (a) Show that $\langle\Phi|M|\Psi\rangle^* = \langle\Psi|M^\dagger|\Phi\rangle$, for any $|\Phi\rangle$ and $|\Psi\rangle$.
   (b) Show that if $M|\Psi\rangle = \lambda|\Psi\rangle$ then $\langle\Psi|M^\dagger = \lambda^*\langle\Psi|$.
5. Show that the transformation matrix from one basis to another is unitary.
6. (a) Show that the matrix $|\Phi\rangle\langle\Phi|$ is Hermitian.
   (b) Show that the photon spin operator S is Hermitian. Generally all physical quantities are represented by Hermitian matrices.
7. Let $|x'(\theta)\rangle$, $|y'(\theta)\rangle$ denote the basis tilted at an angle θ to the x, y basis. Show that the components of a vector $|\Psi\rangle$ in this basis

$$\langle x'(\theta)|\Psi\rangle, \qquad \langle y'(\theta)|\Psi\rangle$$

obey the differential equations

$$-i\frac{\partial}{\partial\theta}\langle x'(\theta)|\Psi\rangle = \langle x'(\theta)|S|\Psi\rangle$$

$$-i\frac{\partial}{\partial\theta}\langle y'(\theta)|\Psi\rangle = \langle y'(\theta)|S|\Psi\rangle.$$

Solve these equations explicitly for $|\Psi\rangle = |R\rangle$ and $|\Psi\rangle = |L\rangle$.

8. The probability that a photon in state $|\Psi\rangle$ passes through an x-polaroid is the average value of a physical observable which might be called the "x-polarizedness." Write down the operator, P_x, corresponding to this observable. Show that it is Hermitian. What are its eigenvalues and eigenstates? Write down its representation in terms of its eigenvalues and eigenstates [as in (1-71)]. Verify that the probability that a photon in state $|\Psi\rangle$ passes through the x-polaroid is $\langle\Psi|P_x|\Psi\rangle$.
9. Photons polarized at 30° to the x axis are sent through a y-polaroid. An attempt is made to determine how frequently the photons that pass through the polaroid, pass through "as right circularly polarized photons," and how frequently they pass through "as left circularly polarized photons"; this attempt is made as follows:
   First, a prism that passes only right circular polarized light is placed between the source of the 30° polarized photons and the y-polaroid, and it is determined how frequently the 30° photons pass through the y-polaroid. Then this experiment is repeated with a prism that passes only left circular polarized light instead of the one that passes only right. Show by explicit calculation that the sum of the probabilities for passing through the y-polaroid measured in these two experiments is different from the probability that one would measure if there were *no* prism in the path of the photon and only the y-polaroid.
   Relate this experiment to the two-slit diffraction experiment.
10. A beam with a certain number of photons is prepared by reflecting from a mirror for a moment a monochromatic beam from a laser; such a laser beam has, for our purposes, a very well defined phase. The number of photons in the reflected beam is determined by measuring the momentum transferred to the mirror in the reflection. Show that ΔN, the uncertainty in the number of photons in the reflected beam, times $\Delta\varphi$, the uncertainty in the phase of the reflected beam is greater than 2π. Show that this uncertainty relation is independent of the angle of incidence of the laser beam on the mirror. The uncertainty relation $\Delta N \Delta\varphi \gtrsim 2\pi$ between the number of photons in a wave and its phase is quite generally true, and doesn't depend on how the beam was prepared.
11. A photon polarized at an angle θ to the optic axis is sent in the z direction through a slab of calcite 10^{-2} cm thick in the z direction. Assume the optic axis to lie in the x, y plane. Calculate, as a function of θ, the transition probability for the photon to emerge left circularly polarized. Sketch the result. Let the

frequency of the light be given by $c/\omega = 5000$ Å, and let $n_e = 1.50$ and $n_o = 1.65$ for the calcite.

12. Using calcite and polaroid, devise a filter that will pass light of frequency $c/\omega = 5000$ Å only if it is right circularly polarized. Use the same indices as in 11.
13. What is the condition on the length of the calcite that, for frequency ω, $|\Psi_{out}\rangle$ is *always*, to within a phase factor, the same state as $|\Psi_{in}\rangle$.
14. Turpentine is an "optically active" substance. If we shoot plane polarized light into turpentine then it emerges with its plane of polarization rotated. Specifically, turpentine induces a left-hand rotation of about 5° per cm of turpentine that the light traverses. Write down the transition matrix that relates the incident polarization to the emergent polarization. Show that the matrix is unitary. Find its eigenvectors and eigenvalues, as a function of the length of turpentine traversed.
15. Unpolarized light traveling in the z direction is sent through a block of calcite whose optic axis lies in the x, y plane. What is the effect of the calcite on the polarization properties of the beam? What will turpentine do to an unpolarized beam?
16. The *trace* of an $n \times n$ matrix is defined as the sum of its diagonal components, i.e., if

$$A = \begin{pmatrix} a_{11} & a_{12} & \cdots & a_{1n} \\ a_{21} & a_{22} & \cdots & \\ \vdots & \vdots & \ddots & \\ a_{n1} & & & a_{nn} \end{pmatrix}$$

then

$$\text{tr}\, A = a_{11} + a_{22} + a_{33} + \ldots + a_{nn}.$$

Show that the trace of a matrix is independent of the basis in which the matrix is written, i.e., that it is invariant under a unitary transformation of the matrix,

$$A \to U^{\dagger} A U,$$

where U is unitary. Show that $\text{tr}\, AB = \text{tr}\, BA$.

17. (a) Suppose that a photon is in the state $|\Psi\rangle$. Let

$$P_{\Psi} = |\Psi\rangle\langle\Psi|.$$

Show that the expectation value for the photon of a physical quantity represented by the operator Q is tr $P_\Psi Q$. The matrix P_Ψ is called the *density matrix* for the pure state $|\Psi\rangle$.

(b) Suppose now that the photon is in a state that is a mixture of state $|\Psi_1\rangle$ with probability p_1, $|\Psi_2\rangle$ with probability p_2, etc. where $\sum_i p_i = 1$.

Let

$$\rho = \sum_i p_i |\Psi_i\rangle\langle\Psi_i|.$$

Show that the expectation value for this mixed state of a physical quantity Q is

$$\langle Q\rangle = \mathrm{tr}\,\rho Q.$$

The matrix ρ is called the *density matrix* for the mixed state.

(c) Show tr $\rho = 1$.

18. (a) The expectation value of any observable, Q, can be calculated in terms of the density matrix as tr ρQ. The density matrix thus contains all information available about the state, pure or mixed, of a photon. Suppose that we have a beam of photons whose polarization state is unknown to us; it might be pure or mixed. We would like to perform several measurements that will completely determine the state. What measurements will determine the polarization state, and what is the minimum number necessary?

(b) What is ρ for unpolarized light? What measurements would determine that the light was unpolarized?

(c) Find ρ for a *mixed* state consisting of 50% x polarized light and 50% right circularly polarized light. Find two orthogonal states that give the same density matrix.

Chapter 2
NEUTRAL K MESONS

The formalism we have developed to discuss the polarization of the photon is applicable to many other problems. We shall consider, as an example, neutral K mesons. These mesons are produced in strong interaction processes like

$$\pi^- + p \to \Lambda^0 + K^0. \tag{2-1}$$

Notice that charge is conserved in this reaction; the left side has total charge zero, and so does the right side. Similarly, there exists another "quantum number," called *strangeness,* that we can assign to the particles that participate in strong interactions, and it is an experimental fact that strangeness is conserved in strong interactions. Protons, neutrons, and π mesons have strangeness zero, whereas Λ^0 particles have strangeness -1, and K^0 mesons have strangeness $+1$. The left side of (2-1) has total strangeness zero, and so does the right side.

Corresponding to the K^0 meson, there is an antiparticle, called the $\overline{K^0}$ meson. It must have the opposite charge and strangeness of the K^0 meson, and thus it is neutral and has strangeness -1. A typical reaction for a $\overline{K^0}$ is the absorption process.

$$\overline{K^0} + p \to \Lambda^0 + \pi^+. \tag{2-2}$$

[Verify that strangeness is conserved in this strong interaction.]

One of the interesting facts about neutral K mesons is that they can be in a linear superposition of K^0 and $\overline{K^0}$ states, exactly as photons can be in a linear superposition of right and left circularly polarized states. Let us represent the states of neutral K mesons by two-dimensional complex vectors, the same type of vectors we used to represent photon polarization. Thus we let $|K^0\rangle$ denote the state in which the meson is a K^0, and we let $|\overline{K}^0\rangle$ denote the state in which the meson is a $\overline{K}^0$. These states are orthogonal, and we choose them to be normalized to one. Thus $|K^0\rangle$ and $|\overline{K}^0\rangle$ form an orthonormal basis. The general state of a neutral K meson is a linear combination of $|K^0\rangle$ and $|\overline{K}^0\rangle$. These basis states are independent of time; we will put all time dependence in the coefficients in this linear combination. One may think of $|K^0\rangle$ as analogous to the state $|R\rangle$ and $|\overline{K}^0\rangle$ as analogous to $|L\rangle$.

We can define a strangeness operator, S, for K mesons, by writing

$$S|K^0\rangle = |K^0\rangle, \qquad S|\overline{K}^0\rangle = -|\overline{K}^0\rangle. \tag{2-3}$$

Thus in the $|K^0\rangle$, $|\overline{K}^0\rangle$ basis

$$S = \begin{pmatrix} 1 & 0 \\ 0 & -1 \end{pmatrix}. \tag{2-4}$$

The eigenvectors of S are $|K^0\rangle$ and $|\overline{K}^0\rangle$, the states that have a definite strangeness, and the corresponding eigenvalues are the strangeness values of the K^0 and $\overline{K}^0$ mesons. Another way of writing S that doesn't involve writing out the components explicitly is

$$S = |K^0\rangle\langle K^0| - |\overline{K}^0\rangle\langle \overline{K}^0|, \tag{2-5}$$

which we find from operating with S on both sides of the completeness relation for the K^0, $\overline{K}^0$ basis.

We can also define a *charge conjugation* operation for neutral K mesons that changes particles into antiparticles and vice versa. This operation, denoted by CP, is defined by

$$CP|K^0\rangle = |\overline{K}^0\rangle, \qquad CP|\overline{K}^0\rangle = |K^0\rangle. \tag{2-6}$$

In the K^0, $\overline{K}^0$ basis we have

$$CP = \begin{pmatrix} 0 & 1 \\ 1 & 0 \end{pmatrix}. \tag{2-7}$$

The eigenstates of CP are

$$|K_S\rangle = \frac{1}{\sqrt{2}}(|K^0\rangle + |\overline{K}^0\rangle), \qquad |K_L\rangle = \frac{1}{\sqrt{2}}(|K^0\rangle - |\overline{K}^0\rangle), \tag{2-8}$$

as may be verified directly from (2-6):

$$CP|K_S\rangle = |K_S\rangle, \qquad CP|K_L\rangle = -|K_L\rangle. \tag{2-9}$$

The eigenvalues of CP are +1 and −1, and the corresponding physical quantity is called *charge conjugation parity*. Thus a neutral K meson in the $|K_S\rangle$ state has positive charge conjugation parity, while a neutral K meson in the $|K_L\rangle$ state has negative charge conjugation parity. We cannot assign a definite value of strangeness to the K_S and K_L states; they are linear combinations, Eq. (2-8), of a state with strangeness +1 and a state with strangeness −1.

It is trivial to verify that $|K_S\rangle$ and $|K_L\rangle$ form an orthonormal basis. We can express $|K^0\rangle$ and $|\overline{K}^0\rangle$ in terms of them as

$$|K^0\rangle = \frac{1}{\sqrt{2}}(|K_S\rangle + |K_L\rangle), \qquad |\overline{K}^0\rangle = \frac{1}{\sqrt{2}}(|K_S\rangle - |K_L\rangle). \tag{2-10}$$

We cannot assign a definite value of charge conjugation parity to the $|K^0\rangle$ and $|\overline{K}^0\rangle$ states; they are linear combinations of a state with CP = +1 and a state with CP = −1.

The neutral K mesons decay by weak interactions. As strangeness is conserved in strong interactions, charge conjugation parity is conserved in weak interactions. [Actually, CP is not perfectly conserved in weak interactions, but we shall only examine the consequences of exact conservation; see Problem 2.] One sees two kinds of decays of neutral K mesons. The first type of decay is into a $\pi^+ + \pi^-$ or $2\pi^0$ state that has CP = +1. Because CP is conserved, the state before the decay must have had CP = +1, and thus the state that decays into $\pi^+ + \pi^-$ or $2\pi^0$'s must be K_S,

$$K_S \to \pi^+ + \pi^-, \qquad K_S \to \pi^0 + \pi^0. \tag{2-11}$$

These processes occur in a time $\tau_S \approx 0.9 \times 10^{-10}$ sec. The other type of decay is into states with CP = −1, such as $\pi\pi\pi$, $\pi\mu\nu$, and $\pi e\nu$. The K state before the decay must have been, by CP conservation, a K_L state:

$$K_L \to \pi\pi\pi, \qquad K_L \to \pi e\nu, \qquad K_L \to \pi\mu\nu. \tag{2-12}$$

These processes occur in a time $\tau_L \approx 518 \times 10^{-10}$ sec, a very much longer time than that for K_S decay. [The subscripts S and L stand for short- and long-lived.]

Let us consider how the states of neutral K mesons change in time. First recall that when we discussed calcite in Chapter 1, we found that a photon with frequency ω in the $|o\rangle$ state had a definite wavenumber, $\omega n_o/c$, a photon with frequency ω in the $|e\rangle$ state also had a definite wavenumber, $\omega n_e/c$, but that we couldn't assign a definite wavenumber to a photon that was in a linear combination of $|o\rangle$ and $|e\rangle$ states. A very similar situation occurs with neutral K mesons. If the meson is short lived then its state changes in time with a definite frequency,

$$\omega_S = \frac{E_S}{\hbar} \tag{2-13}$$

where

$$E_S = (p^2c^2 + m_S^2c^4)^{1/2} \tag{2-14}$$

is the energy of the K_S meson, m_S is its mass, and p is its momentum. If the meson is long lived then its state changes in time with a frequency

$$\omega_L = \frac{E_L}{\hbar} \tag{2-15}$$

where

$$E_L = (p^2c^2 + m_L^2c^4)^{1/2}; \tag{2-16}$$

m_L is the mass of the K_L meson. [This connection between frequency, i.e., the time rate of change of the phase, and the energy, is a fundamental law of quantum mechanics.] If the state of the K is a linear combination of K_S and K_L, such as K^0, then it doesn't have a single frequency; rather it varies as a sum of two frequencies, one for its K_S component, and one for its K_L component.

Suppose that at t = 0, the state $|\Psi(t)\rangle$ of a neutral K meson is pure K_S

$$|\Psi(t=0)\rangle = |K_S\rangle. \tag{2-17}$$

We expect that at a later time, the probability for finding the particle in this state should decrease by a factor e^{-t/τ_S}, because of the exponential decay law for the decay of the K_S into $\pi\pi$. Now this probability is simply $|\langle K_S|\Psi(t)\rangle|^2$. Thus the amplitude $\langle K_S|\Psi(t)\rangle$ must vary in time with a factor $e^{-t/2\tau_S}$, and the state $|\Psi(t)\rangle$ must also have this time dependence, since the basis state $|K_S\rangle$ given by (2-8) is constant in time. Furthermore, as we have discussed, the state

$|\Psi(t)\rangle$ varies in time by a phase factor $e^{-i\omega_S t}$, where $\hbar\omega_S$ is the energy of the K_S meson. Thus putting these two factors together we have

$$|\Psi(t)\rangle = e^{-i\omega_S t - t/2\tau_S}|K_S\rangle \tag{2-18}$$

as the state of the K_S after time t. Similarly, if we start out with a K meson that is in a pure K_L state at $t = 0$, then at time t its state will be

$$|\Psi(t)\rangle = e^{-i\omega_L t - t/2\tau_L}|K_L\rangle. \tag{2-19}$$

QUANTUM INTERFERENCE EFFECTS

Some rather striking quantum mechanical effects can occur with K mesons. As a first example, we consider how K^0 particles can turn into their antiparticles, $\overline{K}^0$. Suppose that at $t = 0$ we produce, by process (2-1) for example, a neutral K meson in the $|K^0\rangle$ state. [We know that if strangeness is conserved in strong interactions the neutral K produced must have a definite strangeness value, +1, and hence be a K^0..] Let us consider how the state $|\Psi(t)\rangle$ of this meson changes in time. At $t = 0$,

$$|\Psi(t=0)\rangle = |K^0\rangle = \frac{1}{\sqrt{2}}(|K_S\rangle + |K_L\rangle). \tag{2-20}$$

The change of this state in time is governed by the change of its K_S and K_L components in time; these are the components that have a well-defined frequency. From (2-18) and (2-19) we find that at time t,

$$|\Psi(t)\rangle = \frac{1}{\sqrt{2}}\left[e^{-i\omega_S t - t/2\tau_S}|K_S\rangle + e^{-i\omega_L t - t/2\tau_L}|K_L\rangle\right]. \tag{2-21}$$

The probability amplitude that the meson in this state is a $\overline{K}^0$ at time t is $\langle\overline{K}^0|\Psi(t)\rangle$. From Eq. (2-8), we find $\langle\overline{K}^0|K_S\rangle = -\langle\overline{K}^0|K_L\rangle = 1/\sqrt{2}$. Thus

$$\langle\overline{K}^0|\Psi(t)\rangle = \frac{1}{2}\left(e^{-i\omega_S t - t/2\tau_S} - e^{-i\omega_L t - t/2\tau_L}\right). \tag{2-22}$$

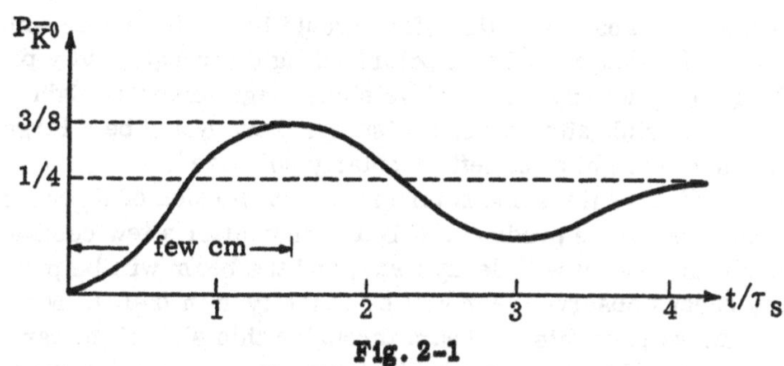

Fig. 2-1
Probability of observing a $\overline{K}^0$ at time t, when the state at time 0 is a K^0. Time is measured in units of τ_S.

The probability, then, that the meson in state $|\Psi(t)\rangle$ will behave as a $\overline{K}^0$ is

$$P_{\overline{K}^0}(t) = |\langle \overline{K}^0 | \Psi(t) \rangle|^2$$

$$= \frac{1}{4}\left[e^{-t/\tau_S} + e^{-t/\tau_L} - 2e^{-t(\tau_S^{-1} + \tau_L^{-1})/2} \cos(\omega_S - \omega_L)t \right]. \tag{2-23}$$

The frequency difference $\omega_S - \omega_L$ is in order of magnitude just $c^2/\hbar$ times the mass difference, $m_S - m_L$, and experimentally $m_L - m_S \approx \hbar/2c^2\tau_S$. Thus the plot of $P_{\overline{K}^0}(t)$ looks something like Fig. 2-1 for $t \ll \tau_L$. It is perhaps more useful to think of the abcissa of Fig. 2-1 as being the distance $l = vt$, that the K meson has traveled from its point of creation; v is its velocity. The picture then shows about the first 10 cm after the creation. Experimentally one observes the creation of the K^0 by looking for the Λ^0 (or rather its decay products) in (2-1). When the K^0 is created, it has no amplitude for being a $\overline{K}^0$, since $\langle \overline{K}^0 | K^0 \rangle = 0$. However, as the particle travels along, its K_S component decays away, leaving just K_L, which contains a large $\overline{K}^0$ component. Of course, after a very large distance, $\sim 10^2$ meters [$= p\tau_L/m_L$ where p is the *momentum* of the meson], the K_L component will also have decayed, but that is too far away to show on Fig. 2-1. The $\overline{K}^0$ component is observed when it undergoes a reaction like (2-2); one looks for the Λ^0 that is created in this reaction. The wiggle in the curve is a result of the $\cos(\omega_S - \omega_L)t$ term in (2-23). This term is due to quantum mechanical interference between the K_S and K_L amplitudes of the particle, and it provides an experimental determination of the mass difference $|m_S - m_L|$.

A good optical analogy to this effect would be a situation in which we had a crystal that absorbed x polarized light strongly, but y polarized light only weakly. Then if we shined right circular light through this crystal, after a small distance there would be a large amplitude for the light to be left circularly polarized.

Another effect of the same nature is the conversion of K_L mesons into K_S mesons. If we produce a K beam, then after a few centimeters the K_S component will decay away, and the beam will be pure K_L. Any decays observed then will be of the type in (2-12). Now suppose that we pass this K_L beam through a thin slab of matter. The K mesons will undergo strong interactions with the protons and neutrons in the matter. The K^0 component of the K_L can be scattered, possibly out of the beam, while the $\overline{K}^0$ component can be both scattered and also absorbed via the reaction (2-2).

Mathematically, the effect of the matter is to multiply the $|K^0\rangle$ component of the beam by a factor $\alpha e^{i\varphi}$, and the $|\overline{K}^0\rangle$ component by a factor $\beta e^{i\varphi'}$. The numbers α and β are positive and generally will be smaller than one, representing loss of particles from the beam due to scattering or absorption. The phase shifts φ and φ' are real. Thus if the state of the beam before the absorber is pure K_L

$$|\Psi_{\text{before}}\rangle = \frac{1}{\sqrt{2}}(|K^0\rangle - |\overline{K}^0\rangle), \tag{2-24}$$

then after the absorber it will be

$$|\Psi_{\text{after}}\rangle = \frac{1}{\sqrt{2}}(\alpha e^{i\varphi}|K^0\rangle - \beta e^{i\varphi'}|\overline{K}^0\rangle). \tag{2-25}$$

The amplitude for the beam to be in the K_S state is therefore

$$\langle K_S|\Psi_{\text{after}}\rangle = \frac{1}{2}(\alpha e^{i\varphi} - \beta e^{i\varphi'}), \tag{2-26}$$

which is nonzero if $\alpha \neq \beta$ or $\varphi \neq \varphi'$.

Thus the piece of matter has the effect of transforming the pure K_L beam into a beam with a K_S component, and it then becomes possible to see K_S decays (2-11) on the far side of the matter. This effect, which is well established experimentally, is called the *regeneration* of K_S mesons; after all the K_S's in the initial beam have decayed away, more can be made by passing the pure K_L beam through matter.

PROBLEMS

1. Suppose that a pure K_L beam is sent through a thin absorber whose only effect is to change the relative phase of the K^0 and $\overline{K}^0$ amplitudes by 10°. Calculate the number of K_S decays, relative to the incident number of particles, that will be observed in the first 5 cm beyond the absorber. Assume for simplicity that the particles have momentum = mc.
2. If CP is not conserved in the decay of neutral K mesons, then the states of definite energy are no longer the K_L, K_S states, but are slightly different states $|K_L'\rangle$ and $|K_S'\rangle$. One can write for example $|K_L'\rangle \sim (1+\varepsilon)|K^0\rangle - (1-\varepsilon)|\overline{K}^0\rangle$ where ε is a very small complex number ($|\varepsilon| \sim 2\times 10^{-3}$) that is a measure of the lack of CP conservation in the decays. The amplitude for a particle to be in $|K_L'\rangle$ (or $|K_S'\rangle$) varies as $e^{-i\omega_L t - t/2\tau_L}$ (or $e^{-i\omega_S t - t/2\tau_S}$) where $\hbar\omega_L = (p^2c^2 + m_L{}^2c^4)^{1/2}$, etc. As before $\tau_L \gg \tau_S$.
   (a) Write out normalized expressions for the states $|K_S'\rangle$ and $|K_L'\rangle$ in terms of $|K^0\rangle$ and $|\overline{K}^0\rangle$.
   (b) Calculate the ratio of (i) the amplitude for a long-lived K to decay into two pions (a CP = +1 state) to (ii) the amplitude for a short-lived K to decay into two pions. What does a measurement of the ratio of these decay rates tell one about ε?
   (c) Suppose that a beam of purely long-lived K mesons is sent through an absorber whose only effect is to change the relative phase of the K^0 and $\overline{K}^0$ components by δ. Derive an expression for the number of two pion events observed as a function of the time of travel from the absorber. How well would such a measurement (given δ) enable one to determine the phase of ε and the short-long mass difference?

Chapter 3

THE MOTION OF PARTICLES IN QUANTUM MECHANICS

The motion of a particle in quantum mechanics is described by a (complex) wave function, $\psi(\mathbf{r}, t)$, that gives the probability amplitude for finding the particle at point $\mathbf{r}$ at time t. The absolute value squared $|\psi(\mathbf{r}, t)|^2$ of the wave function times a volume element d^3r is the probability of finding the particle at time t in the volume element d^3r about t. Because the wave $\psi(\mathbf{r}, t)$ is a probability amplitude, it doesn't tell us how any one particle *will* behave, but rather it tells us the behavior of a large statistical sample of particles subjected to identical conditions. We found the same situation in discussing the polarization state of the photon; we could not say with certainty how any one photon would behave when, for example, passed through a polaroid. We could only give the fraction of a large number of identical photons that passed through the polaroid, and hence only the probability that any one would pass through.

THE SCHRÖDINGER EQUATION[1]

Suppose that we know the wave function for a particle at a certain time t. How will it change over the course of time? Let us begin by answering this question for a particle that moves in one dimension only. It will be most convenient for us to divide the line along which the particle moves into very small intervals, each of length λ (Fig. 3-1). We

[1] This discussion, similar to material in *The Feynman Lectures in Physics*, Vol. III, should not be regarded as a derivation of the Schrödinger equation; rather, it is an attempt to dissect it to see how it works.

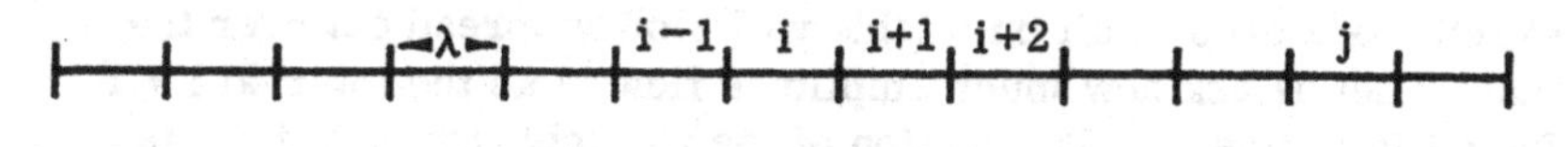

Fig. 3-1
One dimension divided into intervals of length λ.

label the intervals by letters, i, j, ..., and let $\psi_i(t)$ be the probability amplitude for finding the particle in the interval i at time t. Then $|\psi_i(t)|^2$ is the probability of finding the particle in the interval i at time t_1 and since the particle must be somewhere along the line, the total probability, summed over all the intervals must be one,

$$\sum_i |\psi_i(t)|^2 = 1. \tag{3-1}$$

The amplitudes $\psi_i(t)$ are like the components of a giant vector

$$|\Psi\rangle = \begin{pmatrix} \psi_1(t) \\ \psi_2(t) \\ \vdots \\ \psi_{i-1}(t) \\ \psi_i(t) \\ \psi_{i+1}(t) \\ \vdots \end{pmatrix} \tag{3-2}$$

very analogous to the state vector that described the polarization state of the photon. The photon vector had only two components, since there were only two independent possible polarizations, e.g., every vector could be written as a superposition of x and y polarization vectors. On the other hand, this vector (3-2), has an *infinite* number of components, since there are an infinite number of different intervals along the line in which the particle can be.

Let us suppose that at time t, there is some amplitude for the particle to be in the interval i. Then because the particle is free to move about, this amplitude will "leak" into the neighboring intervals, exactly as a drop of oil on a table will slowly spread out over the table. Let us ask how much amplitude flows into the interval $i + 1$ from i in a time Δt. The motion of the particle from i to $i + 1$ is described by a probability amplitude, in very much the same way as we described the passage of photon through a polaroid by a probability amplitude [e.g., Eq. (1-74)]. The increase in the probability amplitude in $i + 1$ is just the probability amplitude for the particle if it is in i to move to $i + 1$ in time Δt times the probability amplitude, $\psi_i(t)$, that it is in i to begin with. The probability amplitude for the particle to move from i to $i + 1$ in time Δt will, for sufficiently small Δt, be proportional to Δt. Let us call the constant of proportionality $-iw_{i+1,i}$, where the i before the w stands for $\sqrt{-1}$. Thus the total amount of amplitude that flows from i to $i + 1$ in time Δt will be

$$-i\Delta t w_{i+1,i}\,\psi_i(t). \tag{3-3}$$

The amplitude in interval i at time $t + \Delta t$ will be increased by amplitude flowing in from the neighboring intervals, and also decreased by amplitude flowing out. We can write the amplitude $\psi_i(t + \Delta t)$ as

$$\psi_i(t+\Delta t) = \psi_i(t) - i\Delta t w_{i,i-1}\psi_{i-1}(t) - i\Delta t w_{i,i+1}\psi_{i+1}(t) - i\Delta t w_{i\,i}\psi_i(t). \tag{3-4}$$

The first term on the right is the amplitude that was in i at time t. The second term represents the amplitude that has flowed in from $i - 1$, and the third term represents the amplitude that has flowed in from $i + 1$. If Δt is sufficiently small, we needn't worry about amplitude flowing into i from intervals further away than the nearest neighbors. The last term in (3-4) includes the decrease in amplitude in i due to its flowing out; this is proportional to Δt and to $\psi_i(t)$. There is another possible change in the amplitude $\psi_i(t)$ that is included in this last term. Even if there were no flow from interval to interval, the amplitude in interval i could change in time Δt by a phase factor of absolute value one, since this wouldn't change the probability of the particle being in the interval i. We shall return to this point shortly.

The w coefficients are not completely arbitrary, but are restricted by the fact that the total probability for finding the particle somewhere must remain one at all times. Thus

$$\sum_i |\psi_i(t+\Delta t)|^2 = \sum_i |\psi_i(t)|^2. \tag{3-5}$$

Substituting $\psi_i(t + \Delta t)$ from (3-4) on the left side and keeping terms to first order in Δt we find

$$\sum_i |\psi_i(t)|^2 (w_{ii} - w_{ii}^*) + \sum_i \psi_i^*(t)\psi_{i+1}(t)(w_{i,i+1} - w_{i+1,i}^*)$$
$$+ \sum_i \psi_i^*(t)\psi_{i-1}(t)(w_{i,i-1} - w_{i-1,i}^*) = 0, \tag{3-6}$$

regardless of what the various $\psi_i(t)$ are. This can only be so if the coefficients in parentheses individually vanish. Thus

$$w_{ii} = w_{ii}^*, \qquad w_{i,i+1} = w_{i+1,i}^*. \tag{3-7}$$

The amplitude to hop one way is the complex cónjugate of the amplitude to do the reverse hop. If we think of the w's as elements of a giant $(\infty \times \infty)$ matrix W, then this matrix has the form

$$W = \begin{pmatrix} w_{11} & w_{12} & & & \bigcirc \\ w_{12}^* & w_{22} & w_{23} & & \\ & w_{23}^* & w_{33} & w_{34} & \\ & & w_{34}^* & w_{44} & w_{45} \\ \bigcirc & & & \cdots & \cdots & \cdots \end{pmatrix} \tag{3-8}$$

It equals the complex conjugate of its transpose, and is therefore Hermitian. Equation (3-4) is the vector equation

$$|\psi(t + \Delta t)\rangle = (1 - i\Delta t W)|\psi(t)\rangle \tag{3-9}$$

written out in components. Compare this equation with (1-99), derived for the space development of the polarization vector of a photon in calcite. Recall that K was also Hermitian as a consequence of conservation of probability.

If we subtract $\psi_i(t)$ from both sides of (3-4), divide both sides by $-i\Delta t$, and then take the limit $\Delta t \to 0$, we find

$$i \frac{\partial \psi_i(t)}{\partial t} = w_{i,i-1}\psi_{i-1}(t) + w_{i,i+1}\psi_{i+1}(t) + w_{ii}\psi_i(t) \tag{3-10}$$

Let us introduce the number v_i by writing

$$w_{ii} = -w_{i,i+1} - w_{i,i-1} + \frac{v_i}{\hbar}. \qquad (3\text{-}11)$$

Then (3-10) takes the form

$$i\frac{\partial\psi_i(t)}{\partial t} = w_{i,i-1}[\psi_{i-1}(t) - \psi_i(t)] + w_{i,i+1}[\psi_{i+1}(t) - \psi_i(t)] + \frac{v_i}{\hbar}\psi_i(t) \qquad (3\text{-}12)$$

or

$$i\frac{\partial}{\partial t}|\Psi(t)\rangle = W|\Psi(t)\rangle.$$

Aside from the last term, this equation looks very much like a *diffusion equation* — the rate of flow of amplitude into interval i from $i \pm 1$ is proportional to the difference of the amplitudes in i and $i \pm 1$. One important difference between this equation and an ordinary diffusion equation, such as the Fourier equation for heat flow,

$$\frac{\partial T(r,t)}{\partial t} = \frac{\varkappa}{c_V}\nabla^2 T(r,t),$$

is that the diffusion constant, $\varkappa/c_V$, is real whereas Eq. (3-12), as we shall see, has an *imaginary* diffusion constant.

To see the significance of the last term in Eq. (3-12), let us suppose that there is no diffusion of amplitude from interval to interval, i e., that the first two terms on the right are zero. This would be the situation if the particle didn't move, if all its energy were potential and none kinetic. Then (3-12) has the solution

$$\psi_i(t) = e^{-iv_i t/\hbar}\psi_i(o). \qquad (3\text{-}13)$$

Since $\hbar$ times the frequency with which the amplitude oscillates in time is the energy of the particle, in this case only potential, we are led to identify v_i with the potential energy of a particle in the interval i. In general, there is diffusion of amplitude and the oscillation frequencies of the amplitude are more complicated than in (3-13), but they still contain a contribution from the potential energy term.

Let us now pass to the continuum limit by letting λ, the size of the intervals, tend to zero. Then ψ_i tends to $\sqrt{\lambda}\,\psi(x_i)$, where $\psi(x)$ is the wave function in the continuum limit and x_i is the position of, say, the center of the interval i. The reason for the $\sqrt{\lambda}$ is that $|\psi_i|^2$ is the probability of the particle being in the interval i, whereas $|\psi(x_i)|^2$ is the probability per unit length for the particle to be at the point x_i, and thus $\lambda|\psi(x_i)|^2$ is the probability for the particle to be in the interval i. The amplitude ψ_i is dimensionless whereas $\psi(x)$ has dimensions of $(\text{length})^{-1/2}$.

Writing (3-12) in terms of $\psi(x_i)$, the $\sqrt{\lambda}$ cancels from both sides, and we have

$$i\frac{\partial}{\partial t}\psi(x_i,t) = w_{i,i-1}[\psi(x_{i-1},t)-\psi(x_i,t)] \\ +w_{i,i+1}[\psi(x_{i+1},t)-\psi(x_i,t)]+\frac{v(x_i,t)}{\hbar}\psi(x_i,t). \tag{3-14}$$

We have indicated explicitly that v may also depend on time. Now $x_{i+1} = x_i + \lambda$, and for small λ we can expand $\psi(x_{i+1}, t)$ about x_i. Thus

$$\psi(x_{i+1},t) = \psi(x_i,t)\pm\lambda\frac{\partial\psi(x_i,t)}{\partial x_i}+\frac{\lambda^2}{2}\frac{\partial^2\psi(x_i,t)}{\partial x_i^2}+\ldots, \tag{3-15}$$

so that (3-14) becomes

$$i\frac{\partial}{\partial t}\psi(x_i,t) = \lambda[w_{i,i+1}-w_{i,i-1}]\frac{\partial\psi(x_i,t)}{\partial x_i} \\ +\frac{\lambda^2}{2}[w_{i,i+1}+w_{i,i-1}]\frac{\partial^2\psi(x_i,t)}{\partial x_i^2}+\frac{v(x_i,t)}{\hbar}\psi(x_i,t). \tag{3-16}$$

Let us write

$$w_{i,i+1} = w_L(x_i), \qquad w_{i,i-1} = w_R(x_i); \tag{3-17}$$

these are the coefficients in the amplitudes for the particle to move to the left or to the right, respectively, into the interval i. From (3-7) we have, to order λ^0,

$$w_R(x) = w_L(x)^*. \tag{3-18}$$

Dropping the now superfluous subscript i, we find the equation

$$i\frac{\partial\psi(x,t)}{\partial t} = \frac{\lambda^2}{2}[w_L(x)+w_R(x)]\frac{\partial^2\psi(x,t)}{\partial x^2}+\lambda[w_L(x)-w_R(x)]\frac{\partial\psi(x,t)}{\partial x} \\ +\frac{1}{\hbar}v(x,t)\psi(x,t). \tag{3-19}$$

In Cartesian coordinates, with no velocity dependent forces on the particle, such as a magnetic field,[2] one has

$$w_R(x) = w_L(x) = w$$

where w is independent of position, and by Eq. (3-18) must be real. Furthermore, as λ tends to zero, it is a fact that

$$\lambda^2 w \to -\frac{\hbar}{2m}, \tag{3-20}$$

where m is the mass of the particle. [We shall see that this identification of w is amply verified experimentally.] Equation (3-19) then becomes

$$i\hbar \frac{\partial \psi(x,t)}{\partial t} = -\frac{\hbar^2}{2m}\frac{\partial^2}{\partial x^2}\psi(x,t) + v(x,t)\psi(x,t), \tag{3-21}$$

which we recognize as the Schrödinger equation in one dimension. Were we to repeat the same argument in three dimensions, and make the same identification of w for the case of no velocity dependent forces, we would find the Schrödinger equation in three dimensions

$$i\hbar \frac{\partial \psi(\mathbf{r},t)}{\partial t} = \left[-\frac{\hbar^2}{2m}\nabla^2 + v(\mathbf{r},t)\right]\psi(\mathbf{r},t). \tag{3-22}$$

While we can give no "logically rigorous" argument for the identification of w we have made in (3-20), we can argue that it is reasonable. $-1/w$ is essentially the characteristic time, τ, for the amplitude to diffuse from one interval to the neighboring interval; from (3-20),

$$\tau = \frac{2m\lambda^2}{\hbar}. \tag{3-23}$$

First of all, we might expect that the more massive the particle is, the more difficult it is for it to hop from interval to interval, and thus the diffusion time τ should increase with m, as in (3-23). Second, as $\hbar$ tends to zero, and we approach the classical limit, then the diffusion process, which is a quantum mechanical process, should become relatively slower, whence the $\hbar$ in the denominator. Lastly,

---

[2] We treat the case of a magnetic field later in this chapter.

the λ^2 in the numerator of τ is characteristic of a *random walk process* — the distance one wanders in a time τ is proportional to $\tau^{1/2}$. In other words, to go a distance λ, from one interval to the next, requires a time proportional to λ^2.

Let us solve the Schrödinger equation (3-22) for a *free particle*, that is, for $v \equiv 0$. [Notice that this equation has the form of a diffusion equation, but with an imaginary diffusion constant.] If we look for a solution of the form

$$\psi(\mathbf{r}, t) = e^{i(\mathbf{p} \cdot \mathbf{r} - Et)/\hbar} \tag{3-24}$$

we find, on substituting (3-24) into (3-22) and carrying out the derivatives, that

$$E = \frac{p^2}{2m}. \tag{3-25}$$

Now E, being $\hbar$ times the frequency of (3-24), is identified as the energy of the particle described by the wave function (3-22), and similarly, $\mathbf{p}$, being $\hbar$ times the wave number of the plane wave (3-24), is identified as the momentum of the particle described by (3-24). Thus (3-25) is just the classical result for the energy of a free particle of mass m; this is a good indication that we have correctly identified m, p, and E.

Notice that the solution (3-24) implies that the probability density, $|\psi(\mathbf{r}, t)|^2$, is independent of space and time. This means that if a particle has a well-defined value of momentum it is completely nonlocalized in space — it has equal probability of being anywhere.

STATE VECTORS

Before proceeding to a discussion of consequences of the Schrödinger equation, let us spend a little time making contact between the description of the polarization of the photon, developed in Chapter 1, and the description of the states of a particle in terms of wave functions. Suppose that a particle is described, at a certain time, by a wave function, $\psi(\mathbf{r})$. Let us define a state vector $|\Psi\rangle$, that is essentially a continuum generalization of the vector (3-2). Like (3-2), it is infinite dimensional. We may think of this vector as looking like

$$|\Psi\rangle = \begin{pmatrix} \psi(\mathbf{r}_1) \\ \psi(\mathbf{r}_2) \\ \psi(\mathbf{r}_3) \\ \vdots \\ \psi(\mathbf{r}_i) \\ \vdots \end{pmatrix}, \tag{3-26}$$

where the index $\mathbf{r}_i$ runs over all the points of space. Of course, (3-26) should be regarded as no more than a way of visualizing this vector, since there is really a *continuum* of points in space, not a denumerable number of points.

The scalar product of two such vectors, $|\Psi\rangle$ corresponding to the wave function $\psi(\mathbf{r})$, and $|\Phi\rangle$ corresponding to the wave function $\varphi(\mathbf{r})$, we define to be

$$\langle\Phi|\Psi\rangle = \int d^3r\ \varphi^*(\mathbf{r})\psi(\mathbf{r}) = \langle\Psi|\Phi\rangle^*. \tag{3-27}$$

This is a continuum generalization of the scalar product (1-21) defined for the two-dimensional photon states.

In order for $\psi(\mathbf{r})$ to represent the probability amplitude for the particle to be at $\mathbf{r}$, then the total probability calculated from $\psi(\mathbf{r})$ must be one, that is,

$$\int d^3r\ |\psi(\mathbf{r})|^2 = 1. \tag{3-28}$$

In terms of the scalar product this normalization condition is

$$\langle\Psi|\Psi\rangle = 1. \tag{3-29}$$

Let us introduce a state $|\mathbf{r}\rangle$ that corresponds to the particle being localized at the point $\mathbf{r}$. The wave function $\psi(\mathbf{r})$ corresponding to the state $|\Psi\rangle$ is then the component of the state $|\Psi\rangle$ along the state $|\mathbf{r}\rangle$,

$$\psi(\mathbf{r}) = \langle\mathbf{r}|\Psi\rangle, \tag{3-30}$$

since it is the amplitude for finding the particle in the state $|\Psi\rangle$ at the point $\mathbf{r}$. The set of all these localized states, $|\mathbf{r}\rangle$, forms a basis since the set of all amplitudes $\langle\mathbf{r}|\Psi\rangle$, for all $\mathbf{r}$, completely specify the state.

In terms of the notation, (3-30), the scalar product, $\langle\Phi|\Psi\rangle$, can be written as

$$\langle \Phi | \Psi \rangle = \int d^3r \, \langle \mathbf{r} | \Phi \rangle^* \langle \mathbf{r} | \Psi \rangle = \int d^3r \, \langle \Phi | \mathbf{r} \rangle \langle \mathbf{r} | \Psi \rangle$$
$$= \langle \Phi | \left(\int d^3r \, | \mathbf{r} \rangle \langle \mathbf{r} | \right) | \Psi \rangle . \tag{3-31}$$

Since $|\Phi\rangle$ and $|\Psi\rangle$ can be any states, the matrix in parentheses must be the unit operator for these infinite-dimensional state vectors:

$$\int d^3r \, |\mathbf{r}\rangle \langle \mathbf{r}| = 1, \tag{3-32}$$

where the 1 is like an $\infty \times \infty$ unit matrix. This equation is the completeness relation for the $|\mathbf{r}\rangle$ basis.

Acting with both sides of (3-32) on a vector $|\Psi\rangle$ gives us

$$|\Psi\rangle = \int d^3r' \, |\mathbf{r}'\rangle \langle \mathbf{r}' | \Psi \rangle = \int d^3r' \, |\mathbf{r}'\rangle \psi(\mathbf{r}'), \tag{3-33}$$

which is the expansion of an arbitrary vector $|\Psi\rangle$ in terms of the $|\mathbf{r}\rangle$ vectors.

What is the value of the scalar product $\langle \mathbf{r} | \mathbf{r}' \rangle$? This is the same as asking what is the wave function corresponding to the state, $|\mathbf{r}'\rangle$, in which the particle is localized at $\mathbf{r}'$. We may find the answer by taking the scalar product of both sides of (3-33) with $\langle \mathbf{r}|$. This gives

$$\langle \mathbf{r} | \Psi \rangle = \int d^3r' \, \langle \mathbf{r} | \mathbf{r}' \rangle \langle \mathbf{r}' | \Psi \rangle$$

or

$$\psi(\mathbf{r}) = \int d^3r' \, \langle \mathbf{r} | \mathbf{r}' \rangle \psi(\mathbf{r}'). \tag{3-34}$$

Thus, since $\psi(\mathbf{r})$ is an arbitrary function,

$$\langle \mathbf{r} | \mathbf{r}' \rangle = \delta(\mathbf{r} - \mathbf{r}'), \tag{3-35}$$

where $\delta(\mathbf{r} - \mathbf{r}')$ is the Dirac delta function in three dimensions.

DIGRESSION ON THE DELTA FUNCTION

For the benefit of those who are not familiar with the delta function, let me spend a moment describing its properties. In one dimension, the function $\delta(x - x')$ is equal to zero unless $x = x'$, when it equals infinity; it has the further property that

$$\int_{-\infty}^{\infty} dx' \, \delta(x - x') f(x') = f(x), \tag{3-36}$$

where $f(x)$ is any arbitrary function. For the special case $f(x) = 1$, we see that

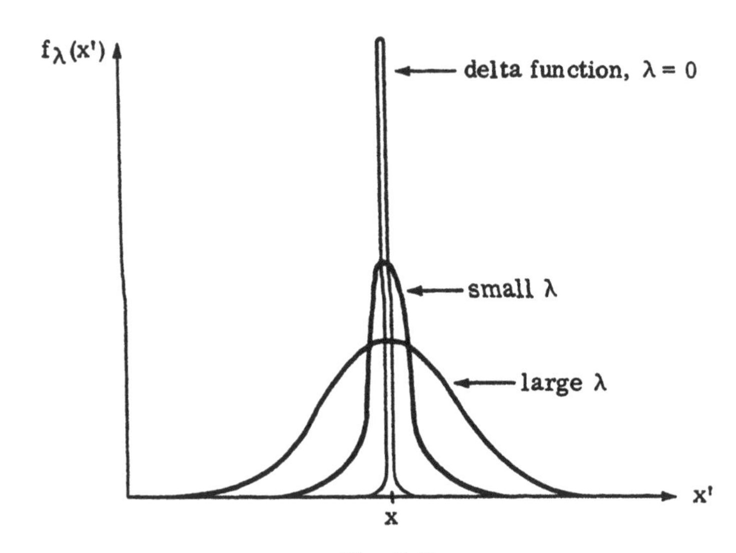

Fig. 3-2
The delta function $\delta(x' - x)$ as the limit of the sequence of functions $f_\lambda(x')$ as $\lambda \to 0$.

$$\int_{-\infty}^{\infty} dx' \; \delta(x - x') = 1; \tag{3-37}$$

the delta function has unit area under it. Also,

$$\delta(x - x') = \delta(x' - x) \tag{3-38}$$

and

$$\delta(ax) = \frac{1}{|a|}\,\delta(x), \tag{3-39}$$

as may be verified by changing the integration variable in (3-36) to ax'.

The delta function can be regarded as the limit of a sequence of functions, each of which has unit area, that become more and more sharply peaked as a function of x' about the point x. For example, the Gaussian

$$f_\lambda(x') = \frac{1}{\lambda\sqrt{2\pi}}\, e^{-(x' - x)^2/2\lambda^2}, \tag{3-40}$$

has unit area

$$\int_{-\infty}^{\infty} dx' \; f_\lambda(x') = 1,$$

and is peaked about $x' = x$ with a width $\sim\lambda$. As λ becomes smaller, $f_\lambda(x')$ becomes more and more sharply peaked as in Fig. 3-2, and finally as $\lambda \to 0$

$$f_\lambda(x') \to \delta(x - x'). \tag{3-41}$$

The delta function in three dimensions $\delta(\mathbf{r} - \mathbf{r}')$ has the property that

$$\int d^3r'\, \delta(\mathbf{r} - \mathbf{r}')f(\mathbf{r}') = f(\mathbf{r}), \tag{3-42}$$

and may be written, in Cartesian coordinates, as

$$\delta(\mathbf{r} - \mathbf{r}') = \delta(x - x')\delta(y - y')\delta(z - z'). \tag{3-43}$$

From (3-30) and (3-35) we see that the wave function, $\psi(\mathbf{r})$, for a particle in the state $|\mathbf{r}_0\rangle$, localized at $\mathbf{r}_0$, is

$$\psi(\mathbf{r}) = \delta(\mathbf{r} - \mathbf{r}_0). \tag{3-44}$$

This wave function is not normalized to one as in (3-28). This is not really a difficulty, since the only use we shall have for these perfectly localized states is as basis states; one never encounters the situation in which a particle is perfectly localized, since such a particle would have, by the uncertainty principle, infinite kinetic energy.

Let us return again to the particle described by a wave function, $\psi(\mathbf{r})$, at a certain time. We can write the Fourier transform of $\psi(\mathbf{r})$ as

$$f(\mathbf{p}) = \int d^3r\, e^{-i\mathbf{p}\cdot\mathbf{r}/\hbar}\psi(\mathbf{r}). \tag{3-45}$$

The $\hbar$ in the exponent implies that $\mathbf{p}$ has the dimensions of momentum. Inverting this Fourier transform, we have

$$\psi(\mathbf{r}) = \int \frac{d^3p}{(2\pi\hbar)^3}\, e^{i\mathbf{p}\cdot\mathbf{r}/\hbar} f(\mathbf{p}). \tag{3-46}$$

This equation says that any state is a linear combination of the plane wave states $e^{i\mathbf{p}\cdot\mathbf{r}/\hbar}$, in which the momentum of the particle has a definite value $\mathbf{p}$. Notice the $(2\pi\hbar)^3$ in (3-46) — this is a consequence of our definition of the Fourier transform.

Let us write the state vector corresponding to the plane wave function $e^{i\mathbf{p}\cdot\mathbf{r}/h}$ as $|\mathbf{p}\rangle$. Then

$$\langle \mathbf{r}|\mathbf{p}\rangle = e^{i\mathbf{p}\cdot\mathbf{r}/\hbar}. \tag{3-47}$$

The Fourier transform of $\psi(\mathbf{r})$ can thus be written

$$f(\mathbf{p}) = \int d^3r\, \langle \mathbf{p}|\mathbf{r}\rangle\langle \mathbf{r}|\Psi\rangle = \langle \mathbf{p}|\Psi\rangle, \tag{3-48}$$

making use of the completeness relation (3-32) for the $|\mathbf{r}\rangle$ vectors; $f(\mathbf{p})$ is the component of the $|\Psi\rangle$ vector along the $|\mathbf{p}\rangle$ vector. Equation (3-46) can therefore be written as

$$\langle \mathbf{r}|\Psi\rangle = \int \frac{d^3p}{(2\pi\hbar)^3} \langle \mathbf{r}|\mathbf{p}\rangle\langle \mathbf{p}|\Psi\rangle, \tag{3-49}$$

which is the vector equation

$$|\Psi\rangle = \int \frac{d^3p}{(2\pi\hbar)^3} |\mathbf{p}\rangle\langle \mathbf{p}|\Psi\rangle \tag{3-50}$$

written out in components in the $|\mathbf{r}\rangle$ basis, or as this basis is more commonly called, the *position representation.*

Equation (3-50) says that the $|\mathbf{p}\rangle$ vectors form a basis; any $|\Psi\rangle$ is a linear combination of $|\mathbf{p}\rangle$ vectors (this is just the Fourier transformation representation). Clearly, from (3-50),

$$\int \frac{d^3p}{(2\pi\hbar)^3} |\mathbf{p}\rangle\langle \mathbf{p}| = 1, \tag{3-51}$$

the infinite-dimensional unit operator, the same unit operator as in (3-32). Equation (3-51) is the completeness relation for the $|\mathbf{p}\rangle$ basis. Taking $|\Psi\rangle = |\mathbf{r}\rangle$ in (3-50) we have

$$|\mathbf{r}\rangle = \int \frac{d^3p}{(2\pi\hbar)^3} |\mathbf{p}\rangle\langle \mathbf{p}|\mathbf{r}\rangle = \int \frac{d^3p}{(2\pi\hbar)^3} |\mathbf{p}\rangle\, e^{-i\mathbf{p}\cdot\mathbf{r}/\hbar} \tag{3-52}$$

which says that a state localized at $\mathbf{r}$ is a linear combination of momentum states with amplitudes $e^{-i\mathbf{p}\cdot\mathbf{r}/\hbar}$.

Taking the scalar product of both sides of (3-52) with $\langle \mathbf{r}'|$ we find

$$\int \frac{d^3p}{(2\pi\hbar)^3} \langle \mathbf{r}'|\mathbf{p}\rangle\langle \mathbf{p}|\mathbf{r}\rangle = \delta(\mathbf{r}-\mathbf{r}')$$

or

$$\int \frac{d^3p}{(2\pi\hbar)^3} e^{i\mathbf{p}\cdot(\mathbf{r}'-\mathbf{r})/\hbar} = \delta(\mathbf{r}-\mathbf{r}') \tag{3-53}$$

which is the completeness relation (3-51), written out in components in the position representation. The matrix elements of the unit operator in the position representation are

$$\langle \mathbf{r}'|1|\mathbf{r}\rangle = \langle \mathbf{r}'|\mathbf{r}\rangle = \delta(\mathbf{r}-\mathbf{r}'). \tag{3-54}$$

We can also interpret (3-49) as the equation that gives the components of $|\Psi\rangle$ in the position representation in terms of the components, $\langle \mathbf{p}|\Psi\rangle$, in the *momentum representation* (as the p basis is often called). Thus $\langle \mathbf{r}|\mathbf{p}\rangle$ is the analogue of the components of the transformation matrix (1-36) introduced for photon states. $\langle \mathbf{r}|\mathbf{p}\rangle$ is called the *transformation function* from the momentum representation to the position representation.

If the state, $|\Psi\rangle$, of the particle is normalized to one, then from the completeness relation, (3-51), for the p basis we find

$$1 = \langle \Psi|\Psi\rangle = \langle \Psi|\left(\int \frac{d^3p}{(2\pi\hbar)^3}|\mathbf{p}\rangle\langle \mathbf{p}|\right)|\Psi\rangle = \int \frac{d^3p}{(2\pi\hbar)^3}|\langle \mathbf{p}|\Psi\rangle|^2; \tag{3-55}$$

this is the normalization condition on $|\Psi\rangle$ in terms of its components in the momentum representation.

A particle in the state $|\mathbf{p}\rangle$ has the definite value $\mathbf{p}$ for its momentum. Thus $\langle \mathbf{p}|\Psi\rangle$, the component of $|\Psi\rangle$ along $|\mathbf{p}\rangle$, has the physical interpretation of being the probability amplitude that a particle in the state $|\Psi\rangle$ [with wave function $\psi(\mathbf{r})$] will have momentum $\mathbf{p}$. Therefore, because of the normalization condition (3-55), $[d^3p/(2\pi\hbar)^3]|\langle \mathbf{p}|\Psi\rangle|^2$ is the probability that the particle will behave as if it had a value of momentum in the interval d^3p about $\mathbf{p}$.

We see then that the wave function, $\psi(\mathbf{r})$, gives the probability amplitude for finding the particle at the various points of position space, whereas the Fourier transform, (3-45), of the wave function gives the probability amplitude for finding the particle at the various points in momentum space.

OPERATORS FOR PHYSICAL QUANTITIES

Corresponding to every physical quantity is an operator. A particle whose state is an eigenstate of the operator has a definite value of the physical quantity. This value is just the eigenvalue belonging to the eigenstate. For instance, if a particle is in the state $|\mathbf{p}\rangle$, whose wave function is the plane wave, $e^{i\mathbf{p}\cdot\mathbf{r}/\hbar}$, then it has the well-defined value, $\mathbf{p}$, of its momentum. What is the operator, $\mathbf{p}_{op}$, that corresponds to the momentum? We can find out if we use the fact that $|\mathbf{p}\rangle$ is an eigenstate of the operator with eigenvalue $\mathbf{p}$. Thus

$$\mathbf{p}_{op}|\mathbf{p}\rangle = \mathbf{p}|\mathbf{p}\rangle, \tag{3-56}$$

and from (3-50), we find that the momentum operator acting on an arbitrary state $|\Psi\rangle$ gives

$$\mathbf{p}_{op}|\Psi\rangle = \int \frac{d^3p}{(2\pi\hbar)^3}\, \mathbf{p}|\mathbf{p}\rangle\langle\mathbf{p}|\Psi\rangle. \tag{3-57}$$

Thus we can represent the momentum operator in terms of its eigenstates and eigenvalues as

$$\mathbf{p}_{op} = \int \frac{d^3p}{(2\pi\hbar)^3}\, \mathbf{p}|\mathbf{p}\rangle\langle\mathbf{p}|. \tag{3-58}$$

If $\psi(\mathbf{r})$ is the wave function corresponding to the state $|\Psi\rangle$, what wave function corresponds to the state $\mathbf{p}_{op}|\Psi\rangle$? To answer this, let us note that from (3-56) and (3-47),

$$\langle\mathbf{r}|\mathbf{p}_{op}|\mathbf{p}\rangle = \mathbf{p}\langle\mathbf{r}|\mathbf{p}\rangle = \frac{\hbar}{i}\nabla\langle\mathbf{r}|\mathbf{p}\rangle. \tag{3-59}$$

Thus for any arbitrary state $|\Psi\rangle$,

$$\langle\mathbf{r}|\mathbf{p}_{op}|\Psi\rangle = \frac{\hbar}{i}\nabla\langle\mathbf{r}|\Psi\rangle. \tag{3-60}$$

The effect of the momentum operator on the wave function of a state is to produce the gradient of the wave function times $\hbar/i$.

We can write the free particle Schrödinger equation simply in terms of the momentum operator. Let us write the time-dependent wave function $\psi(\mathbf{r}, t)$ as the component of a time-dependent state vector $|\Psi(t)\rangle$ along $|\mathbf{r}\rangle$. Then

$$\begin{aligned} i\hbar\frac{\partial}{\partial t}\langle\mathbf{r}|\Psi(t)\rangle &= \frac{1}{2m}\left(\frac{\hbar}{i}\nabla\right)^2\langle\mathbf{r}|\Psi(t)\rangle = \frac{1}{2m}\left(\frac{\hbar}{i}\nabla\right)\langle\mathbf{r}|\mathbf{p}_{op}|\Psi(t)\rangle \\ &= \langle\mathbf{r}|\frac{\mathbf{p}_{op}^2}{2m}|\Psi(t)\rangle. \end{aligned} \tag{3-61}$$

This is the vector equation

$$i\hbar \frac{\partial}{\partial t} |\Psi(t)\rangle = \frac{p_{op}^2}{2m} |\Psi(t)\rangle \tag{3-62}$$

written out in components in the position representation. We recognize that $p_{op}^2/2m$ is just the classical Hamiltonian, $p^2/2m$, of a free particle, with the classical momentum coordinate replaced by the quantum mechanical momentum operator. It is a consequence of the fact that $-\hbar^2\nabla^2/2m$ is the operator representation of the classical Hamiltonian, that the frequency of a free particle momentum eigenstate (3-24), is just the classical energy, $p^2/2m$, divided by $\hbar$.

Let us next determine the operator that corresponds to the position of the particle. If the particle were in the state $|\mathbf{r}'\rangle$, perfectly localized at $\mathbf{r}'$, then it would have a definite value, $\mathbf{r}'$, of its position. Thus

$$\mathbf{r}_{op}|\mathbf{r}'\rangle = \mathbf{r}'|\mathbf{r}'\rangle,$$

and we can write,

$$\mathbf{r}_{op} = \int d^3r' \; \mathbf{r}'|\mathbf{r}'\rangle\langle \mathbf{r}'|, \tag{3-63}$$

as the analog of (3-58).

Let us take the matrix elements of the position operator between $\langle \mathbf{r}|$ and $|\Psi\rangle$. Using (3-63) and (3-35), we find

$$\langle \mathbf{r}|\mathbf{r}_{op}|\Psi\rangle = \int d^3r' \; \mathbf{r}'\langle \mathbf{r}|\mathbf{r}'\rangle\langle \mathbf{r}'|\Psi\rangle = \mathbf{r}\langle \mathbf{r}|\Psi\rangle. \tag{3-64}$$

The effect of the position operator, in terms of the wave function, is simply to multiply the wave function, $\psi(\mathbf{r})$, by $\mathbf{r}$.

In fact, the operator corresponding to a function of position, such as $v(\mathbf{r})$, simply acts to multiply the wave function by $v(\mathbf{r})$. That is, the wave function corresponding to the state $v(\mathbf{r}_{op})|\Psi\rangle$ is

$$\langle \mathbf{r}|v(\mathbf{r}_{op})|\Psi\rangle = v(\mathbf{r})\langle \mathbf{r}|\Psi\rangle. \tag{3-65}$$

For example, $\langle \mathbf{r}|\mathbf{r}_{op}^2|\Psi\rangle = \mathbf{r}\cdot\langle \mathbf{r}|\mathbf{r}_{op}|\Psi\rangle = \mathbf{r}\cdot\mathbf{r}\langle \mathbf{r}|\Psi\rangle = r^2\langle \mathbf{r}|\Psi\rangle$, etc.

We can now write the full Schrödinger equation (3-22) in operator language;

$$i\hbar \frac{\partial \psi(\mathbf{r},t)}{\partial t} = \left[\frac{1}{2m}\left(\frac{\hbar}{i}\nabla\right)^2 + v(\mathbf{r},t)\right]\psi(\mathbf{r},t)$$

or

$$i\hbar \frac{\partial \langle \mathbf{r}|\Psi(t)\rangle}{\partial t} = \left[\frac{1}{2m}\left(\frac{\hbar}{i}\nabla\right)^2 + v(\mathbf{r},t)\right]\langle \mathbf{r}|\Psi(t)\rangle$$

$$= \langle \mathbf{r}|\left(\frac{p_{op}^2}{2m} + v(\mathbf{r}_{op}, t)\right)|\Psi(t)\rangle. \tag{3-66}$$

This is the vector equation

$$i\hbar \frac{\partial |\Psi(t)\rangle}{\partial t} = H_{op}|\Psi(t)\rangle, \tag{3-67}$$

written out in components in the $\mathbf{r}$ basis, where the Hamiltonian operator,

$$H_{op} = \frac{p_{op}^2}{2m} + v(\mathbf{r}_{op}, t), \tag{3-68}$$

is the classical Hamiltonian, $H(\mathbf{r}, \mathbf{p})$, with the classical momentum and position coordinates replaced by the corresponding quantum mechanical operators. The Schrödinger equation in position space thus has the form

$$i\hbar \frac{\partial}{\partial t}\psi(\mathbf{r}, t) = H(\mathbf{r}, \frac{\hbar}{i}\nabla)\psi(\mathbf{r}, t). \tag{3-69}$$

This appearance of the classical Hamiltonian guarantees that the Schrödinger equation yields the correct classical limit; this is the essential reason that v_i in (3-11) is identified as the potential energy.

THE FREE PARTICLE

Let us turn now to discussing the motion of a free particle in some detail. If at time $t = 0$, the state of the particle is $\psi(\mathbf{r}) = e^{i\mathbf{p}\cdot\mathbf{r}/\hbar}$, then from (3-24), the state at a later time t will be

$$\psi(\mathbf{r}, t) = e^{i(\mathbf{p}\cdot\mathbf{r} - Et)/\hbar},$$

where $E = p^2/2m$. Thus a linear combination of plane waves, as in (3-46), develops in time as

$$\psi(\mathbf{r}, t) = \int \frac{d^3p}{(2\pi\hbar)^3} e^{i\mathbf{p}\cdot\mathbf{r}/\hbar} e^{-iEt/\hbar} f(\mathbf{p}). \tag{3-70}$$

In our fancy notation this becomes

$$\langle \mathbf{r}|\Psi(t)\rangle = \int \frac{d^3p}{(2\pi\hbar)^3} \langle \mathbf{r}|\mathbf{p}\rangle\, e^{-iEt/\hbar} \langle \mathbf{p}|\Psi(0)\rangle, \tag{3-71}$$

where $f(\mathbf{p}) = \langle \mathbf{p}|\Psi(0)\rangle$ is the momentum amplitude at time $t = 0$. Clearly,

$$\langle \mathbf{p}|\Psi(t)\rangle = e^{-iEt/\hbar} \langle \mathbf{p}|\Psi(0)\rangle;$$

the momentum amplitude changes in time by a phase factor only, since a free particle with a definite momentum, $\mathbf{p}$, also has a definite energy, $p^2/2m$.

Equation (3-71) in terms of vectors is

$$|\Psi(t)\rangle = \int \frac{d^3p}{(2\pi\hbar)^3} \left(|\mathbf{p}\rangle\, e^{-iEt/\hbar} \langle \mathbf{p}| \right) |\Psi(0)\rangle, \tag{3-72}$$

which is the solution of (3-62). [Don't be alarmed by the exponential factor being written between $|\mathbf{p}\rangle$ and $\langle \mathbf{p}|$. It means the same thing as $e^{-iEt/\hbar}|\mathbf{p}\rangle\langle \mathbf{p}|$.] This equation expresses the state at time t in terms of the state at time zero.

In a state in which the momentum is certainly $\mathbf{p}$ the probability $|\langle \mathbf{r}|\mathbf{p}\rangle|^2$ of finding the particle at point $\mathbf{r}$ is independent of $\mathbf{r}$; the position is then completely uncertain. Similarly if the position of the particle is certainly at $\mathbf{r}$ the probability $|\langle \mathbf{p}|\mathbf{r}\rangle|^2$ of finding the particle to have momentum $\mathbf{p}$ is independent of $\mathbf{p}$, as so the momentum

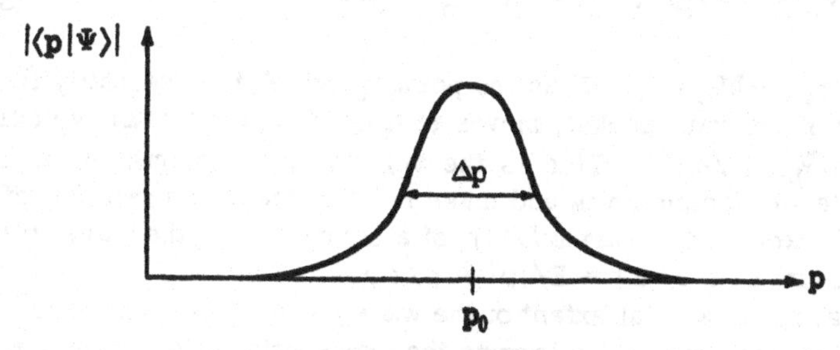

Fig. 3-3
The absolute value of the momentum space amplitude $\langle p|\Psi\rangle$ for a wave packet of mean momentum p_0 and uncertainty in momentum $\sim\Delta p$.

is completely uncertain. Let us consider the intermediate case of a state in which the particle is reasonably well localized in position space, and at the same time has a fairly well defined momentum. We can construct such a state, called a *wave packet,* as follows. Suppose that $|\langle \mathbf{p}|\Psi\rangle|$ is peaked about the value $\mathbf{p}_0$, with a width $\sim \Delta p$, as in Fig. 3-3. At t = 0, let $\langle \mathbf{p}|\Psi\rangle = g(\mathbf{p})e^{i\alpha(\mathbf{p})}$, where $g(\mathbf{p})$ is a real, nonnegative function and $\alpha(\mathbf{p})$ is real. Then

$$\psi(\mathbf{r}, t) = \int \frac{d^3p}{(2\pi\hbar)^3} g(\mathbf{p})\, e^{i[\mathbf{p}\cdot\mathbf{r} - Et + \hbar\alpha(\mathbf{p})]/\hbar} \tag{3-73}$$

is the wave function of a particle whose momentum is approximately $\mathbf{p}_0$, i.e., most of the time that we measure the momentum of a particle in this state we would come out with a value within Δp of the value $\mathbf{p}_0$.

Where is this particle, and how does it move in time? At time t, for most of the points $\mathbf{r}$, the exponential factor in (3-73) oscillates so many times in the region where $g(\mathbf{p})$ is large that the $\mathbf{p}$ integral averages to zero. However, if $\mathbf{r}$ is a point such that the phase of the exponential remains fairly constant over the range of $\mathbf{p}$ for which $g(\mathbf{p})$ is large, then we will get a contribution to the integral. In particular, if at time t, the point $\mathbf{r}$ makes the phase stationary around $\mathbf{p}_0$, then we will get strong constructive interference, i.e., at the point $\mathbf{r}_t$ at which

$$\nabla_p [\mathbf{p}\cdot\mathbf{r}_t - Et + \hbar\alpha(\mathbf{p})]_{(\mathbf{p} = \mathbf{p}_0)} = 0, \tag{3-74}$$

$\psi(\mathbf{r}, t)$ will be relatively large. Solving for $\mathbf{r}_t$ in (3-74) we find

$$\mathbf{r}_t = [(\nabla_p E)t - \hbar\nabla_p \alpha(\mathbf{p})]_{(\mathbf{p} = \mathbf{p}_0)} = \frac{\mathbf{p}_0}{m} t + \mathbf{r}_0 , \tag{3-75}$$

where $\mathbf{r}_0 = -\hbar\nabla_p \alpha(\mathbf{p}_0)$. Thus the point $\mathbf{r}_t$, which is essentially the center of the wave packet, moves in time with a constant velocity $\mathbf{p}_0/m = \nabla_{p_0}(p_0{}^2/2\,m)$. This is the velocity one expects for a free particle of momentum $\mathbf{p}_0$ and mass m; it is the *group velocity* of the wave packet. The *phase velocity* of a component of the wave with momentum $\mathbf{p}$ is $\omega/|k| = E/|p| = p/2m$.

What is the spatial extent of the wave packet? We will have constructive interference as long as the exponential in (3-73) undergoes less than one oscillation as $\mathbf{p}$ varies over the region for which $g(\mathbf{p})$ is large. Now the change in the phase of the exponential as we vary, say, the x component of $\mathbf{p}$ is approximately

$$\frac{1}{\hbar}\Delta p_x \frac{\partial}{\partial p_x}(\mathbf{p}\cdot\mathbf{r} - Et + \hbar\alpha)_{(\mathbf{p}=\mathbf{p}_0)} = \frac{1}{\hbar}\Delta p_x(x - x_t), \tag{3-76}$$

making use of Eq. (3-75). As long as this phase variation is $\leqslant 2\pi$, we will get constructive interference when we do the p_x integral. This tells us the extent of the wave packet in the x direction about the point $\mathbf{r}_t$. As long as $|(1/\hbar)\Delta p_x(x - x_t)| \leqslant 2\pi$ or

$$|x - x_t| \leqslant \frac{2\pi\hbar}{\Delta p_x} \tag{3-77}$$

then we get constructive interference and the point will be in the wave packet. Hence the extent, Δx, of the wave packet in the wave packet in the x direction must be *at least* $2\pi\hbar/\Delta p_x$, and we therefore can write $\Delta x \geqslant 2\pi\hbar/\Delta p_x$, or

$$\Delta x \Delta p_x \geqslant 2\pi\hbar. \tag{3-78a}$$

Carrying out the same argument for the y and z directions tells us that the spatial extent of the wave packet in these directions must obey

$$\Delta y \Delta p_y \geqslant 2\pi\hbar, \qquad \Delta z \Delta p_z \geqslant 2\pi\hbar. \tag{3-78b}$$

These relations between the spatial extent of the wave packet, i.e., the uncertainty in the position of the particle, and the uncertainty in the momentum of the particle, are known as the Heisenberg *uncertainty relations.* They form the cornerstone on which the logical consistency of quantum mechanics rests. [Equation (1-79) is one example of this.]

Notice that there is no uncertainty relation between, for example, Δx and Δp_y. It is possible for the x coordinate of the particle to be well localized at the same time as the particle has a well defined value of its momentum in the y direction; it can't, however, have a well defined value of its momentum in the x direction if its x coordinate is well localized, and vice versa.

In the classical region we can measure both x and p to an accuracy that on a classical scale is arbitrarily fine, and thus a particle in a wave packet state in the classical limit appears to have both a well defined position and a well defined momentum. Classical measurements are never so fine that $\Delta x \Delta p \sim 2\pi\hbar$.

What is the absolutely most "certain" wave packet one can construct, the one with the least uncertainty in x and p_x? This can be answered by first noticing several useful facts. The *commutator* of two operators A and B is defined by

$$[A, B] \equiv AB - BA. \tag{3-79}$$

If the commutator vanishes the operators are said to *commute*. The commutator of x_{op} (the x position operator) with $p_{x,op}$ is given by

$$[x_{op}, p_{x,op}] = i\hbar. \tag{3-80}$$

To prove this we note that for an arbitrary state $|\Psi\rangle$ and basis state $|\mathbf{r}\rangle$:

$$\begin{aligned}\langle \mathbf{r}|(x_{op}p_{x,op} - p_{x,op}x_{op})|\Psi\rangle &= \left(x\frac{\hbar}{i}\frac{\partial}{\partial x} - \frac{\hbar}{i}\frac{\partial}{\partial x}x\right)\langle \mathbf{r}|\Psi\rangle \\ &= i\hbar\langle \mathbf{r}|\Psi\rangle,\end{aligned} \tag{3-81}$$

since the $\partial/\partial x$ acts on *everything* to its right. Equation (3-80) follows at once from (3-81), because $|\Psi\rangle$ and $|\mathbf{r}\rangle$ are arbitrary.

Next we observe that if $|\Theta\rangle$ and $|\Phi\rangle$ are two normalized state vectors, then

$$|\langle\Theta|\Phi\rangle| \leq 1 \tag{3-82}$$

The proof is based on the fact that if $|\zeta\rangle \equiv |\Theta\rangle - e^{i\alpha}|\Phi\rangle$, where α is an arbitrary real number, then

$$0 \leq \langle\zeta|\zeta\rangle = (\langle\Theta| - e^{-i\alpha}\langle\Phi|)(|\Theta\rangle - e^{i\alpha}|\Phi\rangle)$$

or

$$e^{i\alpha}\langle\Theta|\Phi\rangle + e^{-i\alpha}\langle\Phi|\Theta\rangle \leq 2; \tag{3-83}$$

equality holds only if $|\zeta\rangle = 0$. If we choose α so that

$$\langle\Theta|\Phi\rangle = e^{-i\alpha}|\langle\Theta|\Phi\rangle|$$

then (3-82) is an immediate consequence of (3-83). More generally if $|\Theta\rangle$ and $|\Phi\rangle$ are not normalized, then the states $|\Theta\rangle/\langle\Theta|\Theta\rangle^{1/2}$ and $|\Phi\rangle/\langle\Phi|\Phi\rangle^{1/2}$ are normalized to one. Applying (3-82) and (3-83) to these states we derive:

$$\begin{aligned}\sqrt{\langle\Theta|\Theta\rangle}\sqrt{\langle\Phi|\Phi\rangle} &\geq |\langle\Theta|\Phi\rangle| \\ &\geq \frac{1}{2}[e^{i\alpha}\langle\Theta|\Phi\rangle + e^{-i\alpha}\langle\Phi|\Theta\rangle]\end{aligned} \tag{3-84}$$

The first line of (3-84) is called the *Schwarz inequality*.

The uncertainties Δx and Δp_x for a state $|\Psi\rangle$ can be defined precisely by the standard relations

$$\Delta x = \sqrt{\langle (x - \langle x \rangle)^2 \rangle}$$

$$\Delta p_x = \sqrt{\langle (p_{x,\,op} - \langle p_x \rangle)^2 \rangle} \tag{3-85}$$

where the expectation values are in the normalized state $|\Psi\rangle$; e.g.,

$$\langle x^2 \rangle = \langle \Psi | x_{op}{}^2 | \Psi \rangle = \int d^3r\, x^2 |\langle x | \Psi \rangle|^2.$$

Let us assume the particle to be in state $|\Psi\rangle$ and write

$$|\Theta\rangle = (x_{op} - \langle x \rangle)|\Psi\rangle, \qquad |\Phi\rangle = (p_{x,\,op} - \langle p_x \rangle)|\Psi\rangle. \tag{3-86}$$

Then $\langle \Theta | \Theta \rangle = (\Delta x)^2$ and $\langle \Phi | \Phi \rangle = (\Delta p)^2$, and from (3-84) we have, choosing $e^{i\alpha} = -i$,

$$\Delta x \Delta p \geq -\frac{i}{2} \langle \Psi | [x_{op} - \langle x \rangle,\, p_{x,\,op} - \langle p_x \rangle] | \Psi \rangle.$$

Thus from the commutation relation (3-80) we find

$$\Delta x \Delta p \geq \frac{\hbar}{2}. \tag{3-87}$$

$\hbar/2$ is the absolute lower limit on the product of the uncertainties of x and p; in most states $\Delta x \Delta p$ is greater than $\hbar/2$.

The state for which the equality in (3-87) holds is the one for which

$$0 = |\Theta\rangle - e^{i\alpha}|\Phi\rangle = \frac{x_{op} - \langle x \rangle}{\Delta x}|\Psi\rangle + i\,\frac{p_{x,\,op} - \langle p_x \rangle}{\Delta p}|\Psi\rangle, \tag{3-88}$$

where $|\Theta\rangle$ and $|\Phi\rangle$ are here normalized to unity.

Taking the components of (3-88) in the position basis and using (3-60) and (3-64) we find a differential equation for $\langle \mathbf{r} | \Psi \rangle$, the solution to which is

$$\langle \mathbf{r} | \Psi \rangle = \exp\left[\frac{ix\langle p \rangle}{\hbar} - \frac{(x - \langle x \rangle)^2}{4(\Delta x)^2} \right] \cdot g(y, z). \tag{3-89}$$

Thus the *minimum wave packet,* the one that obeys $\Delta x \Delta p = \hbar/2$, is in the form of a Gaussian. The derivation of (3-89) is left as an exercise. Note that these results do not in any way require the particle to be free.

If the wave function of a free particle at time t' is $\psi(\mathbf{r}', t')$, what will be the wave function at later time t? To answer this, let us write (3-72) as

$$\langle \mathbf{r}|\Psi(t)\rangle = \int \frac{d^3p}{(2\pi\hbar)^3} \langle \mathbf{r}|\mathbf{p}\rangle \, e^{-iE(t-t')/\hbar} \langle \mathbf{p}|\Psi(t')\rangle, \tag{3-90}$$

using (3-71). Let us substitute

$$\langle \mathbf{p}|\Psi(t')\rangle = \int d^3r' \, \langle \mathbf{p}|\mathbf{r}'\rangle\langle \mathbf{r}'|\Psi(t')\rangle.$$

into (3-90). Thus, taking the $\mathbf{r}'$ integral out in front we find

$$\langle \mathbf{r}|\Psi(t)\rangle = \int d^3r' \; K(\mathbf{r}t, \mathbf{r}'t')\langle \mathbf{r}'|\Psi(t')\rangle, \tag{3-91}$$

where

$$K(\mathbf{r}t, \mathbf{r}'t') = \int \frac{d^3p}{(2\pi\hbar)^3} \langle \mathbf{r}|\mathbf{p}\rangle \, e^{-iE(t-t')/\hbar} \langle \mathbf{p}|\mathbf{r}'\rangle. \tag{3-92}$$

This function, which is called the *free particle propagator,* gives us the wave function at time t in terms of the wave function at time t'. $K(\mathbf{r}t, \mathbf{r}'t')$ is the amplitude for finding the particle at $\mathbf{r}$ at time t, *if it was* at $\mathbf{r}'$ at time t'. Multiplying K by $\psi(\mathbf{r}', t')$ and summing over all $\mathbf{r}'$ gives us the total amplitude for finding the particle at $\mathbf{r}$ at time t, as in (3-91).

It is easy to calculate K explicitly:

$$\begin{aligned} K(\mathbf{r}t, \mathbf{r}'t') &= \int \frac{d^3p}{(2\pi\hbar)^3} \exp\left\{\frac{i}{\hbar}\left[\mathbf{p}\cdot(\mathbf{r}-\mathbf{r}') - \frac{p^2}{2m}(t-t')\right]\right\} \\ &= \left(\frac{m}{2\pi i\hbar(t-t')}\right)^{3/2} \exp\left[\frac{im}{2\hbar}\frac{(\mathbf{r}-\mathbf{r}')^2}{t-t'}\right]. \end{aligned} \tag{3-93}$$

This tells us how amplitude localized at $\mathbf{r}'$ at time t will spread out in time. As $t - t' \to 0$, $K(\mathbf{r}t, \mathbf{r}'t') \to \delta(\mathbf{r} - \mathbf{r}')$.

The exponent of K has a very simple interpretation. The classical Lagrangian for a free particle is just

$$\mathcal{L} = \frac{1}{2}\,\mathrm{m}\mathrm{v}^2, \tag{3-94}$$

where v is its velocity. Recall that the classical action $S_{cl}(\mathbf{r}t, \mathbf{r}'t')$ for a path of a particle that went from a point $\mathbf{r}'$ at time t' to a point $\mathbf{r}$ at time t was defined as the integral of the Lagrangian between t' and t

$$S_{cl}(\mathbf{r}t, \mathbf{r}'t') = \int_{t'}^{t} \mathcal{L}(t'')\,dt''. \tag{3-95}$$

Now what is the actual value of S for a free particle? The velocity remains constant, so that $\mathcal{L}$ remains constant in time. Thus

$$S_{cl}(\mathbf{r}t, \mathbf{r}'t') = \frac{1}{2}\,m(t-t')v^2.$$

Also the velocity required to go from $\mathbf{r}'$ to $\mathbf{r}$ in time $t - t'$ must be

$$\mathbf{v} = \frac{\mathbf{r}-\mathbf{r}'}{t-t'},$$

so that

$$S_{cl}(\mathbf{r}t, \mathbf{r}'t') = \frac{1}{2}\,m\,\frac{(\mathbf{r}-\mathbf{r}')^2}{t-t'}. \tag{3-96}$$

This is exactly what appears in the exponent of K. Therefore

$$K(\mathbf{r}t, \mathbf{r}'t') = \left(\frac{m}{2\pi i\hbar(t-t')}\right)^{3/2} e^{(i/\hbar)S_{cl}(\mathbf{r}t,\,\mathbf{r}'t')}. \tag{3-97}$$

Thus, as the free particle amplitude propagates from $\mathbf{r}'$ at time t' to $\mathbf{r}$ at time t, it changes by a phase factor that is simply the exponential of the *classical action* evaluated along the *classical path* from $\mathbf{r}'t'$ to $\mathbf{r}t$, times $i/\hbar$.

QUANTUM MECHANICAL MOTION AS A SUM OVER PATHS

It is easy to show that K obeys the composition equation

$$\int d^3r_2\, K(\mathbf{r}_1t_1, \mathbf{r}_2t_2)\, K(\mathbf{r}_2t_2, \mathbf{r}_3t_3) = K(\mathbf{r}_1t_1, \mathbf{r}_3t_3). \tag{3-98}$$

To see this we write

$$\psi(\mathbf{r}_1t_1) = \int K(\mathbf{r}_1t_1, \mathbf{r}_2t_2)\psi(\mathbf{r}_2t_2)\, d^3r_2$$

and

$$\psi(\mathbf{r}_2t_2) = \int K(\mathbf{r}_2t_2, \mathbf{r}_3t_3)\psi(\mathbf{r}_3t_3)\, d^3r_3.$$

Thus

$$\begin{aligned}\psi(\mathbf{r}_1t_1) &= \int K(\mathbf{r}_1t_1, \mathbf{r}_2t_2)\, d^3r_2\, K(\mathbf{r}_2t_2, \mathbf{r}_3t_3)\psi(\mathbf{r}_3t_3)\, d^3r_3\\ &= \int K(\mathbf{r}_1t_1, \mathbf{r}_3t_3)\psi(\mathbf{r}_3t_3)\, d^3r_3.\end{aligned}$$

Since $\psi(\mathbf{r}_3t_3)$ is arbitrary, (3-98) follows immediately.

There is a rather interesting consequence of this composition relation and the expression for K in terms of the classical action. In classical mechanics, the principle of least action states that a particle moving from $\mathbf{r}_3$ at time t_3 to $\mathbf{r}_1$ at time t_1 takes the path that makes the action stationary. How though does the particle ever "know" that the path it takes is the best one, if it doesn't also somehow sample neighboring paths? In fact, a quantum mechanical particle moving from $\mathbf{r}_3t_3$ to $\mathbf{r}_1t_1$ does try *all* possible paths between these points. Let us see why this is so.

The composition equation (3-98) says that the amplitude for going from $\mathbf{r}_3t_3$ to $\mathbf{r}_1t_1$ is composed of the amplitude for going from $\mathbf{r}_3t_3$ to some point $\mathbf{r}_2$ at time t_2, times the amplitude to go from $\mathbf{r}_2t_2$ to $\mathbf{r}_1t_1$, summed over all $\mathbf{r}_2$. We can write the amplitude to go from $\mathbf{r}_3t_3$ to $\mathbf{r}_1t_1$ as the sum of the amplitudes for the two-legged paths as shown in Fig. 3-4. The amplitude for the two-legged path from $\mathbf{r}_3t_3$ to $\mathbf{r}_2t_2$ to $\mathbf{r}_1t_1$ is

$$\begin{aligned}K(\mathbf{r}_1t_1, \mathbf{r}_2t_2)\, K(\mathbf{r}_2t_2, \mathbf{r}_3t_3)\\ &= \left(\frac{m}{2\pi i\hbar(t_1 - t_2)}\right)^{3/2}\left(\frac{m}{2\pi i\hbar(t_2 - t_3)}\right)^{3/2}\\ &\quad\times e^{(i/\hbar)[S_{cl}(\mathbf{r}_1t_1, \mathbf{r}_2t_2) + S_{cl}(\mathbf{r}_2t_2, \mathbf{r}_3t_3)]}.\end{aligned} \tag{3-99}$$

We see that the exponent is just $i/\hbar$ times the classical action $S_{path}(\mathbf{r}_1t_1, \mathbf{r}_3t_3)$ evaluated along the two-legged path:

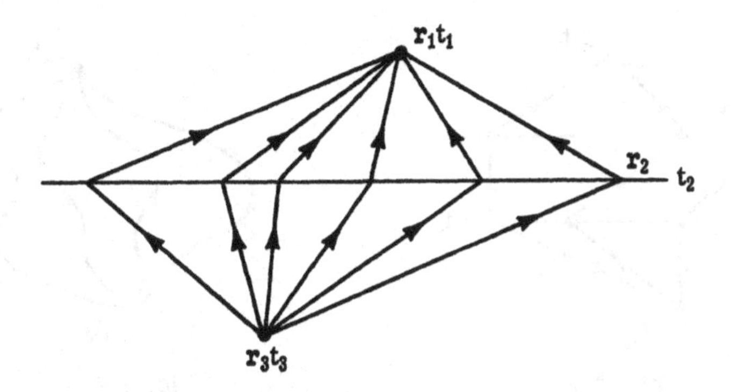

Fig. 3-4
Motion from r_3t_3 to r_1t_1 as a sum over all two-legged paths. The time t_2 is the same for all paths.

$$S_{\text{path}}(\mathbf{r}_1t_1, \mathbf{r}_3t_3) = S_{cl}(\mathbf{r}_1t_1, \mathbf{r}_2t_2) + S_{cl}(\mathbf{r}_2t_2, \mathbf{r}_3t_3). \tag{3-100}$$

We can now think of breaking up these two-legged paths into three-legged paths, using the composition relation again as in

$$K(\mathbf{r}_2t_2, \mathbf{r}_3t_3) = \int d^3r_4\, K(\mathbf{r}_2t_2, \mathbf{r}_4t_4)K(\mathbf{r}_4t_4, \mathbf{r}_3t_3)$$

where t_4 is between t_2 and t_3. Then amplitude for the motion from $\mathbf{r}_3t_3$ to $\mathbf{r}_1t_1$ can be thought of as being the sum of all the amplitudes for motion along three-legged paths, as in Fig. 3-5. The amplitude for each of these three-legged paths is proportional to

$$e^{(i/\hbar)[S_{cl}(\mathbf{r}_1t_1, \mathbf{r}_2t_2) + S_{cl}(\mathbf{r}_2t_2, \mathbf{r}_4t_4) + S_{cl}(\mathbf{r}_4t_4, \mathbf{r}_3t_3)]}.$$

The exponent is again just $i/\hbar$ times the classical action evaluated along the three-legged path.

Let us continue to divide the time interval between t_3 and t_1 into smaller and smaller subintervals. For n intervals, the amplitude $K(\mathbf{r}_1t_1, \mathbf{r}_3t_3)$ can be written as the sum over all n-legged paths from $\mathbf{r}_3t_3$ to $\mathbf{r}_1t_1$. As the intervals become infinitesimally small, the total amplitude $K(\mathbf{r}_1t_1, \mathbf{r}_3t_3)$ becomes the sum over all the amplitudes for going from $\mathbf{r}_3t_3$ to $\mathbf{r}_1t_1$ along *all* alternative possible paths between these two points, as in Fig. 3-6. The amplitude for each of these paths is proportional to

$$e^{(i/\hbar)S_{\text{path}}(\mathbf{r}_1t_1, \mathbf{r}_3t_3)} \tag{3-101}$$

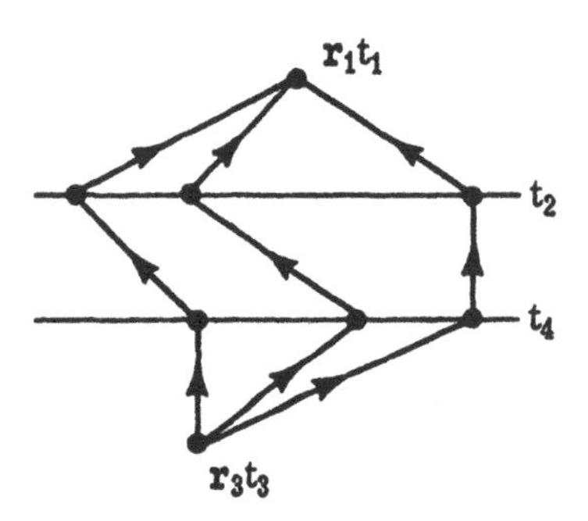

Fig. 3-5
Motion from r_3t_3 to r_1t_1 as a sum over all three-legged paths.

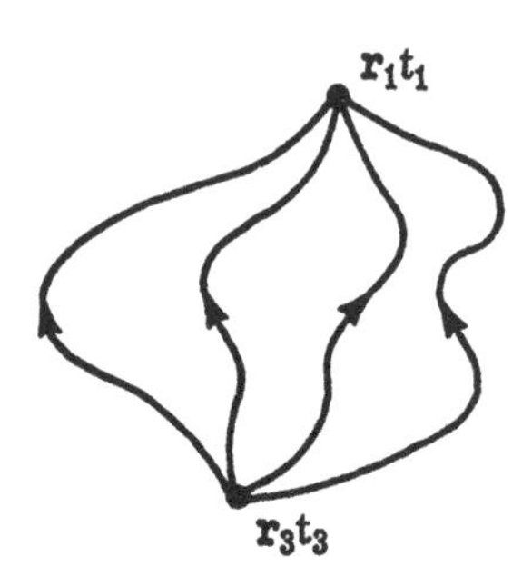

Fig. 3-6
Motion from r_3t_3 to r_1t_1 as a sum over all forward-going paths in space time.

where $S_{path}(\mathbf{r}_1t_1, \mathbf{r}_3t_3)$ is the classical action evaluated along the particular path in question. The total amplitude $K(\mathbf{r}_1t_1, \mathbf{r}_3t_3)$ is thus proportional to

$$\sum_{\text{all paths}} e^{(i/\hbar)S_{path}(\mathbf{r}_1t_1, \mathbf{r}_3t_3)}. \tag{3-102}$$

We have derived this result for free particles where $\mathcal{L} = \frac{1}{2}\ m\ (dr/dt)^2$. The result is true even if there are forces acting on the particle, only we must use the full Lagrangian containing the force terms in evaluating the action integral along the paths; however, the propagator no longer has the simple form (3-97). Thus generally, for particles with no spin, the total amplitude for motion from $\mathbf{r}_3t_3$ to $\mathbf{r}_1t_1$ is proportional to the sum over all alternative paths (3-102) where S_{path} is the classical action evaluated along the path.

We can understand how diffraction works from this point of view. If we put an impermeable obstacle in the way of a particle as in Fig. 3-7(a), then we prevent the particle from moving along the paths that would go through the obstacle. Thus certain terms in the sum (3-102) must be left out, and the total amplitude for arriving at a point $\mathbf{r}_1$ at time t_1 will be different from its value were there no obstacle. This change in the final distribution of amplitude is what is usually called diffraction by the obstacle.

For example, in a one slit diffraction experiment, Fig. 3-7(b), the total amplitude at a point on the screen is the amplitude summed

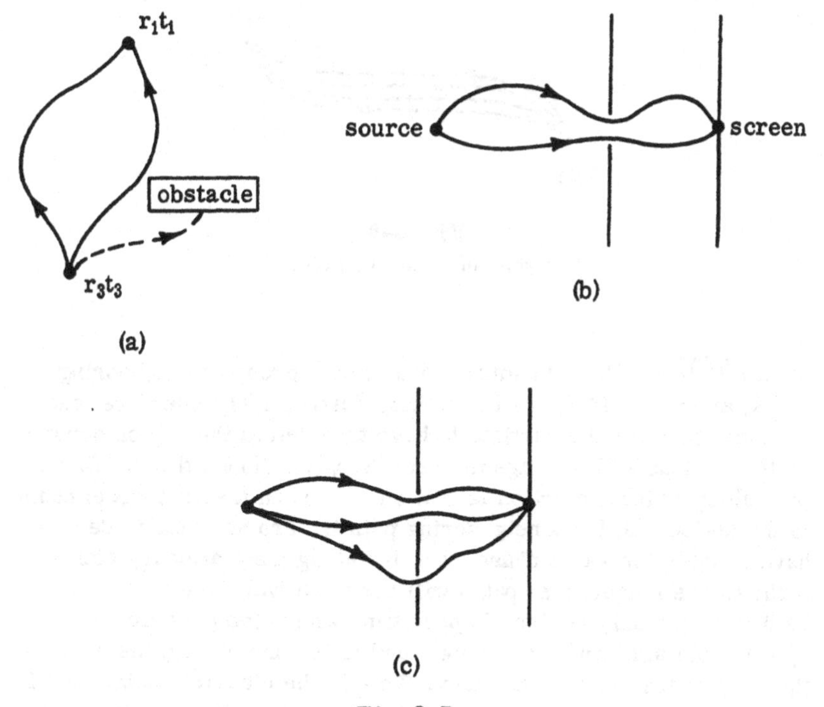

Fig. 3-7

(a) Diffraction caused by an obstacle blocking paths.
(b) One-slit diffraction.
(c) Two-slit diffraction; additional paths change the net amplitude on the screen.

over all the alternative paths from the source through the one slit to the screen. The interferences of the amplitudes for the paths produces a one-slit diffraction pattern. If we open a second slit in addition, as in Fig. 3-7(c), then we make available to the particle additional paths that go through the second slit rather than the first. Including these new alternative paths in the sum over paths (3-102) changes the total amplitude from that for a one-slit diffraction pattern into that for a two-slit pattern.

Now let us examine how the classical principle of least action arises as $\hbar \to 0$, i.e., when the typical variation of the action from path to path is $\ggg \hbar$. In this limit, if the amplitude for a particular path is given by $e^{(i/\hbar)S_0}$, then the action along a neighboring path will be $S_0 + \delta S$, and the amplitude will be $e^{(i/\hbar)(S_0+\delta S)}$. Now because $\hbar$ is so tiny, this amplitude will have a completely different phase

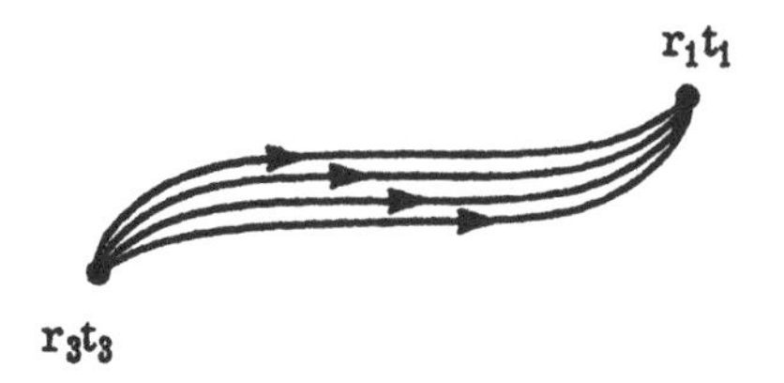

Fig. 3-8
A group of neighboring paths.

from $e^{(i/\hbar)S_0}$. Thus the amplitudes from a group of neighboring paths, as in Fig. (3-8), will generally destructively interfere, and the amplitude for the particle to have traveled in the region occupied by these paths will average to zero. Suppose though that we find a path along which the action is stationary. Then the first-order change in the action, δS, for a neighboring path is zero and amplitude will have roughly the same phase as it had along the stationary action path; all the neighboring paths will constructively interfere. Thus as $\hbar \to 0$, the only region of space time where the particle has an appreciable amplitude for being found is the one around the path that makes the action stationary. This is the classical principle of stationary action for determining the classical path. We see, from a quantum mechanical point of view, that as long as h is nonzero the particle does sample all paths, but as $\hbar$ becomes smaller and smaller, the amplitude concentrates itself about the stationary action path.

This representation of the amplitude, $K(\mathbf{r}_1t_1, \mathbf{r}_3t_3)$ as the sum of amplitudes for each of the alternative paths from $\mathbf{r}_3t_3$ to $\mathbf{r}_1t_1$ while rarely a practical method for doing elementary quantum mechanics calculations, is often very useful for visualizing how amplitudes behave in time,[3] and has become an important technique for calculating in quantum field theory.

PARTICLE IN A MAGNETIC FIELD

The classical Hamiltonian for a particle of charge e in a magnetic field $\mathcal{H}(\mathbf{r}, t)$ is

[3] This description of quantum mechanics in terms of sums over space-time paths was introduced by Feynman in *Rev. Mod. Phys.* **20**, 367 (1948), and is discussed also in *The Feynman Lectures on Physics*, Vol. II, Chapter 19. See in addition R.P. Feynman and A.R. Hibbs, *Quantum Mechanics and Path Integrals* (McGraw-Hill, New York, 1965).

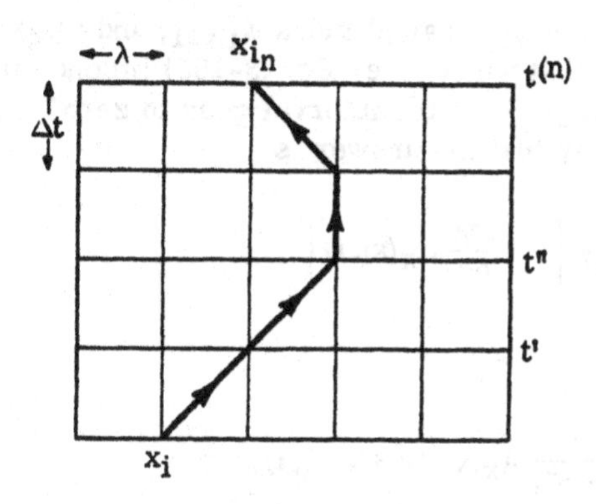

Fig. 3-9

Motion from x_i, t to x_{i_n}, $t^{(n)}$ along an n-legged path with fixed time intervals Δt and space intervals λ.

$$H = \frac{[\mathbf{p} - (e/c)\mathbf{A}(\mathbf{r}, t)]^2}{2m} + V(\mathbf{r}, t), \tag{3-103}$$

where $\mathcal{H}$ is given in terms of the vector potential **A** by $\mathcal{H}(\mathbf{r}, t) = \nabla \times \mathbf{A}(\mathbf{r}, t)$. The potential energy is $V(\mathbf{r}, t)$. Quantum mechanically, the Hamiltonian operator is given by (3-103) with **r** replaced by $\mathbf{r}_{op}$ (in **A** also) and **p** replaced by $\mathbf{p}_{op}$. A state of the particle then develops in time according to

$$i\hbar \frac{\partial}{\partial t}|\Psi(t)\rangle = \left\{\frac{[\mathbf{p}_{op} - (e/c)\mathbf{A}(\mathbf{r}_{op}, t)]^2}{2m} + V(\mathbf{r}_{op}, t)\right\}|\Psi(t)\rangle \tag{3-104}$$

or in terms of the wave function $\langle \mathbf{r}|\Psi(t)\rangle$:

$$i\hbar \frac{\partial \Psi(\mathbf{r}, t)}{\partial t} = \left\{-\frac{\hbar^2\nabla^2}{2m} + \frac{ie\hbar}{2mc}(\mathbf{A}\cdot\nabla + \nabla\cdot\mathbf{A}) + \frac{e^2A^2}{2mc^2} + V(\mathbf{r}, t)\right\}\Psi(\mathbf{r}, t). \tag{3-105}$$

[Note that we must pay attention to the ordering of $\mathbf{p}_{op}$ and $\mathbf{A}(\mathbf{r}_{op}, t)$ in these equations, since they generally do not commute.]

It is very instructive to ask how Eq. (3-105) arises from the discrete point of view we adopted at the beginning of this chapter. For simplicity we consider the one-dimensional case. The question is, what must be the hopping amplitudes $w_{i,i\pm1}$, and $v(x_i)$ in Eqs. (3-16) and (3-11) such that we recover Eq. (3-105) [in one dimension] in the limit that the length λ of the interval goes to zero? It is left as an exercise to verify that the answer is

$$w_{i+1,i} = -\frac{\hbar}{2m\lambda^2}\left[1 + \frac{ie\lambda}{\hbar c} A_x(x_i, t)\right] \tag{3-106}$$

and

$$w_{ii} = \frac{\hbar}{m\lambda^2} + \frac{e^2}{2mc^2\hbar} A_x(x_i, t)^2 + V(x_i, t). \tag{3-107}$$

Notice that in the presence of a magnetic field, a velocity dependent force, the $w_{i,i\pm1}$ are no longer real.

As $\lambda \to 0$, (3-106) can be written as

$$w_{i+1,i} = -\frac{\hbar}{2m\lambda^2} e^{ie A_x(x_i, t)\lambda/\hbar c}; \tag{3-108}$$

this says simply that the effect of a magnetic field is to multiply the hopping amplitude for no field by a phase factor. The amplitude to trace out a path (in the sense of the last section) in this discrete space [Fig. 3-9] from x_i at time t, to x_i' at the next time t', to x_i'' at the next time t'', ..., finally to x_{i_n}, $t^{(n)}$, is then the amplitude with no magnetic field times a factor $1 + (ie\lambda/\hbar c)A_x(x_j, t^{(j)})$ for each leg of the path. The overall extra factor is

$$\prod_j \left(1 + \frac{ie\lambda}{\hbar c} A_x(x_j, t^{(j)})\right) \approx \exp\left[\frac{ie}{\hbar c}\sum_j \lambda A_x(x_j, t^{(j)})\right]. \tag{3-109}$$

In the limit $\lambda \to 0$ the factor becomes simply [going back to three dimensions][4]

$$\exp\left[\frac{ie}{\hbar c}\int_{\text{path}} d\mathbf{l} \cdot \mathbf{A}(\mathbf{r}, t)\right]. \tag{3-110}$$

The general effect of the magnetic field is to multiply the amplitude for a path, as in (3-102), by the phase factor (3-110). Thus the

---

[4] The corrections to the exponent in (3-109) are terms of order λ^2 times the number of intervals; such corrections approach zero as $\lambda \to 0$. Thus (3-110) is the correct limit of the left side of (3-109).

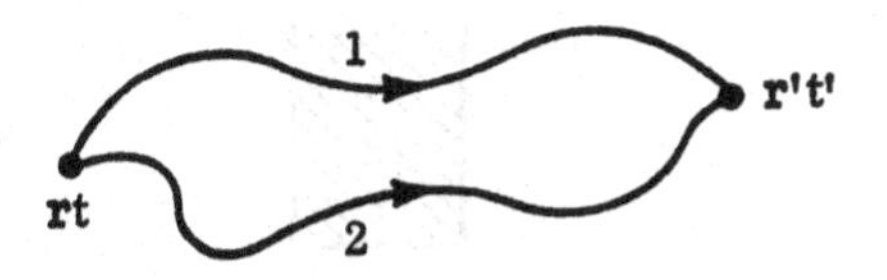

Fig. 3-10
Two paths from rt to r't'. Φ is the flux enclosed by the loop formed by going from r to r' along 1 and back to r along 2.

propagator K_A from rt to r't' is given as a sum over paths by

$$K_A(\mathbf{r}t, \mathbf{r}'t') = \sum_{\text{all paths}} \exp\left[\frac{i}{\hbar} S^{(A=0)}_{\text{path}}(\mathbf{r}t, \mathbf{r}'t')\right] \times \exp\left[\frac{ie}{\hbar c} \int_{\text{path}} d\mathbf{l}'' \cdot \mathbf{A}(\mathbf{r}'', t'')\right] \tag{3-111}$$

where the integral is along the given path from rt to r't', and $S^{(A=0)}_{\text{path}}$ is the classical action along the path with $\mathbf{A} = 0$.

Notice that the combination $S^{(A=0)}_{\text{path}}(\mathbf{r}t, \mathbf{r}'t') + (e/c)\int d\mathbf{l}'' \cdot \mathbf{A}(\mathbf{r}'', t'')$ in the exponents in (3-111) is simply the total action including the magnetic field, since classical Lagrangian $\mathcal{L}_A$ in the presence of a magnetic field is

$$\mathcal{L}_A = \mathcal{L}_{A=0} + \frac{e}{c} \mathbf{v} \cdot \mathbf{A}, \tag{3-112}$$

where **v** is the velocity of the particle.

AHARONOV-BOHM EFFECT

The result (3-111) has some striking physical consequences. Take **A** independent of t and consider the interference in (3-111) between the motion along two paths, as in Fig. 3-10. The amplitude for the particle to take a given path is multiplied by the phase factor (3-110). Thus the relative phase of the amplitudes for paths 1 and 2 in Fig. 3-10 is modified by an amount

$$\frac{e}{\hbar c}\int_1 d\mathbf{l} \cdot \mathbf{A} - \frac{e}{\hbar c}\int_2 d\mathbf{l} \cdot \mathbf{A} = \frac{e}{\hbar c}\Phi, \tag{3-113}$$

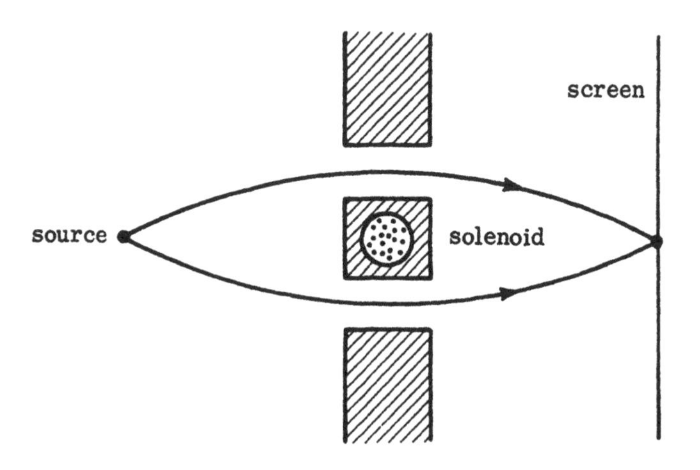

Fig. 3-11

Two-slit diffraction experiment with a solenoid enclosed in the region between the two slits. The magnetic field is out of the paper and confined to the interior of the solenoid; the particles travel only in field free regions.

where Φ is the total magnetic flux passing through the (clockwise) loop formed by the two paths. Thus the relative phase of the two paths is proportional to the flux enclosed by the paths; if the flux is an integral number of flux quanta

$$\Phi = n\varphi_0 \tag{3-114}$$

where $\varphi_0 = 2\pi\hbar c/e = 4.135 \times 10^{-7}$ gauss-cm^2, then the magnetic field produces no change in the relative amplitude.

Now let us look at the two-slit diffraction experiment in Fig. 3-11. Encased in the region between the slits is a solenoid whose magnetic field, pointing out of the paper, is confined to the interior of the solenoid; thus the particles in diffracting never travel in a region where there is a nonzero magnetic field. However, as we increase the flux in the solenoid from zero, the phase of a path that goes through the top slit relative to that of a path that goes through the lower slit changes according to (3-113), and consequently the diffraction pattern produced on the screen shifts, *even though* the particles never feel the magnetic field. This remarkable effect, discovered by Aharonov and Bohm,[5] illustrates most powerfully that the electromagnetic potentials, rather than the fields, are the fundamental quantities in quantum mechanics. In classical mechanics one can always write the equation of motion of a particle directly in terms of $\mathcal{H}$. However, the quantum mechanical wave function of the particle depends directly on the electromagnetic potential, $\mathbf{A}$, and only indirectly on $\mathcal{H}$. The shift of the diffraction pattern depends, however, on $\mathbf{A}$ through Φ and thus is independent of the choice of gauge for the vector potential. Whenever $\Phi = n\varphi_0$, the shift vanishes.

PROBLEMS

1. In the "discrete" Schrödinger equation (3-12) for a free particle, $w_{i,i-1} = w_{i,i+1} \equiv w$, and $v_i = 0$. Assume *periodic boundary conditions*, that is, $\psi_{N+1}(t) = \psi_1(t)$, and generally $\psi_{j+N}(t) = \psi_j(t)$, where N is a large integer.
   (a) Show that there are N linearly independent solutions of this Schrödinger equation.
   (b) Find the N normalized energy eigenstates, that is, those solutions that vary in time with a fixed frequency. What are the possible energy values for the particle? Show that these go over into the free particle energies, $p^2/2m$, in the continuum limit, i.e., as λ, the size of the intervals $\to 0$, but $N\lambda$ remains fixed.
   (c) We can define the *propagator matrix*, K, by $\psi_j(t) = \sum_k K_{jk}(t, t')\psi_k(t')$ in analogy with (3-91). Write down an explicit expression for the matrix elements $K_{jk}(t, t')$.
2. (a) Write out $\delta(\mathbf{r} - \mathbf{r}')$ as the product of three one-dimensional delta functions in spherical coordinates.
   (b) What is the Fourier transform of $\delta(\mathbf{r} - \mathbf{r}')$? Write out $\delta(\mathbf{r} - \mathbf{r}')$ in terms of its Fourier transform.
   (c) Show that $\langle \mathbf{p} | \mathbf{p}' \rangle = (2\pi\hbar)^3 \delta(\mathbf{p} - \mathbf{p}')$.
3. What is the representation of the position operator in the momentum basis, i.e., how is $\langle \mathbf{p} | \mathbf{r}_{op} | \Psi \rangle$ related to $\langle \mathbf{p} | \Psi \rangle$?
4. Suppose that the potential is $v(\mathbf{r}) = (k/2)r^2$. What is the Schrödinger equation written in momentum space; that is, what is the equation of motion of the amplitude $\langle \mathbf{p} | \Psi(t) \rangle$?

[5] Y. Aharonov and D. Bohm, *Phys. Rev.* 115, 485 (1959). Such an effect has been observed experimentally by R.G. Chambers, *Phys. Rev. Letters* 5, 3 (1960).

5. What is $[y_{op}, (p_x)_{op}]$? Compare these commutation relations for position and momentum with the classical Poisson bracket relations satisfied by $\mathbf{p}$ and $\mathbf{r}$.
6. What is the expectation value of the kinetic energy of a particle in terms of its wave function?
7. Derive (3-89) from (3-88).
8. Verify (3-98) by explicit integration using (3-93).
9. (a) Show that the free particle propagator $K(\mathbf{r}t, \mathbf{r}'t')$ obeys the free particle Schrödinger equation

$$\left(i\hbar \frac{\partial}{\partial t} + \frac{\hbar^2 \nabla^2}{2m}\right) K(\mathbf{r}t, \mathbf{r}'t') = 0.$$

(b) Suppose that a particle is acted on by a potential $v(\mathbf{r}, t)$, and that its wave function at t_0 is $\psi_0(\mathbf{r})$. Show that the wave function of the particle at a later time t is given as the solution to the integral equation

$$\psi(\mathbf{r}, t) = \int d^3r' \; K(\mathbf{r}t, \mathbf{r}'t_0)\psi_0(\mathbf{r}')$$

$$+ \frac{1}{i\hbar} \int_{t_0}^{t} dt' \int d^3r' \; K(\mathbf{r}t, \mathbf{r}'t')v(\mathbf{r}', t')\psi(\mathbf{r}', t').$$

10. (a) The probability current density $\mathbf{j}(\mathbf{r}, t)$ is given in terms of the wave function by

$$\mathbf{j}(\mathbf{r}, t) = \frac{1}{2m}\left[\psi^*(\mathbf{r}, t)\frac{\hbar}{i}\nabla\psi(\mathbf{r}, t) - \frac{\hbar}{i}\nabla\psi^*(\mathbf{r}, t)\cdot\psi(\mathbf{r}, t)\right].$$

Show directly from Schrödinger's equation that the probability density, $P(\mathbf{r}, t) = |\Psi(\mathbf{r}, t)|^2$ and the probability current density, obey the continuity equation

$$\frac{\partial}{\partial t} P(\mathbf{r}, t) + \nabla \cdot \mathbf{j}(\mathbf{r}, t) = 0.$$

Do not assume that the potential vanishes.
(b) What is the form for $\mathbf{j}(\mathbf{r}, t)$ when there is a magnetic field present specified by the vector potential $\mathbf{A}(\mathbf{r}, t)$?
11. Consider a particle of charge e traveling in the electromagnetic potentials

$$\mathbf{A}(\mathbf{r}, t) = -\nabla\lambda(\mathbf{r}, t), \qquad \phi(\mathbf{r}, t) = \frac{1}{c}\frac{\partial\lambda(\mathbf{r}, t)}{\partial t}$$

where $\lambda(\mathbf{r}, t)$ is an arbitrary scalar function.

(a) What are the electromagnetic fields described by these potentials?

(b) Show that the wave function of the particle is given by

$$\psi(\mathbf{r}, t) = \exp\left[-\frac{ie}{\hbar c}\lambda(\mathbf{r}, t)\right]\psi^{(0)}(\mathbf{r}, t),$$

where $\psi^{(0)}$ solves the Schrödinger equation with $\lambda \equiv 0$.

(c) Let $v(\mathbf{r}, t) = e\varphi(t)$ be a spatially uniform time varying potential. Show that

$$\psi(\mathbf{r}, t) = \exp\left[-\frac{ie}{\hbar}\int_{-\infty}^{t}\phi(t')\,dt'\right]\psi^{(0)}(\mathbf{r}, t).$$

[Why is the lower limit on the integral $-\infty$?]

12. Consider doing a "two-slit interference" experiment where the slits are replaced by long conducting tubes. The source S emits

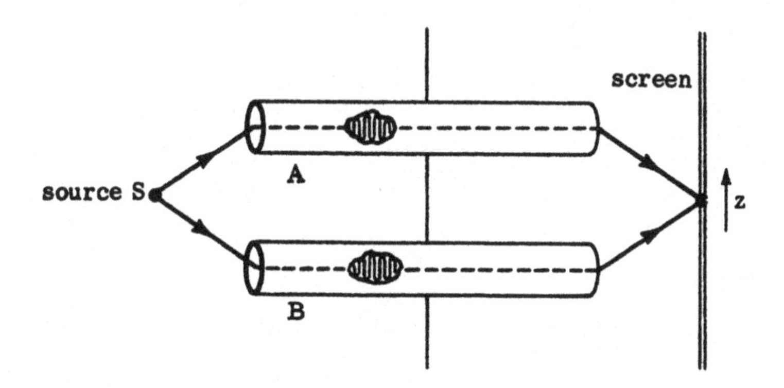

particles in reasonably well defined wave packets, so that one can be sure [if the tubes are long enough] that for a certain time interval, say t_0 to t_1 seconds after emission, the wave packet of the particle is definitely within the tubes. During this time interval, a constant voltage V_A is applied to tube A and a constant voltage V_B is applied to tube B. The rest of the time there is no voltage on the tubes. These voltages produce no fields well within the tubes.

(a) Describe how the interference pattern on the screen depends on V_A and V_B.

(b) If the phase difference between the amplitudes arriving at the screen from the two tubes is Kz, where $K = 10^4 \text{ cm}^{-1}$, by how many cm is the diffraction pattern shifted if $V_A = 0$, $V_B = 10^{-6}$ volts, $t_1 - t_0 = 10^{-9}$ sec?

Chapter 4

POTENTIAL PROBLEMS, MOSTLY IN ONE DIMENSION

PROPERTIES OF HERMITIAN OPERATORS

As a prelude to solving specific problems, let us discuss a few general properties of operators. Recall that the Hermitian adjoint, $B^\dagger$, of an operator B is defined as the operator whose matrix elements are given by

$$\langle\Psi|B^\dagger|\Phi\rangle \equiv \langle\Phi|B|\Psi\rangle^*, \tag{4-1}$$

for all states $|\Phi\rangle$ and $|\Psi\rangle$. The matrix that represents $B^\dagger$ is the complex conjugate of the transpose of the matrix that represents B. It is easy to see from (4-1) that if λ is a number, then

$$(\lambda B)^\dagger = \lambda^* B^\dagger. \tag{4-2}$$

Now what is the Hermitian adjoint of the product, BC, of two operators. Suppose we have a complete set of states $1 = \sum_\alpha |\alpha\rangle\langle\alpha|$. Then we can write

$$\langle\Psi|(BC)^\dagger|\Phi\rangle = \langle\Phi|B(\sum_\alpha|\alpha\rangle\langle\alpha|)C|\Psi\rangle^* = \sum_\alpha \langle\alpha|C|\Psi\rangle^*\langle\Psi|B|\alpha\rangle^*$$

$$= \sum_\alpha \langle\Psi|C^\dagger|\alpha\rangle\langle\alpha|B^\dagger|\Phi\rangle = \langle\Psi|C^\dagger B^\dagger|\Phi\rangle.$$

Thus

$$(BC)^\dagger = C^\dagger B^\dagger. \tag{4-3}$$

If $B^\dagger = B$, then B is Hermitian. It is trivial to see that $\mathbf{r}_{op}$, the position operator, is Hermitian, for in terms of wave functions

$$\begin{aligned} \langle\Phi|\mathbf{r}_{op}|\Psi\rangle^* &= [\int d^3r\ \phi^*(\mathbf{r})\mathbf{r}\ \psi(\mathbf{r})]^* = \int d^3r\ \psi^*(\mathbf{r})\mathbf{r}\phi(\mathbf{r}) \\ &= \langle\Psi|\mathbf{r}_{op}|\Phi\rangle. \end{aligned} \tag{4-4}$$

Similarly, the momentum operator, $\mathbf{p}_{op}$, is Hermitian since, in terms of wave functions,

$$\langle\Phi|\mathbf{p}_{op}|\Psi\rangle = \int d^3r\ \phi^*(\mathbf{r})\frac{\hbar}{i}\nabla\psi(\mathbf{r}).$$

Now integrate by parts[1]; then

$$\begin{aligned} \langle\Phi|\mathbf{p}_{op}|\Psi\rangle &= \int d^3r \left[-\frac{\hbar}{i}\nabla\phi^*(\mathbf{r})\right]\psi(\mathbf{r}) = \left[\int d^3r \left(\frac{\hbar}{i}\nabla\phi(\mathbf{r})\right)\psi^*(\mathbf{r})\right]^* \\ &= \langle\Psi|\mathbf{p}_{op}|\Phi\rangle^*. \end{aligned} \tag{4-5}$$

Thus $\mathbf{p}_{op}^\dagger = \mathbf{p}_{op}$.

Next let us show that the eigenvalues of a Hermitian matrix, B, are real. Let $|b\rangle$ be a normalized eigenstate of B with eigenvalue b. Then $\langle b|B|b\rangle = b\langle b|b\rangle = b$, and from (4-1), $\langle b|B|b\rangle^* = \langle b|B|b\rangle$; thus

$$b^* = b.$$

Another important result is that any two eigenvectors of a Hermitian matrix, having unequal eigenvalues, are orthogonal. Suppose

$$B|b'\rangle = b'|b'\rangle \qquad \text{and} \qquad B|b''\rangle = b''|b''\rangle.$$

Then

$$\langle b''|B|b'\rangle = b'\langle b''|b'\rangle. \tag{4-6}$$

On the other hand

$$\langle b''|B|b'\rangle = \langle b'|B|b''\rangle^* = b''\langle b'|b''\rangle^* = b''\langle b''|b'\rangle, \tag{4-7}$$

that is, B acting to the left gives b''. Thus comparing (4-6) and (4-7)

---

[1] The vanisning of the boundary terms is necessary for $\mathbf{p}_{op}$ to be Hermitian.

we see that

$$b'\langle b''|b'\rangle = b''\langle b''|b'\rangle$$

so

$$\langle b''|b'\rangle = 0 \quad \text{if} \quad b'' \neq b'. \tag{4-8}$$

Finally, we note without proof that the normalized eigenvectors of a Hermitian matrix B form an orthonormal basis. Any $|\Psi\rangle$ can be written as

$$|\Psi\rangle = \sum_{b'} |b'\rangle\langle b'|\Psi\rangle, \tag{4-9}$$

where the sum is over all (linearly independent) eigenvectors of B. If all the eigenvectors of B have different eigenvalues, the sum in (4-9) is over all eigenvectors. But if two or more eigenvectors belong to the same eigenvalue then the sum, for that eigenvalue, is only over a basis of linearly independent orthogonal eigenvectors belonging to the eigenvalue. The set of all eigenstates entering the sum (4-9) is called a *complete set.*

Acting on (4-9) with B we find, since $|\Psi\rangle$ is arbitrary,

$$B = \sum_{b'} b'|b'\rangle\langle b'|. \tag{4-10}$$

Any Hermitian operator can be written in this diagonal form in terms of its eigenstates and eigenvalues.

ENERGY EIGENSTATES

A wide variety of problems in quantum mechanics can be formulated in terms of a particle moving in an external, time-independent potential, $V(\mathbf{r})$. The Schrödinger equation for the particle takes the form

$$i\hbar \frac{\partial \psi(\mathbf{r},t)}{\partial t} = \left[-\frac{\hbar^2}{2m}\nabla^2 + V(\mathbf{r})\right]\psi(\mathbf{r},t). \tag{4-11}$$

Because the Hamiltonian doesn't depend explicitly on time, we can look for solutions of the form

$$\psi(\mathbf{r}, t) = e^{-iEt/\hbar}\psi_E(\mathbf{r}). \tag{4-12}$$

A particle in this state of this type has a well defined energy E, since $E/\hbar$ is the time rate of change of the phase of the wave function. Substituting (4-12) into (4-11) we find the time-independent equation

$$E\psi_E(\mathbf{r}) = \left[-\frac{\hbar^2}{2m}\nabla^2 + V(\mathbf{r})\right]\psi_E(\mathbf{r}). \tag{4-13}$$

For only certain values of E will this equation have solutions. These values are the possible energy values of the particle, and the corresponding $\psi_E(\mathbf{r})$ is the wave function of the particle when it has energy E. Quite often there will be several different $\psi_E(\mathbf{r})$ corresponding to the same energy value E; such states are called *degenerate.*

If we think of $\psi_E(\mathbf{r})$ as a vector $|E\rangle$ written in the position representation,

$$\psi_E(\mathbf{r}) = \langle \mathbf{r}|E\rangle, \tag{4-14}$$

then (4-13) is the eigenvalue equation

$$E|E\rangle = H_{op}|E\rangle, \tag{4-15}$$

written out in the position representation. Finding the possible values E and the states $\psi_E(\mathbf{r})$ is thus equivalent to finding the eigenvectors and eigenvalues of the Hamiltonian operator. Because the vectors written in the $\mathbf{r}$ basis have a continuous infinity of components, the eigenvalue problem in the $\mathbf{r}$ basis takes the form of a differential equation (4-13) rather than a matrix equation, but it is still an eigenvalue problem.

The energy eigenstates $|E\rangle$ play a very important role in quantum mechanics. First of all, we have pointed out that a particle in such a state has a well defined value of its energy. Second, if Q is an operator that does not depend explicitly on time, then its expectation value in an energy eigenstate remains constant in time. To see this we remark that if the state of the particle is $|E\rangle$ at $t = 0$, then at time t it is

$$|E, t\rangle = e^{-iEt/\hbar}|E\rangle, \quad \text{and} \quad \langle E, t| = e^{iEt/\hbar}\langle E|. \tag{4-16}$$

Then the expectation value of Q at time t is

$$\langle Q\rangle_t = \langle E,t|Q|E,t\rangle$$

$$= \langle E|e^{iEt/\hbar}Qe^{-iEt/\hbar}|E\rangle = \langle E|Q|E\rangle,$$

which is independent of time. For this reason, the energy eigenstates are called *stationary states.*

In order that the energy eigenvalues be real, H_{op} must be Hermitian. Then as in (4-9) the normalized eigenstates form an orthonormal basis, and

$$1 = \sum_E |E\rangle\langle E| \tag{4-17}$$

where the sum is over a complete set of orthonormal eigenstates. Any $|\Psi\rangle$ can be written

$$|\Psi\rangle = \sum_E |E\rangle\langle E|\Psi\rangle, \tag{4-18}$$

that is, as a linear superposition of energy eigenstates. Thus if at time zero

$$|\Psi(0)\rangle = \sum_E |E\rangle\langle E|\Psi(0)\rangle, \tag{4-19}$$

at a later time t, because each energy component changes in time by $e^{-iEt/\hbar}$, we have

$$|\Psi(t)\rangle = \sum_E e^{-iEt/\hbar}|E\rangle\langle E|\Psi(0)\rangle. \tag{4-20}$$

[Verify explicitly that (4-20) solves the Schrödinger equation $d|\Psi(t)\rangle/dt = H_{op}|\Psi(t)\rangle$.] This equation tells us how any state develops in time, in terms of the energy eigenstates and eigenvalues. Thus if we know the energy eigenstates and eigenvalues we can use (4-20) to learn all the possible dynamics of the system. Written in the position basis, (4-20) becomes

$$\langle \mathbf{r}|\Psi(t)\rangle = \sum_E e^{-iEt/\hbar}\langle \mathbf{r}|E\rangle\langle E|\Psi(0)\rangle$$

$$= \sum_E e^{-iEt/\hbar}\langle \mathbf{r}|E\rangle \int \langle E|\mathbf{r}'\rangle\, d^3r'\, \langle \mathbf{r}'|\Psi(0)\rangle$$

or

$$\psi(\mathbf{r}, t) = \sum_E e^{-iEt/\hbar}\psi_E(\mathbf{r})\left(\int d^3r'\ \psi_E^*(\mathbf{r}')\psi(\mathbf{r}', 0)\right), \tag{4-21}$$

making use of the completeness relation for the position basis. In the example of free particles in Chapter 3 the energy eigenstates were the momentum eigenstates, and Eqs. (3-51) and (3-71) there were special cases of Eqs. (4-17) and (4-21) here. Remember though that (4-19) works only if the Hamiltonian does not depend explicitly on time.

If we define the propagator function by

$$K(\mathbf{r}t, \mathbf{r}'t') = \sum_E e^{-iE(t-t')/\hbar}\psi_E(\mathbf{r})\psi_E^*(\mathbf{r}') \tag{4-22}$$

then, from (4-21), we have

$$\psi(\mathbf{r}, t) = \int d^3r'\ K(\mathbf{r}t, \mathbf{r}'t')\psi(\mathbf{r}', t'). \tag{4-23}$$

This function is the generalization of the propagator we defined for free particles.

ONE-DIMENSIONAL BARRIERS

Let us turn now to finding the eigenfunctions and eigenvalues for several illustrative problems. We shall begin by solving one-dimensional problems. These are problems in which the potential and the wave function vary only in one direction, which we take to be the x direction. The Schrödinger equation then becomes

$$i\hbar\frac{\partial\psi(x,t)}{\partial t} = -\frac{\hbar^2}{2m}\frac{\partial^2\psi(x,t)}{\partial x^2} + V(x)\psi(x,t), \tag{4-24}$$

and the equation for the stationary states has the form

$$E\psi_E(x) = -\frac{\hbar^2}{2m}\frac{d^2}{dx^2}\psi_E(x) + V(x)\psi_E(x), \tag{4-25}$$

a second-order ordinary differential equation, which has two linearly independent solutions. We see from this equation that the second derivative of ψ must be finite if V, E, and ψ are, and hence the first derivative of ψ, and ψ itself must be continuous.

If V(x) is a constant V, independent of x, then the general solution to (4-25) is clearly of the form

$$\psi_E(x) = Ae^{i\bar{p}x/\hbar} + Be^{-i\bar{p}x/\hbar} \tag{4-26}$$

where A and B are arbitrary and

$$\bar{p} = \sqrt{2m(E-V)}.$$

The solution (4-26) is a linear combination of a wave with momentum $\bar{p}$ (in the x direction) and a wave with momentum $-\bar{p}$.

A somewhat more interesting problem occurs when there is a potential step

$$V(x) = \begin{cases} 0, & x < 0, \\ V = \text{positive constant}, & x > 0, \end{cases} \tag{4-27}$$

as shown in Fig. 4-1. There are then two distinct cases. First if $E > V$, then $\bar{p} = \sqrt{2m(E-V)}$ is real, and we are in the situation where classically the particle can be in the right-hand region as well as the left. If $E < V$, then $\bar{p}$ is imaginary and we have an exponentially damped or growing wave on the right. This corresponds to the physical situation where classically the particle hasn't enough energy to be in the right-hand region. Let us treat the case $E > V$ first.

Then, as can be found in the standard texts, the solution corresponding to no incident wave from the right is

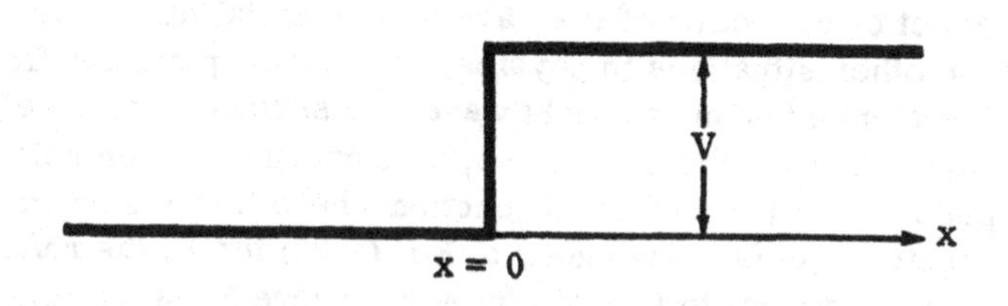

Fig. 4-1
One-dimensional potential step at x=0.

$$\psi(x) = \begin{cases} Ae^{ipx/\hbar} + Be^{-ipx/\hbar}, & x < 0, \\ Ce^{i\bar{p}x/\hbar}, & x > 0, \end{cases} \tag{4-28}$$

where $p = \sqrt{2mE}$ and

$$B = \frac{p - \bar{p}}{p + \bar{p}} A, \qquad C = \frac{2p}{p + \bar{p}} A. \tag{4-29}$$

Physically this solution represents an incident wave $\sim e^{ipx/\hbar}$ on the left, corresponding to the particle moving toward the step, a reflected wave $\sim e^{-ipx/\hbar}$ on the left, corresponding to the particle moving to the left away from the step, and a transmitted wave $\sim e^{i\bar{p}x/\hbar}$ on the right, corresponding to the particle moving to the right away from the step. A is the relative amplitude for the particle to be in the incident wave, B the relative amplitude for it to be in the reflected wave, and C the relative amplitude for it to be in the transmitted wave.

The *current density*, j_L, [see Problem 10, Chapter 3] on the left is the absolute value square of the amplitude $Ae^{ipx/\hbar}$ for the particle to be at x in the incident wave, times its velocity p/m, plus the absolute value square of the amplitude $Be^{-ipx/\hbar}$ for the particle to be at x in the reflected wave, times $-p/m$. Thus

$$j_L = \frac{p}{m} (|A|^2 - |B|^2). \tag{4-30}$$

Similarly the current density on the right is

$$j_R = \frac{\bar{p}}{m} |C|^2. \tag{4-31}$$

It is a trivial calculation to verify from (4-29) that $j_L = j_R$. This says that the *net* current flowing in from the left equals the transmitted current.

The effect of reflection of the wave at a potential step is very familiar in other situations in physics, such as the partial reflection and partial transmission of a light wave at a surface where the index of refraction changes abruptly. Also, a signal in a transmission line will be partially reflected from a junction where there is an impedance mismatch. [In fact analogous to Eq. (4-29) for B, the reflected voltage is proportional to $(Z - Z_0)/(Z + Z_0)$ where Z_0 is the impedance on the left and Z is the impedance on the right.] However from the point of view of particles, the possibility of reflection of a particle at a barrier, when the energy of the particle is sufficient to surmount the barrier, is a strictly quantum mechanical effect.

It is interesting to study the reflection and transmission at a barrier, in terms of wave packets. Let us make up a wave packet of the states of the form (4-28). For $x < 0$ we write

$$\psi(x, t) = \int_0^\infty \frac{dp}{2\pi\hbar} f(p) \left[e^{ipx/\hbar} + \left(\frac{p-\bar{p}}{p+\bar{p}} \right) e^{-ipx/\hbar} \right] e^{-iEt/\hbar} \tag{4-32}$$

and hence for $x > 0$

$$\psi(x, t) = \int_0^\infty \frac{dp}{2\pi\hbar} f(p) \left(\frac{2p}{p+\bar{p}} \right) e^{i\bar{p}x/\hbar} e^{-iEt/\hbar}. \tag{4-33}$$

$\psi(x, t)$, given by (4-32) and (4-33), solves the time-dependent Schrödinger equation with the potential (4-27). We have integrated from 0 to ∞ so that the incident wave will contain only components traveling to the right. This solution is really in the form of two wave packets for $x < 0$

$$\psi_{\text{incident}}(x, t) = \int_0^\infty \frac{dp}{2\pi\hbar} f(p)\, e^{i(px - Et)/\hbar}$$

$$\psi_{\text{reflected}}(x, t) = \int_0^\infty \frac{dp}{2\pi\hbar} f(p) \left(\frac{p-\bar{p}}{p+\bar{p}} \right) e^{i(-px - Et)/\hbar}$$

and one wave packet for $x > 0$,

$$\psi_{\text{transmitted}}(x, t) = \int_0^\infty \frac{dp}{2\pi\hbar} f(p) \left(\frac{2p}{p+\bar{p}} \right) e^{i(\bar{p}x - Et)/\hbar}. \tag{4-34}$$

What is the motion of these wave packets? Let us suppose that $f(p)$ is peaked about some value p_0. We can, with no loss of generality, take $f(p)$ to be real at $p = p_0$; this is equivalent to multiplying $\psi(x, t)$ by an overall phase factor. Then the incident wave packet will be centered about $x = (p_0/m)t$, as long as this value is < 0. For $t > 0$ the incident wave packet becomes negligibly small. The reflected wave packet is centered about $x = -(p_0/m)t$, as long as this value is negative, i.e., when $t > 0$. Similarly the transmitted wave packet is centered about $x = (\bar{p}_0/m)t$, as long as this is positive. For $t < 0$, the transmitted and reflected packets are negligibly small. Thus the incident wave packet "hits" the step at $t = 0$, when it turns into a reflected plus a transmitted packet. For $t < 0$ we then have the

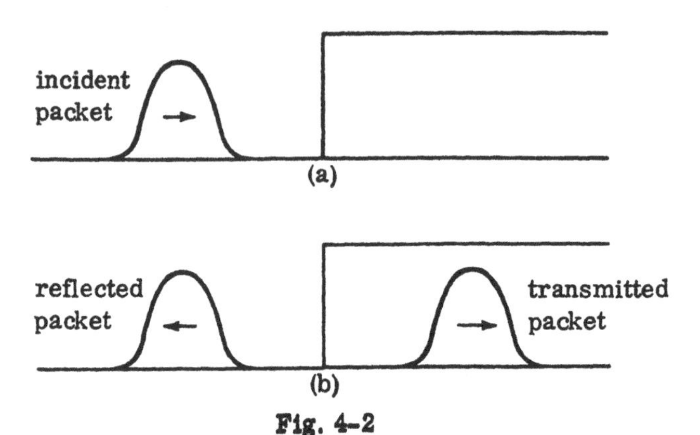

Fig. 4-2

Breakup of incident wave packet (a) into transmitted and reflected packets (b) on hitting the step.

situation in Fig. 4-2(a), and for $t > 0$ that in Fig. 4-2(b). Notice that in no sense does the particle break into two pieces; only its probability amplitude wave divides in two on hitting the step.

This problem with a sharp step has no classical limit. In order to be in the classical regime we must have a smoothly increasing potential and have $lp \gg \hbar$ where l is the distance over which the step rises.

In the case $E < V$, the solution is the analytic continuation of (4-28) and (4-29) to imaginary $\bar{p} = \sqrt{2m(E-V)}$. The correct branch of the square root is $\bar{p} = i\hbar k$ where k is real and positive; then the solution on the right is a decaying exponential. [$\bar{p} = -i\hbar k$ leads to an unphysical solution that grows without bound for $x > 0$.] Since in this case $|B|^2 = |A|^2$, all the incident wave $(\sim e^{ipx/\hbar})$ is reflected; there is no transmission. Let us write

$$\frac{B}{A} = -e^{2i\delta(E)} \equiv N(E) \tag{4-35}$$

where δ is real. Then from (4-29) we find

$$\hbar k = p \cot \delta . \tag{4-36}$$

The reflected wave is shifted in phase, at $x = 0$, from the incident wave at $x = 0$ by a factor $-e^{2i\delta}$. As $p \to 0$, $e^{2i\delta} \to +1$. For the case $E > V$, part of the wave is transmitted and there is no shift in the phase of the reflected wave at $x = 0$, but instead we find a change in its magnitude.

The solution (4-28) in this case indicates that there is a nonzero probability for finding the particle in the classically forbidden region. We are tempted to ask, doesn't it violate conservation of energy for the particle to be in a region where its potential energy exceeds its total energy? The answer is no, there is no real contradiction for the following reason. *If* we observe that the particle is in the forbidden region, then it will no longer be in a state with energy $E < V$; the act of measuring the location of the particle must necessarily introduce an uncertainty in its energy. The particle penetrates the forbidden region with sizeable probability up to a distance $\sim 1/k$. Thus if we determine that the particle is on the right we have localized it within a region $\Delta x \sim 1/k$. This means that we've introduced an uncertainty in its momentum $\Delta p \gtrsim \hbar/\Delta x \sim \hbar k$, and hence an uncertainty in its kinetic energy $\sim[(\Delta p)^2/2m] > (\hbar^2 k^2/2m) = V - E$. Hence its final energy, which is E plus the added kinetic energy from localizing the particle, is just sufficiently uncertain that we are no longer sure that the total energy of the particle is less than V.

In other words, if we know for certain that the particle is on the right, then we cannot say that its energy is less than V. On the other hand, if we definitely know that its energy is less than V then there will be a probability amplitude for observing the particle in the left region where its energy exceeds its potential energy. The particle can't simultaneously have energy $E < V$ and be localized on the right.

TUNNELING

The fact that particles can penetrate into regions that are forbidden classically leads to the very important phenomenon of tunneling. Consider a potential step of height $V > 0$ between $x = 0$ and $x = a$, as in Fig. 4-3. An energy eigenstate corresponding to the

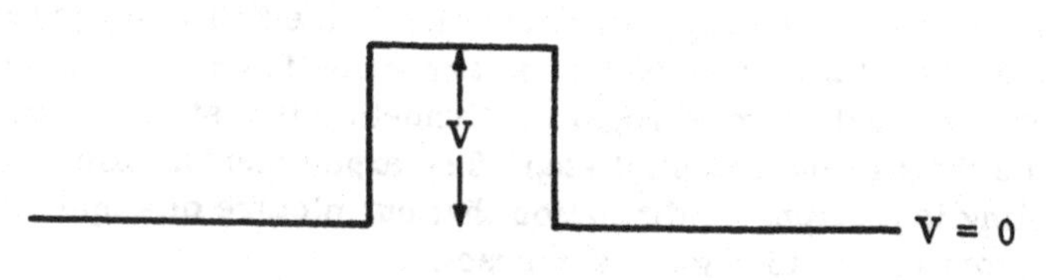

Fig. 4-3

physical situation that a particle of energy $E < V$ is incident from the left has the form

$$\psi(x) = \begin{cases} Ae^{ipx/\hbar} + Be^{-ipx/\hbar}, & x < 0 \\ Ce^{-kx} + De^{kx}, & 0 < x < a \\ AS(E)e^{ip(x-a)/\hbar}, & x > a \end{cases} \tag{4-37}$$

where $p = \sqrt{2mE}$, $\hbar k = \sqrt{2m(V-E)}$. [Because the step is only of length a, the growing exponential solution, for $0 < x < a$, never grows by more than a factor e^{ka}, and thus it is allowable physically. This is in contrast to the case we considered earlier where the step was infinitely long, and the growing exponential solution could grow without bound.] The quantity AS(E) must be nonzero in order to satisfy the boundary conditions at 0 and a. This means that even though a classical particle would be unable to penetrate the step, a quantum mechanical particle incident from the left has a certain amplitude for being found to the right of the step. It is as if the particle digs a tunnel under the step; hence this phenomenon is called *tunneling*.

The function S(E) is called the tunneling matrix element or transmission amplitude; it is essentially the probability amplitude for the process that a particle incident on the left with energy E will tunnel through the step. A detailed calculation yields that for $E < V$

$$S(E) = \frac{2ikp\hbar}{2ikp\hbar \cosh ka + (p^2 - \hbar^2 k^2) \sinh ka}. \tag{4-38}$$

The transmissivity T, the probability that a particle striking the barrier from the left will tunnel through to the right, is given by

$$T(E) = |S|^2 = \left[1 + \frac{\sinh^2 ka}{4(E/V)(1-E/V)}\right]^{-1}. \tag{4-39}$$

T increases monotonically with E, as long as $E < V$.

The possibility of tunneling is built into the Schrödinger equation from the very beginning. In our derivation of the Schrödinger equation, we assumed that if we had some amplitude in one region then it would leak into neighboring regions. Tunneling is just the leaking of amplitude through the potential step. The experimental observation of tunneling is a strong confirmation that our picture of amplitudes leaking from region to region is correct.

It seems as if we have another contradiction lurking in the phenomenon of tunneling. If at early time we have a particle incident from the left with energy $E < V$, and at a later time we observe that it is on the right of the step, then, in tunneling through, hasn't it spent some time in the region of the step, where its total energy was less than the potential V?

Yes, certainly the particle has been in this region, for if we were to place an infinitely high potential barrier anywhere in the region in the way of the particle, we would never observe tunneling. Now in order for there to be a contradiction, we would have to measure the energy of the particle while it was passing through the step, and find a result definitely less than V. However, the accuracy with which we can measure the energy of the particle while it is passing through the step is limited by the uncertainty relation

$$\Delta E \Delta t > \hbar \tag{4-40}$$

where ΔE is the uncertainty in the energy, and Δt is our uncertainty in the times at which the particle is in the step. Were the particle to spend a long time in the step then we could measure E very accurately, and still be sure that the particle was also in the step. What saves us is that the time spent by a particle in traversing the step, defined as the time between the incident wave packet arriving at the step and the transmitted wave packet emerging from the step, is always less than $\hbar/(V-E)$. Thus $\Delta E > V - E$, and we couldn't say for certain that the particle had energy $E < V$ and was simultaneously in the step.

The only situation where this simple argument breaks down is when $V \gg E$ and the barrier is very long. Then the barrier does terrible things to the shape of the wave packet and it is very hard to define a time of travel across the barrier.[2]

The Δt in the uncertainty relation (4-40) is often interpreted as length of time spent measuring the energy. This interpretation is not right since, for example, one can always measure the energy of a free particle arbitrarily accurately by measuring its momentum, and momentum measurements can be performed arbitrarily rapidly. But in an accurate momentum measurement one loses sense of where the particle is, and consequently the time at which it will do things, such as strike screens, becomes uncertain by $\Delta t \gtrsim \hbar/\Delta E$. This loss of the ability to make accurate time predictions for a particle always accompanies accurate energy measurements.

The α decay of nuclei can be qualitatively pictured as a tunneling process. We can think of the nucleus before the decay as consisting

[2] This problem is discussed fully by T. Hartman in *J. Appl. Phys.* **33**, 3422 (1962), in a very interesting paper about the transit times of particles tunneling through potential barriers.

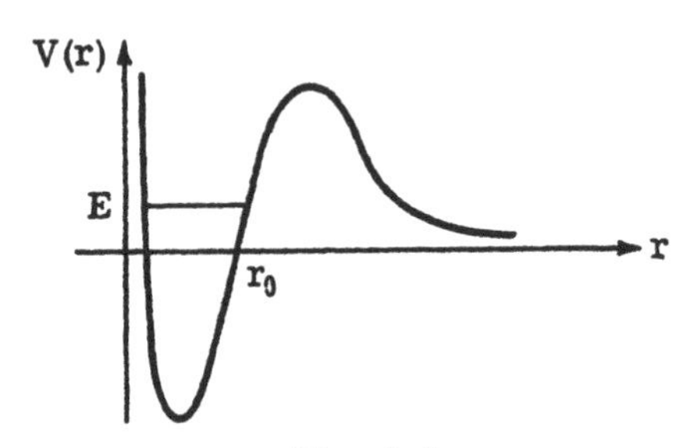

Fig. 4-4
Potential felt by α particle of energy E in the residual nucleus.

of an α particle trapped in the spherically symmetric potential due to the remaining nucleons in the nucleus. This potential, as a function of the distance r from the center of the nucleus, looks roughly as shown in Fig. 4-4. If the α particle has energy > 0 but less than the height of the positive bump, then it will bounce around inside the nucleus. Each time it hits the bump it has a certain probability of tunneling through, and one of the times it hits the bump it does tunnel through to the outside world. This looks then like the emission of an α particle by the nucleus.

If the energy of the α particle in the nucleus is E, then it bounces around in the nucleus $\omega = E/\hbar$ times per second. Hence the probability of the particle tunneling through the barrier in the time $\hbar/E$, equal to one cycle, is T, the transmissivity of the barrier. The lifetime of the particle, which is the inverse of the probability of decay per unit time, is therefore about $[T/(\hbar/E)]^{-1} = \hbar/ET$. For a square barrier this behaves roughly as $\exp[-2a\sqrt{2m(V-E)}/\hbar]$; its actual value is, however, very sensitive to the exact shape of the potential, and can vary enormously from nucleus to nucleus.

A second occurrence of tunneling is in solid state physics. If we separate two metals by a thin insulating layer, about 100 Å thick,

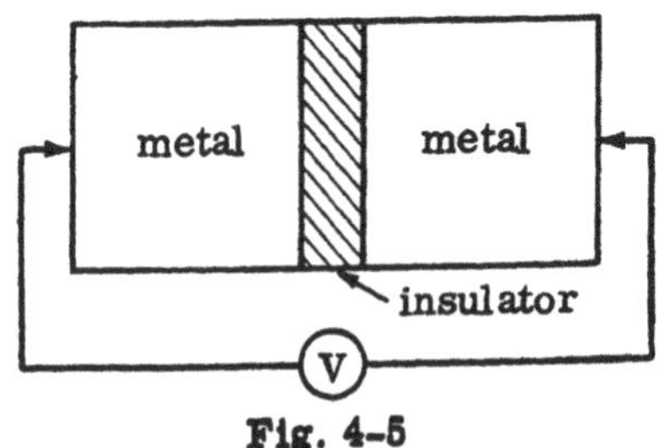

Fig. 4-5
Tunneling junction.

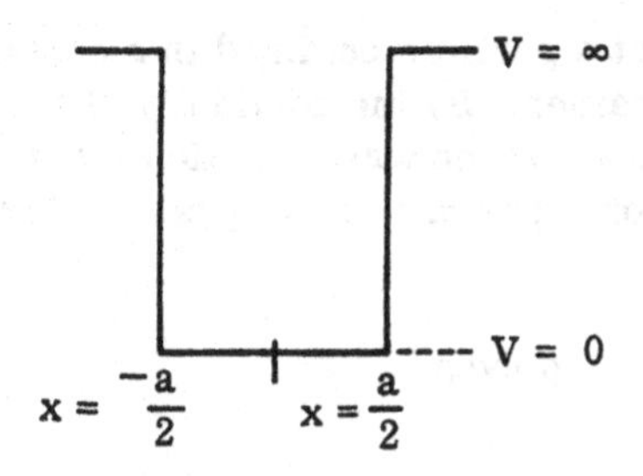

Fig. 4-6
Square well with impermeable walls.

and apply a voltage across the metals (Fig. 4-5), then we will observe a current flowing across the insulator. The insulator acts as a potential barrier between the two metals, and what we are actually observing is the tunneling of electrons from one metal to the other through the potential barrier. By studying the current as a function of the applied voltage one can learn a great deal about the states of the electrons in the metals.

BOUND STATES

In the cases we have been studying so far, the energy could take on any value from zero to infinity. Let us turn now to considering potential wells, where we will find, in addition, bound states of the particle with only discrete energy values. These are the type of states in which, for example, one finds electrons in atoms.

The simplest case is that of a square well in one dimension with infinitely high sides (Fig. 4-6). The wave function inside the well is of the form

$$\psi(x) = A \sin\left(\frac{px}{\hbar} + \varphi\right), \tag{4-41}$$

where φ is some phase angle, and $p^2 = 2mE$. [Why must φ be real?] The wave function outside the well, if the walls were of finite height, $V \gg E$, would be of the form $e^{-k|x|}$ where $\hbar^2 k^2 = 2m(V - E)$, and would join continuously onto the wave function inside the well. Now as $V \to \infty$, $e^{-k|x|} \to 0$, so that the wave function vanishes outside the well, for an infinitely high potential. This agrees of course with our classical notion that a particle confined in a container can be found only inside the container. By the continuity of the wave function at the walls we see that, for the wave function inside, $\psi(\pm a/2) = 0$. The solution can therefore take one of two possible forms:

$$\psi(x) = \begin{cases} \sin \dfrac{n\pi x}{a}, & n \text{ even} \\ \cos \dfrac{n\pi x}{a}, & n \text{ odd}, \end{cases} \tag{4-42}$$

where n is any integer from one to infinity. $n = 0$ is excluded, for then $\psi(x) \equiv 0$. From (4-42) we see that

$$p = n\frac{\pi\hbar}{a}, \tag{4-43}$$

and

$$E = n^2 \frac{\pi^2\hbar^2}{2ma^2}, \qquad n = 1, 2, 3, \ldots \tag{4-44}$$

are the allowable energy values for particles inside the well.

The lowest energy value is $E_0 = \pi^2\hbar^2/2ma^2$. This energy is greater than zero, the lowest value of the potential, because of the uncertainty principle. When the particle is localized in the well, it must have a momentum uncertainty $\Delta p \gtrsim 2\pi\hbar/a$, and hence a finite value of its kinetic energy

$$E \sim \frac{(\Delta p)^2}{2m}. \tag{4-45}$$

We can see this momentum uncertainty explicitly by writing the wave function of the lowest state (called the ground state) as $\frac{1}{2}(e^{i\pi x/a} + e^{-i\pi x/a})$ for $|x| \le a/2$, and 0 for $|x| \ge a/2$; a linear combination of a state with momentum peaked about $\pi\hbar/a$ and one with momentum peaked about $-\pi\hbar/a$.

This minimum amount of kinetic energy coming from the uncertainty principle is called the *zero point energy*. One of its most important manifestations is in keeping helium a liquid (at saturated vapor pressure) down to $T = 0$. If we try to make a solid of He then we have to localize the particles in a lattice. This localization implies that the atoms will have a zero point energy, which is particularly large for small mass particles [cf. (4-45)]. It is so large for He that the particles can't be constrained in a lattice by the interatomic forces, and He instead is a liquid. Only when we apply lots of pressure from the outside can we hold the He atoms in a lattice.

PARITY

The integer n in (4-42) is just the number of nodes of the wave function. We see that

$$\begin{aligned} \psi(-x) &= \psi(x), \quad n \text{ odd} \\ \psi(-x) &= -\psi(x), \quad n \text{ even.} \end{aligned} \tag{4-46}$$

Under the operation of replacing x by −x, called the *parity operation*, these wave functions are multiplied by a constant $(-1)^{n+1}$ Hence the wave functions are eigenstates of the parity operation, and the eigenvalue $(-1)^{n+1}$ is called the *parity* of the state. One says that the states with odd n have even parity (+1), and the states with even n have odd parity (−1).

Let us consider the parity operation from a more general point of view. Suppose that we have a potential $V(\mathbf{r})$ that obeys

$$V(\mathbf{r}) = V(-\mathbf{r}), \tag{4-47}$$

Then if $\psi(\mathbf{r})$ solves the Schrödinger equation

$$\left[-\frac{\hbar^2}{2m}\nabla^2 + V(\mathbf{r})\right]\psi(\mathbf{r}) = E\psi(\mathbf{r})$$

we have, on replacing $\mathbf{r}$ by $-\mathbf{r}$ everywhere

$$\left[-\frac{\hbar^2}{2m}\nabla^2 + V(-\mathbf{r})\right]\psi(-\mathbf{r}) = E\psi(-\mathbf{r}),$$

and since $V(-\mathbf{r}) = V(\mathbf{r})$

$$\left[-\frac{\hbar^2}{2m}\nabla^2 + V(\mathbf{r})\right]\psi(-\mathbf{r}) = E\psi(-\mathbf{r}).$$

Thus if $\psi(\mathbf{r})$ is a solution with energy E, so is $\psi(-\mathbf{r})$ a solution with the same energy value. Hence

$$\psi(\mathbf{r}) + \psi(-\mathbf{r}) \tag{4-48}$$

and

$$\psi(\mathbf{r}) - \psi(-\mathbf{r}), \tag{4-49}$$

are also solutions; (4-48) has even parity, and (4-49) has odd parity. Thus we can always choose the solutions for a potential that obeys $V(\mathbf{r}) = V(-\mathbf{r})$ to have a definite parity value. Notice that we don't automatically have degeneracy; either (4-58) or (4-59) can vanish identically.

We can define a parity operator, P, acting on state vectors, by saying that $P|\Psi\rangle$ has the components

$$\langle \mathbf{r}|P|\Psi\rangle = \langle -\mathbf{r}|\Psi\rangle. \tag{4-50}$$

Note that

$$P^2 = 1 \tag{4-51}$$

since

$$\langle \mathbf{r}|P^2|\Psi\rangle = \langle -\mathbf{r}|P|\Psi\rangle = \langle \mathbf{r}|\Psi\rangle.$$

The energy eigenstates can be chosen to have definite parity if and only if P commutes with H, the Hamiltonian. (From now on we'll drop the subscript "op.") Suppose, first that the energy eigenstates can be chosen to have definite parity. Then

$$PH|E\rangle = PE|E\rangle = EP|E\rangle = \pm E|E\rangle,$$

and

$$HP|E\rangle = \pm H|E\rangle = \pm E|E\rangle,$$

so

$$PH|E\rangle = HP|E\rangle. \tag{4-52}$$

Since the energy eigenstates form a basis we have therefore $PH = HP$, or

$$[P, H] = 0. \tag{4-53}$$

Since $P^2 = 1$, we can write this relation as

$$P^{-1}HP = H. \tag{4-54}$$

In space language

$$P^{-1}H(\mathbf{r}, p)P = H(-\mathbf{r}, -p). \tag{4-55}$$

(4-54) is equivalent to the statement that $V(\mathbf{r}) = V(-\mathbf{r})$.

When the parity operator commutes with H, the energy eigenstates can be assigned definite parity values. To see this, note that $[H, P] = 0$ implies that

$$HP|E\rangle = PH|E\rangle = EP|E\rangle$$

so that $P|E\rangle$ is also an eigenstate of H with the same energy. Thus $(1 \pm P)|E\rangle$ are both eigenstates of H with eigenvalue E and of P with eigenvalue ± 1.

This is a *general rule:* if an operator, Ξ, commutes with the Hamiltonian, then the energy eigenfunctions can be chosen to be simultaneously eigenfunctions of Ξ and H. The proof of this very important theorem goes as follows. Suppose first of all that a given energy level, E, is nondegenerate. If the energy eigenfunction is $|E\rangle$, then

$$H\Xi|E\rangle = \Xi H|E\rangle = E\Xi|E\rangle$$

so that $\Xi|E\rangle$ is an energy eigenfunction with the same energy. But if the level is nondegenerate, then $\Xi|E\rangle$ must be a constant times $|E\rangle$:

$$\Xi|E\rangle = \xi|E\rangle,$$

i.e., $|E\rangle$ is also an eigenfunction of Ξ. Next, suppose that there are n orthonormal states $|1\rangle, |2\rangle, \ldots, |n\rangle$, all with the energy E. We shall show that there are n linearly independent linear combinations of these states each of which is an eigenfunction of Ξ. Since $[\Xi, H] = 0$, the state $\Xi|i\rangle$, $i = 1, \ldots, n$, must also be an eigenstate of H with energy E. Thus it must be a linear combination of $|1\rangle, |2\rangle, \ldots, |n\rangle$,

$$\Xi|i\rangle = \sum_j |j\rangle X_{ji}, \tag{4-56}$$

where $X_{ji} = \langle j|\Xi|i\rangle$, and in general, any linear combination $|C\rangle = \sum_i c_i|i\rangle$ obeys

$$\Xi|C\rangle = \sum_{ji}|j\rangle X_{ji}c_i. \tag{4-57}$$

Thus if we pick the c_i to be eigenvectors of X,

$$\sum_i X_{ji} c_i = \xi c_j, \tag{4-58}$$

then

$$\Xi |C\rangle = \xi |C\rangle. \tag{4-59}$$

There are n linearly independent eigenvectors of the $n \times n$ matrix X. Thus we can choose n linearly independent combinations of the n degenerate energy eigenstates $|1\rangle, \ldots, |n\rangle$, that are simultaneously eigenstates of Ξ. All the energy eigenstates can therefore be chosen to be simultaneously eigenstates of Ξ. This result holds generally for any two (or more) commuting Hermitian operators.

Conversely, if the two (or more) operators can have a complete set of simultaneous eigenstates, they must commute. The proof is identical to that leading to (4-53). Physically this means that if A is the operator for an observable, and B is the operator for another observable, then, in general, the system can have a definite value of both of these observables simultaneously, i.e., be in a simultaneous eigenstate of A and B, if and only if A and B commute. For example, $[x, p_x] = i\hbar$, so that a particle can't have a definite value of its position and its momentum simultaneously. When $[P, H] = 0$, then the particle can have a definite energy value and parity value simultaneously.

Let us use the concept of parity in finding the bound states of the finite square well in one dimension [Fig. 4-7]. Since $V(x) = V(-x)$ the solutions can be chosen to have a definite parity value. Thus the even parity solutions are of the form, for $0 > E > -|V|$,

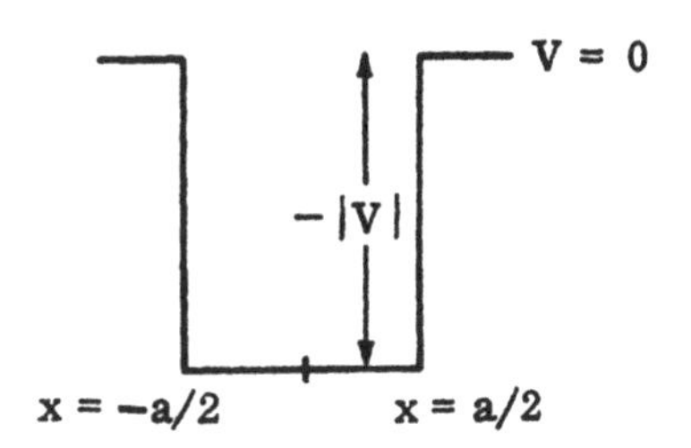

Fig. 4-7
Finite depth square well symmetric about x = o.

$$\psi(x) = \begin{cases} A\cos\frac{px}{\hbar}, & x \text{ in well} \\ Ce^{-kx}, & x > \frac{a}{2} \\ Ce^{kx}, & x < -\frac{a}{2} \end{cases} \tag{4-60}$$

and the odd parity solutions are of the form

$$\psi(x) = \begin{cases} A\sin\frac{px}{\hbar}, & x \text{ in well} \\ Ce^{-kx}, & x > \frac{a}{2} \\ -Ce^{kx}, & x < -\frac{a}{2} \end{cases} \tag{4-61}$$

where $p = \sqrt{2m(E + |V|)}$, $\hbar k = \sqrt{-2mE}$. E cannot be arbitrary, but rather, it is determined by the boundary conditions on the wave function at $x = \pm a/2$. For even parity the boundary conditions are

$$A\cos\frac{pa}{2\hbar} = Ce^{-ka/2} \quad \text{and} \quad -\frac{p}{\hbar}A\sin\frac{pa}{2\hbar} = -kCe^{-ka/2}. \tag{4-62}$$

Dividing the bottom equation by the top equation we find

$$p\tan\frac{pa}{2\hbar} = \hbar k \tag{4-63}$$

as the equation that determines the possible E values for the even parity states. Similarly, for the odd parity solutions the condition on E is

$$p\cot\frac{pa}{2\hbar} = -\hbar k. \tag{4-64}$$

Equations (4-63) and (4-64) are easily solved graphically; the solutions can be found in standard texts. For small $|V|$ there is always one even parity solution. As $|V|$ increases, we find, at higher energy, an odd parity solution also. Then for still higher $|V|$ there is another even parity solution, at a higher energy than either of the first two solutions. For general V the even and odd solutions alternate in energy. There is always a finite number of bound states.

For very small $|V|$

$$E = \frac{-mV^2a^2}{2\hbar^2}. \tag{4-65}$$

The particle is very loosely bound to the well, and its wave function falls off very slowly outside the well.

TRANSMISSION RESONANCES

The energy spectrum for the square well, for $0 > E > -|V|$, consists of a finite number of bound levels. The wave function for these states falls off exponentially as we move away from the well. In addition we can construct solutions for any value of E greater than zero. These states correspond physically to particles shot in from far away which are reflected and transmitted by the potential well.

Let us consider only the case of particles incident from the left. The incident wave is

$$\psi_{\text{inc}}(x) = Ae^{ipx/\hbar}, \quad x < -\frac{a}{2}. \tag{4-66}$$

and the transmitted wave is

$$\psi_{\text{trans}}(x) = Ae^{ip(x-a)\hbar}\, S(E), \quad x > \frac{a}{2}, \tag{4-67}$$

where the transmission amplitude S(E), it turns out, is given by

$$S(E) = \left[\cos\frac{\bar{p}a}{\hbar} - \frac{i}{2}\left(\frac{p}{\bar{p}} + \frac{\bar{p}}{p}\right)\sin\frac{\bar{p}a}{\hbar}\right]^{-1}, \tag{4-68}$$

where $\bar{p} = \sqrt{2m(E + |V|)}$. S(E) vanishes as E approaches zero. The transmissivity, T, of the well, is

$$T(E) = |S(E)|^2 = \left[1 + \frac{\sin^2(\bar{p}a/\hbar)}{4(E/|V|)[1 + (E/|V|)]}\right]^{-1}. \tag{4-69}$$

Compare this result with Eq. (4-39) for a potential barrier. A graph of T(E) is shown in Fig. 4-8. The peaks in T are called *resonances;* they occur whenever

$$\sin\frac{\bar{p}a}{\hbar} = 0, \quad E > 0, \tag{4-70}$$

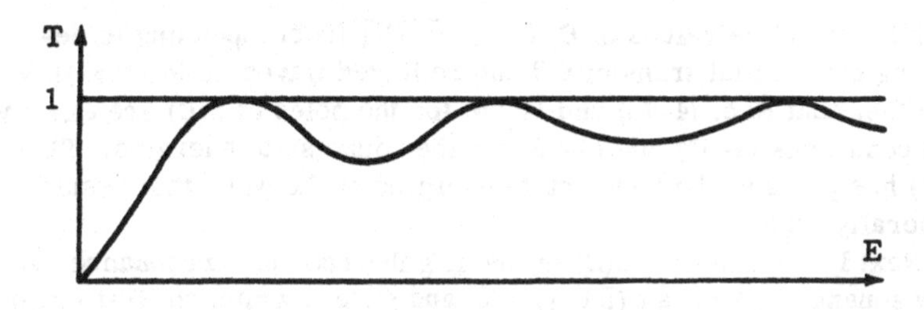

Fig. 4-8
Transmissivity of the square well.

or

$$E = n^2 \frac{\pi^2\hbar^2}{2ma^2} - |V| > 0, \quad \text{and } n = 1, 2, 3, \ldots \tag{4-71}$$

To a particle of this energy, the potential looks perfectly transparent; there is no reflection — all the wave is transmitted.

The transmission amplitude, considered as an analytic function of the energy E, has some very interesting properties. First of all, S(E) has poles at the (negative) values of E that are the bound state energies of the square well. For S(E) to be infinite means that one can have a transmitted wave with no incident wave; this is exactly the condition to have a bound state. Of course, the transmitted (and reflected) wave doesn't propagate; it falls off exponentially. From (4-68) we see that S(E) has a pole whenever

$$\cos\frac{\bar{p}a}{\hbar} = \frac{i}{2}\left(\frac{p}{\bar{p}} + \frac{\bar{p}}{p}\right)\sin\frac{\bar{p}a}{\hbar}. \tag{4-72}$$

Clearly p must be imaginary, or else (4-72) has no solutions. Using $\tan 2x = 2[\cot x - \tan x]^{-1}$, we find that (4-72) is satisfied when

$$\bar{p}\cot\frac{\bar{p}a}{2\hbar} = ip \tag{4-73}$$

or when

$$\bar{p}\tan\frac{\bar{p}a}{2\hbar} = -ip. \tag{4-74}$$

If we choose the branch of $\sqrt{E}$ so that if $E = |E|e^{i\varphi}$ then $\sqrt{E} = |E|^{1/2}e^{i\varphi/2}$, then for negative values of E, $p = i\sqrt{2m|E|}$ [corresponding to decaying exponential transmitted and reflected waves as in (4-67)]. We see then that Eqs. (4-73) and (4-74) for the poles of S(E) are exactly the conditions (4-64) and (4-63) for the bound state energies. Thus S(E) has poles at the bound state energies of the well; this result is generally true.

Next let us examine S(E) in the neighborhood of a resonance. At a resonance $E_0 > 0$, $\tan(\bar{p}a/\hbar) = 0$, and we can write, to first order in $E - E_0$,

$$\left(\frac{p}{\bar{p}}+\frac{\bar{p}}{p}\right)\tan\frac{\bar{p}a}{\hbar} = \frac{4}{\Gamma}(E - E_0), \tag{4-75}$$

where the positive quantity Γ is defined by

$$\frac{4}{\Gamma} = \left[\left(\frac{p}{\bar{p}}+\frac{\bar{p}}{p}\right)\frac{1}{\hbar}\frac{d\bar{p}a}{dE}\right]_{E = E_0}. \tag{4-76}$$

Thus S(E) near a resonance can be written as

$$S(E) = \frac{1}{\cos(pa/\hbar)}\,\frac{i\Gamma/2}{E - E_0 + i\Gamma/2}. \tag{4-77}$$

The $\cos(\bar{p}a/\hbar) \approx \pm 1$. Thus S(E) appears to have a pole at $E = E_0 - i\Gamma/2$, as in Fig. 4-9.

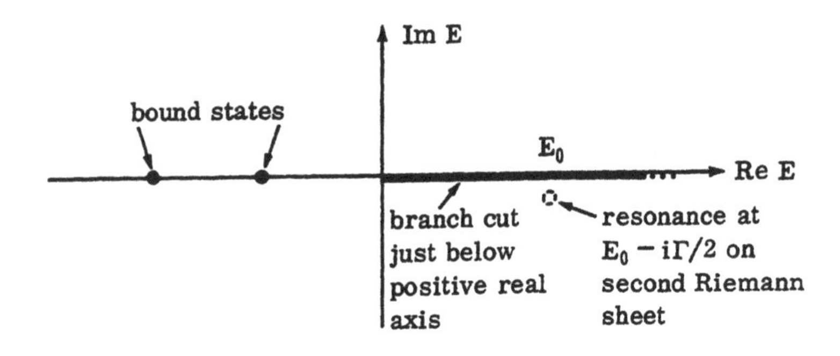

Fig. 4-9

Singularities of the transmission amplitude in the complex E plane.

We have chosen the branch of $\sqrt{E}$ so that there is a discontinuity in $\sqrt{E}$ just below the positive real axis, i.e., if $E = |E|$, then $\sqrt{E} = |E|^{1/2}$ but if $E = |E|e^{2\pi i}$, then $\sqrt{E} = -|E|^{1/2}$. This discontinuity in $\sqrt{E}$ just below the positive real axis implies that S(E), which is a function of $\sqrt{E}$, also has a discontinuity just below the positive real axis. The value of S(E) that we find by analytically continuing S(E) from above to below the positive real axis differs from the value defined by (4-77) below the positive real axis; S(E) has a branch cut just below the real axis. The pole at $E = E_0 - i\Gamma/2$ that we have found in S(E) is not in the function defined by (4-68) but rather is in the analytic continuation of S(E) from above to below the positive real axis. The pole is on the *second Riemann sheet* of S(E).

Thus the bound states of the well correspond to poles of S(E) on the negative real axis; the resonances of the well correspond to bumps on the real axis, or equivalently, correspond to poles that are just below the positive real axis on the second sheet of S(E). We shall see that the closer these poles are to the real axis the more the resonances behave like very long-lived bound states. This structure of S(E) is not peculiar to the square well problem; it occurs for all similar transmission problems in one dimension, and scattering problems in three dimensions.

Let us write S(E) in terms of a phase shift, $\delta(E)$, and the transmissivity, as

$$S(E) = e^{i\delta(E)}|T(E)|^{1/2}. \tag{4-78}$$

Then, from Eq. (4-68), the phase shift is given by

$$\tan\delta(E) = \frac{1}{2}\left(\frac{p}{\bar{p}} + \frac{\bar{p}}{p}\right)\tan\frac{\bar{p}a}{\hbar}. \tag{4-79}$$

Using (4-77) we see that near a resonance $\tan\delta(E) = (2/\Gamma)(E - E_0)$ or

$$\delta(E) = \arctan\frac{2}{\Gamma}(E - E_0). \tag{4-80}$$

Notice that $d\delta(E)/dE$ has a maximum at a resonance. The phase shift varies rapidly with energy at a resonance.

Physically, a resonance is like an "almost bound state" of the well. If we shoot a particle with an energy on resonance at the well from far away, it enters the well and bounces around for a long time, like a bound particle. Finally, however, because its energy is greater than zero, it escapes from the well. This escaping is reminiscent of the decay of an α particle from a nucleus. As we shall see, $\hbar/\Gamma$

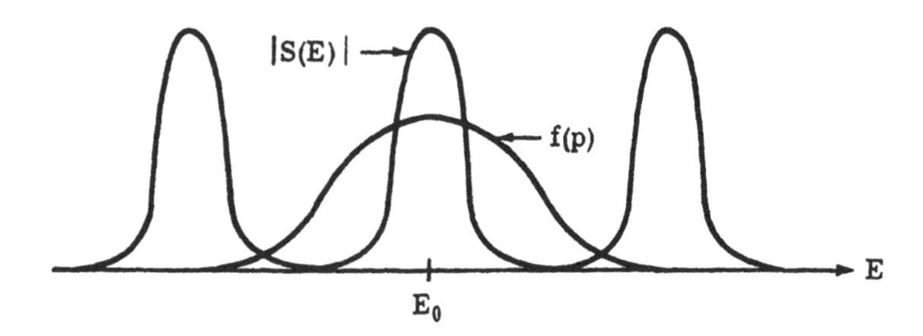

Fig. 4-10
A sharp resonance in S at E_0, and a wave packet whose momentum amplitude varies slowly over the resonance.

is the lifetime of the particle in a resonance in the well.

Because a resonance looks very much like a long-lived bound state, people nowadays refer to resonances in the scattering of elementary particles as new elementary particles with finite lifetimes. For example, in the scattering of π + N there is a resonance at 1238 MeV. It looks like the π and N have formed a temporary bound state at this energy. Now a temporary bound state of π + N is very hard to distinguish from an unstable elementary particle that decays into π + N, so one refers to this resonant state as a new elementary particle, N*, and says that N* decays into π + N. Actually, all of the newly discovered elementary particles are resonances in the scattering of other elementary particles. [In fact, the principle of "nuclear democracy" has been proposed: that *all* elementary particles that interact strongly are either resonances, or true bound states of *each other.* [3]]

Let us look at resonances in more detail — from the point of view of wave packets. We take a potential that vanishes for $x > a/2$ and for $x < -a/2$, like the square well, but we do not make any particular assumptions about its form for $-a/2 < x < a/2$. We shall take an incident wave packet, for $x < -a/2$

$$\psi_{\text{inc}}(x, t) = \int_0^\infty \frac{dp}{2\pi\hbar}\, e^{i(px - Et)/\hbar} f(p) e^{ipa/2\hbar}$$

where f(p) is real. The factor $e^{ipa/2\hbar}$ implies that this packet reaches the well at $t = 0$. The transmitted packet is

[3] See, for example, M. Jacob and G. Chew, *Strong Interaction Physics* [W.A. Benjamin, New York, 1964].

$$\psi_{trans}(x,t) = \int_0^\infty \frac{dp}{2\pi\hbar}\, e^{i(px-Et)/\hbar}\, [S(E)e^{-ipa/\hbar}] f(p) e^{ipa/2\hbar} \tag{4-81}$$

for $x > a/2$, where S(E) is the transmission amplitude of the potential, as defined in (4-67). Let us suppose, as in Fig. 4-10, that S(E) has several sharp resonances, that f(p) is appreciably different from zero at only one resonant energy, E_0, and that f(p) varies slowly over the resonance.

What does $\psi_{trans}(x)$ look like; how does it move in time? Our old argument about the motion of a wave packet [using the method of stationary phase, as in (3-74)] breaks down when the phase of the weighting function, S(E)f(p), varies rapidly with energy. This is exactly the situation at a resonance where $d\delta(E)/dE$ has a maximum. Thus we must proceed differently to find the structure of the wave packet emerging from the well.

Because f(p) varies slowly near the resonance, but extends over only one resonance, we can write approximately

$$\psi_{trans}(x,t) = f(p_0) \int_{\substack{\text{neighborhood} \\ \text{of } E_0}} \frac{dp}{2\pi\hbar}\, e^{i[p(x-a/2)-Et]/\hbar}\, S(E) \tag{4-82}$$

where $p_0 = \sqrt{2mE_0}$, and the integral is over the region of the resonance only. Now in the neighborhood of the resonance at E_0 we can approximate S(E) by (4-77):

$$S(E) \approx \frac{\pm i\Gamma/2}{E - E_0 + i\Gamma/2} \tag{4-83}$$

to within a slowly varying factor. This form, which we derived for the square well, is generally valid near a sharp resonance. Then (4-82) becomes

$$\psi_{trans}(x,t) \approx \pm f(p_0) \int \frac{dp}{2\pi\hbar}\, \frac{e^{i[p(x-a/2)-Et]/\hbar}\, i\Gamma/2}{E - E_0 + i\Gamma/2}\, . \tag{4-84}$$

Actually, this is a rather crude approximation for ψ_{trans} for a square well because the resonances aren't very sharp. It is a much better approximation for potentials with smaller Γ's at the resonances. For example, a square well with high shoulders, Fig. 4-11, has several very sharp resonances. These resonances are the "bound states" of the well (dotted lines) with energy $V > E > 0$. They are not

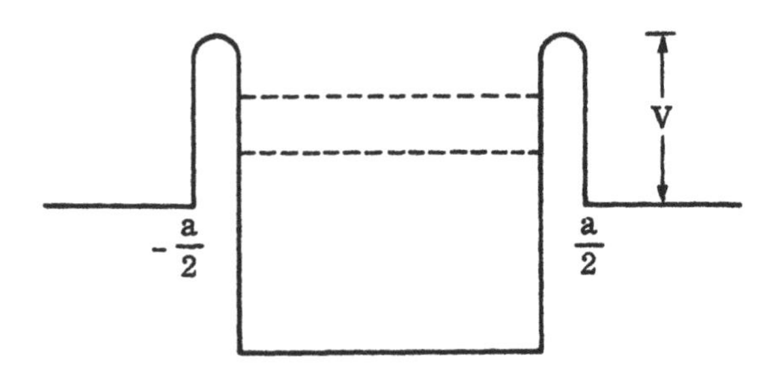

Fig. 4-11
A well with sharp resonances at energies indicated by the dotted lines.

true bound states because a particle in such a state can eventually tunnel its way out through the shoulders. The smaller the probability of tunneling, the smaller will be Γ

Now to get a feeling for how (4-84) behaves let us write

$$dp = \frac{dp}{dE}\, dE = \frac{m}{p}\, dE \approx \frac{m}{p_0} dE$$

and expand the exponent about $E = E_0$ using

$$p \approx p_0 + \left(\frac{dp}{dE}\right)_{E_0} (E - E_0).$$

Thus

$$\psi_{\text{trans}}(x, t) \approx \pm \frac{i\Gamma f(p_0)\, m}{2\, p_0} \exp\left\{\frac{i[p_0(x - a/2) - E_0 t]}{\hbar}\right\}$$
$$\times \int_{E \text{ near } E_0} \frac{dE}{2\pi\hbar} \quad \frac{1}{E - E_0 + i\Gamma/2} \exp\left\{\frac{-i(E - E_0)}{\hbar}\left[t - \frac{m}{p_0}\left(x - \frac{a}{2}\right)\right]\right\} \tag{4-85}$$

Lastly, let us notice that as E becomes far from E_0, the exponent varies rapidly, and the denominator grows as E. Thus we can, as a first approximation, extend the limits on the integral in (4-85) from $-\infty$ to ∞. [But see footnote 4, p. 112.] The integral then becomes

$$\int_{-\infty}^{\infty}\frac{dE}{2\pi\hbar}\frac{1}{E-E_0+i\Gamma/2}\exp\left\{\frac{-i(E-E_0)}{\hbar}\left[t-\frac{m}{p_0}\left(x-\frac{a}{2}\right)\right]\right\}. \tag{4-86}$$

Let us do this integral by the method of contour integrals. If $x > (a/2) + (p_0/m)\,t$, then the exponential goes to zero as $E \to \infty$ in the upper half of the complex E plane. Thus we can close the contour of integration in the upper half-plane. Since the denominator vanishes at $E = E_0 - i\Gamma/2$, in the lower half-plane, the integration surrounds no singularities and therefore the integral vanishes. For $x < (a/2) + (p_0/m)\,t$, the exponential goes to zero as $E \to \infty$ in the lower half-plane, and we can close the contour in the lower half-plane. We thus surround the pole at $E = E_0 - i\Gamma/2$. The value of the integral is thus

$$\frac{-2\pi i}{2\pi\hbar}\exp\left\{-\frac{i}{\hbar}\left[E_0-\frac{i\Gamma}{2}-E_0\right]\left[t-\frac{m}{p_0}\left(x-\frac{a}{2}\right)\right]\right\}$$

Hence we find, for $x > a/2$

$$\psi_{\text{trans}}(x,t) \approx \begin{cases} 0: \quad x > \dfrac{a}{2}+\dfrac{p_0}{m}t \\ \pm\dfrac{m\Gamma f(p_0)}{2p_0\hbar}\exp\left\{\dfrac{i}{\hbar}\left[p_0\left(x-\dfrac{a}{2}\right)-E_0 t\right]\right\} \\ \qquad \times\exp\left\{-\Gamma\left[t-\dfrac{m}{p_0}\left(x-\dfrac{a}{2}\right)\right]\dfrac{1}{2\hbar}\right\}: \quad x<\dfrac{a}{2}+\dfrac{p_0}{m}t. \end{cases} \tag{4-87}$$

Thus for $t > 0$ the (real or imaginary part of the) transmitted wave has the form shown in Fig. 4-12. Actually the wave front isn't perfectly sharp, as we found in (4-87). There is a small contribution for $x > (a/2) + (p_0/m)\,t$ which unfortunately was thrown away when we extended the limits on the integral from $-\infty$ to ∞.

How can we interpret this result? First of all, the front of the wave packet is at $x = (a/2) + (p_0/m)\,t$, which is just the point at which we would find a classical particle leaving $a/2$ at time 0 and having velocity p_0/m. This velocity is essentially that of the incident packet which is centered about $E = E_0$. The wave function is an oscillatory term modulated by the exponential horn

$$\exp\left\{-\Gamma\left[t-\frac{m}{p_0}\left(x-\frac{a}{2}\right)\right]\frac{1}{2\hbar}\right\}.$$

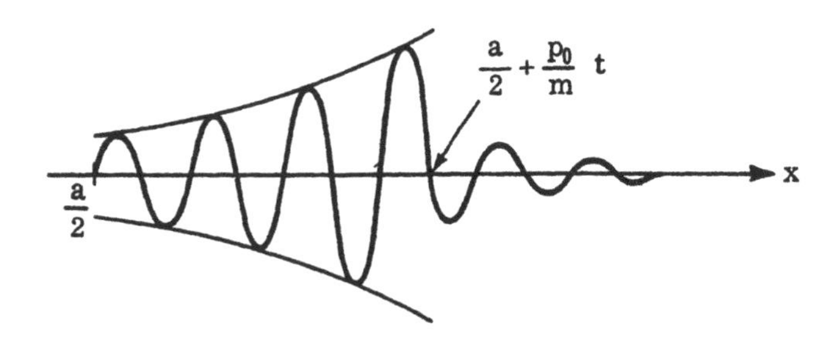

Fig. 4-12

Transmitted wave at a resonance. As t increases the wave moves to the right with constant (absolute value of the) amplitude. The same picture describes the wave function of an α particle escaping from a nucleus: in three dimensions the amplitude falls as $1/r$ as the wave moves away from the origin.

Thus the probability amplitude for finding the particle at a point $x < (a/2) + (p_0/m)\,t$ decreases in magnitude exponentially as $e^{-\Gamma t/2\hbar}$; the probability decreases as $e^{-\Gamma t/\hbar}$. Notice that this behavior corresponds exactly to an exponential decay law for a particle trapped in the potential, such as the α particle in the well in Fig. 4-4; the probability for finding the particle near the well is greatest at $t = 0$, and it decreases exponentially in time. The resonance behaves just like a "bound state" that has a lifetime $\hbar/\Gamma$. The amplitude for finding the particle leaks out of the well at a rate proportional to the amplitude in the well.[4]

In the case of the α particle escaping from the well, its state at $t = 0$, say, when the system is prepared, is a wave packet localized in the well. This wave packet is *not* an eigenstate of the potential, for in that case it would be a stationary state and the α particle would remain forever in the well; the initial state is a linear combination of many different eigenstates of the potential, and the energy of the α particle is uncertain; $\Delta E \sim \Gamma/2$. For $t > 0$ the wave function of the α particle begins to leak out of the well and looks exactly like Fig. 4-12. As time progresses the picture of the wave function in Fig. 4-12 moves bodily to the right. The height of the wavefront remains constant in one dimension, and decreases as r^{-1} in three dimensions. A wave function of the general form (4-87) is the only one consistent with exponential decay.

[4] Actually, the exact behavior of the integral (4-84) turns at very long times from exponented $\sim e^{-\Gamma t/2\hbar}$ to a power law falloff $\sim t^{1/2}$; see L.A. Khalfin, Zh. Eksp. Theor. Fiz. 33, 1371 (1958) [Engl. transl., Sov. Phys. JETP6, 1053 (1958).

The exponential decay behavior of a transmission resonance occurs only when the resonance is sharp. As $\delta(E)$ varies more and more slowly at the resonance, corresponding to a broader and broader resonance, the outcoming wave packet looks more and more like the incoming wave packet, and for very slowly varying $\delta(E)$ we find our old picture of the transmitted packet emerging from the well after a delay time $\sim [d\delta(E)/dE]_{E=E_0}$. The extent to which the well distorts the incoming wave packet in transmitting it depends on how much the well changes the relative phases of the components of the incoming packet. The faster $\delta(E)$ varies the greater the change in the relative phases.[5]

ONE-DIMENSIONAL DELTA FUNCTION POTENTIAL

As a special example of transmission by a potential let us consider a delta function potential in one dimension

$$v(x) = v_0\delta(x). \tag{4-88}$$

v_0 has dimensions of energy times length. The Schrödinger equation is

$$\left(E + \frac{\hbar^2}{2m}\frac{d^2}{dx^2}\right)\psi(x) = v_0\delta(x)\psi(x) = v_0\delta(x)\psi(0), \tag{4-89}$$

and has a solution

$$\psi(x) = \psi_{inc}(x) + \int_{-\infty}^{\infty} \frac{dp}{2\pi\hbar}\,\frac{e^{ipx/\hbar}}{E - p^2/2m}\,v_0\psi(0), \tag{4-90}$$

where $\psi_{inc}(x) \sim e^{\pm i\sqrt{2mE}\,x/\hbar}$ solves the free particle Schrödinger equation. To see that (4-90) is a solution, we act on it with $E + (\hbar^2/2m)(d^2/dx^2)$; this annihilates $\psi_{inc}(x)$ and turns the second term into

$$\int_{-\infty}^{\infty} \frac{dp}{2\pi\hbar}\,\frac{(E - p^2/2m)e^{ipx/\hbar}}{E - p^2/2m}\,v_0\psi(0),$$

[5] A fascinating computer movie of the scattering of a Gaussian wave packet by a square well was prepared by A. Goldberg, H.M. Schey, and J.L. Schwartz; some of the pictures are shown by them in *Am. J. Phys.* 35, 177 (1967).

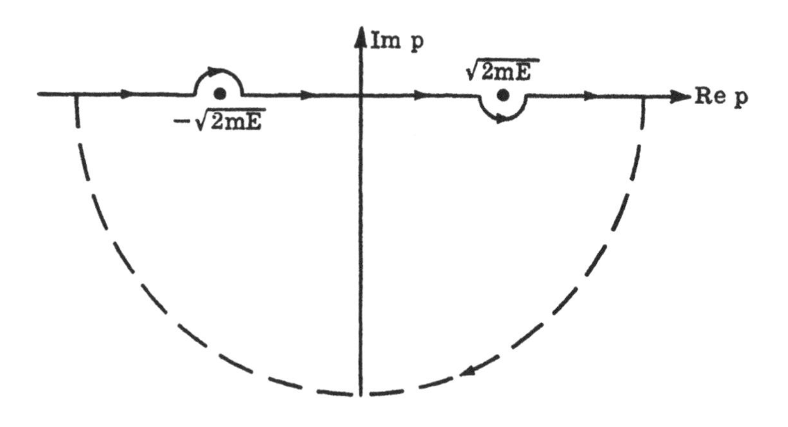

Fig. 4-13

Contour of integration for outgoing waves; the dotted line is the closure of the contour at infinity in the lower half-plane.

which from (3-53) [in one dimension] is just $v_0\psi(0)\delta(x)$.

Now the integral in (4-90) has poles at $p = \pm\sqrt{2mE}$, and to specify completely the solution we must specify what the contour of integration does near these poles; this corresponds to choosing boundary conditions on the solution (4-90). We should like to choose the boundary condition of *outgoing waves,* i.e., that the integral behaves as a wave traveling to the right, for $x > 0$, and a wave traveling to the left, for $x < 0$. We therefore choose the integration contour in Fig. 4-13; then for $x > 0$ we can close the contour of integration in the upper half-plane, surrounding therefore only the pole at $p = \sqrt{2mE}$, and find

$$\int_{-\infty}^{\infty} \frac{dp}{2\pi\hbar} \frac{e^{ipx/\hbar}}{E - p^2/2m} = \frac{me^{i\sqrt{2mE}\,x/\hbar}}{i\hbar\sqrt{2mE}}, \qquad x > 0, \tag{4-91}$$

which is an outgoing wave. For $x < 0$ we can close the contour of integration below, and pick up the contribution from the pole at $p = -\sqrt{2mE}$; this gives

$$\frac{m}{i\hbar} \frac{e^{-i\sqrt{2mE}\,x/\hbar}}{\sqrt{2mE}}, \qquad x < 0, \tag{4-92}$$

which is also an outgoing wave. The solution (4-90) is thus

$$\psi(x) = \psi_{inc}(x) + \frac{mv_0}{ip\hbar} e^{ip|x|/\hbar} \psi(0), \tag{4-93}$$

where $p = \sqrt{2mE}$. Setting x equal to zero and solving for $\psi(0)$ we find

$$\psi(0) = \frac{ip\hbar}{ip\hbar - mv_0} \psi_{inc}(0). \tag{4-94}$$

Thus taking the incident wave to be $e^{ipx/\hbar}$ we have

$$\psi(x) = e^{ipx/\hbar} + \frac{mv_0 e^{ip|x|/\hbar}}{ip\hbar - mv_0}, \tag{4-95}$$

so that

$$\psi_{trans}(x) = \frac{ip\hbar}{ip\hbar - mv_0} e^{ipx/\hbar}. \tag{4-96}$$

The transmission amplitude is then

$$S(E) = \frac{ip\hbar}{ip\hbar - mv_0}, \tag{4-97}$$

and

$$T(E) = |S(E)|^2 = \frac{E}{E + mv_0^2/2\hbar^2}. \tag{4-98}$$

T(E) is linear in E for small E, and has no resonances. The phase shift is given by $\tan \delta(E) = -mv_0/\hbar p$.

Notice that S(E) has a pole at $p\hbar = -imv_0$. Since p is imaginary, this pole occurs for negative energy. But for $E < 0$, $p = i\sqrt{2m|E|}$, so this pole occurs only if $v_0 < 0$. It corresponds to a bound state of the potential at energy given by $\sqrt{-2mE} = -mv_0/\hbar$ or

$$E = \frac{-mv_0^2}{2\hbar^2}. \tag{4-99}$$

As we pointed out, at a pole one can have a transmitted and reflected wave with no incident wave. The wave function of the bound state is given by (4-93) with $\psi_{inc}(x) = 0$,

$$\psi(x) = \frac{mv_0}{ip\hbar} e^{ip|x|/\hbar} \psi(0). \tag{4-100}$$

[We see that this is consistent only if $mv_0/ip\hbar = 1$.] Then the normalized bound state wave function is

$$\psi(x) = \frac{\sqrt{m|v_0|}}{\hbar} e^{-m|v_0 x|/\hbar^2}, \quad v_0 < 0. \tag{4-101}$$

PERIODIC POTENTIALS

As another example of a one-dimensional problem, let us consider, a potential that is periodic in space.

$$V(x) = V(x - \alpha), \tag{4-102}$$

where α is a fixed distance. Such a potential, having the general form shown in Fig. 4-14, is the type of potential felt by an electron in a (one-dimensional) crystal lattice; α is the distance between atoms. The nature of the eigenvalue spectrum of a periodic potential gives one great insight into the question of why some crystals are insulators, and some are conductors.

The Hamiltonian

$$H(x, p) = \frac{p^2}{2m} + V(x) \tag{4-103}$$

remains unchanged if we add a lattice constant, α, onto x:

$$H(x + \alpha, p) = H(x, p). \tag{4-104}$$

As a consequence the energy eigenfunctions can be chosen also to be eigenfunctions of the translation operation

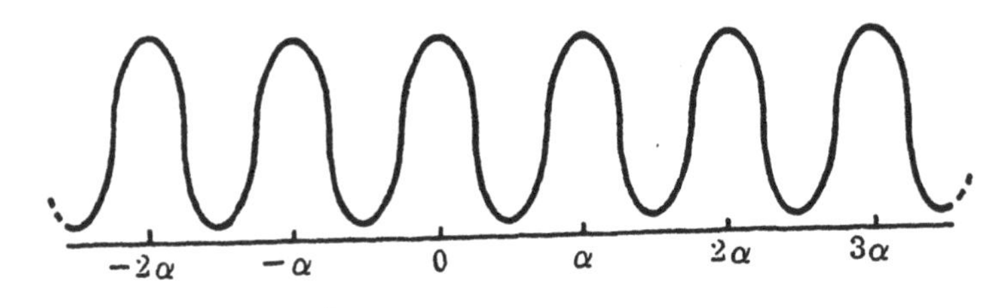

Fig. 4-14

A periodic potential with "lattice constant" a.

$$x \to x + \alpha. \tag{4-105}$$

We have already met a similar situation in the case of parity; when the Hamiltonian was invariant under $\mathbf{r} \to -\mathbf{r}$, the energy eigenfunctions could also be chosen to be eigenfunctions of the parity operation.

The translation operator, T, has the defining property that, if a state $|\Psi\rangle$ has the wave function $\langle x|\Psi\rangle$, then the state $T|\Psi\rangle$ has the wave function

$$\langle x|T|\Psi\rangle = \langle x+\alpha|\Psi\rangle. \tag{4-106}$$

T commutes with the Hamiltonian,

$$TH = HT, \tag{4-107}$$

since for any state $|\Psi\rangle$

$$\langle x|HT|\Psi\rangle = \left[-\frac{\hbar^2}{2m}\frac{d^2}{dx^2} + V(x)\right]\langle x|T|\Psi\rangle = \left[-\frac{\hbar^2}{2m}\frac{d^2}{dx^2} + V(x)\right]\langle x+\alpha|\Psi\rangle$$

$$= \left[-\frac{\hbar^2}{2m}\frac{d^2}{dx^2} + V(x+\alpha)\right]\langle x+\alpha|\Psi\rangle = \langle x+\alpha|H|\Psi\rangle = \langle x|TH|\Psi\rangle.$$

What do the eigenfunctions of T look like? An eigenfunction of T has the property that

$$\langle x|T|\Psi\rangle = \lambda\langle x|\Psi\rangle. \tag{4-108}$$

Let us define k by writing the eigenvalue as $\lambda = e^{ik\alpha}$; we shall see shortly that k must be real. Then (4-108) is

$$\psi(x+\alpha) = e^{ik\alpha}\psi(x). \tag{4-109}$$

If we define

$$u_k(x) = e^{-ikx}\psi(x) \tag{4-110}$$

then it follows from (4-109) that

$$u_k(x+\alpha) = u_k(x), \tag{4-111}$$

i.e., $u_k(x)$ is periodic with period α. Thus the eigenfunctions of T have the form

$$\psi(x) = e^{ikx} u_k(x), \tag{4-112}$$

a plane wave modulated by a function with the periodicity of the lattice. A wave function of this type is called a *Bloch wave.* Notice that (4-112) is an eigenfunction of T for *any* real k. Were k not real, then $\psi(x)$ would become infinitely large as $x \to \infty$ in one direction; such a state would not be allowable physically, so we must restrict ourselves to real k.[6]

As we have seen from the argument of p. 101, the energy eigenfunctions for a periodic potential can be chosen to be simultaneous eigenfunctions of the translation operation, since T and H commute. Thus we can construct a complete set of energy eigenstates in the form of Bloch waves, $e^{ikx}u_k(x)$, and we can construct the eigenstates of T with eigenvalue $e^{ik\alpha}$ also to be eigenstates of H. Thus we can find at least one energy eigenstate for each possible $e^{ik\alpha}$, since all $e^{ik\alpha}$ are eigenvalues of T.

Notice that k and $k + 2\pi n/\alpha$ both give the same eigenvalue $e^{ik\alpha}$. Thus the values of k in the interval

$$\frac{-\pi}{\alpha} < k \leq \frac{\pi}{\alpha} \tag{4-113}$$

give all the eigenvalues of T.

In a three-dimensional crystal the energy eigenstates are also Bloch waves, which look like $e^{i\mathbf{k}\cdot\mathbf{r}}u_{\mathbf{k}}(\mathbf{r})$, where $u_{\mathbf{k}}(\mathbf{r})$ has the same periodicity as the lattice.

As a tractable example of a one-dimensional periodic potential let us take a periodic array of delta functions

$$V(x) = \sum_{n=-\infty}^{\infty} v_0 \delta(x - n\alpha); \tag{4-114}$$

this is a case of the Kronig-Penney model.

Let us look for a solution of the form (4-112) with energy E. Our problem shall be to determine $u_k(x)$ and the energy E. In the interval $0 < x < \alpha$ the potential vanishes and thus

$$\psi(x) = Ae^{iqx} + Be^{-iqx} \tag{4-115}$$

where $\hbar q = \sqrt{2mE}$. Thus, for $0 < x < \alpha$

[6] A finite crystal can have surface states not of this form.

$$u_k(x) = Ae^{i(q-k)x} + Be^{-i(q+k)x}. \tag{4-116}$$

The coefficients A, B, and q are determined by two conditions. First of all $\psi(x)$ must be continuous at the lattice sites; therefore $u_k(x)$ must also be continuous. Hence $u_k(\varepsilon) - u_k(-\varepsilon) \to 0$ as $\varepsilon \to 0$. But from periodicity, $u_k(-\varepsilon) = u_k(\alpha - \varepsilon)$. Thus as $\varepsilon \to 0$, $u_k(\varepsilon) - u_k(\alpha - \varepsilon) \to 0$, or, from (4-116)

$$A + B = Ae^{i(q-k)\alpha} + Be^{-i(q+k)\alpha}. \tag{4-117}$$

A second condition on these coefficients comes from making $\psi(x)$ solve the Schrödinger equation at the lattice sites. The Schrödinger equation is

$$\left(E + \frac{\hbar^2}{2m}\frac{d^2}{dx^2}\right)\psi(x) = \sum_{n=-\infty}^{\infty} v_0 \delta(x - n\alpha)\psi(x); \tag{4-118}$$

integrating both sides from $-\varepsilon$ to ε, where ε is tiny, we find

$$E\int_{-\varepsilon}^{\varepsilon} dx\, \psi(x) + \frac{\hbar^2}{2m}\left[\left(\frac{d\psi(x)}{dx}\right)_{x=\varepsilon} - \left(\frac{d\psi(x)}{dx}\right)_{x=-\varepsilon}\right] = v_0\psi(0). \tag{4-119}$$

Now as $\varepsilon \to 0$, the E term vanishes,

$$\left(\frac{d\psi(x)}{dx}\right)_{x=\varepsilon} \to iq(A - B),$$

$$\left(\frac{d\psi(x)}{dx}\right)_{x=-\varepsilon} = e^{-ik\alpha}\left(\frac{d\psi(x)}{dx}\right)_{x=\alpha-\varepsilon} \to e^{-ik\alpha} iq(Ae^{iq\alpha} - Be^{-iq\alpha}),$$

and $\psi(0) = A + B$. Thus (4-119) becomes

$$\frac{\hbar^2}{2m} iq(A - B - Ae^{i(q-k)\alpha} + Be^{-i(q+k)\alpha}) = v_0(A + B). \tag{4-120}$$

Equations (4-117) and (4-120) are two homogeneous equations for A and B. Solving for B in terms of A, from (4-117), and substituting into (4-120), we find

$$\left[\cos k\alpha - \cos q\alpha - \frac{mv_0}{q\hbar^2}\sin q\alpha\right] A = 0.$$

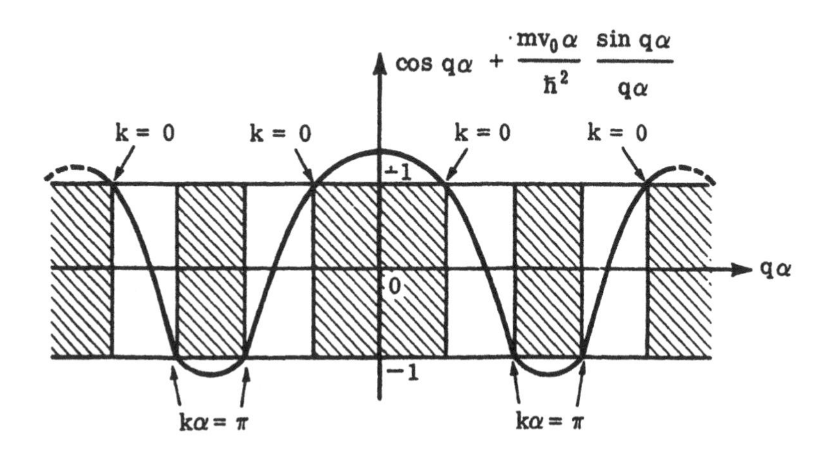

Fig. 4-15

Graphical solution of Eq. (4-121). Allowed values of q are in the unshaded areas on the q axis.

Thus the coefficient in front of A must vanish; this determines q in terms of k:

$$\cos k\alpha = \cos q\alpha + \frac{mv_0\alpha}{\hbar^2}\frac{\sin q\alpha}{q\alpha}. \tag{4-121}$$

This equation can be solved graphically. Plotting the right side as a function of $q\alpha$ we find, for $v_0 > 0$, the curve in Fig. 4-15. For a given value of k, $\cos k\alpha$ is between 1 and -1. Thus q must always lie in the unshaded areas. For each value of $\cos k\alpha$ there are an infinite number of solutions. Taking k to range from $-\pi/\alpha$ to π/α we find the curves for $E = \hbar^2 q^2/2m$ versus $k\alpha$ as shown in Fig. 4-16. The possible energy value lie in *bands*, with gaps between them. This structure of the energy spectrum, bands and gaps, occurs for all periodic potentials, in three dimensions as well as one.

Suppose that we have a crystal of finite length, extending from $x = 0$ to $x = L = N\alpha$; then there are only a finite number of states, i.e., different k values in each band. Let us find that number. In order to avoid complications with end effects, it is most convenient to impose *periodic boundary conditions* on the wave functions, that is, take

$$\psi(0) = \psi(L) \tag{4-122}$$

and

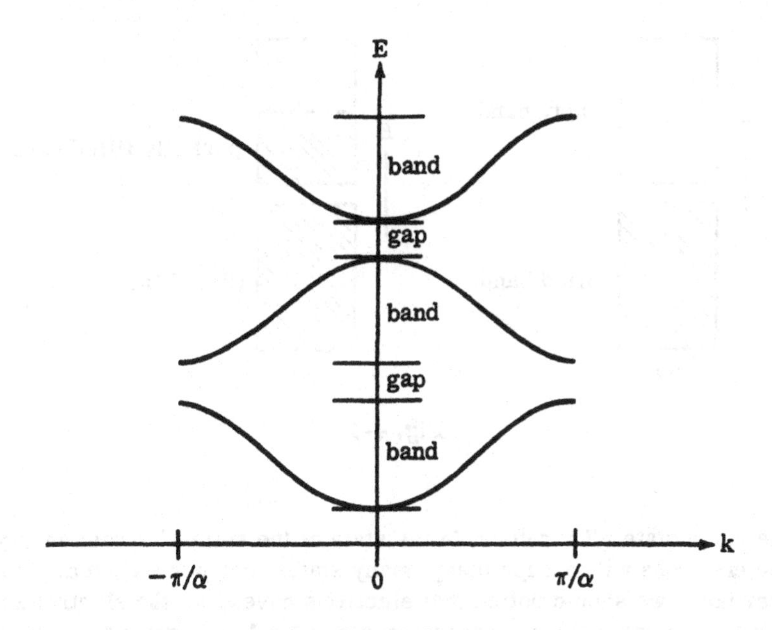

Fig. 4-16
Energy bands for the one-dimensional periodic array of delta functions.

$$\frac{d\psi}{dx}(0) = \frac{d\psi}{dx}(L). \tag{4-123}$$

For a Bloch wave, (4-122) becomes

$$u_k(0) = e^{ikL} u_k(L). \tag{4-124}$$

Since $u_k(L) = u_k(N\alpha) = u_k(0)$, we see that $e^{ikL} = 1$, or k has the N independent values

$$k = \begin{matrix} \pi(2n - N)/L, & N \text{ even}, \\ \pi(2n - N - 1)/L, & N \text{ odd}, \end{matrix} \quad n = 1, 2, \ldots, N \tag{4-125}$$

between $-\pi/\alpha$ and π/α. Thus, for a crystal N atoms long, there are N different states in each band.

How are these states occupied by the several N electrons in a solid? If the electrons behaved as classical particles, then in the ground state they would all sit in the lowest energy state, at the bottom of the lowest band. However electrons must obey the Pauli *exclusion principle,* which says that two electrons cannot occupy

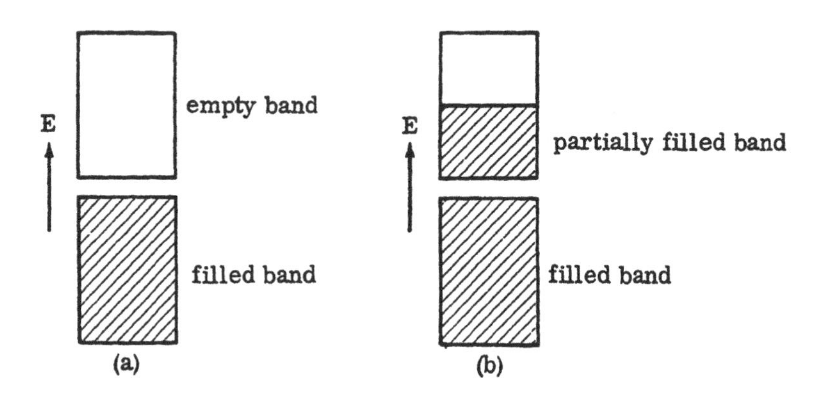

Fig. 4-17

the same state. Therefore, in the state of the solid of lowest energy, the electrons will occupy many, many states, one per electron. Furthermore, we should notice that electrons have *spin;* the electron spin can take on two values, which we can call "up" and "down." An electron with spin up is in a different state from an electron with spin down. Thus each Bloch state can accommodate two electrons, one with spin up and one with spin down, and each band therefore has room for 2N electrons.

The simplest picture of the ground state of a solid is that of bands filled from the lowest Bloch state on up, with two electrons in each state. Briefly, two important situations can occur. In the first, Fig. 4-17(a), the electrons exactly fill to the top of a band, with the next band entirely empty; such a solid behaves as an insulator. In the second, Fig. 4-17(b), the topmost band having occupied states is only partially filled; such a solid will conduct electricity. To see why this is so we note that an electric field applied to a solid tries to accelerate the electrons in the direction opposite to the field; quantum mechanically, it tries to move the electrons to states of higher energy, if they are traveling antiparallel to the field, and to states of lower energy if they are traveling parallel to the field. This causes a net particle current antiparallel to the field. In an insulator, there are no more states available to the electrons without their jumping into the next band; a reasonable strength static electric field can't however cause such interband transitions. Thus the electric field can't accelerate the electrons and no current flows. In a conductor, however, there are many adjacent states available for the field to accelerate the electrons into, and a weak field therefore causes a current to flow.

THE HARMONIC OSCILLATOR

The problem of the harmonic oscillator is one of the most important in quantum mechanics. The Hamiltonian of a particle in a harmonic oscillator potential has the form, in one dimension,

$$H = \frac{p^2}{2m} + \frac{1}{2} m\omega^2 x^2, \tag{4-126}$$

where ω is the classical frequency of the oscillator. Our problem is to determine the states in which the particle has a definite energy value, and also to determine the possible energy values.

We shall solve this problem by a very elegant method due to Dirac. Let us define the operators

$$a = \sqrt{\frac{m\omega}{2\hbar}}\, x + \frac{ip}{\sqrt{2m\hbar\omega}} \tag{4-127}$$

and

$$a^\dagger = \sqrt{\frac{m\omega}{2\hbar}}\, x - \frac{ip}{\sqrt{2m\hbar\omega}}. \tag{4-128}$$

The operator $a^\dagger$ is the Hermitian adjoint of a, since both x and p are Hermitian.

Let us also define the operator

$$N = a^\dagger a. \tag{4-129}$$

The Hamiltonian is expressible in terms of N as

$$H = \hbar\omega\left(N + \frac{1}{2}\right) \tag{4-130}$$

since from (4-127) and (4-128)

$$a^\dagger a = \frac{m\omega}{2\hbar} x^2 + \frac{p^2}{2m\hbar\omega} + \frac{i}{2\hbar}(xp - px) = \frac{H}{\hbar\omega} - \frac{1}{2}. \tag{4-131}$$

We have used $xp - px = i\hbar$. The problem of finding the eigenstates and eigenvalues of H is thus equivalent to finding those of N. If $|n\rangle$ is an eigenstate of N with eigenvalue n,

$$N|n\rangle = n|n\rangle, \tag{4-132}$$

then $|n\rangle$ is an eigenstate of H with eigenvalue $\hbar\omega\ (n + 1/2)$, since

$$H|n\rangle = \hbar\omega\left(N+\frac{1}{2}\right)|n\rangle = \hbar\omega\left(n+\frac{1}{2}\right)|n\rangle. \tag{4-133}$$

Notice that N is Hermitian, since

$$N^\dagger = (a^\dagger a)^\dagger = a^\dagger a = N. \tag{4-134}$$

Thus all the eigenvalues of N, and H, are real. Also, all the eigenvalues of N are greater than or equal to zero, since if $n\rangle$ is an eigenstate, then

$$n = \langle n|N|n\rangle = \langle n|a^\dagger a|n\rangle = \langle \Phi|\Phi\rangle \geq 0 \tag{4-135}$$

where $|\Phi\rangle \equiv a|n\rangle$.

To discover the eigenvalues of N we first need the commutator of a with $a^\dagger$:

$$[a, a^\dagger] = \frac{1}{2\hbar}\{i[p, x] - i[x, p]\} = 1. \tag{4-136}$$

It follows from (4-136) that

$$Na = a^\dagger aa = (aa^\dagger - 1)a = a(N-1) \tag{4-137}$$

and

$$Na^\dagger = a^\dagger(N+1). \tag{4-138}$$

Now let us suppose that $|n\rangle$ is an eigenstate of N with eigenvalue n. Then from (4-137)

$$Na|n\rangle = a(N-1)|n\rangle = (n-1)a|n\rangle. \tag{4-139}$$

Thus $a|n\rangle$ is an eigenstate of N with eigenvalue $n - 1$. If $\langle n|n\rangle = 1$, then $a|n\rangle$ has normalization

$$(\langle n|a^\dagger)(a|n\rangle) = \langle n|a^\dagger a|n\rangle = \langle n|N|n\rangle = n,$$

so that

$$a|n\rangle = \sqrt{n}\,|n-1\rangle, \tag{4-140}$$

where $|n-1\rangle$ is normalized to one. Similarly, $a^2|n\rangle$ is an eigenstate of N with eigenvalue n − 2, and

$$a^2|n\rangle = a\sqrt{n}\,|n-1\rangle = \sqrt{n(n-1)}\;|n-2\rangle. \tag{4-141}$$

Thus if n is an eigenvalue of N, so are n − 1, n − 2, n − 3, . . . , eigenvalues of N. But this sequence can't keep up indefinitely, for after a while we would come to a negative eigenvalue, and we know that all the eigenvalues of N are > 0. The only possibility is that n is an integer, for when we come to the eigenstate $|1\rangle$ we find, by

$$a|1\rangle = |0\rangle \tag{4-142}$$

and

$$a|0\rangle = 0. \tag{4-143}$$

Were we to come to an eigenstate $|\zeta\rangle$ where $0 < \zeta < 1$, then from (4-140), $a|\zeta\rangle = \sqrt{\zeta}\,|\zeta - 1\rangle$, producing an eigenstate with a negative eigenvalue; this contradicts (4-135). Thus the sequence of states produced by acting with a on an eigenstate terminates when we get to $|0\rangle$. The eigenvalues of N are therefore integers going down to zero.

Now let us notice that, from (4-138)

$$Na^\dagger|n\rangle = a^\dagger(N+1)|n\rangle = (n+1)a^\dagger|n\rangle. \tag{4-144}$$

Thus n + 1 is also an eigenvalue of $|n\rangle$, and it is easy to show that

$$a^\dagger|n\rangle = \sqrt{n+1}\;|n+1\rangle, \tag{4-145}$$

where $\langle n+1|n+1\rangle = 1$.

Therefore the eigenvalues of N are all the integers from zero to infinity. Using (4-144) we can write all the normalized eigenstates in terms of $|0\rangle$. Let us assume $\langle 0|0\rangle = 1$. Then

$$|1\rangle = a^\dagger|0\rangle$$

$$|2\rangle = \frac{a^\dagger}{\sqrt{2}}|1\rangle = \frac{(a^\dagger)^2}{\sqrt{2}}|0\rangle$$

$$|3\rangle = \frac{a^\dagger}{\sqrt{3}}|2\rangle = \frac{(a^\dagger)^3}{\sqrt{3!}}|0\rangle$$

and, in general,

$$|n\rangle = \frac{(a^\dagger)^n}{\sqrt{n!}} |0\rangle, \tag{4-146}$$

where $\langle n|n\rangle = 1$. It should be emphasized that all these properties of N follow from the commutation relation $[a, a^\dagger] = 1$.

We have essentially solved the harmonic oscillator problem. The energy eigenstates are those of N, and the energy eigenvalues are, from (4-130)

$$E_n = \hbar\omega\left(n + \frac{1}{2}\right), \qquad n = 0, 1, 2, \ldots . \tag{4-147}$$

The ground state energy is $\hbar\omega/2$ and is greater than zero because of zero point motion.

It is possible to get a fair estimate of the zero point energy using the uncertainty principle alone. Classically, a particle in its ground state would sit still at the origin, since that is the lowest point of potential energy. Quantum mechanically, the more perfectly we try to localize the particle about the origin, the more we increase its kinetic energy. On the other hand, the more we try to decrease its kinetic energy by spreading it out over a large region, the more we increase the amplitude for the particle to be in a region of high potential energy, since $V \sim x^2$. Thus spreading out the wave function increases the potential energy of the particle. The true ground state is a compromise between a not too large kinetic energy and a not too large potential energy. To estimate the ground state energy let us assume that the particle is localized to within $\Delta x = d$ about the origin, i.e., extending a distance $\sim d/2$ in each direction about the origin. Then $\Delta p \sim \hbar/d$ and the kinetic energy is roughly $\hbar^2/2md^2$; this increases as d decreases. Now the potential energy is roughly $\frac{1}{2} m\omega^2 (d/2)^2$, which increases as d increases. The total energy is therefore

$$E(d) \approx \frac{m\omega^2 d^2}{8} + \frac{\hbar^2}{2md^2}, \tag{4-148}$$

and the ground state energy can be estimated as the minimum value of E(d) as we vary d. Thus if $E'(d) = 0$, we have $\frac{1}{4} m\omega^2 d - \hbar^2/md^3 = 0$ or

$$d = \sqrt{\frac{2\hbar}{m\omega}}. \tag{4-149}$$

Putting this in (4-148) we find, as an estimate of the ground state energy, $\hbar\omega/2$, which coincides with the zero point energy we found in the exact calculation. Equation (4-149) is a fairly good measure of the spatial extent of the wave function of the ground state of the harmonic oscillator.

We can think of an excited state, $|n\rangle$ of the oscillator as containing n *quanta* of energy $\hbar\omega$, in addition to the zero point energy. The operator, $a^\dagger$, according to (4-145) creates a quantum of energy; $a^\dagger$ is called a *creation operator*. Similarly, by (4-140) a annihilates a quantum of energy; it is called an *annihilation*, or *destruction operator.*

Let us now construct the wave functions of the oscillator explicitly. The ground state wave function is determined by the condition (4-143). In the x representation

$$0 = \langle x|a|0\rangle = \langle x|\left(\sqrt{\frac{m\omega}{2\hbar}}\,x + \frac{ip}{\sqrt{2m\hbar\omega}}\right)|0\rangle$$

$$= \left(\sqrt{\frac{m\omega}{2\hbar}}\,x + \sqrt{\frac{2\hbar}{m\omega}}\,\frac{d}{dx}\right)\langle x|0\rangle, \tag{4-150}$$

where we have used $\langle x|p|0\rangle = (\hbar/i)d\langle x|0\rangle/dx$. Equation (4-150) is a differential equation to determine $\langle x|0\rangle$, whose solution is

$$\langle x|0\rangle = \left(\frac{m\omega}{\pi\hbar}\right)^{1/4} e^{-m\omega x^2/2\hbar}; \tag{4-151}$$

the constant in front guarantees that $\langle 0|0\rangle = \int_{-\infty}^{\infty} dx|\langle x|0\rangle|^2 = 1$. The ground state wave function has even parity.

Notice that the ground state, being Gaussian, is a *minimum wave packet* as in (3-89). Comparing (4-151) with (3-89) we see at once that $\Delta x = \sqrt{\hbar/2m\omega}$ and $\Delta p = \sqrt{\hbar m\omega/2}$; thus explicitly $\Delta x\Delta p = \hbar/2$.

The first excited state $|1\rangle = a^\dagger|0\rangle$ has a wave function

$$\langle x|1\rangle = \langle x|a^\dagger|0\rangle = \left(\sqrt{\frac{m\omega}{2\hbar}}\,x - \sqrt{\frac{2\hbar}{m\omega}}\,\frac{d}{dx}\right)\langle x|0\rangle$$

$$= \sqrt{\frac{2m\omega}{\hbar}}\,x\langle x|0\rangle; \tag{4-152}$$

it is a state of odd parity. In general, acting with a or $a^\dagger$ on a state of definite parity reverses its parity. Thus all the even n states have even parity, and all the odd n states have odd parity.

The wave function of the state $|n\rangle$ is

$$\langle x|n\rangle = \left(\frac{m\omega}{\pi\hbar}\right)^{1/4} \frac{1}{\sqrt{2^n n!}}\, H_n\left(\sqrt{\frac{m\omega}{\hbar}}\,x\right) e^{-m\omega x^2/2\hbar}, \tag{4-153}$$

where $H_n(\xi)$ is the Hermite polynomial of order n. The first few Hermite polynomials are

$$H_0(\xi) = 1, \quad H_1(\xi) = 2\xi, \quad H_2(\xi) = 4\xi^2 - 2.$$

The wave functions (4-153) obey the orthogonality relation

$$\langle n|m\rangle = \int_{-\infty}^{\infty} dx\, \langle n|x\rangle\langle x|m\rangle = \delta_{nm} \equiv \begin{cases} 1 \text{ if } n = m \\ 0 \text{ if } n \neq m \end{cases}, \tag{4-154}$$

and are called the Hermite orthogonal functions. δ_{nm} is known as the *Kronecker delta symbol.*

PROBLEMS

1. Why are there no solutions for $E < 0$ for the barrier problem,

$$V(x) = \begin{cases} 0, & x < 0 \\ V > 0, & x > 0? \end{cases}$$

2. Consider the potential in Problem 1. Assume that a wave packet with $E_0 = p_0^2/2m < V$ is incident on the barrier from the left. Calculate in terms of E_0 and V the difference in time between the arrival of the incident packet at the step and the departure of the reflected packet from the step. Relate this time delay to a "distance of travel" within the step.
3. (a) Calculate for both $E < V$ and $E > V$ the transmission and reflection amplitudes for the potential barrier of Fig. 4-3.
(b) Calculate the transit time of a wave packet through the barrier; what is its form at resonance?
4. Verify Eqs. (4-30) and (4-31) using the expression for the current density given in Problem 10, Chapter 3.
5. Consider a potential

$$V_a(x) = \begin{cases} 0, & x < -a \\ V(x), & -a < x < a \\ 0, & x > a \end{cases}$$

where $V(x)$ is some arbitrary function. Consider a wave incident from the left. Suppose that the reflection amplitude at $x = -a$ is $N(E)$ [Eq. (4-35)] (where E is positive) and that the transmission amplitude, defined as the ratio of the transmitted wave at $x = a$ to the incident wave at $x = -a$, is $S(E)$. Now consider a wave incident from the right. Show that the transmission amplitude, now defined as the ratio of the transmitted wave at $-a$ to the incident wave at a, is also $S(E)$. Express the reflection amplitude at a in terms of N and S. *(Hint:* the Schrödinger equation in this case is a *real* equation.)

6. Consider a potential step

$$V_I(x) = \begin{cases} 0, & x < a \\ V(x), & a < x < b \\ \infty, & x > b, \end{cases}$$

where $V(x)$ is some function that we don't know and $a > 0$. Suppose however, that we know the reflection amplitude $N(E) = -e^{2i\delta(E)}$ at $x = a$.

(a) Show that $\delta(E)$ must be real.

(b) Consider a particle in the potential well defined by

$$V_{II}(x) = \begin{cases} \infty, & x < 0 \\ V_I(x), & x > 0. \end{cases}$$

Show that to first order in $\delta(E)$ (assumed small), the energy levels of this well are given in terms of the energy levels of the square well

$$V_{III}(x) = \begin{cases} \infty, & x < 0 \\ 0, & 0 < x < a \\ \infty, & x > a \end{cases}$$

by

$$E = E_0 - \frac{p_0\hbar}{ma}\,\delta(E_0)$$

where E_0 is the energy level of the square well and $p_0 = \sqrt{2mE_0}$.

(c) Give a qualitative argument (at least) that $\delta(E)$ must be positive.

7. (a) Show that the energy levels of a double square well

$$V_S(x) = \begin{cases} \infty, & |x| > b \\ 0, & a < |x| < b \\ \infty, & |x| < a \end{cases}$$

are doubly degenerate.

(b) Now suppose that the barrier between $-a$ and a is very high, but finite. Assume that the potential between $-a$ and a is symmetric about the origin. There is now the possibility of tunneling from one well to the other, and this possibility has the effect of splitting the degeneracy of the double well in part (a). Let E_0 be an energy level of the well in part (a), and assume that E_0 is reasonably less than the barrier height. Assume that in the neighborhood of E_0 the reflection amplitude of the barrier at $-a$ is of the form $-e^{2i\delta(E)}$ where $\delta(E)$ is real, positive, and $\ll 1$. (Can $\delta(E)$ be otherwise under these conditions?) Also assume $S(E)$ to be of the form $iJ(E)$ where J is small and positive. (Must the leading term in $S(E)$ as the barrier height becomes infinite be of this form?)

Show that to lowest order in δ and J the well with the finite barrier has two levels E corresponding to each degenerate level of the double square well in (a), given by

$$E_\pm = E_0 - \frac{p_0\hbar}{m(b-a)}\left[\delta(E_0) \mp \frac{J(E_0)}{2}\right].$$

What are the parities of the split levels?

(c) If $\delta(E)$ is positive, as it generally must be, then one level of the double well with tunneling is lower in energy than E_0. This effect provides a qualitative explanation of the "one electron bond" between two ions, e.g., the bond in H_2^+. The ions, when far apart, correspond to the double square well of part (a). When they are close together there is a possibility of an electron tunneling from one ion to the other. Due to the tunneling splitting the electron, and hence the two ions plus electron, can have a lower energy when the two ions are close together; hence they bind. Estimate the binding energy of two ions due to this mechanism.

8. Calculate the transmission amplitude $S(E)$ for the potential

$$V(x) = \begin{cases} 0, & 0 \le |x| < \frac{b}{2} \\ V \ (> 0), & \frac{b}{2} \le |x| < \frac{a}{2} \\ 0, & |x| \ge \frac{a}{2} \end{cases}$$

Calculate the resonances in S(E) when the shoulders are veryhigh.

9. An α particle emitter consisting of N atoms, obeying an exponential decay law ($\sim e^{-\gamma t}$), is put into place at $t = 0$. Assume for simplicity that the particles are all emitted in the positive x direction and with velocity v. Neglect the size of the emitter. Calculate the density n(x, t) of emitted α particles. Sketch the result as a function of x at several successive times, including $t > \gamma^{-1} \log N$. Relate this result to the wave functions of the emitted α particles.

10. (a) The (negative) delta function potential in one dimension can be thought of as the limit of a square well potential as it becomes deep and narrow. Show that the bound state energy and wave function of the delta function are the limiting values of the square well bound states and wave functions.

    (b) Show that S(E) for a delta function potential is the limiting value of S(E) for a square well potential.

11. A particle of mass m is in a potential $V(x) = v\delta(x - a) + v\delta(x + a)$ where $v < 0$.

    (a) What is the wave function for a bound state with even parity?

    (b) Find the expression that determines the bound state energies for even parity states, and determine graphically how many even parity bound states there are.

    (c) Solve for the even parity bound state energy analytically in the case that $m|v|a/\hbar^2 \ll 1$.

    (d) Repeat parts (a) and (b) for odd parity. For what values of $|v|$ are there bound states?

    (e) Find the even and odd parity state binding energies for $a \gg \hbar^2/m|v|$. Explain physically why these energies move closer and closer together as $a \to \infty$.

12. The effective mass m* of an electron in a solid with energy E(k) is defined in one dimension by

$$\frac{1}{m^*} = \frac{1}{\hbar^2}\frac{d^2E}{dk^2}.$$

Calculate m*, in the Kronig-Penney model with delta functions, at the bottom and at the top of the lowest energy band. Explain the physical significance of the results.

13. The energy eigenfunctions of the harmonic oscillator are a complete set, i.e., form a basis. What is the expansion of a delta function, $\delta(x-a)$ in terms of the Hermite orthogonal functions?
14. (a) Express x and p in terms of a and $a^\dagger$.

    (b) Calculate the following matrix elements for the harmonic oscillator

    $\langle n|x|m\rangle$ $\quad\langle n|px|m\rangle$
    $\langle n|p|m\rangle$ $\quad\langle n|xp|m\rangle$
    $\langle n|x^2|m\rangle$ $\quad\langle n|a|m\rangle$
    $\langle n|p^2|m\rangle$ $\quad\langle n|a^\dagger|m\rangle$
    $\langle n|H|m\rangle$

    (c) Show that the expectation value of the potential energy in an energy eigenstate of the harmonic oscillator equals the expectation value of the kinetic energy in that state.
15. Estimate the zero point energy for a particle of mass m in the following potentials:

    (a) $V(x) = vx^4$ in one dimension, where v is a positive constant.

    (b) $V(r) = -e^2/r$ in three dimensions; this is the potential felt by an electron in a hydrogen atom. Express the estimate in eV, taking e and m to be the charge and mass of the electron.

    (c) $V(r) = -q^2e^{-kr}/r$ in three dimensions. Take $k = 1.3\times10^6\text{cm}^{-1}$ and $q^2 = e^2/14$. This problem requires a graphical solution. A potential of this form, called a "screened Coulomb potential," is the type felt by an electron due to an impurity of charge e in a semiconductor such as germanium.
16. A box, containing a particle, is divided in two into a right and a left compartment by a thin partition. Suppose that the amplitude for the particle being on the left side of the box is ψ_1 and the amplitude for it being on the right side of the box is ψ_2. Neglect spatial variations of these amplitudes within the halves of the box. Suppose that the particle can tunnel through the partition and that the rate of change of the amplitude on the right is $K/i\hbar$ times the amplitude on the left, where K is real:

    $$i\hbar\frac{\partial\psi_2}{\partial t} = K\psi_1.$$

    Assume in the absence of tunneling, i.e., an impermeable partition, that $\partial\psi_1/\partial t = 0$.

    (a) What is the equation that determines the rate of change of the amplitude on the left?

(b) Find the normalized energy eigenstates (two-component vectors) of the particle in the box. Have these states definite parities?

(c) Suppose that at time $t = 0$, the amplitude on the right equals $e^{i\delta}$ times the amplitude on the left. Calculate, as a function of time, the time rate of change of the probability of observing the particle on the left?

Chapter 5
EQUATIONS OF MOTION FOR OPERATORS

FUNCTIONS OF OPERATORS

Suppose that the function f(x) has a power series expansion

$$f(x) = f_0 + xf_1 + x^2 f_2 + x^3 f_3 + \cdots . \tag{5-1}$$

Then, if A is an operator, we can define the operator f(A) by

$$f(A) = f_0 + Af_1 + A^2 f_2 + A^3 f_3 + \cdots . \tag{5-2}$$

For example, the operator $e^{\lambda A}$ is

$$e^{\lambda A} = 1 + \lambda A + \frac{\lambda^2}{2!} A^2 + \frac{\lambda^3}{3!} A^3 + \cdots . \tag{5-3}$$

If $|\alpha\rangle$ is an eigenstate of A with eigenvalue α, then operating on $|\alpha\rangle$ with f(A) we find

$$\begin{aligned} f(A)|\alpha\rangle &= (f_0 + Af_1 + A^2 f_2 + \cdots)|\alpha\rangle \\ &= (f_0 + \alpha f_1 + \alpha^2 f_2 + \cdots)|\alpha\rangle = f(\alpha)|\alpha\rangle . \end{aligned} \tag{5-4}$$

Thus $|\alpha\rangle$ is an eigenstate of f(A) with eigenvalue $f(\alpha)$. If the eigenstates of A form a complete set (as they can always be chosen to do if A is Hermitian), i.e.,

$$1 = \sum_\alpha |\alpha\rangle\langle\alpha| \tag{5-5}$$

then operating on both sides with f(A) we find, from (5-4),

$$f(A) = \sum_\alpha f(\alpha)|\alpha\rangle\langle\alpha|. \tag{5-6}$$

In fact, we could take this as the equation defining f(A) in terms of the eigenstates and eigenvalues of A; it isn't necessary for f to have a power series expansion for this to be a usable definition. For example, we can define a delta function of an operator, $\delta(A - x)$, by

$$\delta(A-x) = \sum_\alpha \delta(\alpha - x)|\alpha\rangle\langle\alpha|, \tag{5-7}$$

where x is a number.

One must be very careful manipulating functions of operators, since operators don't generally commute with one another. For example if $[A, B] \neq 0$ then

$$e^{A+B} \neq e^A e^B \neq e^B e^A.$$

In the special case that [A, B] is a number times the unit operator (called a *c-number*), as for example $[x, p] = i\hbar$, then

$$e^{A+B} = e^A e^B e^{-[A,B]/2}; \tag{5-8}$$

this result is known as Weyl's formula. Its proof can be found in standard texts, such as Messiah.

As an application of operator functions, notice that if $|\Psi(0)\rangle$ is the state of a system at time t = 0, then at another time t, we have [Eq. (4-20)]

$$|\Psi(t)\rangle = \sum_E e^{-iEt/\hbar}|E\rangle\langle E|\Psi(0)\rangle \tag{4-20}$$

where the sum is over a complete set of energy eigenstates of the Hamiltonian, H, of the system. Using (5-6) we see that this equation is simply

$$|\Psi(t)\rangle = e^{-iHt/\hbar}|\Psi(0)\rangle. \tag{5-9}$$

$|\Psi(t)\rangle$ in this form clearly solves the Schrödinger equation

$$i\hbar\frac{\partial|\Psi(t)\rangle}{\partial t} = H|\Psi(t)\rangle. \tag{5-10}$$

Equations (5-9) and (4-20) are only valid if H does not explicitly depend on time. The row vector $\langle\Psi(t)|$ is given in terms of $\langle\Psi(0)|$ by

$$\langle \Psi(t)| = \langle \Psi(0)|(e^{-iHt/\hbar})^\dagger = \langle \Psi(0)|e^{iHt/\hbar}, \tag{5-11}$$

using the hermiticity of H.

HEISENBERG REPRESENTATION

We shall often be interested in knowing how expectation values of operators, in the state $|\Psi(t)\rangle$, change in the course of time. If A is an operator, then its expectation value at time t is

$$\langle A\rangle_t = \langle \Psi(t)|A|\Psi(t)\rangle. \tag{5-12}$$

Assume A not to depend on time explicitly. [The operator pt depends on t explicitly.] Using (5-9) and (5-11) we have

$$\langle A\rangle_t = \langle \Psi(0)|e^{iHt/\hbar}Ae^{-iHt/\hbar}|\Psi(0)\rangle. \tag{5-13}$$

If we define a time-dependent operator A(t) by

$$A(t) = e^{iHt/\hbar}Ae^{-iHt/\hbar}, \tag{5-14}$$

then we can also write

$$\langle A\rangle_t = \langle \Psi(0)|A(t)|\Psi(0)\rangle. \tag{5-15}$$

In this equation we can regard the time development of $\langle A\rangle_t$ as occurring because the operator changes in time, while the state remains the same at all times. This way of regarding the time development of the system is called the *Heisenberg representation* or *Heisenberg picture.* Our old way of looking at the time development, as in Eq. (5-12), where the operators remain constant in time, but the states change according to Eq. (5-10), is called the *Schrödinger representation,* or picture. At t = 0, the states and operators are the same in both representations. Both representations give the same results for time dependent expectation values; we can solve for either the time dependence of the states in the Schrödinger representation, or the time dependence of the operators in the Heisenberg representation.

Differentiating (5-14) with respect to t and assuming that A does not depend explicitly on time, we find

$$i\hbar \frac{dA(t)}{dt} = i\hbar \left[\frac{d}{dt} e^{iHt/\hbar}\right] Ae^{-iHt/\hbar} + i\hbar e^{iHt/\hbar} A \frac{d}{dt} e^{-iHt/\hbar}$$
$$= e^{iHt/\hbar}[-HA + AH]e^{-iHt/\hbar}. \tag{5-16}$$

Since H commutes with $e^{-iHt/\hbar}$, (5-16) is simply

$$i\hbar \frac{dA(t)}{dt} = [A(t), H]. \tag{5-17}$$

This is the equation of motion obeyed by operators in the Heisenberg represenation.

In particular, H(t) is independent of time in the Heisenberg representation, if it is independent of time in the Schrödinger representation, since

$$H(t) = e^{-iHt/\hbar} H e^{iHt/\hbar} = e^{-iHt/\hbar} e^{iHt/\hbar} H = H. \tag{5-18}$$

This is the quantum mechanical statement that the energy is a constant of the motion.

The Heisenberg representation position operator for a particle in three dimensions obeys

$$i\hbar \frac{d\mathbf{r}(t)}{dt} = [\mathbf{r}(t), H]. \tag{5-19}$$

Let us assume that H is of the form

$$H = \frac{p^2}{2m} + V(\mathbf{r})$$

Because H is independent of time, we can write it as

$$H = H(t) = \frac{p(t)^2}{2m} + V(\mathbf{r}, t). \tag{5-20}$$

In the commutator in (5-19), $\mathbf{r}(t)$ commutes with $V(\mathbf{r}(t))$ since an operator always commutes with a function of itself. $\mathbf{r}(t)$ fails to commute, however, with $p(t)^2/2m$. To evaluate their commutator, let us notice that, by components,

$$[r_i(t), p_j(t)] = r_i(t)p_j(t) - p_j(t)r_i(t) = e^{iHt/\hbar}(r_i p_j - p_j r_i)e^{-iHt/\hbar}$$
$$= e^{iHt/\hbar} i\hbar\delta_{ij} e^{-iHt/\hbar} = i\hbar\delta_{ij}. \tag{5-21}$$

Thus $\mathbf{r}(t)$ and $\mathbf{p}(t)$ obey the same commutation relation as $\mathbf{r}$ and $\mathbf{p}$ in the Schrödinger representation. [Warning: in general if $t \neq t'$, then $[r_i(t), p_j(t')] \neq i\hbar\delta_{ij}$ since there are factors of $e^{\pm iH(t-t')/\hbar}$ that get in the way.] By simple calculation then

$$\left[\mathbf{r}(t), \frac{\mathbf{p}(t)^2}{2m}\right] = \frac{i\hbar\mathbf{p}(t)}{m},$$

so that

$$\frac{d\mathbf{r}(t)}{dt} = \frac{\mathbf{p}(t)}{m}. \tag{5-22}$$

The position operator in the Heisenberg representation obeys the usual classical equation of motion.

To find the equation of motion of $\mathbf{p}(t)$ we must evaluate $[\mathbf{p}(t), V(\mathbf{r}(t))]$, which equals

$$e^{iHt/\hbar}[\mathbf{p}, V(\mathbf{r})]e^{-iHt/\hbar}.$$

Now

$$[\mathbf{p}, V(\mathbf{r})] = -i\hbar\nabla V(\mathbf{r}), \tag{5-23}$$

since for any $|\Psi\rangle$,

$$\langle\mathbf{r}|[\mathbf{p}, V(\mathbf{r})]|\Psi\rangle = \frac{\hbar}{i}\nabla[V(\mathbf{r})\langle\mathbf{r}|\Psi\rangle] - V(\mathbf{r})\frac{\hbar}{i}\nabla\langle\mathbf{r}|\Psi\rangle$$

$$= -i\hbar\nabla V(\mathbf{r})\langle\mathbf{r}|\Psi\rangle = \langle\mathbf{r}| -i\hbar\nabla V(\mathbf{r})|\Psi\rangle.$$

Thus

$$[\mathbf{p}(t), V(\mathbf{r}(t))] = -i\hbar e^{iHt/\hbar}\nabla V(\mathbf{r})e^{-iHt/\hbar} = -i\hbar\nabla V(\mathbf{r}(t)), \tag{5-24}$$

and $\mathbf{p}(t)$ obeys the equation of motion

$$\frac{d\mathbf{p}(t)}{dt} = -\nabla V(\mathbf{r}(t)). \tag{5-25}$$

Again, this is the usual classical equation of motion; we can interpret $-\nabla V(\mathbf{r}(t))$ as the operator for the force on the particle.

The similarity between the Heisenberg equations of motion and the classical Hamiltonian equations of motion arises from the fact that $(1/i\hbar)[A, B]$, the commutator of the operators for the physical quantities

A and B, divided by $i\hbar$, plays a similar role in quantum mechanics, to that of the Poisson bracket of the classical quantities A and B, in classical mechanics.

Generally, the Heisenberg equations of motion are more difficult to solve than the corresponding classical equations because of the lack of commutation of quantum mechanical operators. There are a few cases, however, that we can easily solve.

For a free particle, the Heisenberg equation of motion for $p(t)$ is

$$\frac{dp(t)}{dt} = 0. \tag{5-26}$$

Thus the momentum operator is a constant of the motion,

$$p(t) = p(0)$$

and the position operator obeys

$$\mathbf{r}(t) = \mathbf{r}(0) + t\frac{p(0)}{m}. \tag{5-27}$$

If $|\Psi(0)\rangle$ is the wave packet of the particle at time 0, then the center of the wave packet, $\langle \mathbf{r} \rangle_t$, is given by

$$\langle \mathbf{r}\rangle_t = \langle \mathbf{r}\rangle_0 + t\frac{\langle p\rangle_0}{m}, \tag{5-28}$$

a familiar result.

The Heisenberg equations of motion for a harmonic oscillator are

$$\frac{dx(t)}{dt} = \frac{p(t)}{m}$$

$$\frac{dp(t)}{dt} = -m\omega^2 x(t), \tag{5-29}$$

and they have the solutions

$$x(t) = x(0)\cos\omega t + \frac{p(0)}{m\omega}\sin\omega t$$

$$p(t) = p(0)\cos\omega t - m\omega x(0)\sin\omega t. \tag{5-30}$$

Taking the expectation value of the first equation in an arbitrary state

$|\Psi\rangle$ of the oscillator, we see that $\langle x\rangle_t = \langle x\rangle_0 \cos \omega t + (\langle p\rangle_0/m\omega)\sin \omega t$; the expectation value of the position oscillates exactly as in a classical oscillator.

Up to now we have assumed that H does not depend explicitly on time. Quite often we meet situations where the Hamiltonian contains terms that do depend explicitly on time; for example,

$$H = \frac{p^2}{2m} + V(\mathbf{r}) - \mathbf{F}(t)\cdot\mathbf{r}, \tag{5-31}$$

where $\mathbf{F}(t)$ is a numerical vector function of time, is the Hamiltonian of a particle in a potential $V(\mathbf{r})$ acted on by an additional time-dependent classical force $\mathbf{F}(t)$, such as a uniform electric field.

The Schrödinger equation, (5-10), does not have the simple solution (5-9) when H depends explicitly on time. To find the solution, let us write

$$|\Psi(t)\rangle = U(t)|\Psi(0)\rangle, \tag{5-32}$$

where U(t) is some operator, and try to solve for U(t). U(t) must be unitary $[U(t)^\dagger U(t) = U(t)U(t)^\dagger = 1]$ to preserve the normalization of the states as time develops. Also, $U(t)^\dagger = U^\dagger(t)$. Substituting (5-32) in (5-10) we see that (5-32) will be a solution to the Schrödinger equation if

$$i\hbar\frac{dU(t)}{dt} = H_S(t)U(t) \tag{5-33}$$

and

$$U(0) = 1,$$

where by $H_S(t)$ we mean the time-dependent Hamiltonian in the Schrödinger representation. [If H_S is independent of time then $U(t) = e^{-iH_S t/\hbar}$.]

We shall construct a formal solution to (5-33) shortly, but now let us discuss the Heisenberg representation when H depends on time. The expectation value of an operator A is given by

$$\langle A\rangle_t = \langle\Psi(t)|A|\Psi(t)\rangle = \langle\Psi(0)|U^\dagger(t)AU(t)|\Psi(0)\rangle.$$

Thus we define the Heisenberg representation operator A(t) by

$$A(t) \equiv U^\dagger(t)AU(t). \tag{5-34}$$

To find the equation of motion for A(t) we differentiate (5-34) with respect to t, assuming that A does not depend explicitly on t:

$$i\hbar \frac{dA(t)}{dt} = i\hbar \frac{dU^\dagger(t)}{dt} AU(t) + U^\dagger(t)A \left(i\hbar \frac{dU(t)}{dt}\right)$$
$$= U^\dagger(t)[-H_s(t)A + AH_s(t)]U(t), \tag{5-35}$$

using (5-34), 5-33 and its Hermitian adjoint:

$$-i\hbar \frac{dU^\dagger(t)}{dt} = U^\dagger(t)H_s(t).$$

Because U(t) is unitary, we can write (5-35) as

$$i\hbar \frac{dA(t)}{dt} = U^\dagger(t)[-H_s(t)U(t)U^\dagger(t)A + AU(t)U^\dagger(t)H_s(t)]U(t)$$
$$= [A(t), H_H(t)] \tag{5-36}$$

where

$$H_H(t) = U^\dagger(t)H_s(t)U(t) \tag{5-37}$$

is the Hamiltonian in the Heisenberg representation.

$H_H(t)$ is no longer a constant of the motion, but rather, it obeys

$$\frac{dH_H(t)}{dt} = \frac{1}{i\hbar}[H_H(t), H_H(t)] + U^\dagger(t)\frac{\partial H_s(t)}{\partial t}U(t)$$
$$= U^\dagger(t)\frac{\partial H_s(t)}{\partial t}U(t) \equiv \frac{\partial H_H(t)}{\partial t},$$

as in classical mechanics.

As an example, the Heisenberg equations of motion for a particle whose Hamiltonian is (5-31) are

$$\frac{d\mathbf{r}(t)}{dt} = \frac{\mathbf{p}(t)}{m} \tag{5-38}$$

and

$$\frac{d\mathbf{p}(t)}{dt} = -\nabla V(\mathbf{r}(t)) + \mathbf{F}(t). \tag{5-39}$$

For a forced harmonic oscillator in one dimension, these equations reduce to

$$m\left[\frac{d^2}{dt^2}+\omega^2\right] x(t) = F(t). \tag{5-40}$$

FORMAL SOLUTION FOR THE TIME DEVELOPMENT OPERATOR

Now let us construct a formal solution to (5-33) for U(t). Suppose for a moment that we didn't have to worry about noncommutation problems with operators. Then we could write the solution to (5-33) as

$$U(t) = \exp\left(-i\int_0^t H_S(t')\, dt'/\hbar\right),$$

as is easily checked by differentiation. However, because H_S at one time fails generally to commute with itself at another time, this is not a valid operator solution. Let us try to find a solution in powers of H_S; we write

$$U(t) = U_0(t) + U_1(t) + U_2(t) + \cdots \tag{5-41}$$

where $U_n(t)$ is proportional to the nth power of H_S. Substituting (5-41) in (5-33) gives

$$i\hbar\frac{d}{dt}(U_0(t) + U_1(t) + \cdots) = H_S(t)(U_0(t) + U_1(t) + \cdots).$$

Comparing powers of H_S on both sides we find

$$i\hbar\frac{d}{dt}U_0(t) = 0$$

$$i\hbar\frac{d}{dt}U_1(t) = H_S(t)U_0(t) \tag{5-42}$$

and generally

$$i\hbar\frac{d}{dt}U_n(t) = H_S(t)U_{n-1}(t). \tag{5-43}$$

At $t = 0$, $U(t) = 1$, so that

$$U_0(t = 0) = 1$$

$$U_1(t = 0) = U_2(t = 0) = \cdots = 0.$$

Thus the solution to (5-42) is

$$U_0(t) = 1$$

while the solution to (5-43) is

$$U_n(t) = \frac{1}{i\hbar}\int_0^t dt' \; H_s(t')U_{n-1}(t') = \frac{1}{(i\hbar)^2}\int_0^t dt'\int_0^{t'} dt'' \; H_s(t')H_s(t'')U_{n-2}(t'')$$

$$= \frac{1}{(i\hbar)^n}\int_0^t dt' \int_0^{t'} dt'' \int_0^{t''} dt''' \cdots \int_0^{t^{(n-1)}} dt^{(n)} \; H_s(t')H_s(t'')H_s(t''') \cdots H_s(t^{(n)}). \tag{5-44}$$

Notice that the H_s at the latest time is on the left, the next to latest is second from the left, and so on, till finally we find the H_s at the earliest time on the right. $U(t)$ is therefore given by

$$U(t) = 1 + \sum_{n=1}^{\infty} \frac{1}{(i\hbar)^n}\int_0^t dt' \int_0^{t'} dt'' \cdots \int_0^{t^{(n-1)}} dt^{(n)} \times H_s(t')H_s(t'') \cdots H_s(t^{(n)}). \tag{5-45}$$

If we introduce the notion of a *time-ordered product* of operators, then (5-44) and (5-45) can be written much more compactly. Let us define $(A(t)B(t')C(t'')\ldots X(t^{(n)}))_+$ to mean the product of the operators in which the operators are written from right to left in order of increasing times, as, for instance, the operators in each $U_n(t)$ occur. For example

$$(A(t)B(t'))_+ = \begin{cases} A(t)B(t'), & t \geq t' \\ B(t')A(t), & t' > t. \end{cases}$$

Such a product is called *time ordered* and occasionally one writes $T(\;)$ instead of $(\;)_+$.

Now, we have

$$\left(\left[\int_0^t H_S(t')\,dt'\right]^2\right)_+ = \left(\int_0^t dt'\;H_S(t')\int_0^t dt''\;H_S(t'')\right)_+$$

$$= \int_0^t dt' \int_0^t dt''\;(H_S(t')H_S(t''))_+ = \int_0^t dt' \int_0^{t'} dt''\;H_S(t')H_S(t'') \tag{5-46}$$

$$+\int_0^t dt'' \int_0^{t''} dt'\;H_S(t'')H_S(t') = 2\int_0^t dt' \int_0^{t'} dt''\;H_S(t')H_S(t''),$$

and in general:

$$\left(\left[\int_0^t H_S(t')\,dt'\right]^n\right)_+ = \left(\int_0^t dt'\;H_S(t')\int_0^t dt''\;H_S(t'')\cdots\int_0^t dt^{(n)}\;H_S(t^{(n)})\right)_+$$

$$= \int_0^t dt' \int_0^t dt'' \cdots \int_0^t dt^{(n)}\;(H_S(t')H_S(t'')\cdots H_S(t^{(n)}))_+ \tag{5-47}$$

$$= n! \int_0^t dt' \int_0^{t'} dt'' \cdots \int_0^{t^{(n-1)}} dt^{(n)}\;H_S(t')H_S(t'')\cdots H_S(t^{(n)})$$

because there are $n!$ possible orderings of the n times $t', t'', \ldots, t^{(n)}$. Thus we find

$$U_n(t) = \frac{1}{(i\hbar)^n}\,\frac{1}{n!}\left(\left[\int_0^t H_S(t')\,dt'\right]^n\right)_+$$

and

$$U(t) = \left(\sum_{n=0}^{\infty} \frac{1}{(i\hbar)^n}\,\frac{1}{n!}\left[\int_0^t H_S(t')\,dt'\right]^n\right)_+. \tag{5-48}$$

The sum inside this time-ordered product is the series expansion of an exponential, so that we may write (5-48) as

$$U(t) = \left(\exp\left[-\frac{i}{\hbar}\int_0^t H_S(t')\,dt'\right]\right)_+. \tag{5-49}$$

This equation is just a convenient shorthand for the long summation (5-45). We can verify directly that (5-49) solves Eq. (5-33), since

$$i\hbar\,\frac{dU(t)}{dt} = \left(H_S(t)\exp\left[-\frac{i}{\hbar}\int_0^t dt'\;H_S(t')\right]\right)_+;$$

in differentiation we don't have to worry about noncommutation of operators inside a time-ordered product, since the order is already dictated. Now t is certainly the latest time in this time-ordered product and therefore all the other operators will be on the right of $H_s(t)$. We can therefore pull it outside the time-ordered product and write

$$i\hbar \frac{dU(t)}{dt} = H_s(t)\left(\exp\left[-\frac{i}{\hbar}\int_0^t dt'\, H_s(t')\right]\right)_+ = H_s(t)U(t). \tag{5-50}$$

It isn't necessary to know U(t) explicitly to find the time-dependent operators; we need only solve the Heisenberg equations of motion. The result for U(t) in Eq. (5-49) proves very useful when considering how the states of a system are changed by time-dependent perturbations.

PROBLEMS

1. (a) Derive and solve the equations of motion for the Heisenberg operators $a(t)$ and $a^\dagger(t)$ for the harmonic oscillator.
   (b) Calculate $[a(t), a^\dagger(t')]$.
2. (a) Show that $[\mathbf{r}, f(\mathbf{p})] = i\hbar\nabla_{\mathbf{p}} f(\mathbf{p})$ where $f(\mathbf{p})$ is an arbitrary function of the momentum operator.
   (b) Using this result show that

   $$e^{i\mathbf{p}\cdot\boldsymbol{\lambda}/\hbar}\,\mathbf{r}\,e^{-i\mathbf{p}\cdot\boldsymbol{\lambda}/\hbar} = \mathbf{r}+\boldsymbol{\lambda}$$

   where $\boldsymbol{\lambda}$ is a numerical vector.
   (c) Show that the wave function of the state

   $$|\Phi\rangle \equiv e^{-i\mathbf{p}\cdot\boldsymbol{\lambda}/\hbar}|\Psi\rangle$$

   is the same as the wave function of the state $|\Psi\rangle$, only shifted a distance $\boldsymbol{\lambda}$. Write out $\langle x|\Phi\rangle$ explicitly if $|\Psi\rangle$ is the ground state of the harmonic oscillator, in one dimension.
   (d) Show that $|\Phi\rangle$ develops in the Schrödinger representation by

   $$|\Phi(t)\rangle = e^{-i\mathbf{p}(-t)\cdot\boldsymbol{\lambda}/\hbar}|\Psi(t)\rangle,$$

   where $\mathbf{p}(-t)$ is the momentum operator in the Heisenberg representation, at time $-t$.

3. Let $|\Psi\rangle$ be an energy eigenstate of the harmonic oscillator in one dimension. Let

$$|\Phi\rangle = e^{ip\lambda/\hbar}|\Psi\rangle.$$

Show that $|\Phi(t)\rangle$ in the Schrödinger representation is a wave packet whose center oscillates from λ to $-\lambda$ with frequency ω, the wave packet never spreading out in time. [In particular, if $|\Psi\rangle = |0\rangle$, then $\Delta x \cdot \Delta p = \hbar/2$ always.] Hint: use Eq. (5-8). This oscillation is precisely what one expects from a classical oscillator displaced a distance λ from the origin and then released.

4. (a) Calculate the correlation function $\langle 0|\, x(t)x(t')\, |0\rangle$ where $|0\rangle$ is the ground state of a one-dimensional harmonic oscillator, and $x(t)$ is the position operator in the Heisenberg representation.

(b) Suppose that a time-dependent force $F(t)$ is applied to the particle in the oscillator potential. Show that $x(t)$ obeys the equation of motion

$$m\left(\frac{d^2}{dt^2} + \omega^2\right) x(t) = F(t),$$

where ω is the frequency of the oscillator.

(c) The function $\eta_+(t-t')$ is defined by the relation

$$\eta_+(t-t') = \begin{cases} 1: & t \geq t' \\ 0: & t < t'. \end{cases}$$

Show that $D(t-t') \equiv -i\hbar^{-1}\langle 0\,|\,[x(t), x(t')]\,|0\rangle\, \eta_+(t-t')$ is the Green's function for the differential equation in part (b), obeying "retarded" boundary conditions, i.e., $D = 0$ if $t < t'$.

(d) Calculate D explicitly and solve for $x(t)$ in terms of F.

5. Suppose that the force applied in Problem 4 is a constant F from time t_0 to time $t_0 + \tau$ and is zero the rest of the time. Assume the initial state to be an energy eigenstate of the unperturbed oscillator.

(a) Calculate the average position of the particle in the oscillator as a function of time, after the force is switched off.

(b) Show that the force does an average amount of work

$$\Delta E = \frac{2F^2}{m\omega^2} \sin^2 \omega\tau/2$$

on the particle. Interpret this formula in the limit $\omega\tau \ll 1$. Why does the average work vanish if τ is an integral number of periods? Must the work done always be positive classically? Explain the difference between the quantum and the classical cases.

6. (a) Solve the Heisenberg equations of motion for the position and momentum of a free particle perturbed by a uniform force $\mathbf{F}(t)$. What is the energy and momentum transferred to the particle by the force after it acts for time t?

   (b) If at time 0 the wave function of the particle is of the form

$$\psi(\mathbf{r}) = e^{i\mathbf{k}\cdot\mathbf{r}}\varphi(\mathbf{r})$$

   where φ is real, show that the uncertainty, $(\Delta r)_t$, in the position of the particle at a later t is given by

$$(\Delta r)_t = \left[(\Delta r)_0^2 + \frac{2t^2}{m}\left(\langle E\rangle - \frac{\hbar^2 k^2}{2m}\right)\right]^{\frac{1}{2}},$$

   regardless of the particular force applied, where $\langle E\rangle$ is the expectation value of the energy of the particle at $t = 0$.

   (c) Find an explicit formula, in this case, for $U(t)$ [not involving time-ordered products]. *Hint:* try a solution of the form

$$U(t) = e^{\boldsymbol{\alpha}\cdot\mathbf{r}}\overline{U}(t).$$

7. (a) Show for the one-dimensional harmonic oscillator that

$$\langle 0|e^{ikx}|0\rangle = e^{-k^2\langle 0|x^2|0\rangle/2}.$$

   (b) Calculate $\langle 0|\delta(x-a)|0\rangle$. Interpret the result.

8. What are the matrix elements of the operator $1/p$ in the position representation, where $p = \sqrt{p_x^2 + p_y^2 + p_z^2}$ and $p_x, \ldots$ are the components of the momentum operator? If a state $|\Psi\rangle$ has a wave function $\psi(\mathbf{r})$, what is the wave function of the state $(1/p)|\Psi\rangle$?

9. In Problem 16, Chapter 4, let Q be the operator whose expectation value is the probability of observing the particle on the left side of the box. Write down explicitly the Hamiltonian, and the equation of motion for Q(t) in the Heisenberg representation, and use this to rederive the results of part (c) of that problem.

Chapter 6
ORBITAL ANGULAR MOMENTUM AND CENTRAL POTENTIALS

The orbital angular momentum of a particle in quantum mechanics is defined exactly as in classical mechanics: if $\mathbf{r}$ is the position operator and $\mathbf{p}$ is the momentum operator for the particle, then the operator, $\mathbf{L}$, for its orbital angular momentum about the origin is defined by

$$\mathbf{L} = \mathbf{r} \times \mathbf{p} \tag{6-1}$$

In components

$$L_x = yp_z - zp_y,$$

$$L_y = zp_x - xp_z,$$

$$L_z = xp_y - yp_x. \tag{6-2}$$

Notice that in these equations, the position operators commute with the momentum operators they multiply, e.g., $[y, p_z] = 0$. The angular momentum operator is therefore Hermitian:

$$L_z{}^\dagger = (xp_y)^\dagger - (yp_x)^\dagger = p_y{}^\dagger x^\dagger - p_x{}^\dagger y^\dagger = p_y x - p_x y = xp_y - yp_x = L_z$$

$$L_y{}^\dagger = L_y, \quad L_z{}^\dagger = L_z.$$

COMMUTATION RELATIONS

Before developing any properties of the angular momentum operators, we must know the commutation relations they satisfy with themselves, with the momentum operator, and with the position operator. To begin with

$$[L_x, x] = [yp_z - zp_y, x] = 0$$

$$[L_x, y] = [yp_z - zp_y, y] = -z[p_y, y] = i\hbar z$$

$$[L_x, z] = [yp_z - zp_y, z] = y[p_z, z] = -i\hbar y. \tag{6-3a}$$

Similarly

$$[L_y, x] = -i\hbar z, \qquad [L_y, y] = 0, \qquad [L_y, z] = i\hbar x \tag{6-3b}$$

and

$$[L_z, x] = i\hbar y, \qquad [L_z, y] = -i\hbar x, \qquad [L_z, z] = 0. \tag{6-3c}$$

In these commutation relations, if reading across the equation x, y, and z occur cyclically, one has $+i\hbar$ on the right, as in $[L_y, z] = i\hbar x$. If x, y, and z occur anticyclically, one has $-i\hbar$, as in $[L_z, y] = -i\hbar x$. Let us use the subscript 1 to denote the x component of a vector, 2 the y component, and 3 the z component. Then the commutation relations (6-3) can be written compactly in terms of ε_{ijk}, the totally antisymmetric unit tensor; this is defined by

$$\varepsilon_{123} = \varepsilon_{231} = \varepsilon_{312} = 1: \text{ cyclic indices}$$

$$\varepsilon_{321} = \varepsilon_{213} = \varepsilon_{132} = -1: \text{ anticyclic indices} \tag{6-4}$$

with all other components equal to zero. Thus if any two indices are equal, $\varepsilon = 0$, and if we interchange any two indices ε changes sign. In terms of ε the commutation relations (6-3) take the form

$$[L_i, r_j] = i\hbar\, \varepsilon_{ijk} r_k \tag{6-5}$$

where a sum over k = 1, 2, 3 is understood. [One always sums over repeated indices in equations with ε_{ijk}.]

Let $\hat{n}$ be a numerical unit vector. Multiplying both sides of (6-5) by $\hat{n}_i$ and summing over i we find

$$[\hat{n} \cdot L, r_j] = i\hbar (r \times \hat{n})_j$$

or

$$[\hat{n} \cdot L, r] = i\hbar (r \times \hat{n}), \tag{6-6}$$

since $\varepsilon_{ijk}a_j b_k = (\mathbf{a} \times \mathbf{b})_i$. Equation (6-6) says that the commutator of the component of the angular momentum along any axis with the $\mathbf{r}$ operator equals $i\hbar$ times the cross product of $\mathbf{r}$ with the unit vector along that axis.

By a short calculation we find that the commutator of the angular momentum vector with the momentum vector is similar to (6-5) and (6-6):

$$[L_i, p_j] = i\hbar\varepsilon_{ijk}p_k \tag{6-7}$$

or

$$[\hat{\mathbf{n}} \cdot \mathbf{L}, \mathbf{p}] = i\hbar\, \mathbf{p} \times \hat{\mathbf{n}}. \tag{6-8}$$

Now what is $[L_i, L_j]$? First recall that

$$[A, BC] = B[A, C] + [A, B]C. \tag{6-9}$$

Then using (6-6) and (6-8) we have

$$[\hat{\mathbf{n}} \cdot \mathbf{L}, \mathbf{L}] = [\hat{\mathbf{n}} \cdot \mathbf{L}, \mathbf{r} \times \mathbf{p}] = \mathbf{r} \times [\hat{\mathbf{n}} \cdot \mathbf{L}, \mathbf{p}] + [\hat{\mathbf{n}} \cdot \mathbf{L}, \mathbf{r}] \times \mathbf{p}$$

$$= i\hbar(\mathbf{r} \times (\mathbf{p} \times \hat{\mathbf{n}})) + i\hbar((\mathbf{r} \times \hat{\mathbf{n}}) \times \mathbf{p}) = i\hbar(\mathbf{r} \times \mathbf{p}) \times \hat{\mathbf{n}} = i\hbar\mathbf{L} \times \hat{\mathbf{n}},$$

which is again of the same form as (6-6) and (6-8). (6-9) can also be written as

$$[L_i, L_j] = i\hbar\, \varepsilon_{ijk}L_k. \tag{6-10}$$

In components this says $L_x L_y - L_y L_x = i\hbar L_z$, etc.; thus (6-10) is the same as the vector relation

$$\mathbf{L} \times \mathbf{L} = i\hbar\mathbf{L}. \tag{6-11}$$

ROTATIONS

The angular momentum operator is very closely related to rotations of the system. We had a glimpse of this connection when we studied the spin angular momentum of the photon. Consider, for example, the operator

$$\mathbf{r}' = \mathbf{r} + \boldsymbol{\alpha} \times \mathbf{r}, \tag{6-12}$$

where $\boldsymbol{\alpha}$ is an infinitesimal vector. If $|\mathbf{r}_0\rangle$ is an eigenstate of $\mathbf{r}$, then

$$\mathbf{r}'|\mathbf{r}_0\rangle = (\mathbf{r}_0 + \boldsymbol{\alpha}\times\mathbf{r}_0)|\mathbf{r}_0\rangle. \tag{6-13}$$

Thus $|\mathbf{r}_0\rangle$ is an eigenstate of the operator $\mathbf{r}'$ with eigenvalue $\mathbf{r}_0 + \boldsymbol{\alpha}\times\mathbf{r}_0$. Now the vector $\mathbf{r}_0 + \boldsymbol{\alpha}\times\mathbf{r}_0$ is just the vector we get if we rotate $\mathbf{r}_0$, in a right-handed sense about the axis $\boldsymbol{\alpha}/|\boldsymbol{\alpha}| \equiv \hat{\boldsymbol{\alpha}}$ by the infinitesimal angle $|\boldsymbol{\alpha}|$. Therefore we can interpret $\mathbf{r}'$ as the position operator rotated about $\hat{\boldsymbol{\alpha}}$ by $|\boldsymbol{\alpha}|$, since its eigenvalues are those of $\mathbf{r}$, only rotated about $\hat{\boldsymbol{\alpha}}$ by $|\boldsymbol{\alpha}|$. From another point of view, $\mathbf{r}'$ is the position operator in a coordinate frame rotated, from our present one, about $\hat{\boldsymbol{\alpha}}$ by an angle $-|\boldsymbol{\alpha}|$.

Making use of the commutation relation between $\mathbf{r}$ and $\boldsymbol{\alpha}\cdot\mathbf{L}$ we can write (6-12) as

$$\mathbf{r}' = \mathbf{r} + \frac{i}{\hbar}[\boldsymbol{\alpha}\cdot\mathbf{L}, \mathbf{r}]. \tag{6-14}$$

This is equivalent to the equation

$$\mathbf{r}' = \exp\left(\frac{i}{\hbar}\boldsymbol{\alpha}\cdot\mathbf{L}\right)\mathbf{r}\exp\left(-\frac{i}{\hbar}\boldsymbol{\alpha}\cdot\mathbf{L}\right), \tag{6-15}$$

to *first* order in α.

In fact, Eq. (6-15) is valid even if the rotation angle $|\alpha|$ is not infinitesimal. To see this we note that if $\mathbf{r}'(|\alpha|)$ is the $\mathbf{r}$ operator rotated about $\hat{\boldsymbol{\alpha}}$ by $|\alpha|$, then if we rotate just a little bit, $\delta|\alpha|$, further about $\hat{\boldsymbol{\alpha}}$, then

$$\mathbf{r}'(|\alpha| + \delta|\alpha|) = \mathbf{r}'(|\alpha|) + \delta|\alpha|\,\hat{\boldsymbol{\alpha}}\times\mathbf{r}'(|\alpha|), \tag{6-16}$$

or equivalently

$$\frac{d\mathbf{r}'(|\alpha|)}{d|\alpha|} = \hat{\boldsymbol{\alpha}}\times\mathbf{r}'(|\alpha|). \tag{6-17}$$

This equation, together with the boundary condition $\mathbf{r}'(0) = \mathbf{r}$, determines $\mathbf{r}'(|\alpha|)$ in terms of $\mathbf{r}$. All we must verify therefore is that (6-15) is, in fact, a solution to (6-17), since certainly (6-15) satisfies the boundary condition. But taking $d/d|\alpha|$ of both sides of (6-15) we find

$$\frac{d\mathbf{r}'}{d|\alpha|} = \exp\left(\frac{i}{\hbar}\boldsymbol{\alpha}\cdot\mathbf{L}\right)\left(\frac{i}{\hbar}\hat{\boldsymbol{\alpha}}\cdot\mathbf{L}\,\mathbf{r} - \mathbf{r}\frac{i}{\hbar}\hat{\boldsymbol{\alpha}}\cdot\mathbf{L}\right)\exp\left(-\frac{i}{\hbar}\boldsymbol{\alpha}\cdot\mathbf{L}\right)$$

$$= \exp\left(\frac{i}{\hbar}\boldsymbol{\alpha}\cdot\mathbf{L}\right)(\hat{\boldsymbol{\alpha}}\times\mathbf{r})\exp\left(-\frac{i}{\hbar}\boldsymbol{\alpha}\cdot\mathbf{L}\right) = \hat{\boldsymbol{\alpha}}\times\mathbf{r}',$$

thus verifying that the transformed operator (6-15) solves (6-17). What we have learned then is that the rotated operator $\mathbf{r}'$ is given in terms of $\mathbf{r}$ and $\mathbf{L}$ by (6-15) for any arbitrary rotation. Notice that we haven't used the fact that $\mathbf{r}$ was actually the position operator; the same argument holds for any vector operator, e.g., $\mathbf{p}, p^2\mathbf{r}$, etc.

What effect does the unitary operator $e^{i\boldsymbol{\alpha}\cdot\mathbf{L}/\hbar}$ have on the states? If $\langle\mathbf{r}_0|$ is an eigenstate of $\mathbf{r}$ then

$$\langle\mathbf{r}_0|e^{i\boldsymbol{\alpha}\cdot\mathbf{L}/\hbar}\mathbf{r}e^{-i\boldsymbol{\alpha}\cdot\mathbf{L}/\hbar} = \langle\mathbf{r}_0|\mathbf{r}' = \mathbf{r}'_0\langle\mathbf{r}_0| \tag{6-18}$$

where $\mathbf{r}'_0$ is the vector $\mathbf{r}_0$ rotated about $\hat{\boldsymbol{\alpha}}$ by $|\alpha|$. Multiplying both sides of (6-18) on the right by $e^{i\boldsymbol{\alpha}\cdot\mathbf{L}/\hbar}$ we find

$$\langle\mathbf{r}_0|e^{i\boldsymbol{\alpha}\cdot\mathbf{L}/\hbar}\mathbf{r} = \mathbf{r}'_0\langle\mathbf{r}_0|e^{i\boldsymbol{\alpha}\cdot\mathbf{L}/\hbar}. \tag{6-19}$$

Thus $\langle\mathbf{r}_0|e^{i\boldsymbol{\alpha}\cdot\mathbf{L}/\hbar}$ is an eigenstate of $\mathbf{r}$ with eigenvalue $\mathbf{r}'_0$, so that we can write

$$\langle\mathbf{r}_0|e^{i\boldsymbol{\alpha}\cdot\mathbf{L}/\hbar} = \langle\mathbf{r}'_0|. \tag{6-20}$$

In particular, for an infinitesimal rotation, $\boldsymbol{\alpha}$,

$$\langle\mathbf{r}_0|\left(1 + i\boldsymbol{\alpha}\cdot\frac{\mathbf{L}}{\hbar}\right) = \langle\mathbf{r}_0 + \boldsymbol{\alpha}\times\mathbf{r}_0|. \tag{6-21}$$

For this reason, one speaks of L as the *generator* of infinitesimal rotations.

If $|\Psi\rangle$ is an arbitrary state of the system, then the state $|\Psi'\rangle = e^{i\boldsymbol{\alpha}\cdot\mathbf{L}/\hbar}|\Psi\rangle$ has the wave function

$$\psi'(\mathbf{r}_0) = \langle\mathbf{r}_0|\Psi'\rangle = \langle\mathbf{r}_0|e^{i\boldsymbol{\alpha}\cdot\mathbf{L}/\hbar}|\Psi\rangle = \langle\mathbf{r}'_0|\Psi\rangle = \psi(\mathbf{r}'_0), \tag{6-22}$$

which is just the wave function of $|\Psi\rangle$ evaluated at the rotated point $\mathbf{r}'_0$.

We should compare the ability of the angular momentum operator to generate infinitesimal rotations, with the corresponding property of the linear momentum $\mathbf{p}$. If $\boldsymbol{\lambda}$ is a *c-number* vector then it is easy to verify that [Problem 2, Chapter 5]

$$e^{i\mathbf{p}\cdot\boldsymbol{\lambda}/\hbar}\mathbf{r}e^{-i\mathbf{p}\cdot\boldsymbol{\lambda}/\hbar} = \mathbf{r}+\boldsymbol{\lambda}, \tag{6-23}$$

and also

$$\langle \mathbf{r}| e^{i\mathbf{p}\cdot\boldsymbol{\lambda}/\hbar} = \langle \mathbf{r}+\boldsymbol{\lambda}|. \tag{6-24}$$

Thus if $|\Psi\rangle$ has the wave function $\psi(\mathbf{r})$, the state $e^{i\mathbf{p}\cdot\boldsymbol{\lambda}/\hbar}|\Psi\rangle$ has the wave function $\psi(\mathbf{r}+\boldsymbol{\lambda})$. The operator p is thus the *generator* of infinitesimal translations.

DIFFERENTIAL OPERATOR REPRESENTATION

We can use Eq. (6-21) to construct explicit differential operator representations of $\mathbf{L}$, analogous to the representation of p by $-i\hbar\nabla$, [Eq. (3-60)]. In spherical coordinates r, θ, φ, where

$$z = r\cos\theta$$

$$y = r\sin\theta\sin\varphi$$

$$x = r\sin\theta\cos\varphi \tag{6-25}$$

we can denote the state $\langle \mathbf{r}|$ by $\langle r, \theta, \varphi|$. Under a rotation of $\mathbf{r}$, the magnitude r is unchanged. First, if we rotate the vector $\mathbf{r}$ about z by α, then

$$\theta \rightarrow \theta, \qquad \varphi \rightarrow \varphi + \alpha.$$

Using (6-21) we see that, for infinitesimal α,

$$\langle r, \theta, \varphi|\left(1 + \frac{i\alpha L_z}{\hbar}\right) = \langle r, \theta, \varphi+\alpha| = \langle r, \theta, \varphi| + \alpha\frac{\partial}{\partial\varphi}\langle r, \theta, \varphi|.$$

Thus

$$\langle r, \theta, \varphi| L_z = \frac{\hbar}{i}\frac{\partial}{\partial\varphi}\langle r, \theta, \varphi|. \tag{6-26}$$

Acting on wave functions, L_z has the representation $-i\hbar\partial/\partial\varphi$.

Now if we do an infinitesimal rotation by α about the x axis, then r is changed by

$$x \rightarrow x, \qquad y \rightarrow y - \alpha z, \qquad z \rightarrow z + \alpha y. \tag{6-27}$$

From (6-26) we see that

$$\delta z = \alpha y = \alpha r\sin\theta\sin\varphi = r\,\delta\cos\theta = -r\sin\theta\,\delta\theta.$$

Thus under this infinitesimal rotation

$$\theta \to \theta + \delta\theta = \theta - \alpha \sin \varphi. \tag{6-28}$$

Since x is unchanged, $\delta(\sin \theta \cos \varphi) = 0$ so that

$$\delta\varphi = \cot \theta \cot \varphi \, \delta\theta$$

or

$$\varphi \to \varphi - \alpha \cot \theta \cos \varphi. \tag{6-29}$$

Thus

$$\langle r, \theta, \varphi| \left(1 + \frac{i\alpha L_x}{\hbar}\right) = \langle r, \theta - \alpha \sin \varphi, \varphi - \alpha \cot \theta \cos \varphi| \tag{6-30}$$

so that expanding the right side to first order in α we find

$$\langle r, \theta, \varphi| L_x = \frac{\hbar}{i}\left(-\sin \varphi \frac{\partial}{\partial \theta} - \cot \theta \cos \varphi \frac{\partial}{\partial \varphi}\right)\langle r, \theta, \varphi|. \tag{6-31}$$

This is the differential operator representation of L_x.

Under an infinitesimal rotation by α about the y axis we can similarly calculate that

$$\langle r, \theta, \varphi| \to \langle r, \theta + \alpha \cos \varphi, \varphi - \alpha \cot \theta \sin \varphi|$$

and therefore

$$\langle r, \theta, \varphi| L_y = \frac{\hbar}{i}\left(\cos \varphi \frac{\partial}{\partial \theta} - \cot \theta \sin \varphi \frac{\partial}{\partial \varphi}\right)\langle r, \theta, \varphi|. \tag{6-32}$$

The operators

$$L_\pm \equiv L_x \pm iL_y \tag{6-33}$$

have the differential operator representation

$$\langle r, \theta, \varphi| L_\pm = \frac{\hbar}{i} e^{\pm i\varphi}\left[\pm i \frac{\partial}{\partial \theta} - \cot \theta \frac{\partial}{\partial \varphi}\right]\langle r, \theta, \varphi|. \tag{6-34}$$

Also the square of the angular momentum, defined by

$$L^2 = L_x{}^2 + L_y{}^2 + L_z{}^2 \tag{6-35}$$

has, by a simple calculation, the representation

$$\langle r, \theta, \varphi | L^2 = -\hbar^2 \left[\frac{1}{\sin^2\theta} \frac{\partial^2}{\partial \varphi^2} + \frac{1}{\sin\theta} \frac{\partial}{\partial\theta}\left(\sin\theta \frac{\partial}{\partial\theta}\right)\right] \langle r, \theta, \varphi | . \qquad (6\text{-}36)$$

We should note that, using (6-36), the Laplacian takes the simple form in spherical coordinates

$$\nabla^2 = \frac{1}{r}\frac{\partial^2}{\partial r^2} r - \frac{L^2/\hbar^2}{r^2}. \qquad (6\text{-}37)$$

where L^2 here means the differential operator in (6-36).

EIGENFUNCTIONS AND EIGENVALUES

Let us now turn to the task of finding the eigenvalues and eigenfunctions of the angular momentum operators. In doing this the only operator properties we shall use are that $\mathbf{L}$ is Hermitian and that it obeys the commutation relation $\mathbf{L} \times \mathbf{L} = i\hbar\mathbf{L}$. *Thus our results will be valid for any Hermitian vector operator obeying this cross product commutation relation.*

First, let us notice that

$$[L^2, \mathbf{L}] = 0. \qquad (6\text{-}38)$$

Now if we have a complete set of eigenstates of L^2, then because $\mathbf{L}$ commutes with L^2, we can construct these states to be simultaneously eigenstates of *one* component of L; we *arbitrarily* choose this component to be L_z. The reason this complete set can't be simultaneous eigenstates of two or three components of $\mathbf{L}$ is that different components fail to commute with each other. For example, if we had an eigenstate $|l_x, l_z\rangle$ of L_x and L_z, then

$$-i\hbar L_y |l_x, l_z\rangle = (L_x L_z - L_z L_x)|l_x, l_z\rangle = (l_x l_z - l_z l_x)|l_x, l_z\rangle = 0$$

since $l_x l_z = l_z l_x$ for numbers. Thus L_y acting on the state $|l_x, l_z\rangle$ gives zero. If we had a complete set of such states then L_y acting on each of them would give zero, implying that L_y was identically zero — there would be no states it didn't annihilate. However, L_y is certainly not the zero operator, and therefore we can't construct a complete set of simultaneous eigenstates of L_x and L_z, or any other two components of $\mathbf{L}$. The best we can do is to find a complete set of simultaneous eigenstates of L^2 and one component, which we take to be L_z.

The eigenvalues of L^2 are all greater than or equal to zero. This follows from the fact that

$$\langle\Psi|L^2|\Psi\rangle = \sum_i \langle\Psi|L_i^2|\Psi\rangle = \sum_i (\langle\Psi|L_i)(L_i|\Psi\rangle) \geqslant 0$$

for any state $|\Psi\rangle$. Thus if $|\Psi\rangle$ is an eigenstate of L^2 with eigenvalue λ, then

$$\langle\Psi|L^2|\Psi\rangle = \lambda\langle\Psi|\Psi\rangle \geqslant 0.$$

We can therefore write the eigenvalues of L^2 as $\hbar^2 l(l+1)$ with $l \geqslant 0$; the reason for writing the eigenvalues this way is that $2l$ shall turn out to be an integer. We also use $\hbar m$ to denote an eigenvalue of L_z, and the joint eigenstates of L^2 and L_z we denote by $|l, m\rangle$; m is called the *azimuthal quantum number.*

The nature of the eigenstates $|l, m\rangle$ follow simply from the properties of the operators $L_\pm = L_x \pm iL_y$. It is easy to verify that

$$[L_z, L_\pm] = \pm\hbar L_\pm \tag{6-39}$$

$$L_+L_- = L_x^2 + L_y^2 - i(L_xL_y - L_yL_x) = L^2 - L_z^2 + \hbar L_z \tag{6-40}$$

$$L_-L_+ = L^2 - L_z^2 - \hbar L_z \tag{6-41}$$

and

$$[L_+, L_-] = 2\hbar L_z. \tag{6-42}$$

Now since $(L_-)^\dagger = L_+$,

$$\langle lm|L_+L_-|lm\rangle \geq 0;$$

using (6-40) we see that

$$\begin{aligned}\langle lm|L_+L_-|lm\rangle &= \langle lm|(L^2 - L_z^2 + \hbar L_z)|lm\rangle \\ &= \hbar^2(l(l+1) - m^2 + m)\langle lm|lm\rangle\end{aligned} \tag{6-43}$$

and therefore

$$l(l+1) - m^2 + m \geq 0$$

or

$$\left(l+\frac{1}{2}\right)^2 \geq \left(m-\frac{1}{2}\right)^2. \tag{6-44}$$

Also, from $\langle lm|L_-L_+|lm\rangle \geq 0$ we find

$$\left(l+\frac{1}{2}\right)^2 \geq \left(m+\frac{1}{2}\right)^2. \tag{6-45}$$

The two equations (6-44) and (6-45) imply that

$$-l \leq m \leq l. \tag{6-46}$$

We now apply the same machinery that we used in solving the harmonic oscillator problem. From the commutation relation (6-39) it follows that

$$\begin{aligned} L_z L_-|l,m\rangle &= L_-(L_z-\hbar)|l,m\rangle = L_-(\hbar m-\hbar)|l,m\rangle \\ &= \hbar(m-1)L_-|l,m\rangle. \end{aligned} \tag{6-47}$$

L_- acting on $|l, m\rangle$ produces an eigenstate of L_z with eigenvalue $\hbar(m-1)$. If we write

$$L_-|l,m\rangle = c|l,m-1\rangle,$$

where $|l, m\rangle$ and $|l, m-1\rangle$ are normalized to one, then from (6-43) we see that

$$\langle l,m|L_+L_-|l,m\rangle = |c|^2\langle l,m-1|l,m-1\rangle = \hbar^2(l(l+1)-m(m-1)).$$

Thus, *choosing* the phase of $|l, m-1\rangle$ so that c is positive we have

$$L_-|l,m\rangle = \hbar\sqrt{l(l+1)-m(m-1)}\;|l,m-1\rangle. \tag{6-48}$$

By a similar argument

$$L_+|l,m\rangle = \hbar\sqrt{l(l+1)-m(m+1)}\;|l,m+1\rangle. \tag{6-49}$$

From (6-48) and (6-49) it appears that if m is an eigenvalue of $L_z/\hbar$, then so are $m+1$, $m+2$, $m+3$, etc. However, (6-46) places an upper limit on $|m|$. Therefore, for some m that doesn't exceed l the coefficient $\hbar\sqrt{l(l+1)-m(m+1)}$ must vanish in order for the series of eigenstates to terminate with $m \leq l$. This occurs clearly for $m = l$; in order for the series to terminate the m values must be smaller than l by integers only, and the largest m value is l itself.

Now let's look at the other end, where $m \geq -l$. From (6-48) we see that the lowest m values satisfies $\hbar\sqrt{l(l+1) - m(m-1)} = 0$, or $m = -l$. The possible m values are greater than $-l$ by integers only, and the least value is $-l$. However, m can differ from both l and $-l$ by integers only if $2l$ is an integer. Putting all this together we have

l = integer or half-integer

$$m = -l, -l+1, -l+2, \ldots, l-2, l-1, l. \qquad (6\text{-}50)$$

Notice that for each l there are $2l+1$ different possible m values.

The picture we have of the angular momentum vector is the following. In an $|l, m\rangle$ state, its z component has length $\hbar m$, it has length $\sqrt{\hbar^2 l(l+1)}$, and its transverse component, $\sqrt{L_x^2 + L_y^2} = \sqrt{L^2 - L_z^2}$ has length $\hbar\sqrt{l(l+1) - m^2}$. We cannot fix the direction of the transverse component, however, because L_x and L_y fail to commute with L_z. If we know that L^2 has the value $\hbar^2 l(l+1)$ then there are $2l+1$ possible values we can find for the projection of L on the z axis,

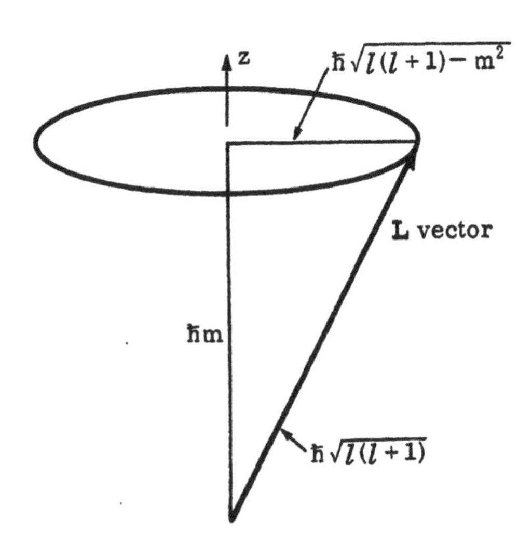

Fig. 6-1

The angular momentum vector in a state in which $L^2 = \hbar^2 l(l+1)$ and $L_z = \hbar m$.

corresponding to the $2l+1$ different m values. This statement is, of course, true for any axis, only we can't know the projection of **L** on two axes simultaneously.

Because L^2 and L_z are Hermitian, the various $|l, m\rangle$ obey the orthogonality relation

$$\langle lm|l'm'\rangle = \delta_{ll'}\delta_{mm'}. \tag{6-51}$$

For the oribital angular momentum l is always an integer. For spin angular momentum, which involves internal degrees of freedom, such as the polarization vector of a photon, l can, for a given type of particle, be either integer or half-integer. The fact that l is an integer for orbital angular monentum is closely related to the single-valuedness of the wave function. Suppose that we have a particle in an eigenstate $|l, m\rangle$ of L^2 and L_z. Then the wave function of this state is $\langle r, \theta, \varphi|lm\rangle$. If we rotate about the z axis by 2π, which brings us back to the same place again, then

$$\langle r, \theta, \varphi|lm\rangle \to \langle r, \theta, \varphi+2\pi|lm\rangle = \langle r, \theta, \varphi|e^{2\pi i L_z/\hbar}|lm\rangle$$

$$= e^{2\pi i m}\langle r\theta\varphi|lm\rangle.$$

Thus in order that $\langle r\theta\varphi|lm\rangle$ equal $\langle r, \theta, \varphi+2\pi|lm\rangle$, i.e., that the wave function be single-valued under a rotation by 2π, then $e^{2\pi i m}$ must equal 1; or m, and therefore l, must be an integer. The reason double-valued *spatial* wave functions must be ruled out, is that for a wave function $\psi(\mathbf{r})$ to be double-valued in three dimensions there must be a line in space, either closed or extending to infinity in both directions, such that when one goes around it 360° and returns to the starting point the wave function changes sign. The existence of such a line, which is not attached to the particle described by the wave function, would, at best, violate rotational invariance.[1]

SPHERICAL HARMONICS

Let us go on now to construct the eigenfunctions of the orbital angular momentum explicitly. The eigenvalue equation

$$L_z|l, m\rangle = \hbar m|l, m\rangle$$

---

[1] See also K. Gottfried, *Quantum Mechanics, Volume I* (W.A. Benjamin, Inc., New York, 1966), p. 86.

becomes in terms of wave functions

$$\langle r\theta\varphi|L_z|lm\rangle = \frac{\hbar}{i}\frac{\partial}{\partial\varphi}\langle r\theta\varphi|lm\rangle = \hbar m\langle r\theta\varphi|lm\rangle \tag{6-52}$$

using (6-26). Solving for the φ dependence, we find

$$\langle r\theta\varphi|lm\rangle = e^{im\varphi}\langle r, \theta, 0|lm\rangle. \tag{6-53}$$

Notice that

$$\langle r, \theta, \varphi+\alpha|lm\rangle = e^{im\alpha}\langle r, \theta, \varphi|lm\rangle. \tag{6-54}$$

To construct the θ dependence we can use the differential operator representation, (6-36) of L^2. We shall proceed instead by using the property $L_+|l, l\rangle = 0$ to find $|l, l\rangle$ and use (6-48) to work our way down to lower m values.

From (6-34) we find that

$$\langle r\theta\varphi|L_+|ll\rangle = \frac{\hbar}{i}e^{i\varphi}\left[i\frac{\partial}{\partial\theta} - \cot\theta\frac{\partial}{\partial\varphi}\right]\langle r\theta\varphi|ll\rangle = 0. \tag{6-55}$$

Thus, from (6-52)

$$\left(\frac{\partial}{\partial\theta} - l\cot\theta\right)\langle r\theta\varphi|ll\rangle = 0, \tag{6-56}$$

and the solution, taking (6-53) into account is,

$$\langle r\theta\varphi|ll\rangle = f(r)e^{il\varphi}(\sin\theta)^l, \tag{6-57}$$

where f(r) is an arbitrary function of r. Now when we manufacture the various $\langle r\theta\varphi|l, m\rangle$ from (6-57) by operating with L_-, it is clear that we won't change f(r). We shall therefore suppress it, and use the notation

$$\langle\theta\varphi|lm\rangle \equiv Y_{lm}(\theta, \varphi) \tag{6-58}$$

to refer to the angular dependence of the wave function of the state $|lm\rangle$. Thus $\langle r\theta\varphi|lm\rangle = f(r)\,Y_{lm}(\theta, \varphi)$. We also shall normalize the radial and angular parts of the wave function separately. Choosing $\int_0^\infty r^2\,dr\,|f(r)|^2 = 1$, the normalization condition on the Y_{lm}'s is

$$\int d\Omega\, Y_{lm}^*(\theta,\varphi) Y_{l'm'}(\theta,\varphi) = \int_0^\pi \sin\theta\, d\theta \int_0^{2\pi} d\varphi\; Y_{lm}^*(\theta,\varphi) Y_{l'm'}(\theta,\varphi) = \delta_{ll'}\delta_{mm'}. \tag{6-59}$$

From (6-57),

$$Y_{ll}(\theta,\varphi) = c e^{il\varphi}(\sin\theta)^l.$$

Now the constant c, using (6-59), obeys

$$2\pi|c|^2 \int_{-1}^{1} du\, (1-u^2)^l = 1.$$

Doing the awful integral we find

$$c = \frac{(-1)^l}{2^l l!}\sqrt{\frac{2l+1}{4\pi}(2l)!}\;; \tag{6-60}$$

the $(-1)^l$ is an arbitrary phase convention.

To find the other $Y_{lm}(\theta,\varphi)$ we use Eqs. (6-48) and (6-34) repeatedly. The general result, correctly normalized, is

$$Y_{lm}(\theta,\varphi) = \frac{(-1)^l}{2^l l!}\sqrt{\frac{2l+1}{4\pi}\frac{(l+m)!}{(l-m)!}}\;\frac{e^{im\varphi}}{(\sin\theta)^m} \times \left(\frac{d}{d\cos\theta}\right)^{l-m}(\sin\theta)^{2l}. \tag{6-61}$$

The Y_{lm} are called *spherical harmonics.*

The first few Y's are

$$Y_{00} = \frac{1}{\sqrt{4\pi}}$$

$$Y_{10} = \sqrt{\frac{3}{4\pi}}\cos\theta, \qquad Y_{1,\pm1} = \mp\sqrt{\frac{3}{8\pi}}\, e^{\pm i\varphi}\sin\theta$$

$$Y_{20} = \sqrt{\frac{5}{16\pi}}\,(3\cos^2\theta - 1), \qquad Y_{2,\pm1} = \mp\sqrt{\frac{15}{8\pi}}\sin\theta\cos\theta\, e^{\pm i\varphi}$$

$$Y_{2,\pm2} = \sqrt{\frac{15}{32\pi}}\sin^2\theta\, e^{\pm 2i\varphi}. \tag{6-62}$$

Generally,

$$Y_{l,-m}(\theta, \varphi) = (-1)^m Y_{l,m}^*(\theta, \varphi). \tag{6-63}$$

We should note the behavior of the Y_{lm}'s under the parity operation, $\mathbf{r} \to -\mathbf{r}$. In spherical coordinates, $\mathbf{r} \to -\mathbf{r}$ is equivalent to

$$r \to r, \qquad \theta \to \pi - \theta, \qquad \varphi \to \varphi + \pi.$$

Thus

$$e^{im\varphi} \to e^{im(\varphi+\pi)} = (-1)^m e^{im\varphi}$$

$$\sin\theta \to \sin\theta, \; \cos\theta \to \cos(\pi - \theta) = -\cos\theta,$$

and we see from (6-61) that

$$Y_{lm}(\theta, \varphi) \to (-1)^l Y_{lm}(\theta, \varphi). \tag{6-64}$$

The even l states have even parity (+1) and the odd l states have odd parity (−1).

Lastly, let us note that because the $|lm\rangle$ are a complete set, the $Y_{lm}(\theta, \varphi)$ form a complete set of functions in the sense that any function $f(\theta, \varphi)$ can be written as a sum of various $Y_{lm}(\theta, \varphi)$:

$$f(\theta, \varphi) = \sum_{l,m} f_{lm} Y_{lm}(\theta, \varphi) \tag{6-65}$$

where from (6-59)

$$f_{lm} = \int d\Omega \, Y_{lm}^*(\theta, \varphi) f(\theta, \varphi). \tag{6-66}$$

CENTRAL FORCES

We have now developed the theory of the orbital angular momentum sufficiently to study the problem of a particle in a central potential $V(|\mathbf{r}|)$. The Hamiltonian of the particle is

$$H = \frac{p^2}{2m} + V(|\mathbf{r}|), \tag{6-67}$$

and since the Hamiltonian is unchanged by a rotation about the origin — it is a scalar function of $\mathbf{r}$ and $\mathbf{p}$ — we have

$$[H, \mathbf{L}] = 0, \tag{6-68}$$

and therefore

$$[H, L^2] = 0. \tag{6-69}$$

This means that we can find a complete set of states that are simultaneous eigenfunctions of H, L^2, and L_z. We denote such a state by $|E, l, m\rangle$.

From (6-68) we see that $[H, L_\pm] = 0$. Thus

$$HL_\pm|E, l, m\rangle = L_\pm H|E, l, m\rangle = EL_\pm|E, l, m\rangle; \tag{6-70}$$

$L_\pm|E, l, m\rangle$ is also an energy eigenstate with the same E and l but with azimuthal quantum number m ± 1. It follows then that all $2l+1$ different values of m will have the same energy. Therefore, from the rotational invariance of the Hamiltonian we conclude that an energy level with angular momentum l must be $2l+1$ -fold degenerate. This makes sense physically; if we change the direction of the angular momentum vector we can't change the energy when the potential is spherically symmetric – the $2l+1$ different m values correspond to different orientations of the angular momentum vector.

From (6-37) we see that the Schrödinger equation for the energy eigenstates takes the form in spherical coordinates

$$\left[-\frac{\hbar^2}{2m}\frac{1}{r}\frac{\partial^2}{\partial r^2}r + \frac{L^2}{2mr^2} + V(r)\right]\psi(\mathbf{r}) = E\psi(\mathbf{r}). \tag{6-71}$$

The first term can be written as $p_r^2/2m$ where

$$p_r = -i\hbar\left(\frac{\partial}{\partial r} + \frac{1}{r}\right). \tag{6-72}$$

This Hermitian operator is essentially the radial component of the linear momentum.

The eigenfunctions are of the form

$$\psi(\mathbf{r}) = f_l(r)Y_{lm}(\theta, \varphi). \tag{6-73}$$

All m values have the same $f_l(r)$, since $L_\pm$ contains only derivatives with respect to angles. Putting (6-73) into (6-71) we see that $f_l(r)$ obeys

$$\left[-\frac{\hbar^2}{2m}\frac{1}{r}\frac{d^2}{dr^2}r + \frac{\hbar^2}{2m}\frac{l(l+1)}{r^2} + V(r) - E\right]f_l(r) = 0. \tag{6-74}$$

If we write $f_l(r) = u_l(r)/r$, then $u_l(r)$ satisfies

$$\left(-\frac{\hbar^2}{2m}\frac{d^2}{dr^2} + \frac{\hbar^2}{2m}\frac{l(l+1)}{r^2} + V(r) - E\right)u_l(r) = 0. \tag{6-75}$$

This equation is in the form of a one-dimensional Schrödinger equation in the region $0 \le r < \infty$, with the added potential term $\hbar^2 l(l+1)/2mr^2$. This term which becomes arbitrarily large as $r \to 0$ is called a *centrifugal barrier*. Acting like a centrifugal force, it prevents particles with nonzero angular momentum from getting too close to the origin.

We can see how this barrier works by looking at the solution to (6-75) for r very close to zero when $l \neq 0$. We assume that $r^2V(r)$ goes to zero as $r \to 0$. Then for sufficiently small r, the centrifugal barrier term is much bigger than the $V(r) - E$ term, and (6-75) becomes

$$\left[-\frac{\hbar^2}{2m}\frac{d^2}{dr^2} + \frac{\hbar^2}{2m}\frac{l(l+1)}{r^2}\right]u_l(r) = 0.$$

The solutions are $u_l(r) \sim r^{l+1}$ and $u_l(r) \sim r^{-l}$. The latter solution diverges too rapidly at $r = 0$ for the wave function to be normalizable, and only the r^{l+1} solution is physical. Thus near $r = 0$, $f_l(r) \sim r^l$; the wave function of a particle with $l \neq 0$ vanishes at the origin because of the barrier. Also the probability of finding the particle near the origin decreases as l increases. For $l = 0$, u_l must vanish at $r = 0$. If it diverged at $r = 0$, it could not solve (6-75) as $r \to 0$. If it approached a constant, then f would go as $1/r$ at the origin. Thus, because $l = 0$, $\psi(r)$ would also behave as $1/r$. However such a ψ wouldn't solve the Schrödinger equation at $r = 0$ since $\nabla^2(1/r) = -4\pi\delta(r)$, and one would have an extra $\delta(r)$ floating around.

If $V \to 0$ at ∞, then for very large r we have

$$-\frac{\hbar^2}{2m}\frac{d^2}{dr^2}u_l(r) = Eu_l(r).$$

Thus for $E > 0$, and r very large, $u_l(r) \sim e^{\pm ikr}$, where $E = \hbar^2k^2/2m$, and for $E < 0$, $u_l(r) \sim e^{-Kr}$, where $E = -\hbar^2K^2/2m$.

A bit of nomenclature: one refers to states with $l = 0$ as *s states*, $l = 1$ as *p states*, $l = 2$ as *d states*, $l = 3$ as *f states*, $l = 4$ as *g states*, and through the alphabet from here on. These letters come from spectroscopy and are the initials of "sharp," "principal," "diffuse," and "fundamental."

Let us consider how one goes about constructing a solution to the three-dimensional "square well" potential

$$V(r) = \begin{cases} V_0, & r < a \\ 0, & r > a. \end{cases} \qquad (6\text{-}76)$$

Since this potential is constant as a function of r in the two regions $r < a$ and $r > a$, we can find the eigenfunctions by patching together the solutions to the radial Schrödinger equation for a constant potential, using the continuity conditions on the wave function and its derivative at $r = a$. We therefore need the solutions to (6-74) when $V(r)$ is a constant V. First consider $E > V$, and let $\hbar k = \sqrt{2m(E - V)}$. Then (6-74) reduces to

$$\left(\frac{d^2}{dr^2} + \frac{2}{r}\frac{d}{dr} - \frac{l(l+1)}{r^2} + k^2\right) f_l(r) = 0; \qquad (6\text{-}77)$$

this is a form of Bessel's equation. It has, as a solution, the spherical Bessel function, $j_l(kr)$, which is related to the ordinary Bessel function $J_n(x)$ by

$$j_l(x) = \left(\frac{\pi}{2x}\right)^{1/2} J_{l+1/2}(x). \qquad (6\text{-}78)$$

Since Eq. (6-77) is symmetric under the interchange $l + {}^1/_2 \to -(l + {}^1/_2)$, a second solution is the spherical Neumann function, $n_l(kr)$, which is

$$n_l(x) = (-1)^{l+1}\left(\frac{\pi}{2x}\right)^{1/2} J_{-l-1/2}(x). \qquad (6\text{-}79)$$

For real x and l, both n_l and j_l are real.

The first few j_l are

$$j_0(x) = \frac{\sin x}{x}$$

$$j_1(x) = \frac{\sin x}{x^2} - \frac{\cos x}{x}$$

$$j_2(x) = \left(\frac{3}{x^3} - \frac{1}{x}\right)\sin x - \frac{3}{x^2}\cos x \qquad (6\text{-}80)$$

and the first few n_l are

$$n_0(x) = -\frac{\cos x}{x}$$

$$n_1(x) = -\frac{\cos x}{x^2} - \frac{\sin x}{x}$$

$$n_2(x) = -\left(\frac{3}{x^3} - \frac{1}{x}\right)\cos x - \frac{3}{x^2}\sin x. \tag{6-81}$$

These functions have the forms, as $x \to 0$,

$$j_l(x) \to \frac{x^l}{1 \cdot 3 \cdot 5 \cdots (2l+1)}$$

$$n_l(x) \to -\frac{1 \cdot 1 \cdot 3 \cdot 5 \cdots (2l-1)}{x^{l+1}}, \tag{6-82}$$

and as $x \to \infty$,

$$j_l(x) \to x^{-1}\cos\left(x - (l+1)\frac{\pi}{2}\right) = x^{-1}\sin\left(x - l\frac{\pi}{2}\right)$$

$$n_l(x) \to x^{-1}\sin\left(x - (l+1)\frac{\pi}{2}\right) = -x^{-1}\cos\left(x - l\frac{\pi}{2}\right). \tag{6-83}$$

Notice that $n_l(kr)$ diverges at $r = 0$; thus if the potential is a constant near $r = 0$, the solution must be $j_l(kr)$.

At infinity, either j_l or n_l are valid solutions. We can also take linear combinations of j_l and n_l — the spherical Hankel functions, defined by

$$h_l(x) = j_l(x) + in_l(x), \qquad h_l(x)^* = j_l(x) - in_l(x). \tag{6-84}$$

From (6-83) we see that $h_l(x)$ behaves as an outgoing wave, $(1/x)e^{i(x-(l+1)\pi/2)}$, at infinity, while $h_l{}^*$ behaves as an incoming wave, $(1/x)e^{-i(x-(l+1)\pi/2)}$, at infinity.

Now, if $E < V$ then let us write $\hbar K = \sqrt{2m(V-E)}$. Then the solutions to the Schrödinger equation must be $j_l(iKr)$ and $n_l(iKr)$. Again $n_l(iKr)$ diverges at the origin and is therefore unallowable there. At infinity, $h_l{}^*(iKr) \sim e^{Kr}/r$, so that only $h_l(iKr)$ is an allowable solution at infinity; it vanishes there asymptotically as e^{-Kr}/r.

Thus the solution to the square well problem, or any more complicated problem where the potential is piecewise constant, is a linear combination of j_l and n_l; only j_l is allowable at $r = 0$, and if $E < V$ at infinity, then only h_l is allowable there.

As an example, we can write the plane wave function, $e^{i\mathbf{k}\cdot\mathbf{r}}$, of a free particle as a linear superposition of the spherical eigenstates, $j_l(kr)Y_{l,m}(\theta, \varphi)$. For fixed k these states all have the same energy, $\hbar^2k^2/2m$. The expansion is

$$e^{i\mathbf{k}\cdot\mathbf{r}} = 4\pi \sum_{l=0}^{\infty} \sum_{m=-l}^{l} [i^l Y_{lm}^*(\theta_k, \varphi_k)] j_l(kr) Y_{lm}(\theta, \varphi) \tag{6-85}$$

where θ, φ are the angles of $\mathbf{r}$ and θ_k, φ_k are the angles of $\mathbf{k}$.

PROBLEMS

1. Show that all scalar functions $S(\mathbf{r}, \mathbf{p})$ of $\mathbf{r}$ and $\mathbf{p}$, e.g., $\mathbf{r}\cdot\mathbf{p}$, commute with $\mathbf{L}$. Then show that all vector functions $\mathbf{v}(\mathbf{r}, \mathbf{p})$ of $\mathbf{r}$ and $\mathbf{p}$, e.g., $(\mathbf{r}\times\mathbf{p})(\mathbf{r}\cdot\mathbf{p})$, satisfy

   $$[\hat{\mathbf{n}}\cdot\mathbf{L}, \mathbf{v}] = i\hbar(\mathbf{v}\times\hat{\mathbf{n}}).$$

   If $t(\mathbf{r}, \mathbf{p})$ transforms under rotation as a tensor of rank 2 what commutation relation does it obey with $\mathbf{L}$?
2. (a) Explicitly construct the three 3×3 matrices that represent L_x, L_y, and L_z in the space of $l = 1$ functions:

   $$(L_i)_{m,m'} = \langle l = 1, m | L_i | l = 1, m' \rangle$$
   $$= \int d\Omega\, Y_{1m}^*(\theta, \varphi) L_i Y_{1,m'}(\theta, \varphi),$$

   where $i = x, y, z$.

   (b) Show by explicit calculation that these three matrices obey the commutation relations of angular momentum.

   (c) Find the matrices that represent L_+, L_-, and L^2.
3. Show that p_r is Hermitian. (Assume that the wave functions are finite at $r = 0$ and $\to 0$ at ∞.)
4. Find the eigenfunctions and energy levels of a two-dimensional circular box with impermeable rigid walls.
5. (a) Find the energy levels and wave functions of a two-dimensional isotropic harmonic oscillator, $V(r) = m\omega_0^2 r^2/2$ (where $r^2 = x^2 + y^2$) by solving the wave equation in Cartesian coordinates.

Find the degeneracy of each level. Write out the wave functions of the ground state, and each of the first excited states.
(b) Write the Schrödinger equation for this problem in polar coordinates. Explicitly construct the wave functions of the first excited states with angular momentum $+\hbar$ and $-\hbar$; these wave functions are linear combinations of the wave functions found in part (a).

6. A particle of charge e and mass m is in a uniform magnetic field $\mathcal{H}$ pointing in the z direction. Let the vector potential be along $\hat{y}$. Find the energy eigenstates and eigenvalues. Are the states eigenstates of L_z? Classically a particle in a uniform magnetic field in the z direction executes uniform motion along z and circular motion in the x, y plane. What is the quantum mechanical motion like for these eigenstates? (Can a particle with a fixed p_y travel in a circle in the x, y plane?)

Chapter 7

THE HYDROGEN ATOM

The hydrogen atom consists of an electron and a proton interacting through an attractive Coulomb potential. In order to discuss the properties of the hydrogen atom, we have to know how to describe two particle systems quantum mechanically.

TWO PARTICLE SYSTEMS

So far we have been talking only about one particle, described by a wave function $\psi(\mathbf{r}, t)$, which is the probability amplitude for observing the particle at $\mathbf{r}$ at time t. A two particle system must be described by a more complicated wave function $\psi(\mathbf{r}_1, \mathbf{r}_2, t)$, which is the probability amplitude for observing, at time t, particle 1 at $\mathbf{r}_1$ and particle 2 at $\mathbf{r}_2$. [The first argument refers to particle 1 and the second to particle 2.] Thus, for example, $\int d^3r_2 |\psi(\mathbf{r}_1, \mathbf{r}_2, t)|^2$ is the probability of observing particle 1 at $\mathbf{r}_1$ at time t, regardless of where particle 2 is. Since the total probability of observing particle 1 somewhere must be one, we have the normalization condition

$$\int d^3r_1 d^3r_2 |\psi(\mathbf{r}_1, \mathbf{r}_2, t)|^2 = 1. \tag{7-1}$$

One can think of the wave function $\psi(\mathbf{r}_1, \mathbf{r}_2, t)$ as being the components of a state vector $|\Psi(t)\rangle$ in the "position of 1, position of 2" basis,

$$\psi(\mathbf{r}_1, \mathbf{r}_2, t) = \langle \mathbf{r}_1, \mathbf{r}_2 | \Psi(t) \rangle. \tag{7-2}$$

Exactly as for a single particle, physical quantities such as position and momentum are associated with operators; we introduce operators $\mathbf{r}_1$, $\mathbf{p}_1$, $\mathbf{r}_2$, $\mathbf{p}_2$ for the position and momentum of particle 1 and position and momentum of particle 2. In the position basis the operators $\mathbf{r}_1$ and $\mathbf{r}_2$ simply multiply $\psi(\mathbf{r}_1, \mathbf{r}_2)$ by the vectors $\mathbf{r}_1$ and $\mathbf{r}_2$, while the momentum operator $\mathbf{p}_1$ acts as $\hbar\nabla_1/i$ and $\mathbf{p}_2$ acts as $\hbar\nabla_2/i$. All operators referring to particle 1 must commute with all operators referring to particle 2; we can in principle specify, for example, the momentum of particle 1 and the position of particle 2, so that $[\mathbf{p}_1, \mathbf{r}_2] = 0$. We can see, explicitly, in the position representation that

$$\left[\mathbf{r}_1, \frac{\hbar\nabla_2}{i}\right] = 0 = \left[\mathbf{r}_2, \frac{\hbar\nabla_1}{i}\right]. \tag{7-3}$$

$\mathbf{r}_1$ and $\mathbf{p}_1$ still obey the usual commutation relation $[(\mathbf{r}_1)_i, (\mathbf{p}_1)_j] = i\hbar\delta_{ij}$, as do $\mathbf{r}_2$ and $\mathbf{p}_2$. The angular momentum of particle 1 is $\mathbf{L}_1 = \mathbf{r}_1 \times \mathbf{p}_1$ and the angular momentum of particle 2 is $\mathbf{L}_2 = \mathbf{r}_2 \times \mathbf{p}_2$.

The time rate of change of the state vector is still given by Schrödinger's equation: $i\hbar(\partial/\partial t)|\Psi(t)\rangle = H|\Psi(t)\rangle$ where H is the Hamiltonian operator for the two-particle system.

Note that all the results up to here are readily generalized to $3, 4, \ldots, 10^{23}, \ldots$ particle systems; one merely introduces more operators and bigger wave functions. There is an important point to notice about the wave function of an N-particle system: it is *not* a function over 3-dimensional space; rather, it is a function over the 3N-dimensional configuration space of the N particles. Wave functions are not in any sense tangible, like real matter fields or electromagnetic fields — they are probability amplitude functions.

Let us now consider two particles interacting via a potential $V(\mathbf{r}_1 - \mathbf{r}_2)$. The Hamiltonian operator is then

$$H = \frac{p_1^2}{2m_1} + \frac{p_2^2}{2m_2} + V(\mathbf{r}_1 - \mathbf{r}_2), \tag{7-4}$$

the analogue of the classical energy. Schrödinger's equation becomes

$$\left(-\hbar^2 \frac{\nabla_1^2}{2m_1} - \hbar^2 \frac{\nabla_2^2}{2m_2} + V(\mathbf{r}_1 - \mathbf{r}_2)\right)\psi(\mathbf{r}_1, \mathbf{r}_2, t) = i\hbar \frac{\partial}{\partial t}\psi(\mathbf{r}_1, \mathbf{r}_2, t), \tag{7-5}$$

a seven-dimensional partial differential equation! We can reduce this problem, exactly as in classical mechanics, by going over to center-of-mass and relative coordinates. Let us define the operators

$\mathbf{r} = \mathbf{r}_1 - \mathbf{r}_2$: relative position

$\mathbf{R} = \dfrac{m_1\mathbf{r}_1 + m_2\mathbf{r}_2}{m_1 + m_2}$: center-of-mass position

$\mathbf{P} = \mathbf{p}_1 + \mathbf{p}_2$: total momentum

$\mathbf{p} = \dfrac{m_2\mathbf{p}_1 - m_1\mathbf{p}_2}{m_1 + m_2}$: "relative momentum"

$m = \dfrac{m_1 m_2}{m_1 + m_2}$: reduced mass

$M = m_1 + m_2$: total mass. (7-6)

Then, as in classical mechanics,

$$\frac{p_1^2}{2m_1} + \frac{p_2^2}{2m_2} = \frac{p^2}{2m} + \frac{P^2}{2M}. \tag{7-7}$$

It is not hard to verify that, acting on a wave function, **p** becomes the differential operator $\hbar\nabla_r/i$, and **P** becomes $\hbar\nabla_R/i$. If we write $\psi(\mathbf{r}_1, \mathbf{r}_2, t) = \varphi(\mathbf{r}, \mathbf{R}, t)$, then (7-5) becomes

$$\left[-\frac{\hbar^2}{2M}\nabla_R^2 - \frac{\hbar^2}{2m}\nabla_r^2 + V(\mathbf{r})\right]\varphi(\mathbf{r}, \mathbf{R}, t) = i\hbar\frac{\partial}{\partial t}\varphi(\mathbf{r}, \mathbf{R}, t), \tag{7-8}$$

which is a separable equation, since the interaction potential is independent of the center-of-mass position. If we write

$$\varphi(\mathbf{r}, \mathbf{R}, t) = f(\mathbf{R}, t)\psi(\mathbf{r}, t), \tag{7-9}$$

then (7-10) becomes the two equations

$$-\frac{\hbar^2}{2M}\nabla_R^2 f(\mathbf{R}, t) = i\hbar\frac{\partial f(\mathbf{R}, t)}{\partial t} \tag{7-10}$$

$$\left(-\frac{\hbar^2}{2m}\nabla_r^2 + V(\mathbf{r})\right)\psi(\mathbf{r}, t) = i\hbar\frac{\partial\psi(\mathbf{r}, t)}{\partial t}. \tag{7-11}$$

The wave function $f(\mathbf{R}, t)$ tells us the probability amplitude for observing the center-of-mass of the two-particle system at point **R**. Equation (7-10) says that the center-of-mass behaves as a free particle of mass M. The center-of-mass eigenfunctions are

$$f(\mathbf{R}, t) = e^{i[\mathbf{K}\cdot\mathbf{R} - (K^2/2M)t]/\hbar}, \tag{7-12}$$

and are eigenstates of the center-of-mass momentum. Furthermore, the problem of the relative motion has reduced to the problem of a single particle of mass m in the external potential V(r).

Suppose that we have found the energy eigenstates for this problem

$$\psi(\mathbf{r}t) = e^{-iEt/\hbar}\psi(\mathbf{r}) \tag{7-13}$$

where

$$\left(-\frac{\hbar^2}{2m}\nabla^2 + V(\mathbf{r})\right)\psi(\mathbf{r}) = E\psi(\mathbf{r}), \tag{7-14}$$

and E is the energy of relative motion. Then the energy eigenstates of the two-particle problem are of the form

$$\psi(\mathbf{r}_1, \mathbf{r}_2, t) = e^{-i(E + K^2/2M)t/\hbar}\, e^{i\mathbf{K}\cdot\mathbf{R}/\hbar}\, \psi(\mathbf{r}); \tag{7-15}$$

the total energy is $E + K^2/2M$. The hard part of the problem is solving (7-14) for the relative motion.

HYDROGEN ATOM WAVE FUNCTIONS

The potential in the hydrogen atom problem is

$$V(r) = -\frac{e^2}{r}, \tag{7-16}$$

which is spherically symmetric. Thus the relative angular momentum, $\mathbf{L} = \mathbf{r} \times \mathbf{p}$, is a constant of the motion. We look, therefore, for simultaneous eigenstates of energy, L^2 and L_z. These are of the form

$$\psi(\mathbf{r}) = Y_{lm}(\theta, \varphi)R(r). \tag{7-17}$$

The radial function obeys the equation

$$\left(-\frac{\hbar^2}{2m}\frac{d^2}{dr^2} + \frac{\hbar^2 l(l+1)}{2mr^2} - \frac{e^2}{r}\right) r\,R(r) = E\,r\,R(r). \tag{7-18}$$

[Notice that $m \approx m_e$ for the hydrogen atom.] The solution is standard and can be found in any quantum mechanics book, so we merely quote the results. For each l there are an infinite number of solutions, labeled $R_{nl}(r)$, where n is an integer $\geq l + 1$, called the *principal quantum number.* The normalized solutions are

$$R_{nl}(r) = - \left\{ \left(\frac{2}{na_0}\right)^3 \frac{(n-l-1)!}{2n[(n+l)!]^3} \right\}^{1/2} e^{-r/na_0} \left(\frac{2r}{na_0}\right)^l L_{n+l}^{2l+1}\left(\frac{2r}{na_0}\right) \tag{7-19}$$

where

$$L_{n+l}^{2l+1}(\rho) = \sum_{k=0}^{n-l-1} \frac{(-1)^{k+1}[(n+l)!]^2 \rho^k}{(n-l-1-k)!\,(2l+1+k)!} \tag{7-20}$$

is an *associated Laguerre polynomial,* and a_0 is the Bohr radius,

$$a_0 = \frac{\hbar^2}{me^2} = 0.529 \times 10^{-8} \text{ cm}. \tag{7-21}$$

The energy value depends only on n, and is

$$E_n = -\frac{e^2}{2a_0}\frac{1}{n^2}; \tag{7-22}$$

this is the Lyman formula. $e^2/2a_0 = 13.6$ eV is the Rydberg. Thus a hydrogen level is specified by three quantum numbers n, l, and m, and

$$\psi_{nlm}(\mathbf{r}) = R_{nl}(r) Y_{lm}(\theta, \varphi). \tag{7-23}$$

The first few R's look like

$$R_{10} = \frac{2}{a_0^{3/2}} e^{-r/a_0} : \text{ ground state}$$

$$R_{20} = \frac{1}{(2a_0)^{3/2}}\left(2 - \frac{r}{a_0}\right) e^{-r/2a_0}$$

$$R_{21} = \frac{1}{(2a_0)^{3/2}} \frac{r}{a_0\sqrt{3}} e^{-r/2a_0}. \tag{7-24}$$

Two R_{nl} with different n but the same l are orthogonal.

For each n we can have $l = 0, 1, 2, 3, \ldots, n-1$, and each l state is $2l + 1$-fold degenerate. Thus for one energy value E_n, there are $\sum_{l=0}^{n-1}(2l + 1) = n^2$ different possible eigenfunctions. This is far more than the $2l + 1$-fold degeneracy one expects on the basis of the rotational symmetry of the potential; we shall explore the reasons for this extra, or "accidental" degeneracy, shortly.

The structure of the energy levels can be most clearly seen by displaying them as follows

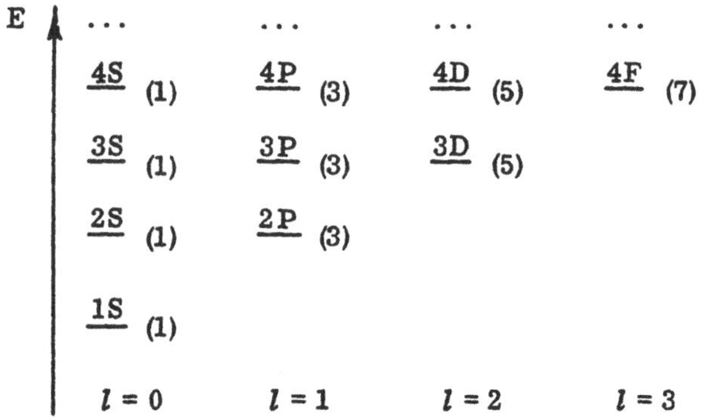

E ↑				
	...	...	...	...
	4S (1)	4P (3)	4D (5)	4F (7)
	3S (1)	3P (3)	3D (5)	
	2S (1)	2P (3)		
	1S (1)			
	$l = 0$	$l = 1$	$l = 2$	$l = 3$

The states are labeled by a number and a letter, e.g., 3P; the number is the n value and the letter is the angular momentum value. The number in parentheses is the degeneracy of the level.

Let us examine the asymptotic form of the wave function. At very large r, the potential term dominates the centrifugal barrier. There

$$\left(-\frac{\hbar^2}{2m}\frac{d^2}{dr^2} - \frac{e^2}{r}\right)u_{nl}(r) = E_n u_{nl}(r), \tag{7-25}$$

where $R_{nl}(r) = r\, u_{nl}(r)$. It is easy to verify that the asymptotic solution to (7-25) is

$$u_{nl}(r) \sim r^n e^{-r/na_0}. \tag{7-26}$$

The long range of the Coulomb potential modifies the usual asymptotic form for a short-range potential, $\sim e^{-\sqrt{-2mE}\, r/\hbar}$ by a factor r^n.

The probability density in the ground state,

$$d^3r e^{-2r/a_0} = (r^2 e^{-2r/a_0}) dr d\Omega$$

is maximum at $r = a_0$. However, because of the long tail of the wave function, the expectation value of r in the ground state is $3a_0/2$, a larger value than the most likely value.

THE SYMMETRY OF THE HYDROGEN ATOM

The extraordinary degeneracy of the hydrogen atom is closely connected with the fact that the bound orbits of the classical Kepler problem, an attractive $1/r$ potential, close on themselves; they do not precess. Thus the vector that points from the origin along the semimajor axis of the orbit (Fig. 7-1) is a constant of the motion, which is peculiar to the Kepler problem. Explicitly, this vector, called Lenz's vector, is

$$\mathbf{R} = \frac{1}{m}\,\mathbf{p}\times\mathbf{L} - \frac{e^2}{r}\,\mathbf{r}, \tag{7-27}$$

when the potential is $-e^2/r$.

In the quantum mechanical hydrogen atom problem, we can similarly form a Hermitian operator

$$\mathbf{R} = \frac{1}{2m}\,(\mathbf{p}\times\mathbf{L} - \mathbf{L}\times\mathbf{p}) - \frac{e^2}{r}\,\mathbf{r} \tag{7-28}$$

which is a constant of the motion; it is a straightforward exercise to verify that $[\mathbf{R}, H] = 0$. As we saw before, the fact that $\mathbf{L}$ was a

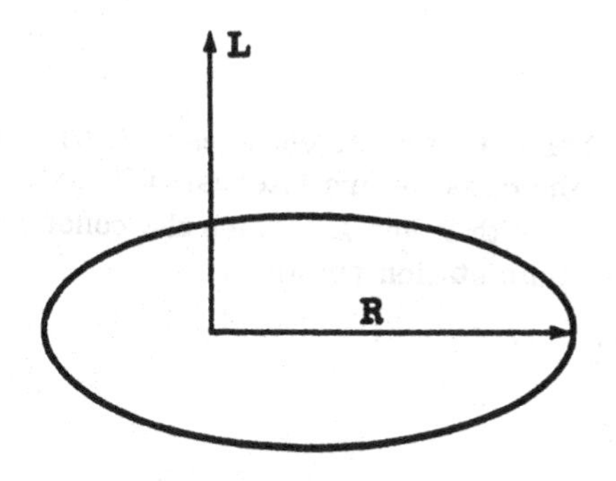

Fig. 7-1
The angular momentum L and Lenz's vector R for an elliptic orbit of the classical Kepler problem.

constant of the motion led to $2l + 1$-fold degeneracy. Similarly, we expect the existence of still another constant of the motion, **R**, to lead to further degeneracy. We shall show that, in fact, the existence of **R** explains the n^2 degeneracy of the hydrogen levels.

As classically, Lenz's vector is perpendicular to the angular momentum,

$$\mathbf{R}\cdot\mathbf{L} = \mathbf{L}\cdot\mathbf{R} = 0. \tag{7-29}$$

Furthermore the square of Lenz's vector is

$$R^2 = e^4 + \frac{2H(L^2 + \hbar^2)}{m}. \tag{7-30}$$

We see the unusual feature of the hydrogen atom problem that the Hamiltonian can be written in terms of other constants of the motion, R^2 and L^2.

Because **R** is a vector, it obeys the commutation relation with **L**

$$[R_i, L_j] = i\hbar\varepsilon_{ijk}R_k. \tag{7-31}$$

The commutation relation of **R** with itself turns out to be

$$[R_i, R_j] = i\hbar\frac{-2H}{m}\varepsilon_{ijk}L_k. \tag{7-32}$$

It is convenient to absorb the factor $-2H/m$ into **R** by defining a new vector

$$\mathbf{K} \equiv \sqrt{\frac{-m}{2H}}\,\mathbf{R}; \tag{7-33}$$

K is Hermitian when acting on eigenstates of H with *negative* eigenvalues; these are the ones we are interested in here. (The operator $\sqrt{-m/2H}$ commutes with **R** and **L**.) Then the constants of the motion **K** and **L** obey the commutation relations

$$[K_i, K_j] = i\hbar\varepsilon_{ijk}L_k \tag{7-34}$$

$$[K_i, L_j] = i\hbar\varepsilon_{ijk}K_k \tag{7-35}$$

and also

$$[L_i, L_j] = i\hbar\varepsilon_{ijk}L_k. \tag{7-36}$$

From (7-30) and (7-33) we see that we can write

$$H = -\frac{me^4}{2(K^2 + L^2 + \hbar^2)}. \tag{7-37}$$

[Note the resemblance to the formula for the energy eigenvalues.] We will now calculate the energy eigenvalues and their degeneracies from this formula and the three commutation relations above.

Let us define

$$\mathbf{M} = \frac{\mathbf{L}+\mathbf{K}}{2} \tag{7-38}$$

$$\mathbf{N} = \frac{\mathbf{L}-\mathbf{K}}{2}. \tag{7-39}$$

Then from (7-34), (7-35), and (7-36) it follows that

$$[M_i, M_j] = i\hbar\varepsilon_{ijk}M_k$$

$$[N_i, N_j] = i\hbar\varepsilon_{ijk}N_k$$

$$[M_i, N_j] = 0. \tag{7-40}$$

Hence **M** obeys the commutation relations of an angular momentum, and so does **N**, and these two "angular momenta" commute with each other. [**M** and **N** are called angular momenta because of the commutation relations they obey; they have nothing to do with freshman physics angular momentum.] In terms of **M** and **N** the Hamiltonian is

$$H = -\frac{me^4}{2(2M^2 + 2N^2 + \hbar^2)}. \tag{7-41}$$

Now because M and N obey angular momentum commutation relations we know all about their eigenstates and eigenvalues. We can find simultaneous eigenstates of M^2, M_z, N^2, and N_z, which we donote by $|\mathcal{M}, \mathcal{N}, \mu, \nu\rangle$, where

$$M^2|\mathcal{M}, \mathcal{N}, \mu, \nu\rangle = \hbar^2\mathcal{M}(\mathcal{M}+1)|\mathcal{M}, \mathcal{N}, \mu, \nu\rangle$$

$$N^2|\mathcal{M}, \mathcal{N}, \mu, \nu\rangle = \hbar^2\mathcal{N}(\mathcal{N}+1)|\mathcal{M}, \mathcal{N}, \mu, \nu\rangle$$

$$M_z|\mathcal{M}, \mathcal{N}, \mu, \nu\rangle = \hbar\mu|\mathcal{M}, \mathcal{N}, \mu, \nu\rangle$$

$$N_z|\mathcal{M}, \mathcal{N}, \mu, \nu\rangle = \hbar\nu|\mathcal{M}, \mathcal{N}, \mu, \nu\rangle. \tag{7-42}$$

$\mathcal{M}$ can take on the values $0, {}^1/_2, 1, {}^3/_2, \ldots$, as can $\mathcal{N}$. [There is no restriction to integer values because these are not orbital angular momenta in any sense.] Also

$$\mu = -\mathcal{M}, -\mathcal{M}+1, \ldots, \mathcal{M}-1, \mathcal{M}$$

$$\nu = -\mathcal{N}, -\mathcal{N}+1, \ldots, \mathcal{N}-1, \mathcal{N}. \tag{7-43}$$

Because $\mathbf{R}\cdot\mathbf{L} = 0$ we have $\mathbf{K}\cdot\mathbf{L} = 0$, and hence

$$M^2 = N^2. \tag{7-44}$$

The only states $|\mathcal{M}, \mathcal{N}, \mu, \nu\rangle$ that are relevant to the problem of finding the bound states of the hydrogen atom are therefore the ones with $\mathcal{M} = \mathcal{N}$, since operating with M^2 on these states gives the same result as operating with N^2. These states $|\mathcal{M} = \mathcal{N}, \mu, \nu\rangle$ are eigenstates of H, since from (7-41)

$$\begin{aligned} H|\mathcal{M} = \mathcal{N}, \mu, \nu\rangle &= -\frac{me^4}{2\hbar^2(4\mathcal{M}(\mathcal{M}+1)+1)}|\mathcal{M} = \mathcal{N}, \mu, \nu\rangle \\ &= -\frac{me^4}{2\hbar^2(2\mathcal{M}+1)^2}|\mathcal{M} = \mathcal{N}, \mu, \nu\rangle. \end{aligned} \tag{7-45}$$

The energy eigenvalue is therefore

$$E = -\frac{me^4}{2\hbar^2(2\mathcal{M}+1)^2}; \tag{7-46}$$

$2\mathcal{M}+1$, which can take on values 1, 2, 3, ... is the principle quantum number n.

Now let us count the degeneracy. For a fixed value of $\mathcal{M} = \mathcal{N}$, there are $2\mathcal{M}+1$ different μ values and $2\mathcal{N}+1$ different ν values. Hence there are $(2\mathcal{M}+1)^2 = n^2$ different states all with the same energy, E_n. Thus we see how the existence of the extra symmetry of the hydrogen atom, the fact that it had the extra constants of the motion $\mathbf{R}$, leads to a complete account of the n^2 degeneracy of the hydrogen levels.

It should be noted that while the states $|\mathcal{M}=\mathcal{N}, \mu, \nu\rangle$ are eigenstates of $L_z = M_z + N_z$ with eigenvalue $\hbar(\mu+\nu)/2$, always an integer, they are not eigenstates of L^2. These states are linear combinations of the $|nlm\rangle$ hydrogen atom states with fixed n and m but different l values.

PROBLEMS

1. (a) Calculate the probability that an electron in the ground state of a hydrogen atom will be found at a greater distance from the nucleus than would be allowed classically.
(b) Show that the uncertainty principle for Δx and Δp_x is obeyed by an electron in the ground state of a hydrogen atom.
(c) Make sketches of the regions in the x, z plane where the probability of finding an electron in a hydrogen atom ground state is large. Repeated for the 2S; the 2P, m = 0; and the 2P, m = 1 states.
2. Calculate for the hydrogen atom all the nonvanishing matrix elements of x between n = 2 states and the ground state.
3. A very elegant method for solving the hydrogen atom problem, due to Schwinger, involves transforming the radial equation of the hydrogen atom into the radial equation of a two-dimensional isotropic harmonic oscillator. To carry out this procedure first replace r by the variable $\lambda\rho^2/2$, where λ is a constant to be determined, and let $R_{n,l}(r)$, the hydrogen atom radial function, equal $F(\rho)/\rho$.
(a) Show that $F(\rho)$ obeys the radial equation of a two-dimensional harmonic oscillator, of frequency $\omega = \sqrt{-2\lambda^2 E/m}$, with angular momentum $2l + 1$ and energy $2e^2\lambda$. E is the energy of the hydrogen level.
(b) Deduce, from the results of Problem 5, Chapter 6, the formula for the energy levels of the hydrogen atom, and the degeneracies of these levels.
(c) Use this procedure to construct explicitly the normalized ground state wave function of the hydrogen atom.
4. (a) Show by explicit calculation that Lenz's vector is constant of the motion of the hydrogen atom, i.e., that $[\mathbf{R}, H] = 0$.
(b) Verify Eqs. (7-30), (7-31), and (7-32).
(c) Show that $\mathbf{L} \cdot \mathbf{R} = \mathbf{R} \cdot \mathbf{L} = 0$.

Chapter 8
COOPER PAIRS

Superconductivity arises from a very complicated set of interactions between the electrons in a metal. I would like to consider a very simplified model of the interaction of a pair of electrons in a metal, since on the one hand, it is an instructive example of how quantum mechanics works, and on the other hand, the main features of this model are at the starting point of the modern theory of superconductivity.

If we neglect all effects of the crystal structure, then a metal of volume V can be looked upon simply as a box filled with electrons. If we neglect all interactions between electrons, the normalized energy eigenstates, using periodic boundary conditions, are

$$\psi(\mathbf{r}) = (V)^{-1/2}\, e^{i\mathbf{k}\cdot\mathbf{r}/\hbar} \tag{8-1}$$

where $\varepsilon_k = k^2/2m$. The possible k vectors in a cubic box of side L, are given by

$$k_x = \frac{2\pi n_x \hbar}{L}, \quad k_y = \frac{2\pi n_y \hbar}{L}, \quad k_z = \frac{2\pi n_z \hbar}{L} \tag{8-2}$$

where n_x, n_y, and n_z are integers ranging from $-\infty$ to ∞.

When we fill up the box with electrons we can put two electrons in each state (two for spin). The configuration of lowest total energy for N electrons will have the states filled up to a certain maximum value, k_f, the *Fermi momentum*, which is given by

$$N = \sum_{k<k_f} 2, \tag{8-3}$$

where the sum is over all k with $k < k_f$. The filled $\mathbf{k}$ states form

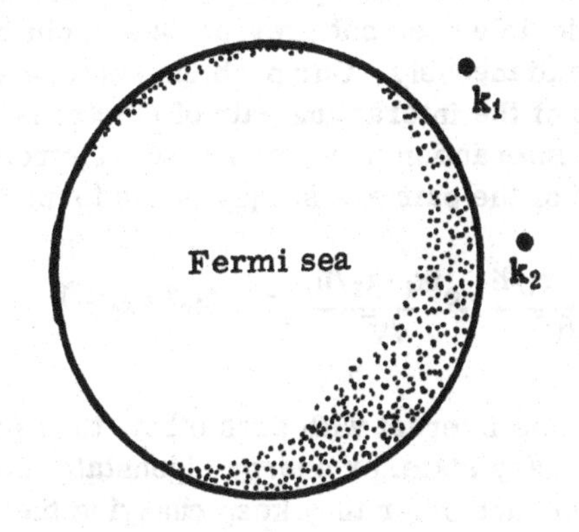

Fig. 8-1

Two noninteracting electrons outside the Fermi sea.

a sphere, called the *Fermi sea* [Fig. 8-1]. For macroscopic L and N, the **k** vectors are spaced closely enough so that we can replace the sum over **k** by an integral. Since the interval between k_x values is $2\pi\hbar/L$,

$$\sum_{k_x} \rightarrow \frac{L}{2\pi\hbar}\int dk_x$$

and

$$\sum_{\mathbf{k}} \rightarrow V\int \frac{d^3k}{(2\pi\hbar)^3}, \tag{8-4}$$

where $V = L^3$. There are $V/(2\pi\hbar)^3$ states per unit volume in momentum space. From (8-3) we then find that k_f is given by

$$k_f = (3\pi^2 n)^{1/3}\hbar \tag{8-5}$$

where $n = N/V$ is the density of particles; k_f is an *intensive* parameter.

The model we want to consider is that of two electrons just outside the surface of the Fermi sea (the Fermi surface) interacting through a weak attractive force. We shall neglect all interactions between the electrons in the Fermi sea, and between the electrons in the Fermi sea and the pair. This is where this model differs from a model of superconductivity. We shall also assume that the electrons have opposite spin values, one up and the other down, and therefore the exclusion principle does not prevent them from being in the same spatial state simultaneously. Our problem shall be to find the energy eigenstates of the interacting pair of electrons.

If there is no interaction between the two electrons, then the energy eigenstates of the pair are simply of the form

$$\psi(\mathbf{r}_1, \mathbf{r}_2, t) = \frac{e^{i\mathbf{k}_1 \cdot \mathbf{r}_1/\hbar}}{\sqrt{V}} \frac{e^{i\mathbf{k}_2 \cdot \mathbf{r}_2/\hbar}}{\sqrt{V}} e^{-i(\varepsilon_{k_1} + \varepsilon_{k_2})t/\hbar}. \tag{8-6}$$

Now if the electrons interact with each other, then (8-6) will no longer be a stationary state, or energy eigenstate, since as the particles scatter from each other they keep changing their momentum values. The states of the interacting pair will be of the form

$$\psi(\mathbf{r}_1, \mathbf{r}_2, t) = \sum_{\mathbf{k}_1\mathbf{k}_2} a_{\mathbf{k}_1\mathbf{k}_2}(t) \frac{e^{i\mathbf{k}_1 \cdot \mathbf{r}_1/\hbar}}{\sqrt{V}} \frac{e^{i\mathbf{k}_2 \cdot \mathbf{r}_2/\hbar}}{\sqrt{V}} \tag{8-7}$$

and for an energy eigenstate

$$a_{\mathbf{k}_1\mathbf{k}_2}(t) = e^{-iEt/\hbar} a_{\mathbf{k}_1\mathbf{k}_2}; \tag{8-8}$$

E is the total energy of the pair. Equation (8-7) expresses ψ as a double Fourier series in $\mathbf{r}_1$ and $\mathbf{r}_2$. The amplitude $a_{\mathbf{k}_1\mathbf{k}_2}(t)$ is the amplitude for finding particle 1 with momentum $\mathbf{k}_1$ and 2 with momentum $\mathbf{k}_2$. Because all the states inside the Fermi sea are already filled, the amplitude for finding either particle in a state inside the Fermi sea must be zero, that is, $a_{\mathbf{k}_1\mathbf{k}_2}$ is zero unless *both* $\mathbf{k}_1$ and $\mathbf{k}_2$ are greater than k_f.

How does the amplitude $a_{\mathbf{k}_1\mathbf{k}_2}(t)$ change in time? If there is no interaction then because (8-6) is a stationary state,

$$i\hbar \frac{\partial}{\partial t} a_{\mathbf{k}_1\mathbf{k}_2}(t) = (\varepsilon_{\mathbf{k}_1} + \varepsilon_{\mathbf{k}_2}) a_{\mathbf{k}_1\mathbf{k}_2}(t); \tag{8-9}$$

only the phase of the amplitude changes. When the particles interact, then if at one instant they have momenta $\mathbf{k}_1$ and $\mathbf{k}_2$, at a slightly later instant they will have an amplitude for having different momenta $\mathbf{k}_1'$ and $\mathbf{k}_2'$, because they can scatter from each other.

Thus we expect a change in the amplitude $a_{\mathbf{k}_1\mathbf{k}_2}(t)$ due to pairs with $\mathbf{k}_1'$ and $\mathbf{k}_2'$ scattering into $\mathbf{k}_1$, $\mathbf{k}_2$ [Fig. 8-2]; this term will be

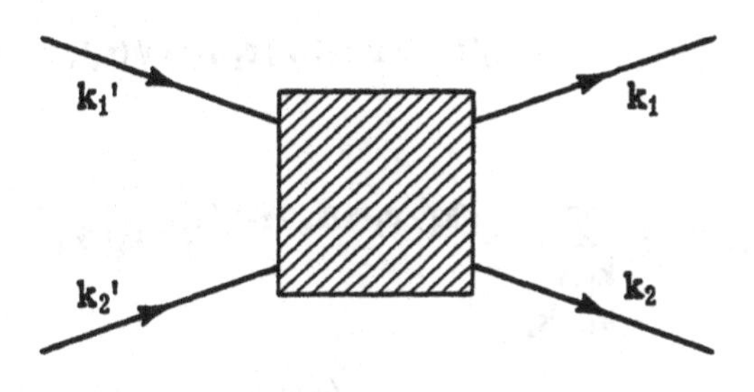

Fig. 8-2
Scattering of a pair from k_1', k_2', to k_1, k_2.

proportional to the amplitude $a_{\mathbf{k}_1'\mathbf{k}_2'}$ for the particles to have momenta $\mathbf{k}_1'$, $\mathbf{k}_2'$. This process will add a term on the right side of (8-9) of the form $\sum_{\mathbf{k}_1'\mathbf{k}_2'} \langle \mathbf{k}_1\mathbf{k}_2 | \mathcal{V} | \mathbf{k}_1'\mathbf{k}_2' \rangle a_{\mathbf{k}_1'\mathbf{k}_2'}$. The quantity $\langle \mathbf{k}_1\mathbf{k}_2 | \mathcal{V} | \mathbf{k}_1'\mathbf{k}_2' \rangle$ is called the matrix element for scattering from $\mathbf{k}_1'\mathbf{k}_2'$ to $\mathbf{k}_1\mathbf{k}_2$. With this term the Schrödinger equation for the rate of change of the amplitude becomes

$$i\hbar \frac{\partial}{\partial t} a_{\mathbf{k}_1\mathbf{k}_2}(t) = (\varepsilon_{\mathbf{k}_1} + \varepsilon_{\mathbf{k}_2}) a_{\mathbf{k}_1\mathbf{k}_2}(t) + \sum_{\mathbf{k}_1'\mathbf{k}_2'} \langle \mathbf{k}_1\mathbf{k}_2 | \mathcal{V} | \mathbf{k}_1'\mathbf{k}_2' \rangle \, a_{\mathbf{k}_1'\mathbf{k}_2'}(t). \tag{8-10}$$

It looks as if we've left out a change in the amplitude due to particles with $\mathbf{k}_1\mathbf{k}_2$ scattering to other states $\mathbf{k}_1'\mathbf{k}_2'$. This term, being proportional to $a_{\mathbf{k}_1\mathbf{k}_2}$, has the same structure as the "diagonal" term $\langle \mathbf{k}_1\mathbf{k}_2 | \mathcal{V} | \mathbf{k}_1\mathbf{k}_2 \rangle a_{\mathbf{k}_1\mathbf{k}_2}$ in (8-10), so we can regard it as being already included in (8-10).

One can regard the amplitudes $\langle \mathbf{k}_1\mathbf{k}_2 | \mathcal{V} | \mathbf{k}_1'\mathbf{k}_2' \rangle$ as being the matrix elements of an interaction operator $\mathcal{V}$ in the "$\mathbf{k}_1$, $\mathbf{k}_2$ basis." In a very complicated system, the matrix elements $\langle \mathbf{k}_1\mathbf{k}_2 | \mathcal{V} | \mathbf{k}_1'\mathbf{k}_2' \rangle$ are hard to determine; usually one has to make educated guesses and see how well the results agree with experiment.

One immediate requirement on the matrix elements is that the total probability $\sum_{\mathbf{k}_1\mathbf{k}_2} |a_{\mathbf{k}_1\mathbf{k}_2}(t)|^2$ not change in time. This implies, as usual, that

$$\langle \mathbf{k}_1\mathbf{k}_2 | \mathcal{V} | \mathbf{k}_1'\mathbf{k}_2' \rangle^* = \langle \mathbf{k}_1'\mathbf{k}_2' | \mathcal{V} | \mathbf{k}_1\mathbf{k}_2 \rangle; \tag{8-11}$$

$\mathcal{V}$ must be a Hermitian matrix.

It is illuminating to write (8-10) in terms of the wave function $\psi(\mathbf{r}_1, \mathbf{r}_2, t)$. The result is

$$i\hbar \frac{\partial}{\partial t}\psi(\mathbf{r}_1,\mathbf{r}_2,t) = \left(-\frac{\hbar^2}{2m}\nabla_1{}^2 - \frac{\hbar^2}{2m}\nabla_2{}^2\right)\psi(\mathbf{r}_1,\mathbf{r}_2,t)$$
$$+\int d\mathbf{r}_1'd\mathbf{r}_2'\langle \mathbf{r}_1\mathbf{r}_2|\mathcal{V}|\mathbf{r}_1'\mathbf{r}_2'\rangle\,\psi(\mathbf{r}_1',\mathbf{r}_2',t), \tag{8-12}$$

where

$$\langle \mathbf{r}_1\mathbf{r}_2|\mathcal{V}|\mathbf{r}_1'\mathbf{r}_2'\rangle = \frac{1}{V^2}\sum_{\substack{\mathbf{k}_1,\mathbf{k}_2\\ \mathbf{k}_1',\mathbf{k}_2'}} e^{(i\mathbf{k}_1\cdot\mathbf{r}_1 + i\mathbf{k}_2\cdot\mathbf{r}_2)/\hbar}\langle \mathbf{k}_1\mathbf{k}_2|\mathcal{V}|\mathbf{k}_1'\mathbf{k}_2'\rangle$$
$$\times\, e^{(-i\mathbf{k}_1'\cdot\mathbf{r}_1' - i\mathbf{k}_2'\cdot\mathbf{r}_2')/\hbar} \tag{8-13}$$

Equation (8-12) is in the form of a Schrödinger equation, only with a *nonlocal* potential $\langle \mathbf{r}_1\mathbf{r}_2|\mathcal{V}|\mathbf{r}_1'\mathbf{r}_2'\rangle$. The rate of change of the amplitude for the particles being at $\mathbf{r}_1$, $\mathbf{r}_2$ depends on the amplitude for the particles being at far away points as well as close points. If the interaction can be represented simply by a potential $v(\mathbf{r}_1 - \mathbf{r}_2)$, as in the hydrogen atom, then

$$\langle \mathbf{r}_1\mathbf{r}_2|\mathcal{V}|\mathbf{r}_1'\mathbf{r}_2'\rangle = \delta(\mathbf{r}_1-\mathbf{r}_1')\delta(\mathbf{r}_2-\mathbf{r}_2')v(\mathbf{r}_1-\mathbf{r}_2). \tag{8-14}$$

In this case

$$\langle \mathbf{k}_1\mathbf{k}_2|\mathcal{V}|\mathbf{k}_1'\mathbf{k}_2'\rangle = \frac{1}{V}v(\mathbf{k}-\mathbf{k}')\delta_{\mathbf{K},\mathbf{K}'} \tag{8-15}$$

where $\mathbf{k} = (\mathbf{k}_1 - \mathbf{k}_2)/2$, $\mathbf{k}' = (\mathbf{k}_1' - \mathbf{k}_2')/2$, $\mathbf{K} = \mathbf{k}_1 + \mathbf{k}_2$ and $\mathbf{K}' = \mathbf{k}_1' + \mathbf{k}_2'$; also $v(\mathbf{q}) = \int d^3r e^{-i\mathbf{q}\cdot\mathbf{r}}v(\mathbf{r})$. The δ symbol guarantees that the total momentum of the pair is conserved in the interaction.

Generally, if the total momentum of the pair of particles is conserved in the interaction the matrix element connects only states with the same total momentum and thus it has the form

$$\langle \mathbf{k}_1\mathbf{k}_2|\mathcal{V}|\mathbf{k}_1'\mathbf{k}_2'\rangle = \mathcal{V}_{\mathbf{k},\mathbf{k}'}(\mathbf{K})\delta_{\mathbf{K},\mathbf{K}'}. \tag{8-16}$$

Then the nonlocal potential has the form

$$\langle \mathbf{r}_1\mathbf{r}_2|\mathcal{V}|\mathbf{r}_1'\mathbf{r}_2'\rangle = \langle \mathbf{r}|\mathcal{V}(\mathbf{R}-\mathbf{R}')|\mathbf{r}'\rangle, \tag{8-17}$$

where $\mathbf{R} = (\mathbf{r}_1 + \mathbf{r}_2)/2$, $\mathbf{R}' = (\mathbf{r}_1' + \mathbf{r}_2')/2$, $\mathbf{r} = \mathbf{r}_1 - \mathbf{r}_2$ and $\mathbf{r}' = \mathbf{r}_1' - \mathbf{r}_2'$ are the center-of-mass and relative coordinates.

In an energy eigenstate, when total momentum is conserved

$$a_{\mathbf{k}_1,\mathbf{k}_2}(t) = a_{\mathbf{k}}(\mathbf{K})e^{-iEt/\hbar}, \tag{8-18}$$

and (8-10) becomes

$$(E - \varepsilon_{\mathbf{k}_1} - \varepsilon_{\mathbf{k}_2})a_{\mathbf{k}}(\mathbf{K}) = \sum_{\mathbf{k}'} \mathcal{V}_{\mathbf{k},\mathbf{k}'}(\mathbf{K})a_{\mathbf{k}'}(\mathbf{K}). \tag{8-19}$$

The total momentum appears in this equation only as a parameter. For a general interaction $\mathcal{V}_{\mathbf{k},\mathbf{k}'}(\mathbf{K})$, Eq. (8-19) is very hard to solve.

In a metal one has a repulsive Coulomb interaction between the electrons, together with an attractive interaction between the electrons and ions. Now when an electron moves, it tends to pull the ions toward it. Of course, the ions don't get very far because they are very heavy and are bound to the vicinity of their lattice sites. However, when an ion moves toward the electron, the other electrons tend to follow the ion, and therefore the first electron. This means that because of the presence of the ions, there is some tendency for electrons to attract each other. The total interaction between any two electrons is a composite of this attraction plus the Coulomb repulsion. In some metals, this total interaction is attractive for electrons near the Fermi surface. To a first approximation we can represent this interaction by

$$\mathcal{V}_{\mathbf{k},\mathbf{k}'}(\mathbf{K}) = \begin{cases} -\dfrac{v_0}{V}: & k_f < k_1, k_2, k_1', k_2' < k_a \\ 0: & \text{otherwise} \end{cases} \tag{8-20}$$

where v_0 is positive, and k_a is a momentum slightly greater than k_f.

With this interaction, the Schrödinger equation becomes

$$(E - \varepsilon_{\mathbf{k}_1} - \varepsilon_{\mathbf{k}_2})a_{\mathbf{k}}(\mathbf{K}) = -\frac{v_0}{V}\sum_{\mathbf{k}'}{}' a_{\mathbf{k}'}(\mathbf{K}), \tag{8-21}$$

where the prime on the sum indicates that only values of k' between k_f and k_a, and such that $k_f < |(\mathbf{K}/2) \pm \mathbf{k}'| < k_a$ are to be summed over.[1] To solve (8-21) we divide both sides by $E - \varepsilon_{\mathbf{k}_1} - \varepsilon_{\mathbf{k}_2}$ and sum over the allowed values of **k**. Then

[1] For an s-state, for which $a_{\mathbf{k}}(\mathbf{K})$ is independent of the direction of **k**, the sum over $\mathbf{k}'$ cannot vanish. Some $a_{\mathbf{k}}(\mathbf{K})$ must be nonzero else $\psi \equiv 0$. Then if the sum vanished, Eq. (8-21) would imply that for this **k**, $E = \epsilon_{k1} + \epsilon_{k2} = (k^2/m) + (K^2/4m)$. However only one **k** can satisfy this condition, so that $a_{\mathbf{k}}(\mathbf{K})$ can be nonzero only for one **k**. Thus the only non-zero terms in the sum are all equal, whereupon the sum over $\mathbf{k}'$ must be nonzero.

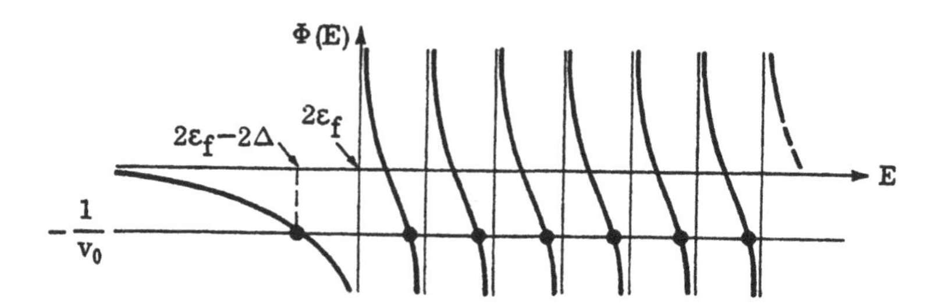

Fig. 8-3
Graphical solution of Eq. (8-24).

$$\sum_{\mathbf{k}}{}' a_{\mathbf{k}}(\mathbf{K}) = -\frac{v_0}{V}\sum_{\mathbf{k}}{}' \frac{1}{E-\varepsilon_{\mathbf{k}_1}-\varepsilon_{\mathbf{k}_2}}\sum_{\mathbf{k}'}{}' a_{\mathbf{k}'}(\mathbf{K}). \tag{8-22}$$

Canceling the sum over $a_{\mathbf{k}}$ from both sides we find a condition to determine the possible eigenvalues E:

$$1 = -\frac{v_0}{V}\sum_{\mathbf{k}}{}' \frac{1}{E-\varepsilon_{\mathbf{k}_1}-\varepsilon_{\mathbf{k}_2}}. \tag{8-23}$$

The nature of the solutions can be seen graphically. Let

$$\Phi(E) = \frac{1}{V}\sum_{\mathbf{k}}{}' \frac{1}{E-\varepsilon_{\mathbf{k}_1}-\varepsilon_{\mathbf{k}_2}}. \tag{8-24}$$

The eigenvalue condition is thus $\Phi(E) = -1/v_0$. $\Phi(E)$ has a pole at each possible energy, $\varepsilon_{\mathbf{k}_1} + \varepsilon_{\mathbf{k}_2}$, of a noninteracting pair of electrons outside the Fermi sea, with total momentum $\mathbf{K}$, and k_1 and k_2 between k_f and k_a; these energies are at least $2\varepsilon_f$, where $\varepsilon_f = k_f^2/2m$. Thus $\Phi(E)$ looks as shown in Fig. 8-3. We see that $\Phi(E)$ intersects $-1/v_0$ at many places above the minimum energy of a noninteracting pair; these energies correspond to states qualitatively like the noninteracting states (8-6). Also, for v_0 positive, there is always one intersection at an energy, E_b, below the minimum. This means that the interaction has produced a "bound state" of the two electrons; this new state is quite different from the noninteracting states (8-6).

Let us solve for E_b in the case $\mathbf{K} = 0$. Then $\mathbf{k}_1 = \mathbf{k}$, $\mathbf{k}_2 = -\mathbf{k}$, and the restriction that k_1 and k_2 be outside the Fermi surface is trivial to handle; replacing the sum by an integral we have

$$\Phi(E) = \int_{k_f}^{k_a} \frac{d^3k}{(2\pi\hbar)^3} \frac{1}{E - 2\varepsilon_k}, \tag{8-25}$$

$$= \frac{m}{2\pi^2\hbar^3} \int_{\varepsilon_f}^{\varepsilon_a} k \, d\varepsilon_k \frac{1}{E - 2\varepsilon_k}, \tag{8-26}$$

where $\varepsilon_a = k_a^2/2m$. For ε_a close to ε_f we can replace the k inside the integrand by k_f, and we find, for $E < 2\varepsilon_f$,

$$\Phi(E) = -\frac{N(0)}{2} \ln \left| \frac{2\varepsilon_a - E}{2\varepsilon_f - E} \right|, \tag{8-27}$$

where

$$N(0) = \frac{mk_f}{2\pi^2\hbar^3} = \int \frac{d^3k}{(2\pi\hbar)^3} \delta(\varepsilon_k - \varepsilon_f) \tag{8-28}$$

is the density of states at the Fermi surface. Equating $\Phi(E)$ to $-1/v_0$, and writing

$$E_b = 2\varepsilon_f - 2\Delta$$

we find

$$\Delta = \frac{\varepsilon_a - \varepsilon_f}{e^{2/v_0 N(0)} - 1} \tag{8-29}$$

The energy $\varepsilon_a - \varepsilon_f$ is on the order of the Debye energy, ω_D, which is $\hbar$ times the maximum frequency of a lattice oscillation in the metal. Typically $\omega_D/\varepsilon_f \sim 1/100$, and $v_0 N(0) \sim 1/4$. Thus the binding energy per electron in this bound state is

$$\Delta \approx \omega_D e^{-2/v_0 N(0)}. \tag{8-30}$$

Notice that Δ is a highly nonanalytic function of the parameter v_0. This mechanism for binding electrons together, which is called "pairing," was discovered by Cooper; electrons in such bound states are called Cooper pairs.[2]

The number of possible **k** values allowed in (8-24) drops sharply as **K** becomes different from zero. The poles of $\Phi(E)$ are at the energy values of the possible noninteracting pairs but because there

[2] L.N. Cooper *Phys. Rev.* **104**, 1189 (1956).

are fewer $\mathbf{k}$ values allowed, $\Phi(E)$ is smaller in magnitude for E less than the minimum singularity. This means that for $\mathbf{K} \neq 0$ the intersection of $\Phi(E)$ with $-1/v_0$ lies closer to the minimum noninteracting energy value of the pair than for $\mathbf{K} = 0$, and thus the binding energy of a bound pair decreases, rapidly in fact, with increasing $\mathbf{K}$. The point is that the fewer the number of states connected together by the attractive interaction $\langle \mathbf{k}_1\mathbf{k}_2 | \mathcal{V} | \mathbf{k}_1'\mathbf{k}_2' \rangle$, the smaller will be the binding energies that result. The biggest binding energy occurs for electrons diametrically opposite each other around the Fermi sea.

To find the wave function of a Cooper pair we notice that from (8-21)

$$a_{\mathbf{k}}(\mathbf{K}) = \frac{1}{E - \varepsilon_{k_1} - \varepsilon_{k_2}} \times \text{constant} \tag{8-31}$$

Thus, from (8-7), keeping $\mathbf{K}$ fixed in the sum

$$\psi(\mathbf{r}_1, \mathbf{r}_2) \sim e^{i\mathbf{K}\cdot(\mathbf{r}_1+\mathbf{r}_2)/2\hbar} \frac{1}{V} \sum_{\mathbf{k}}{}' \frac{e^{i\mathbf{k}\cdot(\mathbf{r}_1-\mathbf{r}_2)/\hbar}}{E - \varepsilon_{\mathbf{k}_1} - \varepsilon_{\mathbf{k}_2}}. \tag{8-32}$$

The relative wave function $\varphi(r)$ is thus

$$\varphi(r) \sim \int' \frac{d^3k}{(2\pi\hbar)^3} \frac{e^{i\mathbf{k}\cdot\mathbf{r}/\hbar}}{E - \varepsilon_{k_1} - \varepsilon_{k_2}}. \tag{8-33}$$

For the $K = 0$ bound pair $\varphi(r)$ behaves roughly as $(1/r)\sin(k_f r/\hbar)$ times a slowly varying function of r similar to $\sin(m\Delta r/2k_f\hbar)$. The length $\xi = 2k_f\hbar/m\Delta$ is essentially the size of bound pair. The wave function $\varphi(r)$ is spherically symmetric, and therefore the pair has angular momentum zero; it is in an s-state.

To explain superconductivity one must take into account the interactions between *all* the electrons at the same time. Then one finds that all the electrons at the Fermi surface form pairs with the same total momentum. When this momentum is different from zero there is a net current – a supercurrent. In order to slow down this current one can't simply slow down the electrons one at a time, as in a normal conductor, because they are all paired together to the same total momentum; one has to slow down all the electrons at the same time. Because this is extraordinarily difficult to do, supercurrents persist indefinitely.[3]

---

[3] See, e.g., J.R. Schrieffer, *Theory of Superconductivity* [W.A. Benjamin, New York, 1964].

A few final comments about the nonlocal potential $\langle \mathbf{r}_1\mathbf{r}_2|\mathcal{V}|\mathbf{r}_1'\mathbf{r}_2'\rangle$: If we assume that the relative angular momentum of the pair of particles is conserved in their interaction, then if in the Schrödinger equation (8-12) $\psi(\mathbf{r}_1, \mathbf{r}_2, t)$ is an eigenstate of angular momentum, so must be the term

$$\int d^3r_1' \, d^3r_2' \, \langle \mathbf{r}_1\mathbf{r}_2|\mathcal{V}|\mathbf{r}_1'\mathbf{r}_2'\rangle \psi(\mathbf{r}_1', \mathbf{r}_2', t),$$

in order that the angular momentum of the pair be constant in time. This implies that $\langle \mathbf{r}_1\mathbf{r}_2|\mathcal{V}|\mathbf{r}_1'\mathbf{r}_2'\rangle$ must be of the form, when total momentum is also conserved

$$\langle \mathbf{r}_1, \mathbf{r}_2|\mathcal{V}|\mathbf{r}_1', \mathbf{r}_2'\rangle = \sum_{l=0}^{\infty} \sum_{m=-l}^{l} Y_{lm}(\Omega_r) Y_{lm}(\Omega_r') \mathcal{V}_l(\mathbf{R}-\mathbf{R}', |\mathbf{r}|, |\mathbf{r}'|) \tag{8-34}$$

where Ω_r denotes the angles of the vector $\mathbf{r}$. Then Fourier transforming, and using formula (6-85) we see that $\mathcal{V}_{\mathbf{k},\mathbf{k}'}(\mathbf{K})$ must be of the form

$$\mathcal{V}_{\mathbf{k},\mathbf{k}'}(\mathbf{K}) = \sum_{l=0}^{\infty} \sum_{m=-l}^{l} Y_{lm}(\Omega_k) Y_{lm}(\Omega_k') \mathcal{V}_{l,k,k'}(\mathbf{K}). \tag{8-35}$$

$\mathcal{V}_{l,k,k'}(\mathbf{K})$ is the matrix element for the scattering from a state with magnitude of relative momentum k' and relative angular momentum l to one of relative momentum k and angular momentum l. Looking back at the interaction (8-20) we see that it is an interaction only between particles in relative s-states, which explains why the wave function of the Cooper pair also was an s-state.

In addition to electrons in metals, pairing takes place in liquid He^3 below 0.0027°K, in p-states. Nucleons outside closed shells in nuclei, as well as bulk nuclear matter in neutron stars, can also undergo pairing.

PROBLEMS

1. Calculate $\langle r^2 \rangle$ for the $K = 0$ bound pair.
2. To see the role played by the Fermi sea in the Cooper pair problem, suppose that $k_f = 0$. What is then the exact condition on v_0 that there be a bound state ($E < 0$) for $K = 0$?

3. Estimate as a function of K the volume of **k**-space entering the sum (8-24) for $K \neq 0$.
4. Find the possible bound state energies, and eigenfunctions for a Cooper pair with total momentum zero, if the potential is attractive and constant in d-states when both particles are in a thin shell about the Fermi surface, and otherwise zero.

Chapter 9
POTENTIAL SCATTERING

WAVE PACKETS[1]

We want to consider the scattering of particles by a fixed short-ranged potential; this excludes the case of a Coulomb potential, which requires separate treatment. Suppose, as in Fig. 9-1, that the source of particles is located far to the left of the target, and that a particle emitted by the source at early time t_0 is in a wave packet state

$$\psi(\mathbf{r}, t_0) = \int \frac{d^3k}{(2\pi)^3} e^{i\mathbf{k}\cdot\mathbf{r}} a_{\mathbf{k}}. \tag{9-1}$$

We assume that this wave packet is localized well out of the range

Fig. 9-1
Wave packet of average momentum $\hbar\mathbf{k}_0$ incident on a target.

---

[1]More detailed descriptions can be found, e.g., in E. Merzbacher, *Quantum Mechanics* (John Wiley and Sons. Inc., New York, 1961) and K. Gottfried. *Quantum Mechanics, Volume I* (W. A. Benjamin, Inc., New York, 1966).

of the potential; on the other hand we assume that $a_{\mathbf{k}}$ is reasonably sharply peaked about a wave vector $\mathbf{k}_0$, as in Fig. 3-3. The wave packet then travels with velocity $\hbar\mathbf{k}_0/m$ toward the target. Our problem is to find the wave function $\psi(\mathbf{r}, t)$ at a time after the particle has interacted with the target.

The method for solving this problem is very similar to that used to solve the analogous one-dimensional problem of the reflection and transmission of wave packets by potential barriers. First we construct the exact eigenstates, $\psi_{\mathbf{k}}(\mathbf{r})$, of the potential problem. These satisfy

$$\left(\frac{\hbar^2}{2m}\nabla^2 + E_k\right)\psi_{\mathbf{k}}(\mathbf{r}) = v(\mathbf{r})\psi_{\mathbf{k}}(\mathbf{r}), \tag{9-2}$$

where $E_k = \hbar^2k^2/2m$. All values of $E_k > 0$ are possible eigenvalues. Then we expand $\psi(\mathbf{r}, t_0)$ in terms of these eigenstates

$$\psi(\mathbf{r}, t_0) = \int \frac{d^3k}{(2\pi)^3}\,\psi_{\mathbf{k}}(\mathbf{r})A_{\mathbf{k}}. \tag{9-3}$$

Only states with energy $E > 0$ enter this expansion, since the wave functions of the bound states of the potential go to zero rapidly at great distances from the potential. Now not all eigenfunctions with $E > 0$ enter the expansion (9-3); the wave function, $\psi(\mathbf{r}, t)$, consists of a part incident on the target plus a part that has been scattered by the target. This latter part is a wave moving outward from the target. Therefore, the eigenfunctions that we need in (9-3) are those in the form of an incident plane wave plus an outgoing scattered wave. Once we have determined the expansion (9-3), then, because the ψ_k are energy eigenfunctions, at any later time ψ is given by

$$\psi(\mathbf{r}, t) = \int \frac{d^3k}{(2\pi)^3}\,\psi_{\mathbf{k}}(\mathbf{r})A_{\mathbf{k}}\,e^{-iE_k(t-t_0)/\hbar}; \tag{9-4}$$

in principle, this is the solution to the scattering problem. Now let us work out the details.

To construct the eigenstates, $\psi_{\mathbf{k}}$, we introduce the Green's function, $G(\mathbf{r}, k)$, for the free particle Schrödinger equation; this is defined by

$$\left(\frac{\hbar^2}{2m}\nabla^2 + E_k\right)G(\mathbf{r}, k) = \delta(\mathbf{r}). \tag{9-5}$$

In terms of this Green's function, the Schrödinger equation (9-2) becomes an integral equation

$$\psi_{\mathbf{k}}(\mathbf{r}) = \varphi_0(\mathbf{r}) + \int d^3r' \, G(\mathbf{r}-\mathbf{r}', k)\, v(\mathbf{r}')\psi_{\mathbf{k}}(\mathbf{r}'), \tag{9-6}$$

as can be verified by operating with $\hbar^2\nabla^2/2m + E_k$ on both sides. The function $\varphi_0(\mathbf{r})$ is a solution to the homogeneous, that is, free particle Schrödinger equation.

Equation (9-6) can be interpreted as follows. The total wave $\psi_{\mathbf{k}}(\mathbf{r})$ is made up of an incident wave $\varphi_0(\mathbf{r})$ plus a scattered wave. In the second term, the amplitude of the wave at $\mathbf{r}'$ times the potential at $\mathbf{r}'$ acts as a source for the scattered wave; $G(\mathbf{r}-\mathbf{r}')$ is the amplitude of the scattered wave at $\mathbf{r}$ due to a unit source at $\mathbf{r}'$. The total scattered wave is given by summing over all source points $\mathbf{r}'$. It is clear from this interpretation that $G(\mathbf{r}-\mathbf{r}')$ must produce outgoing, not incoming, waves.

The Green's function is given explicitly by the integral

$$G(\mathbf{r}, k) = \int \frac{d^3p}{(2\pi)^3} \frac{e^{i\mathbf{p}\cdot\mathbf{r}}}{E_k - \hbar^2p^2/2m} = -\frac{m}{2\pi^2 ir\hbar^2} \int_{-\infty}^{\infty} \frac{p\,dp\, e^{ipr}}{p^2-k^2}, \tag{9-7}$$

for if we operate on the first integral with $\hbar^2\nabla^2/2m + E_k$ we immediately get $\delta(\mathbf{r})$. The integrand is singular at $p = \pm k$, so we must specify how the contour of integration is to behave at these singularities. The contour that will give an outgoing wave in the second term in (9-6) is just that in Fig. 4-13. Then since r is positive, we can close the contour of integration in the upper half-plane, picking up the contribution from the pole at k only. Thus, with this choice of contour

$$G(r, k) = -\frac{m}{2\pi^2 ir\hbar^2}(2\pi i)\frac{ke^{ikr}}{2k} = -\frac{m}{2\pi\hbar^2}\frac{e^{ikr}}{r}, \tag{9-8}$$

which is an outgoing wave. [Verify explicitly that (9-8) solves (9-5).]

Substituting (9-8) into (9-6) we have

$$\psi_{\mathbf{k}}(\mathbf{r}) = e^{i\mathbf{k}\cdot\mathbf{r}} - \frac{m}{2\pi\hbar^2}\int d^3r' \frac{e^{ik|\mathbf{r}-\mathbf{r}'|}}{|\mathbf{r}-\mathbf{r}'|} v(\mathbf{r}')\psi_{\mathbf{k}}(\mathbf{r}'); \tag{9-9}$$

we are interested in the solutions for which $\varphi_0(r) = e^{i\mathbf{k}\cdot\mathbf{r}}$, an incident plane. On multiplying (9-9) by $e^{-iE_kt/\hbar}$, we see explicitly that the wave fronts in the second term move outward from the potential. Note that the magnitude of the outgoing wave vector is also k; energy is conserved in potential scattering.

The detectors of the scattered particle are located far away from the potential. We therefore need the form of the eigenstates in the far field. For $\mathbf{r}$ far from the potential we can write

$$k|\mathbf{r}-\mathbf{r}'| \approx kr - k\mathbf{r}'\cdot\hat{\mathbf{r}} = kr - \mathbf{k}'\cdot\mathbf{r}', \tag{9-10}$$

where $\mathbf{k}' = k\hat{\mathbf{r}}$ is the wave vector as seen in the far field. Thus the asymptotic solutions are of the form

$$\psi_{\mathbf{k}}(\mathbf{r}) = e^{i\mathbf{k}\cdot\mathbf{r}} + \frac{e^{ikr}}{r} f_{\mathbf{k}}(\Omega_r), \tag{9-11}$$

where Ω_r specifies the direction of $\mathbf{r}$ and

$$f_{\mathbf{k}}(\Omega_r) = -\frac{m}{2\pi\hbar^2}\int d^3r'\; e^{-i\mathbf{k}'\cdot\mathbf{r}'}\, v(\mathbf{r}')\psi_{\mathbf{k}}(\mathbf{r}'); \tag{9-12}$$

$f_{\mathbf{k}}(\Omega_r)$ is known as the *scattering amplitude*, and has dimensions of length. The final term in (9-11) is essentially the amplitude $f_{\mathbf{k}}(\Omega_r)$ that an incident particle with momentum $\mathbf{k}$ will be aimed in direction Ω_r by the target times the amplitude e^{ikr}/r that it will reach $\mathbf{r}$ from the region of the target.

We must now expand $\psi(\mathbf{r}, t_0)$ in terms of the eigenstates $\psi_{\mathbf{k}}(\mathbf{r})$. This turns out to be no problem at all, since, as we shall show, the coefficients $A_{\mathbf{k}}$ in (9-3) are in fact equal to the Fourier coefficients $a_{\mathbf{k}}$ in (9-1). To show this, let us use (9-9) to write (9-1) as

$$\psi(\mathbf{r}, t_0) = \int \frac{d^3k}{(2\pi)^3} a_{\mathbf{k}} \left[\psi_{\mathbf{k}}(\mathbf{r}) + \frac{m}{2\pi\hbar^2}\int d^3r' \,\frac{e^{ik|\mathbf{r}-\mathbf{r}'|}}{|\mathbf{r}-\mathbf{r}'|} v(\mathbf{r}')\psi_{\mathbf{k}}(\mathbf{r}')\right]. \tag{9-13}$$

The quantity in the square brackets is just a fancy way of writing $e^{i\mathbf{k}\cdot\mathbf{r}}$. The second term is proportional to

$$\int \frac{d^3k}{(2\pi)^3} a_{\mathbf{k}}\, e^{ik|\mathbf{r}-\mathbf{r}'|}\psi_{\mathbf{k}}(\mathbf{r}'). \tag{9-14}$$

Since $a_{\mathbf{k}}$ is peaked about $\mathbf{k}_0$, we can, except at a sharp scattering resonance where $\psi_{\mathbf{k}}$ varies rapidly with $\mathbf{k}$, replace $\psi_{\mathbf{k}}(\mathbf{r}')$ by $\psi_{\mathbf{k}_0}(\mathbf{r}')$. Also, since the $\mathbf{k}$ vectors are all near $\mathbf{k}_0$ we can approximate k by $\mathbf{k}\cdot\hat{\mathbf{k}}_0$. Then (9-14) becomes

$$\int \frac{d^3k}{(2\pi)^3} a_{\mathbf{k}}\, e^{i\mathbf{k}\cdot\hat{\mathbf{k}}_0|\mathbf{r}-\mathbf{r}'|}\psi_{\mathbf{k}_0}(\mathbf{r}') = \psi(\hat{\mathbf{k}}_0|\mathbf{r}-\mathbf{r}'|, t_0)\; \psi_{\mathbf{k}_0}(\mathbf{r}'). \tag{9-15}$$

But $\hat{\mathbf{k}}_0\,|\mathbf{r}-\mathbf{r}'|$ is a vector to the right of the potential, and we know that the initial wave packet vanishes there. Thus (9-15) equals zero, and (9-13) becomes simply

$$\psi(\mathbf{r}, t_0) = \int \frac{d^3k}{(2\pi)^3}\, a_{\mathbf{k}}\, \psi_{\mathbf{k}}(\mathbf{r}). \tag{9-16}$$

The expansion coefficients for our initial wave packet are the same whether we use plane waves, or exact energy eigenstates. This is the point in the argument where the fact is used that the incident particle is approaching, rather than receding from, the target.

At a later time the wave packet is given by

$$\psi(\mathbf{r}, t) = \int \frac{d^3k}{(2\pi)^3}\, a_{\mathbf{k}} \psi_{\mathbf{k}}(\mathbf{r})\, e^{-iE_k(t-t_0)/\hbar}. \tag{9-17}$$

For $\mathbf{r}$ far from the potential we can use the asymptotic form (9-11) for $\psi_{\mathbf{k}}$ in this expression, and find

$$\psi(\mathbf{r}, t) = \psi_0(\mathbf{r}, t) + \int \frac{d^3k}{(2\pi)^3}\, a_{\mathbf{k}} \frac{e^{i(kr - E_k(t-t_0)/\hbar)}}{r} f_{\mathbf{k}}(\Omega_r), \tag{9-18}$$

where

$$\psi_0(\mathbf{r}, t) = \int \frac{d^3k}{(2\pi)^3}\, a_{\mathbf{k}}\, e^{i\mathbf{k}\cdot\mathbf{r} - iE_k(t-t_0)/\hbar} \tag{9-19}$$

is the wave packet that we would have at time t were there no potential present. Also in the second term of (9-18) we assume $f_{\mathbf{k}}(\Omega_r)$ to be slowly varying around $\mathbf{k}_0$ and replace it by $f_{\mathbf{k}_0}(\Omega_r)$. Also we replace k by $\mathbf{k}\cdot\hat{\mathbf{k}}_0$ here. Then

$$\psi(\mathbf{r}, t) = \psi_0(\mathbf{r}, t) + \frac{f_{\mathbf{k}_0}(\Omega_r)}{r}\, \psi_0(\hat{\mathbf{k}}_0 r, t). \tag{9-20}$$

This equation says that the total wave function after scattering is given by the wave packet one would have at the later time t if there were no scattering, plus a scattered term. The structure of the scattered term is very intuitive, $\psi_0(\hat{\mathbf{k}}_0 r, t)$ is the value the wave function would have at point $\mathbf{r}$, if all the potential did was to bend the path of the particle from the forward direction, toward $\mathbf{r}$. This amplitude is multiplied by $f_{\mathbf{k}_0}(\Omega_r)/r$, which is the probability amplitude that the potential did just this bending of the path of the particle.

Equation (9-20) is *not* a valid expression for the wave function in the asymptotic region in two situations: the first is if there is a sharp scattering resonance at energy $\hbar^2k^2/2m$; then there is considerable distortion of the wave packet in scattering, similar to the distortion we found at a transmission resonance for a one-dimensional potential. The outgoing wave then has the exponential structure of Fig. (4-12), times a factor $1/r$ for three dimensions. The second is in the case of the Coulomb potential which is so long ranged that the asymptotic form (9-11) is incorrect.

CROSS SECTIONS

It is usual to express the results of a scattering experiment in terms of the differential cross section, $d\sigma/d\Omega$, which is defined for a beam of particles as the number of particles *scattered* into a unit solid angle per unit time, divided by the number of particles in the incident beam crossing a unit area per unit time. This cross section is equivalently, in terms of one particle, the total probability that the particle is scattered into a unit solid angle, divided by the total probability that crosses a unit area in front of the target. [We assume that the wave packet is much broader than the target and that the detector is out of the incident beam.] The total probability of being scattered into an infinitesimal solid angle $d\Omega$ at $\mathbf{r}$ is the rate that probability strikes an area $r^2 d\Omega$ in the detector plane, integrated over all time. This rate is the velocity $\hbar k_0/m$ times $r^2 d\Omega$ times $(|f_{\mathbf{k}_0}(\Omega_r)|^2/r^2)\,|\psi_0(\hat{\mathbf{k}}_0 r, t)|^2$, the square of the scattered amplitude. Thus the total probability of being scattered into $d\Omega$ is

$$|f_{\mathbf{k}_0}(\Omega_r)|^2 d\Omega\left(\frac{\hbar k_0}{m}\int_{-\infty}^{\infty} dt\,|\psi_0(\hat{\mathbf{k}}_0 r, t)|^2\right). \tag{9-21}$$

Furthermore the total probability that crosses a unit area, at a point $\mathbf{r}_0$ in front of the target in the incident beam, is the probability flux $(\hbar k_0/m)\,|\psi_0(\mathbf{r}_0, t)|^2$, integrated over all t:

$$\frac{\hbar k_0}{m}\int_{-\infty}^{\infty} dt\,|\psi_0(\mathbf{r}_0, t)|^2. \tag{9-22}$$

If we neglect spreading of the wave packet between $\mathbf{r}_0$ and $\hat{\mathbf{k}}_0 r$, (9-22) equals the latter term in (9-21). The ratio of (9-21) to (9-22) is the cross section $d\sigma/d\Omega$ times the little solid angle $d\Omega$. Thus we find the simple result that

$$\frac{d\sigma}{d\Omega} = |f_{\mathbf{k}_0}(\Omega_r)|^2. \tag{9-23}$$

This result is independent of the detailed shape of the incident wave packet. The total cross section, σ, which is $d\sigma/d\Omega$ integrated over all angles, is the total probability of a particle being scattered divided by the total probability that crossed a unit area in front of the target. From (9-23)

$$\sigma = \int d\Omega \, |f_{\mathbf{k}_0}(\Omega_r)|^2. \tag{9-24}$$

Experimentalists give their measurements in terms of cross sections. The theorist's problem is to calculate the differential cross section from the potential; to do this one must calculate the wave functions in the asymptotic region far from the potential.

PARTIAL WAVES

The task of calculating cross sections is considerably simplified when the potential is spherically symmetric, that is, it depends only on $|\mathbf{r}|$, for then the angular momentum of the incident particle is a constant of the motion. In (9-1) we represented the incident wave packet as a linear superposition of plane waves. We can go one step further and represent each of these plane waves as a linear combination of angular momentum eigenfunctions. The expansion, taking the z axis along $\mathbf{k}$, is given by (6-85).

$$\begin{aligned} e^{i\mathbf{k}\cdot\mathbf{r}} &= \sum_{l=0}^{\infty} i^l (2l+1) P_l(\cos\theta)\, j_l(kr) \\ &= \frac{1}{2} \sum_{l=0}^{\infty} i^l (2l+1) P_l(\cos\theta)\, (h_l(kr) + h_l^*(kr)) \end{aligned} \tag{9-25}$$

where

$$P_l(\cos\theta) = \sqrt{\frac{4\pi}{2l+1}}\; Y_{l0}(\theta, \varphi) \tag{9-26}$$

is the Legendre polynomial of order l. Only $m = 0$ eigenfunctions enter the expansion (9-25), because clearly the incident particle can have no angular momentum along its direction of motion. Put another way, the phase of the plane wave $e^{i\mathbf{k}\cdot\mathbf{r}}$ remains constant as $\mathbf{r}$ rotates about $\mathbf{k}$. Consequently the scattered wave must also be rotationally symmetric about $\mathbf{k}$. This means that $f_{\mathbf{k}}(\Omega_r)$ can only depend on θ, and not on the aximuthal angle φ. (f also depends on the energy E of the incident particle.)

From the rotational invariance of the potential we see that if the incident particle is in an angular momentum state l, with $m = 0$, then the scattered particle is also in an angular momentum eigenstate with the same l, and with $m = 0$. Thus we can study the scattering of each angular momentum component of (9-25) separately.

Let us therefore expand the wave function $\psi_{\mathbf{k}}(\mathbf{r})$ in angular momentum eigenstates:

$$\psi_{\mathbf{k}}(\mathbf{r}) = \sum_{l=0}^{\infty} i^l (2l+1) P_l(\cos\theta) R_l(r). \tag{9-27}$$

The radial wave functions R_l then obey

$$\left[\frac{d^2}{dr^2} + k^2 - \frac{l(l+1)}{r^2}\right] r R_l(r) = \frac{2m}{\hbar^2} v(r) r R_l(r). \tag{9-28}$$

At distances r far beyond the range of the potential [assuming $r^2 v(r) \to 0$ as $r \to \infty$] the right side of this equation becomes vanishingly small and (9-28) reduces to a simple Bessel equation. Thus in the far field the wave function $R_l(r)$ must be a linear combination of the Hankel function $h_l(kr)$ and its complex conjugate:

$$R_l(r) = B_l[h_l^*(kr) + S_l(E) h_l(kr)]. \tag{9-29}$$

B_l and S_l are functions of k which must be determined. In the absence of scattering, i.e., $v(r) = 0$, we see, by comparing (9-28) and (9-25) that

$$R_l = j_l = \frac{1}{2}(h_l + h_l^*);$$

in this case $B_l = 1/2$, $S_l = 1$. Now h_l^* is an incoming spherical wave, and h_l is an outgoing spherical wave. The effect of the potential, as can be seen from (9-11) is to modify only the outgoing part of the incident wave. Thus even in the presence of scattering, $B_l = 1/2$, and the scattering only modifies the coefficient of the h_l part of the solution.

It is easy to see that $|S_l(E)| = 1$ for elastic scattering. This is because the radial component of the total probability current

$$j_r(r) = \frac{\hbar}{2im}\left[R_l^*(r)\frac{\partial}{\partial r}R_l(r) - R_l(r)\frac{\partial}{\partial r}R_l^*(r)\right] \tag{9-30}$$

must vanish; the potential is neither a sink nor a source of particles. Using (9-29) in (9-30), we see that $j_l(r)$ vanishes only if $|S_l| = 1$. Thus we can write

$$S_l(E) = e^{2i\delta_l(E)} \tag{9-31}$$

where δ_l is real. The angle δ_l is known as the *phase shift*; $2\delta_l$ is the difference in phase of the outgoing parts of the actual wave function,

$\psi_{\mathbf{k}}(\mathbf{r})$, and the plane wave, $e^{i\mathbf{k}\cdot\mathbf{r}}$.

The scattering process is completely described in terms of the phase shifts. Asymptotically (that is, far outside the range of the potential) we can write, using (9-29)

$$\psi_{\mathbf{k}}(\mathbf{r}) = \frac{1}{2}\sum_{l} i^{l}\,(2l+1)\,P_{l}(\cos\theta)\,[h_{l}^{*}(kr) + e^{2i\delta_{l}}h_{l}(kr)]. \tag{9-32}$$

$$\psi_{\mathbf{k}}(\mathbf{r}) = e^{i\mathbf{k}\cdot\mathbf{r}} + \frac{1}{2}\sum_{l} i^{l}\,(2l+1)\,P_{l}(\cos\theta)\,(e^{2i\delta_{l}} - 1)\,h_{l}(kr). \tag{9-33}$$

using (9-25). Since $h_{l}(kr)$ is asymptotically equal to $e^{i(kr-l\pi/2)}/ikr$ we find that the scattering amplitude is given in terms of the δ_{l} by

$$f(\theta) = \frac{1}{2ik}\sum_{l}(2l+1)\,P_{l}(\cos\theta)\,(e^{2i\delta_{l}} - 1)$$
$$= \frac{1}{k}\sum_{l}(2l+1)\,P_{l}(\cos\theta)\,e^{i\delta_{l}}\sin\delta_{l}. \tag{9-34}$$

The quantity $e^{2i\delta_{l}} - 1$ is called the partial wave scattering amplitude. To find the total cross section, σ, we integrate $|f(\theta)|^{2}$ over all angles. The P_{l}'s obey the orthogonality relation

$$\int_{-1}^{1}\frac{dx}{2}\,P_{l}(x)\,P_{l'}(x) = \frac{\delta_{l,l'}}{2l+1} \tag{9-35}$$

and therefore

$$\sigma = \frac{4\pi}{k^{2}}\sum_{l}(2l+1)\sin^{2}\delta_{l}. \tag{9-36}$$

We may regard σ as the sum of terms

$$\sigma_{l} = \frac{4\pi}{k^{2}}(2l+1)\sin^{2}\delta_{l}, \tag{9-37}$$

which are the partial cross sections for scattering of particles in angular momentum l states. The scattering in each angular momentum channel is clearly limited by

$$\sigma_{l} < \frac{4\pi(2l+1)}{k^{2}}. \tag{9-38}$$

In the total cross section there is no interference between different partial waves. There is, however, such interference in the expression for $d\sigma/d\Omega$.

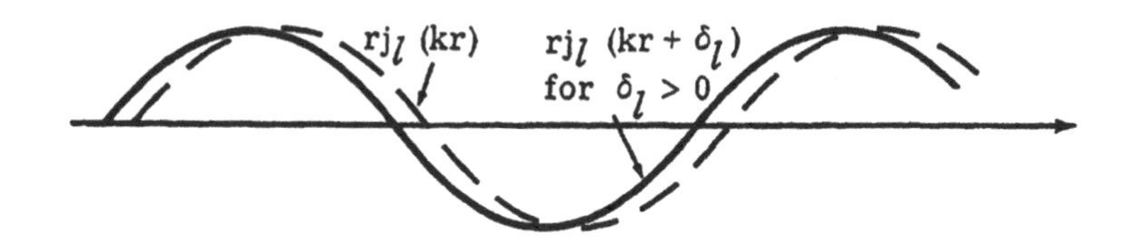

Fig. 9-2

$\delta_l > 0$ means that the asymptotic wave is pulled in; for $\delta_l < 0$ it is pushed out.

The wave function $R_l(r)$ is also given asymptotically by

$$R_l(r) \sim e^{i\delta_l} j_l(kr + \delta_l). \tag{9-39}$$

Except for s waves, the right-hand side is not rigorously equal to $\frac{1}{2}(h_l^* + e^{2i\delta_l} h_l)$; it is only asymptotically equal; thus (9-39) does not solve the Schrödinger equation for large r where the potential vanishes. Since for no scattering,

$$R_l(r) = j_l(kr),$$

the potential has the effect of either "pulling in" the wave function, for $\delta_l > 0$ (see Fig. 9-2), or pushing it out, for $\delta_l < 0$. Imagine turning on the potential slowly from zero to its actual value. If the potential is predominantly attractive, then as the potential is increased in magnitude the wave function has to oscillate more rapidly in the region of the potential. This has the effect of pulling in the wave, as can be seen in Fig. 9-3. Thus a positive phase shift corresponds to a predominantly attractive potential, and a negative phase shift corresponds to a predominantly repulsive potential.

The phase shift representation is useful for a short-range potential since, for a given energy, only a limited number of δ_l are

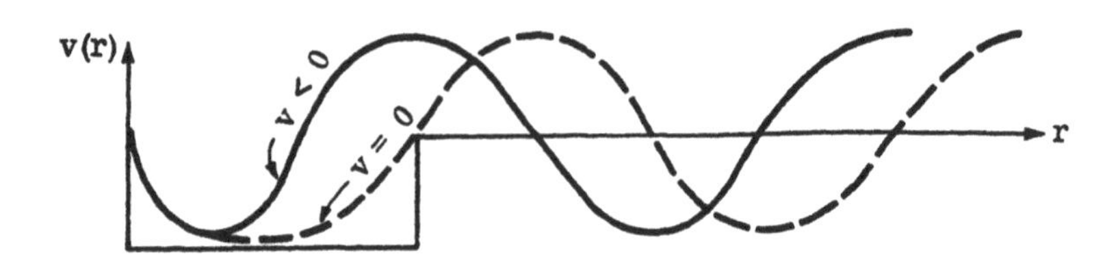

Fig. 9-3

An attractive potential pulling in the asymptotic wave.

appreciably different from zero. Roughly speaking, if the classical impact parameter of the incident particle, $\hbar l/\hbar k$, is much larger than the range b of the potential, then there will only be small scattering of the particle Thus δ_l will usually be small for $l \gg kb$.

THE OPTICAL THEOREM

There is a simple relation between the total cross section and the imaginary part of the forward scattering amplitude $f(\theta = 0)$. The imaginary part of (9-34) is

$$\operatorname{Im} f(\theta) = \frac{1}{k}\sum_l (2l+1)\, P_l(\cos\theta)\sin^2\delta_l$$

and since $P_l(1) = 1$, we find

$$\sigma = \frac{4\pi}{k}\operatorname{Im} f(0). \tag{9-40}$$

This relation is known as the *optical theorem;* its origin is the conservation of particles (or probability). The total scattered current, $\hbar k\sigma/m$, must be equal to the amount by which the current in the incident beam is depleted by the scattering. This depletion is due to the interference of the incident beam with the scattered beam and is therefore proportional to the scattering amplitude in the forward direction. The physics of the optical theorem is most easily seen by calculating the radial current $j_r(\mathbf{r}, t)$ for the wave function (9-20). We have [cf. (9-30)]

$$j_r(\mathbf{r},t) = \frac{\hbar}{m}\operatorname{Im}\left[\psi^*(\mathbf{r},t)\frac{\partial\psi(\mathbf{r},t)}{\partial r}\right]. \tag{9-41}$$

To calculate j_r in the far field we substitute (9-20), write $\partial\psi_0(\hat{\mathbf{k}}_0 r, t)/\partial r \approx ik_0\psi_0(\hat{\mathbf{k}}_0 r, t)$, $\partial\psi_0(\mathbf{r},t)/\partial r \approx i\mathbf{k}_0\cdot\hat{\mathbf{r}}\,\psi_0(\mathbf{r},t)$, and drop the r^{-3} term, which is small in the far field; then

$$j_r(\mathbf{r},t) = j_{r,0}(\mathbf{r},t) + j_{r,\text{scatt}}(\mathbf{r},t) + j_{r,\text{int}}(\mathbf{r},t), \tag{9-42}$$

where the incident current, j_0, is that of the wave packet (9-19), j_{scatt} is the scattered current we've already calculated, and the current due to the interference between the incident and scattered beam is

$$j_{r,\text{int}}(\mathbf{r},t) = \frac{\hbar k_0}{mr}\operatorname{Im}\left[i f_{\mathbf{k}_0}(\Omega_r)\psi_0^*(\mathbf{r},t)\psi_0(\hat{\mathbf{k}}_0 r,t)(1+\hat{\mathbf{k}}_0\cdot\hat{\mathbf{r}})\right]. \tag{9-43}$$

Next we observe that $\psi_0(\hat{\mathbf{k}}_0 r, t)$ is nonzero only after the particle reaches the target; then $\psi_0^*(\mathbf{r}, t)$ and hence the interference current are nonzero *only* in the forward direction; there $\mathbf{k}_0 \cdot \hat{\mathbf{r}} = k_0$ and $f_{\mathbf{k}_0}(\Omega_r) = f_{\mathbf{k}_0}(0)$. The interference term essentially produces the shadow of the target; it reduces the net current in the forward direction from $\mathbf{j}_0$, to account for the scattered current.

In order that the total scattered current exactly equal that taken out of the incident beam, the integral of (9-43) over all angles and times must equal minus the integral of (9-21) over all angles and times. Now

$$\int d\Omega\, j_{r,\,\mathrm{int}}(\mathbf{r}, t) = \frac{2\hbar k_0}{r} \operatorname{Im}\left[i f_{\mathbf{k}_0}(0)\psi_0(\hat{\mathbf{k}}_0 r, t) \int d\Omega\, \psi_0^*(\mathbf{r}, t)\right]; \tag{9-44}$$

in the angular integration of $\psi_0(\mathbf{r}, t)$, (9-19), we find

$$\begin{aligned} \int d\Omega\, e^{i\mathbf{k}\cdot\mathbf{r}} &= \frac{2\pi(e^{ikr} - e^{-ikr})}{ikr} \\ &\approx \frac{2\pi(e^{i\mathbf{k}\cdot(\hat{\mathbf{k}}_0 r)} - e^{-i\mathbf{k}\cdot(\hat{\mathbf{k}}_0 r)})}{ik_0 r} \end{aligned} \tag{9-45}$$

for $\mathbf{k}$ near $\mathbf{k}_0$. Thus

$$\int d\Omega\, \psi_0(\mathbf{r}, t) = \frac{2\pi}{ik_0 r}\left[\psi_0(\hat{\mathbf{k}}_0 r, t) - \psi_0(-\hat{\mathbf{k}}_0 r, t)\right], \tag{9-46}$$

and (9-44) becomes

$$-\frac{4\pi\hbar}{mr^2} \operatorname{Im} f_{\mathbf{k}_0}(0)\, |\psi_0(\hat{\mathbf{k}}_0 r, t)|^2, \tag{9-47}$$

since $\psi_0(\hat{\mathbf{k}}_0 r, t)\psi_0(-\hat{\mathbf{k}}_0 r, t) = 0$ in the far field. Comparing minus the integral of (9-47) over all t with the integral of (9-21) over all angles we find simply $\sigma = (4\pi/k_0) \operatorname{Im} f(0)$, the optical theorem. This argument does not depend on the potential being spherically symmetric.

THE BORN APPROXIMATION

Except in special cases, it is possible to calculate the phase shifts only by numerically integrating (9-28). However, if the potential has only a small effect on a given partial wave, as in the case of large l, or small k for p-wave and higher, we can then write a simple approximate expression for δ_l. We derive this as follows: substituting

(9-27) for $\psi_{\mathbf{k}}$ into Eq. (9-12) for f yields

$$f = -\frac{2m}{\hbar^2}\sum_{l=0}^{\infty}(2l+1)\,P_l(\cos\theta)\int_0^{\infty} r^2 dr\, v(r) j_l(kr) R_l(r); \tag{9-48}$$

in doing the integration over the angles of $\mathbf{r}'$ we have used (6-85) for $e^{-i\mathbf{k}'\cdot\mathbf{r}'}$. Comparing (9-48) with (9-34) we see that

$$e^{i\delta_l}\sin\delta_l = -\frac{2mk}{\hbar^2}\int_0^{\infty} r^2 dr\, v(r) j_l(kr) R_l(r). \tag{9-49}$$

Now if the potential has a small effect on $R_l(r)$, then δ_l will be small, and we may, to a first approximation, replace $R_l(r)$ on the right by $j_l(kr)$, its value for no potential. Keeping terms to first order in δ_l on the left gives

$$\delta_l \approx -\frac{2mk}{\hbar^2}\int_0^{\infty} r^2 dr\, v(r)[j_l(kr)]^2. \tag{9-50}$$

This expression is known as the *Born approximation* for δ_l. [One must be more careful in approximating $R_l(r)$ than we have been if the potential is very large in magnitude anywhere, as in the case of a hard sphere. This is because R_l will be small in regions of large potential, but j_l won't necessarily also be small.]

If the potential has a small effect on all partial waves, then we can write a Born approximation for f directly. When the scattered wave is small compared with the incident wave we can approximate $\psi_{\mathbf{k}}(\mathbf{r})$ in Eq. (9-12) by the incident wave alone, $e^{i\mathbf{k}\cdot\mathbf{r}}$. This gives

$$f(\Omega_r) = -\frac{m}{2\pi\hbar^2}\int d^3r'\, e^{i(\mathbf{k}-\mathbf{k}')\cdot\mathbf{r}'}\, v(\mathbf{r}') = -\frac{m}{2\pi\hbar^2}\, v_{\mathbf{k}'-\mathbf{k}} \tag{9-51}$$

where $v_{\mathbf{k}'-\mathbf{k}}$ is the Fourier transform of $v(\mathbf{r})$. In this expression $\mathbf{k}$ is the incident wave vector, $\mathbf{k}'$ is the wave vector of the scattered particle, and

$$(\mathbf{k}-\mathbf{k}')^2 = k^2 + k'^2 - 2\mathbf{k}\cdot\mathbf{k}' = 2k^2(1-\cos\theta) = \left(2k\sin\frac{\theta}{2}\right)^2 \tag{9-52}$$

where θ is the scattering angle. [For Eq. (9-51) to be valid it is necessary that $v(\mathbf{r})$ have a Fourier transform. For a Yukawa potential $v(\mathbf{r}) = ae^{-Kr}/r$, we have $v_q = 4\pi a/(q^2 + K^2)$, and

$$\frac{d\sigma}{d\Omega} = \frac{a^2}{\left[4E_k\sin^2\frac{\theta}{2} + \frac{\hbar^2K^2}{2m}\right]^2} \tag{9-53}$$

It is amusing to note that if $a = e^2$ then (9-53) reduces to the classical Rutherford cross section as $K \to 0$. This happens to be the correct quantum mechanical result, as we shall see, but its derivation from the Born approximation is strictly accidental; (9-51) is invalid for a Coulomb potential because of its long range.

PROPERTIES OF THE SCATTERING AMPLITUDE

Let us turn now to considering some general properties of the phase shifts. These properties will be easiest to see if we make the assumption that $v(r)$ is strictly zero for r greater than some radius b, though our results will be valid as long as the potential approaches zero at infinity as rapidly as r^{-2}. Since $v(r) = 0$ for $r > b$, the wave function is given there by

$$R_l(r) = \frac{1}{2}(h_l^*(kr) + e^{2i\delta_l} h_l(kr)), \tag{9-54}$$

while for $r < b$ it is necessary to solve (9-28) to find $R_l(r)$. At $r = b$ both the wave function and its derivative must be continuous. Thus, from (9-54)

$$\left[\frac{\frac{\partial}{\partial r}(h_l^*(kr) + e^{2i\delta_l} h_l(kr))}{h_l^*(kr) + e^{2i\delta_l} h_l(kr)}\right]_{r=b} = \alpha_l \quad . \tag{9-55}$$

where

$$\alpha_l \equiv \frac{1}{R_l(b)}\left(\frac{dR_l(r)}{dr}\right)_{r=b} \tag{9-56}$$

is the logarithmic derivative of R_l at $r = b$. Rearranging (9-55) we find

$$e^{2i\delta_l} - 1 = \left[\frac{2(\partial j_l/\partial r - \alpha_l j_l)}{\alpha_l h_l - \partial h_l/\partial r}\right]_{r=b} \tag{9-57}$$

or equivalently

$$\cot\delta_l = \left[\frac{\partial n_l/\partial r - \alpha_l n_l}{\partial j_l/\partial r - \alpha_l j_l}\right]_{r=b}, \tag{9-58}$$

since

$$e^{2i\delta_l} - 1 = \frac{2i}{\cot \delta_l - i}. \tag{9-59}$$

Thus the phase shift depends on the potential only through the logarithmic derivative α_l.

As an example, if v(r) represents a hard sphere of radius b, then $R_l(b) = 0$. Consequently at b, $\alpha_l = \infty$ and from (9-58)

$$\cot \delta_l = \frac{n_l(kb)}{j_l(kb)}.$$

For s waves this becomes

$$\delta_0 = -kb.$$

We see here explicitly a repulsive potential yielding a negative phase shift.

Let us first examine the behavior of δ_l for small values of k. Using (6-82) we find, for kb ≪ 1:

$$\cot \delta_l \approx (kb)^{-2l-1} (2l-1)!!\,(2l+1)!!\,\frac{l+1+b\alpha_l(E)}{l-b\alpha_l(E)} \tag{9-60}$$

where

$$(2l+1)!! = 1 \cdot 3 \cdot 5 \cdots (2l+1).$$

Thus as $kb \to 0$, $\alpha_l(E) \to \alpha_l(0)$, and we see that,

$$\sin \delta_l \sim k^{2l+1}. \tag{9-61}$$

This has one immediate consequence; for sufficiently low incident energy, the contribution to $f(\theta)$ of all partial waves with $l \geq 1$ can be neglected, and

$$\frac{d\sigma}{d\Omega} = \frac{4\pi \sin^2 \delta_0}{k^2}. \tag{9-62}$$

Thus at low enough energy, the scattering is entirely s-wave and therefore isotropic. We can also derive this result from (9-12); for sufficiently long wavelengths the wave function $\psi_{\mathbf{k}}(\mathbf{r})$ in the region of the potential is essentially isotropic, and therefore the scattered wave is also isotropic. For example, for a hard sphere at low energy, $\sigma = 4\pi b^2$, a value four times the classical cross section.

A second feature of the partial wave scattering amplitude, $e^{2i\delta_l} - 1$, is that, considered as an analytic function of the energy E, it has poles at the energies of the bound states of the potential with angular momentum l. To see this, recall that the wave function of a bound state with angular momentum l and energy E_b is given by

$$R_l(r) \sim h_l(i\kappa r), \quad r > b \tag{9-63}$$

where $E = -\hbar^2\kappa^2/2m$. Thus matching logarithmic derivatives at b implies that at a value of κ corresponding to a bound state

$$\alpha_l(E) = \left[\frac{1}{h_l(i\kappa r)}\frac{\partial}{\partial r}h_l(i\kappa r)\right]_{r=b} \tag{9-64}$$

But from (9-58), this is the same condition as

$$\cot\delta_l(E) = i. \tag{9-65}$$

Thus we see either from (9-59), or directly from (9-57), that $e^{2i\delta_l(E)}$ has a pole at each bound state energy. For $\kappa b \ll 1$, (9-64) reduces to

$$l + 1 + b\alpha_l(E_b) = 0, \tag{9-66}$$

to within terms of order[2] κb, as the condition for having a bound state very close to zero energy ($|E_b| \ll \hbar^2/2mb^2$). We see that if there is such a bound state, then by continuity $l + 1 + \alpha_l(0)$ will be small, and this in turn will, from (9-60), increase the relative values of $\sin\delta_l$ (and hence the low-energy partial wave cross section) over its value were there no such bound state.

As we have seen, at sufficiently low energy the scattering is entirely s-wave. Let us study the behavior of the low-energy cross section in more detail. From (9-60), for $kb \ll 1$,

$$k\cot\delta_0 = -\frac{1 + b\alpha_0(0)}{b^2\alpha_0(0)} \equiv -\frac{1}{a}; \tag{9-67}$$

the length a is called the *scattering length.* In terms of a the zero energy cross section is

$$\sigma = 4\pi a^2; \tag{9-68}$$

at very low energy the potential scatters as if it were a hard sphere of radius a.

In terms of the scattering length,

[2]These terms make the difference between $\cot\delta_l = 0$, as (9-65) together with (9-60) seems to imply, and $\cot\delta_l = i$, the actual condition for a bound state.

$$e^{2i\delta_0} - 1 = \frac{2i}{\cot\delta_0 - i} = \frac{2ka}{i - ka}, \qquad (9\text{-}69)$$

for $|kb| \ll 1$. Thus the scattering amplitude has a pole at $k = i/a$, or at $\kappa = -ik = 1/a$, as long as[3] $b \ll |a|$. This corresponds then to a bound state if κ is positive and small. A pole at a negative value of κ corresponds to a solution of the Schrödinger equation that grows exponentially at large distances, and is therefore not a physically allowable state. [Such a pole is on the second Riemann sheet of $e^{2i\delta_0(E)}$.] Therefore we may conclude that the scattering length is positive and large in magnitude if there is an s-wave bound state near zero energy. [As an exercise, one should study the detailed relation between a and the nearly zero energy bound states of an attractive square well potential.] The energy of the bound state is given by

$$E_b = -\frac{\hbar^2\kappa^2}{2m} = -\frac{\hbar^2}{2ma^2}. \qquad (9\text{-}70)$$

Since the wave function of the bound state behaves as $e^{-\kappa r}/r$ at large distances, a is essentially the size of the bound state. Furthermore, the low energy cross section is given by

$$\sigma = \frac{4\pi}{k^2}\,\frac{1}{\cot^2\delta_0 + 1} = \frac{2\pi\hbar^2/m}{E - E_b}. \qquad (9\text{-}71)$$

This result is very remarkable; if there is an s-wave bound state near zero energy it completely determines the low-energy cross section. The only parameter in (9-71) is the energy of the bound state.

An important application of (9-71) is to the scattering of neutrons from protons. The deuteron is a bound state of a neutron and a proton with binding energy $|E_b| = 2.23$ MeV, or a $\sim 4.3 \times 10^{-13}$ cm. The interaction between neutrons and protons corresponds to a potential about 25 MeV deep, and of range $\sim 2 \times 10^{-13}$ cm. Thus the deutron is effectively a bound state close to zero energy, and (9-71) should predict the p-n scattering.

Unfortunately, things are somewhat more complicated than this. Both the neutron and the proton have spin 1/2, like the electron, and it turns out that the interaction between neutrons and protons depends on which way their spins are pointed. In the deuteron the spins of the neutron and proton are "parallel" (triplet state), and therefore

---

[3] For the hard sphere b = a and $e^{2i\delta_0}$ has no poles.

(9-71) should be the cross section for parallel spin neutron-proton scattering. The cross section for "antiparallel" (singlet state) spin n-p scattering has the same form as (9-71),

$$\sigma_{\text{antiparallel}} = \frac{2\pi\hbar^2/m}{E - E_V} \tag{9-72}$$

where $E_V \approx -100$ KeV. However there is no bound state of a neutron and a proton at energy E_V, since the scattering length for antiparallel spins is negative. One says that the neutron and proton have a *virtual state* at energy E_V. If the potential were slightly more attractive, the inverse of this scattering length would go through zero and become slightly positive, giving a true n-p bound state.

LOW ENERGY RESONANCES

At very low energy the scattering is entirely s-wave. Let us now examine the structure of the scattering at somewhat higher energies, but still in the small kb regime, where Eq. (9-60) for $\cot \delta_l$ is valid, We see from (9-60) that in general $\cot \delta_l$ is quite large, being proportional to k^{-2l-1}. Hence the partial cross section

$$\sigma_l = \frac{4\pi}{k^2}(2l+1)\,\frac{1}{\cot^2\delta_l + 1} \tag{9-73}$$

will be proportional to k^{4l}, and with the exception of s-waves, σ_l will be quite small. However, if we are near an energy E_r where $l + 1 + b\alpha_l(E)$ passes through zero,

$$l + 1 + b\alpha_l(E_r) = 0, \tag{9-74}$$

then $\cot \delta_l$ passes through zero and σ_l will be proportional to k^{-2} and hence quite large, as in Fig. 9-4. Such a bump in σ_l is known as a scattering *resonance;* it is very similar to the resonances we found in transmission through a barrier in one dimension. Resonances can occur at low energy for $l \geq 1$.

In the neighborhood of E_r we can write

$$l + 1 + b\alpha_l(E) \approx (E - E_r)b\left(\frac{\partial \alpha_l}{\partial E}\right)_{E=E_r}, \tag{9-75}$$

$$l - b\alpha_l(E) \approx 2l + 1, \tag{9-76}$$

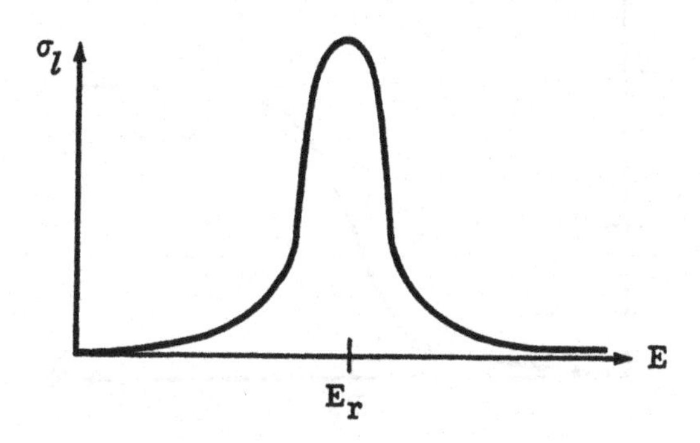

Fig. 9-4
A low-energy resonance at energy E_r in a partial wave cross section for $l \geq 1$.

and therefore

$$\cot \delta_l \approx \frac{-2(E - E_r)}{\Gamma_k} \tag{9-77}$$

where

$$\Gamma_k = -\frac{2k^{2l+1}b^{2l}}{[(2l-1)!!]^2(\partial\alpha_l/\partial E)_{E_r}}. \tag{9-78}$$

Since α_l is a decreasing function of energy,[4] Γ_k is positive. Then near the resonance, the partial cross section σ_l is given by

$$\sigma_l = \frac{4\pi\Gamma_k{}^2/k^2}{4(E - E_r)^2 + \Gamma_k{}^2}. \tag{9-79}$$

When $E - E_r = \pm\Gamma_k/2$, the partial cross section drops to half its value. Thus Γ, equal to Γ_k evaluated at E_r, is essentially the width of the

---

[4]To show this, divide both sides of (9-28) by $u_l = R_l/r$, and differentiate both sides with respect to k^2. This yields $(\partial/\partial k^2)[(1/u_l)(\partial/\partial r)^2 u_l] = -1$, which can easily be juggled into the form $-(\partial/\partial r)[u_l{}^2(\partial^2 \ln u_l/\partial(k^2)\partial r)] = u_l{}^2$. Integrating both sides from zero to r gives $-u_l{}^2(\partial^2 \ln u_l/\partial(k^2)\partial r) = \int_0^r dr' u_l(r')^2$, since $u_l(0) = 0$. Thus $\partial^2 \ln u_l/\partial(k^2)\partial r = (\hbar^2/2m)(\partial\alpha_l(r)/\partial E) < 0$.

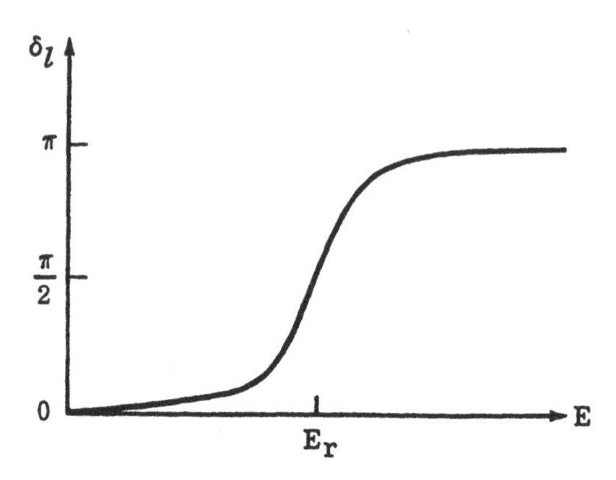

Fig. 9-5
Behavior of the phase shift near a sharp resonance.

resonance at half height. We see from (9-79) that the more rapidly the logarithmic derivative, α_l, varies with E at E_r, the sharper will be the resonance. Also, the smaller the value of kb and the larger the value of l at which the resonance occurs the sharper it will be.

For s-waves, σ_0 does not have a maximum when $1 + b\alpha_0(E) = 0$, unless $\partial\alpha_0/\partial k^2 > 1/\sqrt{2k}$ at E_r. Even then the maximum of σ_0 is much less pronounced than it is at a higher angular momentum resonance.

A resonance occurs (for $l \geq 1$) whenever $\cot \delta_l$ vanishes, or equivalently, whenever

$$\delta_l = \left(n + \frac{1}{2}\right)\pi. \tag{9-80}$$

Near the resonance, δ_l is given by

$$\delta_l = \frac{\pi}{2} + \tan^{-1}\left(\frac{E - E_r}{\Gamma/2}\right), \tag{9-81}$$

as can be checked from (9-77). Thus as E passes through E_r the phase shift climbs suddenly from a value near zero (plus $n\pi$) to a value near π (plus $n\pi$); it jumps by π in a small energy interval $\sim\Gamma$, as in Fig. 9-5. The slope of δ_l equals $2/\Gamma$ at E_r. Thus the sharper the climb, the sharper is the resonance.

The condition (9-74) is very similar to the condition (9-66) for there to be a bound state very near zero energy. A resonance is in a sense an almost bound state that hasn't quite gotten below zero energy to be a true bound state. If the potential v(r) is an attractive

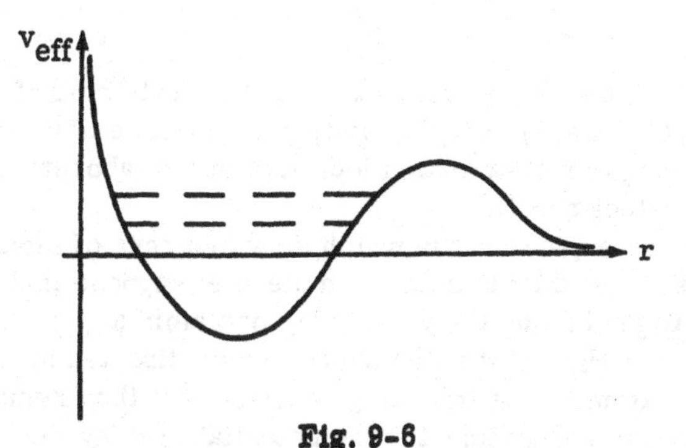

Fig. 9-6
Resonances corresponding to almost bound levels, the dotted lines, of the potential.

well, then the combination $v(r) + \hbar^2 l(l+1)/mr^2$ is an effective potential of the form in Fig. 9-6. Resonances are the "bound states" of the well at positive energy, indicated by dotted lines. A particle in such a state is held in by the centrifugal barrier term. Eventually, however, the particle tunnels through the barrier, with a time constant Γ, and leaves. The mathematics of this process are identical to those we studied in the one-dimensional case. It is clear that the lower the energy of the particle, or the higher the barrier, the longer the time the particle spends in the well. What happens in scattering at a resonant energy is that the incident particle has a large probability of becoming temporarily trapped in such a quasi-bound state of the well; this possibility increases the scattering cross section.

We have noted that the true bound states appear as poles in the scattering amplitude $e^{2i\delta_l} - 1$ at negative energies. Near a resonance

$$e^{2i\delta_l} - 1 = -\frac{i\Gamma_k}{E - E_r + i\Gamma_k/2}, \tag{9-82}$$

so we see that a sharp resonance is related to a pole in the scattering amplitude at the slightly complex energy

$$E = E_r - \frac{i\Gamma}{2} \tag{9-83}$$

in the lower half-plane (on the second Riemann sheet).

PROBLEMS

1. If only $\delta_0(E)$ and $\delta_1(E)$ are nonzero in the scattering of a particle of mass m_1 from an initially stationary particle of mass m_2, what does the angular distribution look like in the laboratory frame? Do not neglect recoil.
2. Consider an experiment in which slow neutrons of momentum $\hbar k$ are scattered by a diatomic molecule; suppose that the molecule is aligned along the y axis with one atom at $y = -b$ and the other at $y = +b$, and that the neutrons are directed along $\hat{z}$. Assume the atoms to be infinitely heavy so that they remain fixed throughout the experiment. The potential seen by the neutron from each atom can be adequately represented by a delta function; thus

$$V(r) = a\delta(y-b)\delta(x)\delta(z) + a\delta(y+b)\delta(x)\delta(z).$$

(a) Calculate the scattering amplitude, and differential cross section, in the Born approximation.
(b) In what ways does the quantum result differ from what one would expect classically?

Chapter 10
COULOMB SCATTERING

Our entire discussion in the last chapter was based on the assumption that the potential was short ranged. The Coulomb potential, which governs the scattering of one charged particle by another, is so long ranged, however, that all our assumptions about the wave functions becoming asymptotically free particle wave functions break down. The Coulomb force makes itself felt no matter how far apart the colliding particles are from each other.

We can see difficulties arising by looking at the radial wave functions of the hydrogen atom. At large r these behave as

$$R_{nl}(r) \sim r^{n-1}e^{-\kappa r}, \tag{10-1}$$

as opposed to the behavior

$$r_l(r) \sim \frac{e^{-\kappa r}}{r} \tag{10-2}$$

for a bound state of a short-range potential. Since, for the hydrogen atom,

$$E = -\frac{me^4}{2\hbar^2 n^2} = -\frac{\hbar^2\kappa^2}{2m}, \tag{10-3}$$

we have

$$n = \frac{me^2}{\hbar^2\kappa}. \tag{10-4}$$

Thus we can write (10-1) as

$$R_{nl}(r) \sim \frac{1}{r}\, e^{-\kappa r + (me^2/\hbar^2\kappa)\ln r}. \tag{10-5}$$

We therefore expect that, asymptotically, the continuum ($E > 0$) solutions to the Schrödinger equation[1]

$$\left(-\frac{\hbar^2}{2m}\nabla^2 - \frac{e^2}{r}\right)\psi(r) = E\psi(r), \tag{10-6}$$

(these solutions describe the nonrelativistic scattering of an electron from a proton, for example) have the same form as (10-5), only with κ replaced by $\pm ik$; that is,

$$R(r) \sim \frac{1}{r} e^{\pm i(kr - \gamma \ln r)}, \quad E > 0, \tag{10-7}$$

where

$$\gamma = -\frac{me^2}{\hbar^2 k} = -\frac{1}{a_0 k} \tag{10-8}$$

a_0 is the Bohr radius. Thus the Coulomb potential induces a phase shift, $-\gamma \ln r$, which grows arbitrarily large as r approaches infinity.

The simplest way of finding the Coulomb scattering amplitude is to solve (10-6) in parabolic coordinates; the method can be found in the standard texts. If we choose the z axis along **k**, the incident wave vector, the solution to (10-6) is

$$\psi(r) = \Gamma(1 + i\gamma)e^{-\pi\gamma/2}\, e^{ikz}\, {}_1F_1(-i\gamma; 1; ik(r - z)), \tag{10-9}$$

where the confluent hypergeometric function ${}_1F_1$ is defined by the power series expansion

$${}_1F_1(a; b; x) = 1 + \frac{ax}{b\cdot 1} + \frac{a(a+1)x^2}{b(b+1)\cdot 1\cdot 2} + \cdots . \tag{10-10}$$

As $-ix$ becomes very large, and positive,

$${}_1F_1(-i\gamma; 1; x) \to \frac{e^{\pi\gamma/2}}{\Gamma(1+i\gamma)} \left\{ e^{i\gamma \ln(-ix)}\left(1 - \frac{\gamma^2}{x} + \cdots\right) + \frac{\Gamma(1+i\gamma)}{\Gamma(-i\gamma)}\frac{e^x}{x} e^{-i\gamma \ln(-ix)} \right\}, \tag{10-11}$$

where $\Gamma(z)$ is the gamma function, defined by

[1]We shall restrict ourselves to the potential $-e^2/r$; to find the scattering of a particle of charge q_1 from a particle of charge q_2 we simply replace $-e^2$ by $q_1 q_2$ in the final results.

$$\Gamma(z) = \int_0^\infty da\, a^{z-1} e^{-a}, \tag{10-12}$$

in the region of the complex z plane, Re z > 0. We see from (10-11) that for large values of $r - z$,

$$\psi(\mathbf{r}) = \psi_i(\mathbf{r}) + \psi_{sc}(\mathbf{r}), \tag{10-13}$$

where

$$\psi_i(\mathbf{r}) = e^{ikz + i\gamma \ln k(r-z)} \left(1 - \frac{\gamma^2}{ik(r-z)}\right) \tag{10-14}$$

and

$$\psi_{sc}(\mathbf{r}) = \frac{e^{ikr - i\gamma \ln k(r-z)}}{ik(r-z)} \frac{\Gamma(1+i\gamma)}{\Gamma(-i\gamma)}. \tag{10-15}$$

ψ_i has roughly the form of an incident wave; in fact, the probability current from ψ_i is $\hbar k/m$, in the limit $r - z \to \infty$. Thus we shall interpret ψ_i as the incident part of the wave function. Furthermore, $\psi_{sc}(\mathbf{r})$ has roughly the form of an outgoing radial times a function of θ; the current from ψ_{sc} is entirely radial, for large $r - z$. [Note that for any $\theta \neq 0$, the condition "large $r - z$" is essentially equivalent to "large r."] Thus we can interpret ψ_{sc} as the scattered wave.

Since $r - z = r(1 - \cos\theta)$, we can write (10-15) as

$$\psi_{sc}(\mathbf{r}) = \frac{e^{i(kr - \gamma \ln 2kr)}}{r} f_c(\theta), \tag{10-16}$$

where the Coulomb scattering amplitude is given by

$$f_c(\theta) = \frac{e^{-i\gamma \ln \sin^2 \theta/2}}{2ik \sin^2 \theta/2} \frac{\Gamma(1+i\gamma)}{\Gamma(-i\gamma)}. \tag{10-17}$$

Notice that the radial dependence of (10-17) has the same form that we anticipated in (10-7). The gamma function obeys

$$\Gamma(z+1) = z\Gamma(z) \tag{10-18}$$

and

$$\Gamma(z^*) = \Gamma(z)^*. \tag{10-19}$$

Therefore we can write $f_c(\theta)$ as

$$f_c(\theta) = -\frac{\gamma}{2k} e^{2i\delta_0} \left(\sin^2\frac{\theta}{2}\right)^{-i\gamma-1} \tag{10-20}$$

where

$$e^{2i\delta_0} = \frac{\Gamma(1+i\gamma)}{\Gamma(1-i\gamma)}; \tag{10-21}$$

δ_0 is a real number.

For large values of r, the current from ψ_{sc} is

$$j_{sc} = \frac{\hbar k}{m} |\psi_{sc}(r)|^2 = \frac{\hbar k}{m} \frac{1}{r^2} |f_c(\theta)|^2, \tag{10-22}$$

and is in the radial direction. Thus the differential cross section is

$$\frac{d\sigma}{d\Omega} = |f_c(\theta)|^2 = \frac{e^4}{16E^2} \frac{1}{\sin^4\frac{\theta}{2}}, \tag{10-23}$$

which is precisely the classical Rutherford cross section.

The cross section (10-23) becomes arbitrarily large in the forward direction — so much so that the cross section is infinite. Classically speaking, the reason for this enormous amount of small angle scattering is that no matter how large the impact parameter of the incident particle, the long range of the Coulomb potential will cause a slight deflection, that is, a small angle scattering of the particle.

It is very hard to do scattering experiments using isolated point charges as targets; generally there are just enough charges near the target of the opposite sign to create overall charge neutrality. Consider, for example, the scattering of fast charged particles from a nucleus of charge Z. Such a nucleus is usually surrounded by a cloud of Z orbiting electrons. Particles that pass in close to the nucleus are mostly scattered through finite angles by the nucleus, as described by the Rutherford cross section away from the forward direction. However, particles that pass by very far from the nucleus will see only the electric field of the nucleus plus the electrons, a neutral system. This field is of much shorter range than that due to the nucleus alone; one says that the electrons *screen out* the field of the nucleus. As a consequence, the amount of small angle scattering of the incident particles is greatly reduced from the pure Coulomb value, so much so as to make the cross section finite.

ANALYTIC PROPERTIES OF THE SCATTERING AMPLITUDE

Let us spend a moment discussing the poles of the scattering amplitude (10-17) considered as a function of energy. The function $\Gamma(z)$ has simple poles at $z = 0, -1, -2, \ldots$ To see this we can use (10-18) to write

$$\Gamma(z) = \frac{\Gamma(z+1)}{z} = \frac{\Gamma(z+2)}{z(z+1)} = \frac{\Gamma(z+3)}{z(z+1)(z+2)} \quad \text{etc.} \tag{10-24}$$

Clearly Γ has a pole at $z = 0$ with residue $\Gamma(1) = 1$, at $z = -1$ with residue $\Gamma(1)/(-1) = -1$, at $z = -2$ with residue $\Gamma(1)/(-1 \cdot -2) = 1/2$, etc. Also $\Gamma(z)$ is never zero in the complex plane. Thus we see from Eq. (10-17) that $f_C(\theta)$ will have a simple pole whenever

$$1 + i\gamma = 0, -1, -2, \ldots,$$

or

$$\gamma = -\frac{1}{ka_0} = in; \quad n = 1, 2, 3, \ldots \tag{10-25}$$

These points correspond to k values

$$k = \frac{i}{na_0}, \tag{10-26}$$

or energy values

$$E = \frac{\hbar^2 k^2}{2m} = -\frac{me^4}{2\hbar^2 n^2}. \tag{10-27}$$

These are precisely the bound state energies of the attractive Coulomb potential. However, only if k in (10-26) is on the positive imaginary axis, corresponding to positive real $\kappa = -ik$ values and therefore to wave functions that become exponentially small at infinity, do these poles indicate actual bound states. For particles of the same charge, we must replace e^2 in (10-26) by $-e^2$; the poles then occur at negative imaginary k, and therefore negative κ.

The radial part of the asymptotic form of (10-9), at the k values (10-21), is just that in (10-1); the incident part of ψ is infinitely smaller than the scattered part. However the angular dependence is a linear combination of all angular moment components l less than n.

It is possible to solve the Schrödinger equation (10-6) directly in spherical coordinates. The solution looks like

$$\psi_c(\mathbf{r}) = \frac{1}{kr} \sum_{l=0}^{\infty} (2l+1) i^l P_l(\cos\theta)\, e^{i\delta_l} F_l(\gamma, kr) \tag{10-28}$$

where asymptotically ($kr \gg l$),

$$F_l \sim \sin\left(kr - \gamma \ln 2kr - \frac{l\pi}{2} + \delta_l\right) \tag{10-29}$$

and

$$e^{2i\delta_l} = \frac{\Gamma(l+1+i\gamma)}{\Gamma(l+1-i\gamma)}. \tag{10-30}$$

The δ_l are like phase shifts for a short ranged interaction. It is the additional l-independent phase shift, $-\gamma \ln 2kr$, that distinguishes the Coulomb solution from that for a short ranged potential.

How is the scattering amplitude $f_c(\theta)$ related to the δ_l? To expand $f_c(\theta)$ in partial waves

$$f_c(\theta) = \sum_l (2l+1) P_l(\cos\theta) f_l \tag{10-31}$$

we note that

$$\begin{aligned} f_l &= \frac{1}{2} \int_{-1}^{1} d(\cos\theta)\, P_l(\cos\theta)\, f_c(\theta) \\ &= -\frac{\gamma e^{2i\delta_0}}{4k} \int_{-1}^{1} d(\cos\theta)\, P_l(\cos\theta) \left(\sin^2\frac{\theta}{2}\right)^{-i\gamma-1} \\ &= -\frac{\gamma e^{2i\delta_0}}{4k} \int_{-1}^{1} dx\, P_l(x) \left(\frac{1-x}{2}\right)^{-i\gamma-1}, \end{aligned} \tag{10-32}$$

using (10-20) for f_c. Unfortunately the integral in (10-32) doesn't converge for real values of γ. This is because f_c is infinite in the forward direction. What we must do is find the expansion coefficients f_l in the region $\mathrm{Re}(-i\gamma) > 0$ of the complex γ plane where (10-32) converges, and then analytically continue the result to pure imaginary γ. Using

$$P_l(x) = \frac{1}{l!\,2^l} \frac{d^l}{dx^l} (x^2-1)^l \tag{10-33}$$

we can then integrate (10-32) by parts l times and find

$$f_l = \frac{1}{2ik} \frac{(l+i\gamma)\cdots(1+i\gamma)}{(l-i\gamma)\cdots(1-i\gamma)} e^{2i\delta_0} = \frac{1}{2ik} e^{2i\delta_l}. \tag{10-34}$$

Thus

$$f_c(\theta) = \sum_l (2l+1) P_l(\cos\theta) \frac{e^{2i\delta_l}}{2ik}. \tag{10-35}$$

This result is valid for all γ; but for $\mathrm{Re}(-i\gamma) < 0$, (10-35) does not converge in the forward direction. We can equivalently write (10-35) as

$$f_c(\theta) = \sum_l (2l+1) P_l(\cos\theta) \frac{(e^{2i\delta_l} - 1)}{2ik} \tag{10-36}$$

as for a short ranged potential; the added term $-{}^1/_{2ik} \sum_l (2l+1) P_l(\cos\theta)$ equals $(i/2k)\delta(1-\cos\theta)$ [because $\int_{-1}^{1} (dx/2) P_l(x)\delta(1-x) = \frac{1}{4}$], and since f_c is already very infinite at $\theta = 0$ it doesn't change matters to add on a delta function in the forward direction.

From (10-30) we see that, considered as an analytic function of k, f_l has poles at

$$l+1+i\gamma = 0, -1, -2, \ldots \tag{10-37}$$

or

$$k = \frac{i}{na_0}, \quad n = l+1, l+2, \ldots \tag{10-38}$$

For an attractive Coulomb potential, these are precisely the k values corresponding to the energies of bound states with angular momentum l.

REGGE POLES

We can look at the analytic properties of f_l from a different point of view by asking: for real E, where in the *complex l plane* does f_l have poles? These still occur whenever (10-37) is satisfied, that is, at

$$l = -p - i\gamma = -p + \frac{ime^2}{\hbar^2 k}\;; \; p = 1, 2, 3, \ldots \tag{10-39}$$

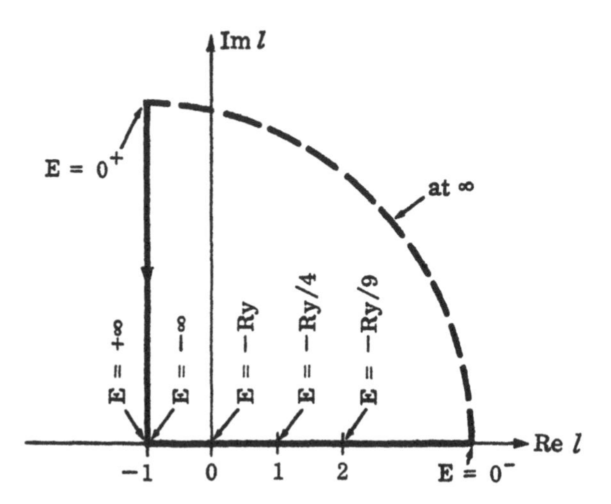

Fig. 10-1
Regge trajectory (p = 1) in the complex l plane for an attractive Coulomb potential. The energies for which $l = 0, 1, 2, \ldots$ are the energies of the bound states.

These poles in the complex l plane are known as *Regge poles.* As we vary the energy these poles move around in the l plane. The path they follow is called a Regge trajectory. The trajectory for $p = 1$ and an attractive Coulomb interaction is shown in Fig. 10-1, where Ry is the Rydberg; the $p = 1$ trajectory for a repulsive Coulomb interaction is shown in Fig. 10-2. For $E < 0$, $k = +i\sqrt{-E}$. Clearly, whenever a trajectory passes through a nonnegative integer value of l the system has a bound state.

For an attractive Coulomb interaction the Regge trajectory remains along the positive real axis all the way out to plus infinity. This is because this system has bound states for arbitrarily high angular momentum. A short ranged potential, on the other hand, can't have bound states for arbitrarily large l, due to the centrifugal barrier. Thus, for a short ranged potential, we expect the Regge trajectories to leave the positive real axis at some finite value of Re l.

Consider for example a bound state with angular momentum l and energy E. We showed in the last chapter [Eq. (9-65)] that at this energy

$$\cot \delta_l(E) = i. \tag{10-40}$$

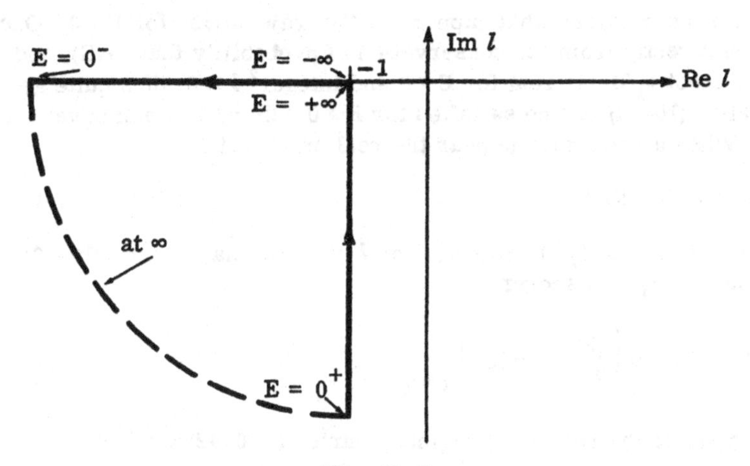

Fig. 10-2

Regge trajectory (p = 1) in the complex l plane for a repulsive Coulomb potential. The trajectory never passes through a non-negative integer l.

From the Regge point of view, this equation determines the Regge trajectories $l = l(E)$ for *all* real E, since whenever (10-40) is satisfied $e^{2i\delta_l(E)}$ has a pole. [There are generally many solutions $l(E)$ to (10-40), as in the Coulomb case.] Whenever, as E increases from $-\infty$ to zero, a trajectory passes, for some $E = E_b$, through an integer $l_b \geq 0$, the system will have a bound state at energy E_b with angular momentum l_b. Fig. 10-3 shows a possible situation where a trajectory passes through $l = 0$, 1, and 2 as E is increased from $-\infty$ to 0, indicating bound s, p, and d states.

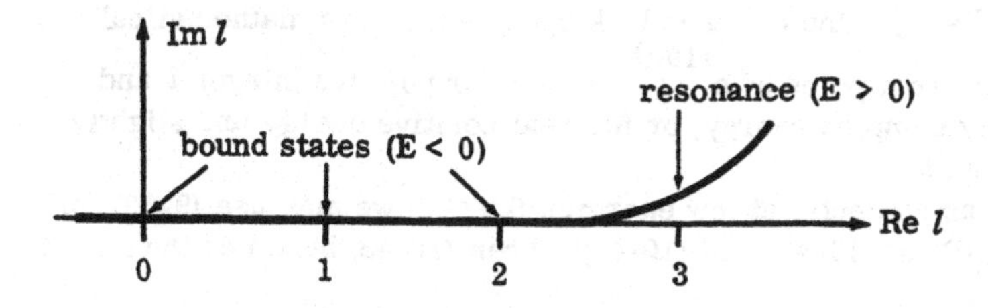

Fig. 10-3

Regge trajectory for a short ranged potential on which lie bound states for $l = 0, 1$, and 2 and a resonance at $l = 3$.

Let us consider what happens to the trajectories for $E > 0$. One sees directly from the conservation of probability that $\delta_l(E)$, and hence $\cot \delta_l(E)$, is real for $E > 0$ and integer[2] $l \geq 0$; thus quite generally, (10-40) can be satisfied for $E > 0$ only by a complex value of l. When a trajectory is near the real axis, that is,

$$l(E) = l_0(E) + il_1(E), \tag{10-41}$$

where $l_0(E)$ and $l_1(E)$ are real, and $l_1 \ll 1$, we may expand (10-40) about $l = l_0$ to discover

$$\cot \delta_{l_0}(E) + il_1 \left[\frac{\partial}{\partial l} \cot \delta_l(E) \right]_{l=l_0} = i. \tag{10-42}$$

Comparing the real and imaginary parts of (10-42) we find

$$\cot \delta_{l_0}(E) = 0, \tag{10-43}$$

as the equation determining the real part of $l(E)$ as a function of E; and

$$l_1 = \left[\frac{\partial}{\partial l} \cot \delta_l(E) \right]^{-1}_{l=l_0}, \tag{10-44}$$

which expresses the imaginary part of $l(E)$ in terms of l_0 and E.

We see from (9-73) (for integer l) that a resonance corresponds to a solution of (10-43) for integer l_0 and $E > 0$; that is, if the real part of a Regge trajectory passes through an integer l_0 at energy $E > 0$, as in Fig. 10-3 where $l_0 = 3$, there is a resonance at that energy and angular momentum[3] l_0. [An exception is s-waves at very low energy where $\cot \delta_0 = 0$ does not necessarily lead to a resonance in σ_0.] The closer l_1 is to zero, the sharper is the resonance. Basically, if $e^{2i\delta_l} - 1$ has a pole at a slightly complex value of $l = l_0 + il_1$ then by continuity it will be particularly large at the nearby real point $l = l_0$. One can thus look upon resonances mathematically as arising from poles of $e^{2i\delta_l(E)}$ either for positive integer l and slightly complex energy, or for real positive energy and slightly complex l.

As an example, at low energies [$kb \ll 1$] we may use (9-60) for $\cot \delta_{l_0}(E)$ in (10-43) and (10-44). Then (10-43) becomes the familiar

---

[2]The integer condition is not necessary.

[3]R. Newton, in his book *The Complex j Plane* [W.A. Benjamin, New York, 1964], gives many examples of calculated Regge trajectories in potential scattering.

equation

$$l_0 + 1 + b\alpha_{l_0}(E) = 0, \tag{10-45}$$

and (10-44) becomes, on using (10-43),

$$l_1 = \frac{(kb)^{2l_0+1}}{[(2l_0-1)!!]^2[1+b\partial\alpha_{l_0}(E)/\partial l_0]}$$

$$= -\frac{\Gamma_k}{2}\, b\, \frac{\partial\alpha_{l_0}(E)}{\partial E}\, \frac{1}{1+b\partial\alpha_{l_0}(E)/\partial l_0} \tag{10-46}$$

where Γ_k is given by (9-78). The derivation of (10-46), which at a resonance relates the imaginary part of $l(E)$ to the width of the resonance, is left as an exercise.

PROBLEMS

1. To a first approximation, the potential that a charged particle feels from a hydrogen atom can be thought of as that due to a positive point charge at the origin plus a uniform cloud of negative charge occupying a sphere of radius a_0 about the origin.
   (a) Calculate, in the Born approximation, the differential and total cross section for the scattering of a charged particle from the hydrogen atom. Neglect recoil of the hydrogen atom.
   (b) What is the form of the differential cross section for low energy? Compare this result with the pure Coulomb cross section.
   (c) Show that the differential cross section becomes more and more like a pure Coulomb cross section as the energy of the incident particle increases. Explain why this happens.
   (d) Calculate $\delta_0(E)$ and $\delta_1(E)$ in the Born approximation. Calculate the scattering length. Compare the results with those of part (b).
2. Consider the following somewhat oversimplified model of an atom: Assume the nucleus to have charge Ze, and the Z electrons to be uniformly distributed in a thin *shell* about the nucleus of radius $a_0/Z^{1/3}$, where a_0 is the Bohr radius. Assume that the atom is sufficiently heavy that $Z^{1/3}$ is $\gg 1$.
   (a) Calculate *exactly,* including p waves, the low energy scattering amplitude for the scattering of electrons from this atom. What is the scattering length? Sketch the total cross section versus energy.

(b) Discuss the possible bound states near zero energy.

3. The scattering from a uniform sphere of charge, with a radius $a \ll a_0$ and total charge e, is very much like that from a pure Coulomb potential ($a = 0$) if $ka \ll 1$. Calculate, to lowest order in ka, the deviations of the scattering amplitude of this potential from the Coulomb case.
4. Verify Eq. (10-46).

Chapter 11
STATIONARY STATE PERTURBATION THEORY

Often one meets a problem in quantum mechanics that can't be solved exactly, but looks very similar to a problem that can be solved exactly. For example, suppose that we have to find the energies and eigenstates of a Hamiltonian of the form

$$H = H_0 + V \tag{11-1}$$

where the eigenstates $|n\rangle$ and energies ε_n of H_0 are known exactly

$$H_0|n\rangle = \varepsilon_n|n\rangle . \tag{11-2}$$

If V is in some sense a small correction to H_0 then we might expect that the eigenstates $|N\rangle$ of H are only slightly different from the eigenstates $|n\rangle$ of H_0, and the energies E_n of H are only slightly different from the ε_n. In other words, if we put a parameter λ in front of V, so that

$$H = H_0 + \lambda V \tag{11-3}$$

then we expect that as the coupling constant λ is varied from zero to one each eigenstate $|n\rangle$ will smoothly become an eigenstate $|N\rangle$ of $H_0 + V$ and the energy ε_n will smoothly become the energy of the state $|N\rangle$.

Let us then look for expansions of the eigenstates and energies of $H_0 + \lambda V$ in powers of λ [eventually we shall set $\lambda = 1$, but for now it is a convenient way of keeping track of powers of V]:

$$|N\rangle = |n\rangle + \lambda|N^{(1)}\rangle + \lambda^2|N^{(2)}\rangle + \cdots \tag{11-4}$$

$$E_n = \varepsilon_n + \lambda E_n^{(1)} + \lambda^2 E_n^{(2)} + \cdots . \tag{11-5}$$

It should be emphasized right away that the existence of such expansions is a rather strong assumption. For example, the energy of the Cooper pair, Eq. (8-29), can certainly not be expanded in powers of v_0, the interaction strength. The bound states of a weak attractive potential never come out of doing perturbation theory on free particle states. They can't: expanding the states, say, in a power series implies that the physics with a small positive V is only slightly different from the physics with a small negative V. But this is certainly not the case for a bound state; a weak attractive potential can have a bound state, but a weak repulsive potential can't — a much different situation. Look, for example, at Eq. (4-65) for the bound state energy of a shallow square well:

$$E = \frac{-mV^2a^2}{2\hbar^2}.$$

This is definitely not the first (or second) term in a power series in V; were that the case, this formula would have to give us the correct result for either sign of V, but a state with $E < 0$ for a repulsive potential is clearly impossible. With this warning we return to the expansions.

Let us assume that the eigenstates of H_0 are normalized to 1, $\langle n|n\rangle = 1$, and let us choose the normalization of $|N\rangle$ so that

$$\langle n|N\rangle = 1. \tag{11-6}$$

Therefore

$$\langle n|N\rangle = 1 = \langle n|n\rangle + \lambda\langle n|N^{(1)}\rangle + \lambda^2\langle n|N^{(2)}\rangle + \cdots .$$

The coefficients in this equation of each power of λ must vanish individually, and therefore

$$\langle n|N^{(i)}\rangle = 0, \quad i = 1, 2, \ldots . \tag{11-7}$$

To find the coefficients in the power series for $|N\rangle$ and E_n we put the expansions (11-4) and (11-5) into the Schrödinger equation

$$(H_0 + \lambda V)|N\rangle = E_n|N\rangle \tag{11-8}$$

and compare the coefficients of powers of λ on both sides of (11-8). This yields the following equations:

$$H_0|n\rangle = \varepsilon_n|n\rangle \tag{11-9}$$

as the coefficient of λ^0. The λ^1 term is

$$H_0|N^{(1)}\rangle + V|n\rangle = \varepsilon_n|N^{(1)}\rangle + E_n^{(1)}|n\rangle \tag{11-10}$$

and in general for λ^k $(k \geq 1)$,

$$H_0|N^{(k)}\rangle + V|N^{(k-1)}\rangle = \varepsilon_n|N^{(k)}\rangle + E_n^{(1)}|N^{(k-1)}\rangle + \cdots + E_n^{(k)}|n\rangle. \tag{11-11}$$

Taking the scalar product of both sides of (11-10) with $\langle n|$ gives

$$\langle n|H_0|N^{(1)}\rangle + \langle n|V|n\rangle = \varepsilon_n \langle n|N^{(1)}\rangle + E_n^{(1)}, \tag{11-12}$$

since $\langle n|n\rangle = 1$. Using (11-2) in the first term on the left-hand side of (11-12) we find

$$E_n^{(1)} = \langle n|V|n\rangle. \tag{11-13}$$

Thus to first order (in λ, and therefore V), the energy shift is

$$E_n - \varepsilon_n = \lambda\langle n|V|n\rangle, \tag{11-14}$$

which is just the expectation value of the perturbation λV.

Taking the scalar product of both sides of (11-11) with $\langle n|$ and using (11-7) we find

$$E_n^{(k)} = \langle n|V|N^{(k-1)}\rangle. \tag{11-15}$$

Thus once we know the change in the state to a certain order, we can find the energy change in the next order from (11-15). The remaining problem is to find the change in the state $|n\rangle$ due to the perturbation. Notice that if we multiply both sides of (11-15) by λ^k and sum over k from one to infinity we find the simple result

$$E_n = \varepsilon_n + \lambda\langle n|V|N\rangle. \tag{11-16}$$

The kth-order change in the state $|N^{(k)}\rangle$ can be expanded in terms of the complete set of eigenstates of H_0:

$$|N^{(k)}\rangle = \sum_m{}' |m\rangle\langle m|N^{(k)}\rangle, \quad k = 1, 2, \ldots \tag{11-17}$$

where the prime on the sum means that, because of the orthogonality relation (11-7), we shouldn't include the state $|n\rangle$ in the sum. Now to find the coefficient $\langle m|N^{(i)}\rangle$ we take the scalar product of both sides of Eq. (11-11) with $\langle m|$. Thus

$$\langle m|H_0|N^{(k)}\rangle + \langle m|V|N^{(k-1)}\rangle = \varepsilon_n\langle m|N^{(k)}\rangle + \cdots + E_n^{(k-1)}\langle m|N^{(1)}\rangle.$$

However, $\langle m|H_0 = \varepsilon_m \langle m|$, so that if $\varepsilon_m \neq \varepsilon_n$,

$$\langle m|N^{(k)}\rangle = \frac{1}{\varepsilon_n - \varepsilon_m}\Big(\langle m|V|N^{(k-1)}\rangle - E_n^{(1)}\langle m|N^{(k-1)}\rangle - \cdots - E_n^{(k-1)}\langle m|N^{(1)}\rangle\Big). \tag{11-18}$$

This formula enables us to find the kth-order correction to the state in terms of lower-order corrections to $|N\rangle$ and ε_n, as long as the state $|n\rangle$ is nondegenerate. Later we shall consider what one must do if the level $|n\rangle$ being perturbed is degenerate.

Now let us examine the first few orders. Letting $k = 1$ in (11-18) we find

$$\langle m|N^{(1)}\rangle = \frac{1}{\varepsilon_n - \varepsilon_m}\langle m|V|n\rangle \tag{11-19}$$

since $|N^{(0)}\rangle = |n\rangle$ and $\langle m|n\rangle = 0$. Hence to *first order* in λ

$$|N\rangle = |n\rangle + \lambda \sum_m{}' |m\rangle \frac{\langle m|V|n\rangle}{\varepsilon_n - \varepsilon_m} \tag{11-20}$$

and

$$E_n = \varepsilon_n + \lambda \langle n|V|n\rangle. \tag{11-21}$$

[One may set $\lambda = 1$ in these formulas at this point; they clearly are first order in V.]

Now that we have found the first-order change in the state, we can find the second-order energy immediately from (11-15). Thus

$$E_n^{(2)} = \sum_m{}' \langle n|V|m\rangle \frac{1}{\varepsilon_n - \varepsilon_m}\langle m|V|n\rangle = \sum_m{}' \frac{|\langle n|V|m\rangle|^2}{\varepsilon_n - \varepsilon_m}. \tag{11-22}$$

Notice that if $|n\rangle$ is the ground state of H_0, then $\varepsilon_m - \varepsilon_n > 0$, and therefore $E_n^{(2)}$ is negative. We find from (11-18) that the second-order term $|N^{(2)}\rangle$ is

$$\sum_m{}'\sum_k{}' |m\rangle \frac{\langle m|V|k\rangle\langle k|V|n\rangle}{(\varepsilon_n - \varepsilon_k)(\varepsilon_n - \varepsilon_m)} - \sum_m{}' |m\rangle \frac{\langle m|V|n\rangle\langle n|V|n\rangle}{(\varepsilon_n - \varepsilon_m)^2}. \tag{11-23}$$

Consider as an example a one-dimensional harmonic oscillator perturbed by a constant force

$$V = -Fx. \tag{11-24}$$

Then the only nonvanishing matrix elements of V are

$$\langle n|V|n+1\rangle = -F\langle n|x|n+1\rangle = -F \left[\frac{\hbar(n+1)}{2m\omega}\right]^{1/2} \tag{11-25}$$

and

$$\langle n|V|n-1\rangle = -F \left[\frac{\hbar n}{2m\omega}\right]^{1/2}. \tag{11-26}$$

The first-order energy correction $E_n^{(1)}$ vanishes since $\langle n|x|n\rangle = 0$. The second-order change in the energy is

$$E_n^{(2)} = \sum_m{}' \frac{|\langle n|V|m\rangle|^2}{\varepsilon_n - \varepsilon_m} = \frac{|\langle n|V|n+1\rangle|^2}{-\hbar\omega} + \frac{|\langle n|V|n-1\rangle|^2}{\hbar\omega} = \frac{-F^2}{2m\omega^2}. \tag{11-27}$$

The energy of each state decreases by the same amount.

We could have learned this by a more direct argument, since

$$H = \frac{p^2}{2m} + \frac{m\omega^2 x^2}{2} - Fx = \frac{p^2}{2m} + \frac{m\omega^2}{2}\left(x - \frac{F}{m\omega^2}\right)^2 - \frac{F^2}{2m\omega^2},$$

which is again a Hamiltonian for simple harmonic motion, now centered about the point $F/m\omega^2$. The frequencies are the same and the only change in the energy is the constant term (11-27). One expects then that the states are just the old states translated in space by the distance $F/m\omega^2$, and are therefore given by[1]

$$|\bar{N}\rangle = e^{-ipF/m\omega^2\hbar}|n\rangle, \tag{11-28}$$

where p is the momentum operator. It is left as exercise to show from the perturbation theory that

$$|\bar{N}^{(1)}\rangle = -\frac{iF}{\hbar m\omega^2}\,p|n\rangle$$

$$|\bar{N}^{(2)}\rangle = -\left(\frac{F}{\hbar m\omega^2}\right)^2 p^2|n\rangle, \tag{11-29}$$

etc.

THE WAVE FUNCTION RENORMALIZATION CONSTANT

We have constructed the perturbed state $|N\rangle$ to have the normalization $\langle n|N\rangle = 1$. If we want the perturbed state to be normalized to

[1] Note that the perturbed state $|\bar{N}\rangle$ in (11-28) is already normalized to one; this is indicated by the bar. See Eq. (11-31).

one, we must multiply $|N\rangle$ by

$$(\langle N|N\rangle)^{-1/2} \equiv Z^{1/2}; \tag{11-30}$$

The state

$$|\overline{N}\rangle = Z^{1/2}|N\rangle \tag{11-31}$$

obeys

$$\langle\overline{N}|\overline{N}\rangle = 1. \tag{11-32}$$

The quantity Z is known as the *wave function renormalization constant.* Multiplying both sides of (11-31) by $\langle n|$ we see that

$$Z^{1/2} = \langle n|\overline{N}\rangle. \tag{11-33}$$

Let us calculate Z to second order. Using (11-6) we find

$$\begin{aligned}\langle N|N\rangle &= (\langle n| + \lambda\langle N^{(1)}| + \lambda^2\langle N^{(2)}|)(|n\rangle + \lambda|N^{(1)}\rangle + \lambda^2|N^{(2)}\rangle)\\ &= 1 + \lambda^2\langle N^{(1)}|N^{(1)}\rangle + \cdots .\end{aligned}$$

Thus

$$\langle N|N\rangle = 1 + \lambda^2 \sum_m{}' \frac{|\langle m|V|n\rangle|^2}{(\varepsilon_n - \varepsilon_m)^2}. \tag{11-34}$$

We see therefore that the state $|N\rangle$ is, to first order, normalized to 1, with the first correction occurring in second order. To this order

$$\begin{aligned}Z &= 1 - \lambda^2 \sum_m{}' \frac{|\langle m|V|n\rangle|^2}{(\varepsilon_n - \varepsilon_m)^2}\\ &= \frac{\partial}{\partial\varepsilon_n}\left(\varepsilon_n + \lambda\langle n|V|n\rangle + \lambda^2 \sum_m{}' \frac{|\langle m|V|n\rangle|^2}{\varepsilon_n - \varepsilon_m}\right) = \frac{\partial E_n}{\partial \varepsilon_n}.\end{aligned} \tag{11-35}$$

This result is true to all orders: if the system is in the eigenstate $|N\rangle$, the probability $Z = |\langle n|\overline{N}\rangle|^2$ of observing it in the unperturbed state $|n\rangle$ is given by the partial derivative of the perturbed energy with respect to the unperturbed energy, keeping fixed the matrix elements of the perturbation as well as the other ε_m.

DEGENERATE PERTURBATION THEORY

It is clear from Eqs. (11-20), (11-22), and (11-23) that the above formulation of perturbation theory is essentially an expansion in quantities like $\lambda\langle n|V|m\rangle/(\varepsilon_n - \varepsilon_m)$. We may expect it to be more rapidly converging the smaller are the matrix elements of V compared with the level spacing of the unperturbed system. On the other hand if there are any states for which $\varepsilon_n = \varepsilon_m$ but $\langle n|V|m\rangle \neq 0$, then the theory breaks down. Let us consider what must be done in this case.

Suppose that we have a group of states

$$|n_a\rangle, |n_b\rangle, \ldots, |n_k\rangle$$

that are degenerate states of the unperturbed Hamiltonian H_0:

$$H_0|n_i\rangle = \varepsilon_n|n_i\rangle \quad \text{for } i = a, b, c, \ldots, k. \tag{11-36}$$

Then if $\langle n_a|V|n_b\rangle$ is nonzero for $a \neq b$, the previous perturbation method will fail. However, any linear combination of the states $|n_a\rangle$, $|n_b\rangle, \ldots, |n_k\rangle$ is also an eigenstate with the same energy ε_n. Thus if we can choose a set of k orthogonal states

$$|n_\alpha\rangle = \sum_{i=a}^{k} C_{\alpha i}|n_i\rangle \tag{11-37}$$

such that

$$\langle n_\alpha|V|n_\beta\rangle = 0 \quad \text{if } \alpha \neq \beta, \tag{11-38}$$

we can use the perturbation procedure as given, for then vanishing energy denominators will always be accompanied by vanishing numerators and we will have no trouble. The correct choice of basis states to use in doing the perturbation expansion is therefore the one that diagonalizes V within each group of degenerate states, i.e., $\langle n_\alpha|V|n_\beta\rangle = 0$ if $\alpha \neq \beta$ and $\varepsilon_{n_\alpha} = \varepsilon_{n_\beta}$.

Should we be interested in the perturbation of only one group of degenerate states of H_0 with energy ε_n, then it is only necessary to diagonalize V within that one group of states. This is because all the energy denominators that occur in the perturbation expansion are energy differences between ε_n and other states, and thus other degeneracies of H_0 don't lead to vanishing denominators.

The problem of diagonalizing V within a group of states $|n_a\rangle$, $|n_b\rangle, \ldots$ is just that of finding the eigenvector of the $k \times k$ matrix

$$\begin{pmatrix} \langle n_a|V|n_a\rangle & \langle n_a|V|n_b\rangle & \langle n_a|V|n_c\rangle & \cdots \\ \langle n_b|V|n_a\rangle & \langle n_b|V|n_b\rangle & \cdots & \cdots \\ \cdots & \cdots & \cdots & \cdots \end{pmatrix}. \tag{11-39}$$

We shall now show that if the $C_{\alpha i}$ are the components of the eigenvectors $\mathbf{C}_\alpha$ of this matrix, then (11-38) is satisfied. To see this, suppose that

$$\sum_i \langle n_j|V|n_i\rangle\, C_{\alpha i} = E_{n_\alpha}{}^{(1)} C_{\alpha j}, \tag{11-40}$$

where $E_{n_\alpha}{}^{(1)}$ is the name of the eigenvalue, and that C_α is normalized to one:

$$\sum_i |C_{\alpha i}|^2 = 1. \tag{11-41}$$

Multiplying on the left by $C_{\beta j}{}^*$, the components of another eigenvector of unit length, we have

$$\left(\sum_i C_{\beta j}{}^*\langle n_j|\right) V \left(\sum_i C_{\alpha i}|n_i\rangle\right) = E_{n_\alpha}{}^{(1)} \sum_j C_{\beta j}{}^* C_{\alpha j}.$$

However,

$$\sum_j C_{\beta j}{}^* C_{\alpha j} = \delta_{\alpha\beta} \sum_j C_{\beta j}{}^* C_{\beta j} = \delta_{\alpha\beta} \tag{11-42}$$

since the eigenvectors of the $k \times k$ Hermitian matrix $\langle n_a|V|n_b\rangle$ can be taken to form an orthonormal set. Thus the vectors

$$|n_\alpha\rangle = \sum_i C_{\alpha i}|n_i\rangle \tag{11-43}$$

obey

$$\langle n_\beta|V|n_\alpha\rangle = E_{n_\alpha}{}^{(1)}\, \delta_{\alpha\beta}. \tag{11-44}$$

The eigenvalue $E_{n_\alpha}{}^{(1)}$ is clearly λ^{-1} times the first-order energy shift of the state $|n_\alpha\rangle$.

Thus the group of states $|n_a\rangle$, $|n_b\rangle, \ldots$ in the presence of the perturbation splits into k states $|N_\alpha\rangle$, $|N_\beta\rangle, \ldots$ which are given in first order by

$$|N\alpha\rangle = |n\alpha\rangle + \sum_m{}' \frac{|m\rangle\langle m|V|n\alpha\rangle}{\varepsilon_n - \varepsilon_m} \tag{11-45}$$

where the prime on the sum over m means that the sum runs over all states *except* the k states $|n_\alpha\rangle$, $|n_\beta\rangle, \ldots$ The energy shift to second order is

$$E_{n\alpha} = \varepsilon_n + \lambda\langle n_\alpha|V|n_\alpha\rangle + \lambda^2 \sum_m{}' \frac{|\langle m|V|n_\alpha\rangle|^2}{\varepsilon_n - \varepsilon_m} \tag{11-46}$$

where $\langle n_\alpha|V|n_\alpha\rangle$ is the eigenvalue $E_{n\alpha}^{(1)}$ of the matrix $\langle n_a|V|n_b\rangle$.

As an example let us consider the splitting by a uniform electric field $\mathscr{E}$, of the fourfold degenerate n = 2 levels of the hydrogen atom This phenomenon is known as the *Stark effect.* The four n = 2 states are

$$|2S_0\rangle, \quad |2P_1\rangle, \quad |2P_0\rangle, \quad |2P_{-1}\rangle.$$

Taking the field to be in the z direction, the perturbation λV is $e\mathscr{E}z$. First of all we notice that the matrix elements of λV between states with different eigenvalues of L_z all vanish. This is because $e\mathscr{E}z$ commutes with L_z. Thus, for example

$$\begin{aligned} 0 = \langle 2P_{-1}|[e\mathscr{E}z, L_z]|2P_1\rangle &= \langle 2P_{-1}|e\mathscr{E}z(L_z|2P_1\rangle) - (\langle 2P_{-1}|L_z)e\mathscr{E}z|2P_1\rangle \\ &= 2\hbar\langle 2P_{-1}|e\mathscr{E}z|2P_1\rangle \end{aligned}$$

so that

$$\langle 2P_{-1}|e\mathscr{E}z|2P_1\rangle = 0.$$

This is a quite general rule: if the perturbation commutes with a constant of the motion, A, of the unperturbed Hamiltonian, then matrix elements of the perturbation between eigenstates of A with different eigenvalues must vanish. This rule, an example of a *selection rule,* often simplifies the problem of selecting the correct linear combination of degenerate states with which to do the perturbation expansion. The only nonvanishing off-diagonal matrix elements $\langle n_a|V|n_b\rangle$ are $\langle 2S_0|e\mathscr{E}z|2P_0\rangle$ and $\langle 2P_0|e\mathscr{E}z|2S_0\rangle$. We must only worry about the degeneracy between the two m = 0 states.

Because the hydrogen atom is invariant under the parity operation: $\mathbf{r} \to -\mathbf{r}$, the diagonal matrix elements of $e\mathscr{E}z$ are zero; since $|\psi_{nlm}(\mathbf{r})|^2 = |\psi_{nlm}(-\mathbf{r})|^2$ for unperturbed hydrogen atom wave functions, the electron in the state $|nlm\rangle$ is as likely to be found at $\mathbf{r}$ as at $-\mathbf{r}$. Thus $\langle nlm|z|nlm\rangle$, the expectation value of z, must vanish.

We must therefore find the eigenvalues of the matrix

$$\begin{pmatrix} 0 & \langle 2P_0|e\mathcal{E}z|2S_0\rangle \\ \langle 2S_0|e\mathcal{E}z|2P_0\rangle & 0 \end{pmatrix}. \tag{11-47}$$

Since

$$\langle 2P_0|z|2S_0\rangle = \langle 2S_0|z|2P_0\rangle = -3a_0, \tag{11-48}$$

the eigenvectors are $\frac{1}{\sqrt{2}}\begin{pmatrix} 1 \\ \pm 1 \end{pmatrix}$ with eigenvalues $\mp 3a_0$. Hence the state which in zero order is $(1/\sqrt{2})\,(|2S_0\rangle + |2P_0\rangle)$ will have a first-order energy shift $-3e\mathcal{E}a_0$, while the state which in zero order is $(1/\sqrt{2})\,(|2S_0\rangle - |2P_0\rangle)$ will have a first-order energy shift $+3e\mathcal{E}a_0$. In order of magnitude the energy shift is a dimensionless factor $\mathcal{E}/(e/a_0{}^2)$ times the spacing of the low-lying hydrogen levels; the "atomic field" $e/a_0{}^2$ is the electric field of a proton at one Bohr radius;

$$\frac{e}{a_0{}^2} = 5.15 \times 10^9 \text{ volts/cm}. \tag{11-49}$$

Thus the perturbation expansion converges rapidly for everyday electric field strengths.

If the atom is originally, for example, in a $|2S_0\rangle$ state before the field is turned on, then in the presence of the electric field this state becomes a linear combination of the two perturbed eigenstates that in lowest order are

$$\frac{1}{\sqrt{2}}(|2S_0\rangle + |2P_0\rangle), \qquad \frac{1}{\sqrt{2}}(|2S_0\rangle - |2P_0\rangle).$$

If we measure the energy of the perturbed $|2S_0\rangle$ state we find either

$$-\frac{e^2}{2a_0}\left(\frac{1}{4} + \frac{6\mathcal{E}}{e/a_0{}^2}\right) \quad \text{or} \quad -\frac{e^2}{2a_0}\left(\frac{1}{4} - \frac{6\mathcal{E}}{e/a_0{}^2}\right)$$

with equal probability.

Notice that in the nondegenerate case if the diagonal matrix element $\langle n|V|n\rangle$ vanishes then the first energy shift is second order; however, as we see in the Stark effect one can still get a first-order energy shift in the degenerate case even if the diagonal matrix elements of V vanish in the original basis.

VAN DER WAALS INTERACTION

An important application of perturbation theory is the derivation of the energy of interaction of two widely separated atoms. To see

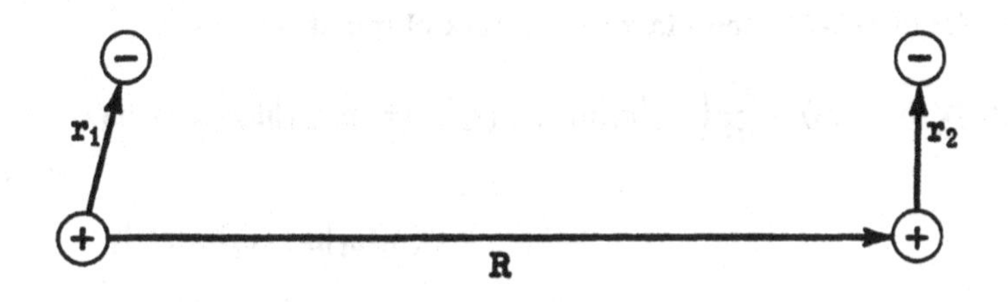

Fig. 11-1

how this works let's consider the interaction between two hydrogen atoms, treating the two protons as fixed point charges separated by a vector $\mathbf{R}$. Let $\mathbf{r}_1$ be the vector from the first proton to its electron and $\mathbf{r}_2$ the vector from the second proton to its electron [Fig. 11-1]. Then the interaction between the two atoms is the sum of the various Coulomb interactions between the charges on the first atom and the second atom:

$$V = e^2 \left[\frac{1}{R} + \frac{1}{|\mathbf{R} + \mathbf{r}_2 - \mathbf{r}_1|} - \frac{1}{|\mathbf{R} + \mathbf{r}_2|} - \frac{1}{|\mathbf{R} - \mathbf{r}_1|} \right]. \tag{11-50}$$

If R is many Bohr radii we may expand the denominators in (11-50) in powers of $\mathbf{r}_1/R$ and $\mathbf{r}_2/R$. The first nonvanishing terms are second order:

$$V = -e^2 (\mathbf{r}_1 \cdot \nabla)(\mathbf{r}_2 \cdot \nabla) \frac{1}{R} = e^2 \left[\frac{\mathbf{r}_1 \cdot \mathbf{r}_2}{R^3} - \frac{3(\mathbf{r}_1 \cdot \mathbf{R})(\mathbf{r}_2 \cdot \mathbf{R})}{R^5} \right]. \tag{11-51}$$

Thus, at large separation the interaction between the atoms is simply that between two dipoles $e\mathbf{r}_1$ and $e\mathbf{r}_2$ separated by $\mathbf{R}$. If we choose the z axis along $\mathbf{R}$ then the interaction operator (11-51) becomes

$$V = \frac{e^2}{R^3} (x_1 x_2 + y_1 y_2 - 2 z_1 z_2). \tag{11-52}$$

Suppose that in the absence of this interaction, atom 1 is in state $|n_1\rangle$ and atom 2 is in state $|n_2\rangle$. The unperturbed energy of the two atoms is $\varepsilon_{n_1} + \varepsilon_{n_2}$. To find the interaction energy of the two atoms we calculate the energy of two atoms including V by perturbation theory. This energy depends on R, and hence behaves as an effective potential energy of interaction. To first order

$$E(R) = \varepsilon_{n_1} + \varepsilon_{n_2} + \langle n_1 n_2 | V(R) | n_1 n_2 \rangle. \tag{11-53}$$

Now from (11-52) the diagonal matrix element of V is

$$\langle n_1n_2|V(R)|n_1n_2\rangle = \frac{e^2}{R^3}\Big(\langle n_1|x_1|n_1\rangle\langle n_2|x_2|n_2\rangle + \langle n_1|y_1|n_1\rangle\langle n_2|y_2|n_2\rangle - 2\langle n_1|z_1|n_1\rangle\langle n_2|z_2|n_2\rangle\Big), \tag{11-54}$$

since the wave function of the state $|n_1n_2\rangle$ is a simple product of the wave functions of the states $|n_1\rangle$ and $|n_2\rangle$. If either of the atoms is in its ground state then from the reflection symmetry of the ground state wave function, (11-54) vanishes and the first correction to the energy appears in second order. On the other hand, if both of the atoms are in degenerate states we must carry out the necessary diagonalization procedure for V; then we find a nonvanishing interaction energy in first order. This interaction energy behaves as $1/R^3$.

To second order, assuming $|n_1\rangle$ and $|n_2\rangle$ to be nondegenerate,

$$E^{(2)}(R) = \frac{e^4}{R^6}\sum_{m_1m_2}{}' \frac{|\langle n_1n_2|(x_1x_2 + y_1y_2 - 2z_1z_2)|m_1m_2\rangle|^2}{\varepsilon_{n_1} + \varepsilon_{n_2} - \varepsilon_{m_1} - \varepsilon_{m_2}}. \tag{11-55}$$

The state with $m_1 = n_1$ and $m_2 = n_2$ is excluded from the sum. We see two immediate features of this interaction energy; first, if both sums are in their ground states then $E^{(2)}(R)$ is negative. Furthermore, the interaction energy falls off as the inverse sixth power of the separation of the atoms. Let us write

$$\zeta = \frac{e^2}{a_0{}^5}\sum_{m_1m_2}{}' \frac{|\langle n_1n_2|(x_1x_2 + y_1y_2 - 2z_1z_2)|m_1m_2\rangle|^2}{\varepsilon_{m_1} + \varepsilon_{m_2} - \varepsilon_{n_1} - \varepsilon_{n_2}}; \tag{11-56}$$

ζ is a dimensionless constant whose value, from detailed calculations for the case of two hydrogen atoms in their ground states, is 6.5.

The total energy of the two atoms is thus

$$E(R) = \varepsilon_{n_1} + \varepsilon_{n_2} - \frac{e^2}{a_0}\left(\frac{a_0}{R}\right)^6 \zeta; \tag{11-57}$$

the second-order energy acts as an effective potential energy for the interaction between two widely separated hydrogen atoms in their ground states. This effective potential is known as a *van der Waals* interaction. The physics behind this interaction is that even though a hydrogen atom in its ground state has no permanent electric dipole moment ($\langle -er\rangle = 0$), the presence of a second atom in the vicinity of this first atom induces a dipole moment in the first whose components are proportional to the dipole moment operator of the second atom times R^{-3}. The van der Waals interaction is essentially the interaction ($\sim R^{-3}$) of this induced dipole moment with the dipole moment operator of the second atom (and vice versa). [The interaction is certainly not that between the induced dipole moment of the first atom ($\sim R^{-3}$) and the induced dipole moment of the second atom ($\sim R^{-3}$) since that goes as R^{-9}.] Atoms in their ground states attract each other; however atoms in excited states can either attract or repel each other.

The interaction at short distances is very complicated, because it is then necessary to take into account the exclusion principle between the electrons on the two atoms. Noble gas atoms – helium, argon, etc. – tend to repel each other at short distances because the electrons refuse to overlap each other. This repulsion leads to an energy behaving roughly as R^{-12}. The combination of a repulsive R^{-12} potential plus an attractive R^{-6} van der Waals potential is known as a *6-12 potential*. We shall study this short ranged interaction when we look at molecules.

ALMOST DEGENERATE PERTURBATION THEORY

Let us consider how to do perturbation theory when two energy levels of H_0 are very closely spaced. This is a troublesome case because the perturbation theory expands in quantities like $\langle n|V|m\rangle/(\varepsilon_n - \varepsilon_m)$ and if $|\varepsilon_n - \varepsilon_m|$ is ever small compared with $|\langle n|V|m\rangle|$ then the expansions for the effects of the perturbation on the states $|n\rangle$ and $|m\rangle$ will converge very slowly.

Suppose for simplicity that only two eigenstates $|n\rangle$ and $|m\rangle$ of H_0 are very closely spaced in energy. To develop a more rapidly convergent perturbation expansion we can proceed in two steps, analogous to the procedure for degenerate states. First we handle the effects of the matrix elements of the perturbation, $\langle n|V|n\rangle$, $\langle m|V|m\rangle$, $\langle n|V|m\rangle$, and $\langle m|V|n\rangle$ exactly, and then treat the rest of the perturbation by ordinary perturbation theory. Let us therefore write V as

$$V = \sum_{ij} |i\rangle\langle i|V|j\rangle\langle j| \tag{11-58}$$

where the sums are over eigenstates of H_0. Then let us separate out the part of V referring to $|n\rangle$ and $|m\rangle$ alone by writing

$$V = V_1 + V_2 \tag{11-59}$$

where

$$V_1 = |m\rangle\langle m|V|m\rangle\langle m| + |n\rangle\langle n|V|n\rangle\langle n| + |m\rangle\langle m|V|n\rangle\langle n| + |n\rangle\langle n|V|m\rangle\langle m|. \tag{11-60}$$

Our procedure shall be to find the exact eigenstates and eigenvalues of the Hamiltonian

$$H_1 = H_0 + V_1 \tag{11-61}$$

and then to treat V_2 by ordinary perturbation theory. Since $\langle n|V_2|n\rangle = \langle m|V_2|m\rangle = \langle n|V_2|m\rangle = \langle m|V_2|n\rangle = 0$, the closeness of the levels ε_n and ε_m will prove no problem in applying the standard perturbation theory to V_2.

Clearly if $|i\rangle$ is an eigenstate of H_0 other than $|m\rangle$ or $|n\rangle$ then it is also an eigenstate of H_1, since $V_1|i\rangle = 0$. However, neither $|m\rangle$ nor $|n\rangle$ is an eigenstate of H_1; we must therefore find the linear combinations $\alpha|n\rangle + \beta|m\rangle$ of these two states that diagonalize H_1:

$$H_1(\alpha|n\rangle + \beta|m\rangle) = E'(\alpha|n\rangle + \beta|m\rangle). \tag{11-62}$$

Now

$$H_1|n\rangle = E_n^{(1)}|n\rangle + \langle m|V|n\rangle|m\rangle$$

$$H_1|m\rangle = E_m^{(1)}|m\rangle + \langle n|V|m\rangle|n\rangle \tag{11-63}$$

where $E_n^{(1)} = \varepsilon_n + \langle n|V|n\rangle$, etc.; thus substituting (11-60) into (11-62) and comparing coefficients of $|n\rangle$ and $|m\rangle$ we see that α and β must satisfy

$$\begin{pmatrix} E_n^{(1)} & \langle n|V|m\rangle \\ \langle m|V|n\rangle & E_m^{(1)} \end{pmatrix} \begin{pmatrix} \alpha \\ \beta \end{pmatrix} = E' \begin{pmatrix} \alpha \\ \beta \end{pmatrix}. \tag{11-64}$$

There are two sets of solutions which, to within a normalizing constant, are

$$\alpha = \langle n|V|m\rangle$$

$$\beta_\pm = \frac{E_m^{(1)} - E_n^{(1)}}{2} \pm \sqrt{\left(\frac{E_m^{(1)} - E_n^{(1)}}{2}\right)^2 + |\langle n|V|m\rangle|^2} \tag{11-65}$$

corresponding to the energy eigenvalues of H_1

$$E_{\pm}' = \frac{E_m^{(1)} + E_n^{(1)}}{2} \pm \sqrt{\left(\frac{E_m^{(1)} - E_n^{(1)}}{2}\right)^2 + |\langle n|V|m\rangle|^2}. \qquad (11\text{-}66)$$

Thus we have found the exact eigenstates and energies of H_1. The next step is to handle V_2 by perturbation theory in the usual manner.

As an example of this type of problem let us consider an electron in a one-dimensional lattice, and assume that the periodic potential, $v(x)$, is weak. Let d be the lattice constant, so that $v(x) = v(x + d)$; let us take the lattice to be of length $L = Nd$ and apply periodic boundary conditions to the wave functions:

$$\psi(x) = \psi(x+L). \qquad (11\text{-}67)$$

The unperturbed wave functions are plane waves

$$\psi_k(x) = \frac{e^{ikx}}{L^{1/2}} \qquad (11\text{-}68)$$

where k must be an integer times $2\pi/L$ in order to satisfy the periodic boundary conditions. The unperturbed energies are $\varepsilon_k = \hbar^2 k^2/2m$.

Because v is periodic, we can expand it in a Fourier series

$$v(x) = \sum_{n=-\infty}^{\infty} e^{iKnx}\, v_n \qquad (11\text{-}69)$$

where n runs over the integers and $K = 2\pi/d$. The only nonvanishing matrix elements of v are therefore

$$\langle k+nK|v|k\rangle = v_n. \qquad (11\text{-}70)$$

The potential only joins together states differing in wave number by an integer multiple of the "reciprocal lattice vector" K.

In general ε_k and ε_{k+nK} are unequal; however, if k is near $-nK/2$ then these energies can be very close to each other. Thus if k is not near a point $-nK/2$, $n \neq 0$, the first-order wave functions are

$$\Psi_k(x) = \frac{e^{ikx}}{L^{1/2}} + \sum_{n \neq 0} \frac{e^{i(k+nK)x}}{L^{1/2}} \frac{v_n}{\varepsilon_k - \varepsilon_{k+nK}} \qquad (11\text{-}71)$$

and the second-order energies are

$$E_k = \varepsilon_k + v_0 + \sum_{n \neq 0} \frac{|v_n|^2}{\varepsilon_k - \varepsilon_{k+nK}}. \qquad (11\text{-}72)$$

[Note that (11-71) is explicitly in the form of a Bloch wave, e^{ikx} times a periodic function of x with period d.] If k is near $-nK/2$ we must apply the procedure for almost degenerate energy levels, since the states ψ_k and ψ_{k+nK} are then strongly mixed by the perturbation. To calculate this mixing we must diagonalize the matrix

$$\begin{pmatrix} \varepsilon_k + v_0 & v_n^* \\ v_n & \varepsilon_{k+nK} + v_0 \end{pmatrix} .$$

The eigenvalues are, from (11-66),

$$E_{k\pm} = v_0 + \frac{\varepsilon_k + \varepsilon_{k+nK}}{2} \pm \sqrt{\left(\frac{\varepsilon_k - \varepsilon_{k+nK}}{2}\right)^2 + |v_n|^2}. \tag{11-73}$$

For $|\varepsilon_k - \varepsilon_{k+nK}| \gg |v_n|$, (11-73) reduces in first order to the two solutions

$$E_{k+} = \varepsilon_k + v_0$$

$$E_{k-} = \varepsilon_{k+nK} + v_0 \tag{11-74}$$

which are the first-order energy shifts of the states k and k + nK. However as k approaches $-nK/2$ we find that $E_{k\pm}$ becomes $\varepsilon_{nK/2} \pm |v_n|$. Thus in this order the energy spectrum develops gaps of magnitude $2|v_n|$ at the "zone boundaries" $k = nK/2$. [See Fig. 11-2.] These

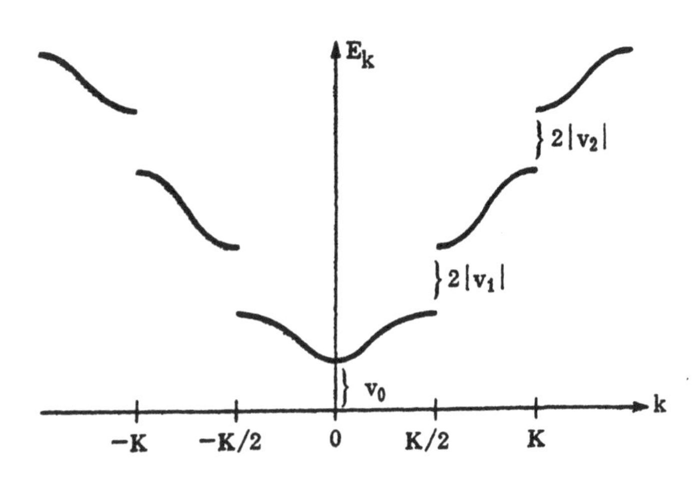

Fig. 11-2
Structure of the energy bands in the weak binding approximation.

energy gaps are just like those we found in the Kronig-Penney model and, as we remarked then, they are characteristic of periodic potentials. Treating a weak periodic potential by "almost-degenerate" perturbation theory is known in solid state physics as the *weak binding approximation.*

BRILLOUIN-WIGNER PERTURBATION THEORY

The Rayleigh-Schrödinger perturbation theory — the one we have developed — yields rather complicated expressions for the change in the states in second order and beyond. There is a somewhat different form of the perturbation theory, due to Brillouin and Wigner, that enables us to see more clearly the structure of the perturbed states. Let us multiply the Schrödinger equation

$$(E_n - H_0)|N\rangle = \lambda V|N\rangle \tag{11-75}$$

on the left by $\langle m|$. Thus

$$(E_n - \varepsilon_m)\langle m|N\rangle = \lambda \langle m|V|N\rangle. \tag{11-76}$$

Let us choose the normalization $\langle n|N\rangle = 1$. If $|m\rangle = |n\rangle$ this gives us the simple formula (11-16) for the perturbed energy. Now writing

$$|N\rangle = \sum_m |m\rangle\langle m|N\rangle = |n\rangle\langle n|N\rangle + \sum_m{}' |m\rangle\langle m|N\rangle, \tag{11-77}$$

and dividing both sides of (11-76) by $E_n - \varepsilon_m$, we have

$$|N\rangle = |n\rangle + \sum_m{}' |m\rangle \frac{1}{E_n - \varepsilon_m} \lambda\langle m|V|N\rangle. \tag{11-78}$$

This is the basic formula of the Brillouin-Wigner perturbation theory. From it we can develop a series expansion of $|N\rangle$ in powers of λ with coefficients depending on the perturbed energy E_n rather than on ε_n. Iterating (11-78) we find

$$\begin{aligned}|N\rangle = |n\rangle &+ \lambda \sum_m{}' |m\rangle \frac{1}{E_n - \varepsilon_m}\langle m|V|n\rangle \\ &+ \lambda^2 \sum_{jm}{}' |j\rangle \frac{1}{E_n - \varepsilon_j}\langle j|V|m\rangle \frac{1}{E_n - \varepsilon_m}\langle m|V|n\rangle \\ &+ \lambda^3 \sum_{kjm}{}' |k\rangle \frac{1}{E_n - \varepsilon_k}\langle k|V|j\rangle \frac{1}{E_n - \varepsilon_j}\langle j|V|m\rangle \frac{1}{E_n - \varepsilon_m}\langle m|V|n\rangle + \cdots .\end{aligned} \tag{11-79}$$

This is not a simple power series expansion in λ because E_n also depends on λ. However, if we expand the terms $(E_n - \varepsilon_m)^{-1}$ in powers of λ, we then recover the Rayleigh-Schrödinger series for $|N\rangle$. The λ^k term for $|N\rangle$ in the Brillouin-Wigner series is the same as the leading term of order λ^k in the Rayleigh-Schrödinger series only with E_n instead of ε_n; the extra terms in the Rayleigh-Schrödinger series of order λ^k are just the corrections to the energy denominators of the lower order, $(\lambda^{k-1}, \lambda^{k-2}, \ldots)$ terms. For example, the Rayleigh-Schrödinger expression for the states to second order can be rewritten as

$$|N\rangle = |n\rangle + \lambda \sum_m{}' |m\rangle \left(\frac{1}{\varepsilon_n - \varepsilon_m} - \frac{\lambda\langle n|V|n\rangle}{\varepsilon_n - \varepsilon_m} \right) \langle m|V|n\rangle$$

$$+ \lambda^2 \sum_{mj}{}' |j\rangle \frac{1}{\varepsilon_n - \varepsilon_j} \langle j|V|m\rangle \frac{1}{\varepsilon_n - \varepsilon_j} \langle m|V|n\rangle . \tag{11-80}$$

The coefficient in the λ term is $(E_n - \varepsilon_m)^{-1}$ expanded to first order in λ.

The Brillouin-Wigner series is not straightforward to evaluate, since it involves the perturbed energy E_n rather than the unperturbed energy ε_n. However, this is often an advantage. For instance, if we explicitly evaluate (11-79) as a function of E_n up to a given order in λ and use this expression for $|N\rangle$ in (11-16) we find a nonlinear equation for E_n. The solution of this equation is often much more accurate than a simple power series expansion. As an example, this nonlinear equation, taking the first two terms on the right in (11-79) is

$$E_n = \varepsilon_n + \lambda\langle n|V|n\rangle + \lambda^2 \sum_m{}' \frac{|\langle m|V|n\rangle|^2}{E_n - \varepsilon_m} . \tag{11-81}$$

Generally, if we have a good idea of the value of E_n, the Brillouin-Wigner perturbation series for $|N\rangle$ should converge more rapidly than the Rayleigh-Schrödinger series.

NONPERTURBATIVE METHODS

Let us consider a very important nonperturbative (i.e., no series expansions) method of finding approximate ground state energies and wave functions — the *Rayleigh-Ritz variational principle.* This principle is based on the simple observation that the expectation value of H in any state $|\psi\rangle$ is always greater than or equal to the ground state energy, E_0, i.e.,

$$\frac{\langle\psi|H|\psi\rangle}{\langle\psi|\psi\rangle} \geq E_0. \tag{11-82}$$

This relation is an equality only if $|\psi\rangle$ is the true ground state. To prove (11-82) we write H in terms of its eigenstates

$$H = \sum_N |N\rangle E_n \langle N|.$$

Thus

$$\langle\psi|H|\psi\rangle = \sum_N \langle\psi|N\rangle E_N \langle N|\psi\rangle \geq \sum_N \langle\psi|N\rangle E_0 \langle N|\psi\rangle = E_0\langle\psi|\psi\rangle. \tag{11-83}$$

As long as $|\psi\rangle$ has a component along any state that is not the ground state then clearly (11-82) is an inequality.

The principle works as follows. First of all, if we pick any $|\psi\rangle$, then $\langle\psi|H|\psi\rangle/\langle\psi|\psi\rangle$ always gives an upper bound to the ground state energy. Furthermore, the closer $|\psi\rangle$ is to being the true ground state, the more closely (11-82) will be an equality. Thus if we pick a "trial" $|\psi\rangle$ containing a parameter α, for example, $\psi(x) = e^{-\alpha x^2}$, in one dimension, then if we choose the value of α that minimizes $\langle\psi|H|\psi\rangle/\langle\psi|\psi\rangle$, the $|\psi\rangle$ for that α is the closest of all the $|\psi(\alpha)\rangle$ to the true ground state. Generally, the person whose $|\psi\rangle$ gives the least value for $\langle\psi|H|\psi\rangle/\langle\psi|\psi\rangle$, has the most accurate approximation to the ground state energy and wave function.

The variational principle can also work occasionally for excited states. For example, if H is spherically symmetric then if we limit the trial functions to p states, the minimum of $\langle\psi|H|\psi\rangle/\langle\psi|\psi\rangle$ will be the lowest energy p state; d state trial functions will give the lowest energy d state, etc.

In the special case that V (in $H_0 + V$) is a positive operator, in the sense that

$$\langle\psi|V|\psi\rangle \geq 0 \tag{11-84}$$

for *all* $|\psi\rangle$, then the ground state energy ε_0 of H_0 is a *lower bound* on the ground state energy E_0 of $H_0 + V$. To see this we simply remark that if $|0\rangle$ is the normalized ground state of $H_0 + V$ then

$$E_0 = \langle 0|(H_0 + V)|0\rangle \geq \langle 0|H_0|0\rangle$$

using (11-84). But from the Rayleigh-Ritz variational principle $\langle 0|H_0|0\rangle \geq \varepsilon_0$, so that

$$E_0 \geq \varepsilon_0. \tag{11-85}$$

Throwing away a positive term from H always lowers the ground state energy.

PROBLEMS

1. Show that in general the energy to second order is the expectation value of the Hamiltonian in the first-order normalized states.
2. Calculate the van der Waals interactions in lowest order between two hydrogen atoms in $n = 2$ states. Interpret the results.
3. (a) Obtain for two hydrogen atoms a lower bound on the coefficient ξ in the second-order van der Waals interaction by keeping only $n = 2$ states in the sum over states. Show that it is a lower bound.
   (b) Obtain an upper bound on ξ by replacing the energy denominators in the sum over states by an appropriate constant and using the completeness relation to evaluate the sum over states.
4. (a) A hydrogen atom originally in its ground state is placed in a uniform electric field. Write down an expression (to lowest order in the field) for the electric dipole moment induced in the atom by the field.
   (b) An alternative method for calculating the dipole moment is to differentiate the energy with respect to the electric field. Show that this method yields the same expression found in (a).
   (c) Applying the procedures of Problem 3, find upper and lower bounds on the induced dipole moment.
5. In the weak binding approximation (in one dimension) calculate, to first order V, the wave functions near the zone boundaries, $k = nK/2$.
6. (a) Using Brillouin-Wigner perturbation theory show that $(\partial E_n/\partial \varepsilon_n) = Z$ is exact to all orders in the perturbation, where Z is the wave function renormalization constant, E_n the exact energy, ε_n the unperturbed energy, and the derivative is at constant matrix elements and ε_m for $m \neq n$.
   (b) For the one-dimensional harmonic oscillator perturbed by a constant force F, calculate Z to second order directly from (11-28) and compare with $\partial E_n/\partial \varepsilon_n$ calculated from (11-27).
7. Estimate the ground state energy of the hydrogen atom using a three-dimensional harmonic oscillator ground state wave function as a trial function.
8. Find upper and lower bounds for the ground state energy of the (three-dimensional) Yukawa potential $V(r) = -Ae^{-qr}/r$, $A > 0$.
9. Estimate the ground state energy of an electron in a three-dimensional potential

$$V(\mathbf{r}) = V_0 \sin\frac{2\pi x}{a} \sin\frac{2\pi y}{a} \sin\frac{2\pi z}{a}$$

in a cubic box whose side is of length Na, $N \gg 1$. Assume periodic boundary conditions.

10. Estimate the ground state energy of a particle in the one-dimensional well

$$V(x) = \begin{cases} \frac{1}{2} m\omega_0^2 (x-a)^2, & \text{for } x > 0 \\ \infty, & \text{for } x < 0 \end{cases}$$

where $a > 0$. Is the energy greater or less than $\hbar\omega_0/2$?

11. A hydrogen atom in its ground state is placed in a uniform magnetic field $\mathcal{H}$. Calculate to first order in $\mathcal{H}$ the magnetization **M** induced in the atom by the field. Recall that the magnetization of a system is given by the negative of the derivative of the energy of the system with respect to the field. Neglect the effects of spin. *Hint:* Use the gauge in which $\mathbf{A} = -\mathbf{r} \times \mathcal{H}/2$.

Chapter 12
TIME-DEPENDENT PERTURBATION THEORY

In the last chapter we studied how to find approximately the energy eigenstates and eigenvalues of a system whose Hamiltonian is of the form $H_0 + V$, where V is "small" and time independent. Let us now consider a somewhat different situation. Suppose that we begin with a system whose Hamiltonian is H_0, and then we proceed to act on the system with a time-dependent external force, describable by an interaction term V_t added to the Hamiltonian. The question is, what effect does the force have on the system? For example, we might shine light on an atom, and ask what are the chances that the light ionizes the atom. Or we might apply a voltage to a piece of metal, and ask how much current is produced.

The general problem then is that we begin, at some early time t_0, with the system in a state $|\psi_t^0\rangle$ that solves

$$i\hbar \frac{\partial}{\partial t} |\psi_t^0\rangle = H_0 |\psi_t^0\rangle, \quad t < t_0 \tag{12-1}$$

and then apply a perturbation V_t, for $t > t_0$. To answer any question about the behavior of system at a later time we must find its state vector $|\psi_t\rangle$ at that time; that is, we must solve

$$i\hbar \frac{\partial}{\partial t} |\psi_t\rangle = (H_0 + V_t) |\psi_t\rangle, \quad t > t_0 \tag{12-2}$$

subject to the boundary condition that $|\psi_t\rangle = |\psi_t^0\rangle$ for $t \le t_0$.

It is usually impossible to solve (12-2) in closed form exactly. Often though, V_t is a small perturbation, and it is adequate to find the lowest-order effects of V_t on the system. Let us then develop the solution to (12-2) in powers of V_t. To do this we note that a large portion of the time dependence of $|\psi_t\rangle$ comes from the H_0. Let us take this dependence out explicitly by writing $|\psi_t\rangle$ as

$$|\psi_t\rangle = e^{-iH_0t/\hbar}\,|\psi(t)\rangle . \tag{12-3}$$

Substituting (12-3) into (12-2) we see that the H_0 drops out of the equation and $|\psi(t)\rangle$ obeys

$$i\hbar \frac{\partial}{\partial t}|\psi(t)\rangle = V(t)\,|\psi(t)\rangle \tag{12-4}$$

where

$$V(t) \equiv e^{iH_0t/\hbar}\,V_t\,e^{-iH_0t/\hbar}. \tag{12-5}$$

The state $|\psi(t)\rangle$ and the operator $V(t)$ are said to be in the *interaction representation* [(12-5) defines the interaction representation for *any* operator], since the time dependence of the state is due only to the interaction, while the time dependence of the operator is due to its explicit time dependence plus the dependence on H_0. If $V_t = 0$, then the interaction representation reduces to the Heisenberg representation.

Next, let us integrate both sides of (12-4) with respect to t from t_0 to t:

$$|\psi(t)\rangle = |\psi(t_0)\rangle + \frac{1}{i\hbar}\int_{t_0}^{t} dt'\; V(t')\,|\psi(t')\rangle . \tag{12-6}$$

From this form of the equation we can develop $|\psi(t)\rangle$ in "powers" of V by iteration. To first order in V we have

$$|\psi(t)\rangle = |\psi(t_0)\rangle + \frac{1}{i\hbar}\int_{t_0}^{t} dt'\; V(t')\,|\psi(t_0)\rangle ; \tag{12-7}$$

to find the second-order correction, we substitute the right side of (12-7) for $|\psi(t)\rangle$ in the integral in (12-6). The second-order change in $|\psi(t)\rangle$ is thus

$$\frac{1}{(i\hbar)^2}\int_{t_0}^{t} dt' \int_{t_0}^{t'} dt''\; V(t')V(t'')\,|\psi(t_0)\rangle . \tag{12-8}$$

Notice that in (12-8), and in all subsequent iterations the operators $V(t')$, $V(t'')$, etc., always occur in order of increasing times from right to left. In fact, it is easy to see that the general solution to (12-7) can be written in terms of the time ordered product [cf. Eq. (5-50)] as

$$|\psi(t)\rangle = \left(e^{-(i/\hbar)\int_{t_0}^{t} V(t')dt'}\right)_+ |\psi(t_0)\rangle. \tag{12-9}$$

FIRST-ORDER TRANSITIONS: GOLDEN RULE

We can now examine the following very important problem: suppose that the system initially is in an eigenstate $|0\rangle$ of H_0 (with energy ε_0, *not* necessarily the ground state); what is the probability that the system will be observed, after the interaction V_t has had time to act, in a different (orthogonal) eigenstate $|n\rangle$ of H_0? In other words, what is the probability that the interaction causes the system to make a transition from $|0\rangle$ to $|n\rangle$? The amplitude for observing the system in the state $|n\rangle$ at time t is simply

$$\langle n|\psi_t\rangle = e^{-i\varepsilon_n t/\hbar} \langle n|\psi(t)\rangle. \tag{12-10}$$

Since

$$|\psi(t_0)\rangle = |0\rangle,$$

we see from (12-7), that to first order in V_t, the transition amplitude is

$$\langle n|\psi_t\rangle = \frac{1}{i\hbar} e^{-i\varepsilon_n t/\hbar} \int_{t_0}^{t} dt' \langle n|V(t')|0\rangle. \tag{12-11}$$

Since from (12-5)

$$\langle n|V(t')|0\rangle = e^{i(\varepsilon_n - \varepsilon_0)t'/\hbar} \langle n|V_{t'}|0\rangle$$

the transition probability is given in lowest order by

$$P_{0\to n}(t) = |\langle n|\psi_t\rangle|^2 = \left|\frac{1}{i\hbar}\int_{t_0}^{t} dt'\, e^{i(\varepsilon_n - \varepsilon_0)t'/\hbar} \langle n|V_{t'}|0\rangle\right|^2. \tag{12-12}$$

Let us explore the consequences of this formula.

As a first example, suppose that the interaction potential V_t is turned on sharply at t_0, but is independent of time thereafter. Let us also take $t_0 = 0$ for convenience. Then

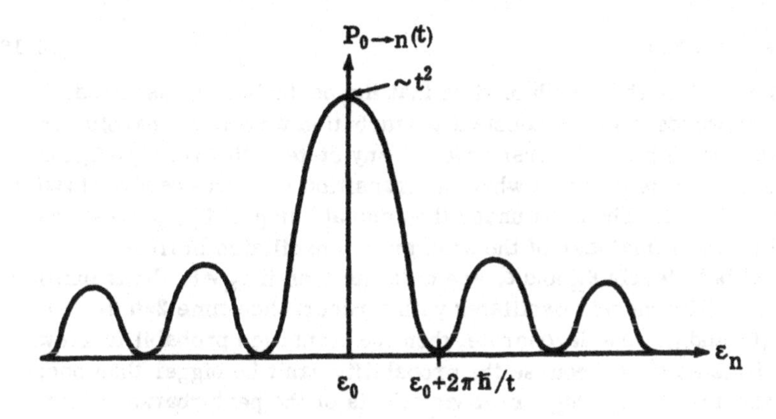

Fig. 12-1

Probability of a first-order transition from $|0\rangle$ to $|n\rangle$ after the potential has been on a time t, assuming $|\langle n|V|0\rangle|^2$ to be a slowly varying function of ϵ_n.

$$V_t = \begin{cases} 0 : t < 0 \\ V : t > 0 \end{cases} \tag{12-13}$$

Putting (12-13) into (12-12) we find

$$\begin{aligned} P_{0\to n}(t) &= \left| \frac{e^{i(\varepsilon_n - \varepsilon_0)t/\hbar} - 1}{\varepsilon_n - \varepsilon_0} \langle n|V|0\rangle \right|^2 \\ &= \left[\frac{\sin(\varepsilon_n - \varepsilon_0)t/2\hbar}{(\varepsilon_n - \varepsilon_0)/2} \right]^2 |\langle n|V|0\rangle|^2. \end{aligned} \tag{12-14}$$

This probability is shown as a function of ε_n in Fig. 12-1. For very short times the probability grows from zero as t^2, for all ε_n. However, as time goes on, the probability is largest for those states whose energy lies under the bump around ε_0. The height of the bump grows as t^2 while its width decreases as $1/t$. Thus ε_n will lie under the bump as long as

$$|\varepsilon_n - \varepsilon_0| < \frac{2\pi\hbar}{t}. \tag{12-15}$$

This means that the likely "spread in energy" $\Delta\varepsilon$ of the system caused by the perturbation obeys the uncertainty relation

$$\Delta\varepsilon \cdot \Delta t \gtrsim 2\pi\hbar \tag{12-16}$$

where Δt is the length of time that the perturbation has acted. In other words, while a constant perturbation turned on sharply can cause transitions in first order to any state with $\langle n|V|0\rangle \neq 0$, the probability is greatest when the transition conserves energy to within $\Delta\varepsilon = 2\pi\hbar/t$. The area under the central bump of $P_{0\to n}$ grows as t, while the remainder of the area simply oscillates in time.

If both levels ε_n and ε_0 are discrete then if $\varepsilon_n \neq \varepsilon_0$ the transition probability merely oscillates with a recurrence time $2\pi\hbar/|\varepsilon_0 - \varepsilon_n|$; if $|0\rangle$ and $|n\rangle$ are degenerate, then the transition probability grows in time as t^2. Of course the probability can't be bigger than one; after a while, the higher-order effects of the perturbation which we've neglected so far, become important, and prevent the transition probability from exceeding one.

Suppose, though, that the state $|n\rangle$ is one of a continuum of energy states, or a group of very closely spaced levels, such as those of a free particle. Then we would normally ask for the probability that the system makes a transition to a small group of states about $|n\rangle$. For example, for a free particle we would ask for the transition probability to an element of phase space $d^3k/(2\pi)^3$. As we have remarked the area under the bump grows as t so that we expect the transition probability to such a set of states, if they have the same energy as $|0\rangle$, to grow linearly in time, i.e., that there is a constant transition rate to the group of final states.

To calculate this transition rate we must sum (12-14) over the final states of interest. If we suppose that $|\langle n|V|0\rangle|^2$ doesn't change violently within the group of final states, we can write the sum over the final states as an integral over energies ε_n times $\rho(\varepsilon_n)$, the number of states per unit energy. Thus the total transition probability is

$$\sum_{\substack{n \\ \text{in group}}} P_{0\to n}(t) \approx |\langle n|V|0\rangle|^2 \int_{\text{group}} d\varepsilon_n\, \rho(\varepsilon_n) \left\{ \frac{\sin[(\varepsilon_n - \varepsilon_0)t/2\hbar]}{[(\varepsilon_0 - \varepsilon_n)/2]} \right\}^2 \tag{12-17}$$

where $\langle n|V|0\rangle$ is a typical matrix element for the group. Now when t is long enough, the central bump falls entirely within the group of states, and we can take the density of states factor outside the integral. The area under the bump is essentially

$$\int_{-\infty}^{\infty} d\varepsilon_n \left\{ \frac{\sin[(\varepsilon_n - \varepsilon_0)t/2\hbar]}{[(\varepsilon_0 - \varepsilon_n)/2]} \right\}^2 = \frac{2\pi}{\hbar}\, t \tag{12-18}$$

so that we find

$$\sum_n P_{0\to n}(t) = \Gamma t \tag{12-19}$$

where the *transition rate* Γ is given by

$$\Gamma = \frac{2\pi}{\hbar}[|\langle n|V|0\rangle|^2 \rho(\varepsilon_n)]_{\varepsilon_n = \varepsilon_0}. \tag{12-20}$$

This formula for the transition rate was named by Fermi – as some measure of its importance – the *golden rule.*

Another way of writing the golden rule comes from noticing that as t grows, $4(\sin^2[(\varepsilon_n - \varepsilon_0)t/2\hbar]/(\varepsilon_n - \varepsilon_0)^2$ becomes more and more peaked about $\varepsilon_n = \varepsilon_0$, has total area $2\pi t/\hbar$, and therefore approaches $(2\pi t/\hbar)\delta(\varepsilon_0 - \varepsilon_n)$ [except for the small wiggles in the wings]. Thus we can write

$$P_{0\to n}(t) = \Gamma_{0\to n} t \tag{12-21}$$

where

$$\Gamma_{0\to n} = \frac{2\pi}{\hbar}|\langle n|V|0\rangle|^2 \delta(\varepsilon_n - \varepsilon_0). \tag{12-22}$$

Remember, though, that to get actual numbers from this formula we must sum $\Gamma_{0\to n}$ over a continuous group of final states

$$\Gamma = \sum_{\substack{n \\ \text{in group}}} \Gamma_{0\to n}. \tag{12-23}$$

The golden rule is not valid for all times. First of all, in order that the central bump of $P_{0\to n}$ fall within the group of final states that we are looking at, the range of energies $\Delta\varepsilon$ of these states must be larger than $2\pi\hbar/t$, i.e., we need

$$t > \frac{2\pi\hbar}{\Delta\varepsilon}. \tag{12-24a}$$

On the other hand, the time must be short enough so that many states fall within the bump, i.e., the level spacing $\delta\varepsilon$ must be small compared with $2\pi\hbar/t$, or

$$t \ll \frac{2\pi\hbar}{\delta\varepsilon}. \tag{12-24b}$$

Furthermore one must keep in mind the depletion of the initial state after long times.

As a simple example of the golden rule, let us suppose that we have a particle in a momentum state in a large box of volume l^3 and that we turn on a potential $V(\mathbf{r})$ inside the box. What is the rate at which the particle makes transitions to other momentum states? The matrix element of V between an initial state of momentum $\hbar\mathbf{k}$ and a final state of momentum $\hbar\mathbf{k}'$ is

$$\langle \mathbf{k}'|V|\mathbf{k}\rangle = \int d^3r \, \frac{e^{-i\mathbf{k}'\cdot\mathbf{r}}}{l^{3/2}} \, V(\mathbf{r}) \, \frac{e^{i\mathbf{k}\cdot\mathbf{r}}}{l^{3/2}} = \frac{V_{\mathbf{k}'-\mathbf{k}}}{l^3} \tag{12-25}$$

where $V_{\mathbf{k}'-\mathbf{k}}$ is the Fourier transform of $V(\mathbf{r})$. Then from (12-22) the rate of transition from $\mathbf{k}$ to $\mathbf{k}'$ is

$$\Gamma_{\mathbf{k}\to\mathbf{k}'} = \frac{2\pi}{\hbar} \, \frac{|V_{\mathbf{k}'-\mathbf{k}}|^2}{l^6} \, \delta(\varepsilon_{k'} - \varepsilon_k) \tag{12-26}$$

where $\varepsilon_k = \hbar^2k^2/2m$. Suppose that we ask for the rate of scattering $d\Gamma$ into a small solid angle $d\Omega'$. Then we must sum (12-26) over all momentum states lying in this solid angle:

$$d\Gamma = \sum_{\mathbf{k}' \text{ in } d\Omega} \Gamma_{\mathbf{k}\to\mathbf{k}'}. \tag{12-27}$$

Now there are

$$l^3 \, \frac{d^3\mathbf{k}'}{(2\pi)^3} = \frac{l^3 mk'}{(2\pi)^3\hbar^2} \, d\Omega' d\varepsilon_{k'} \tag{12-28}$$

states in the volume d^3k' of phase space and therefore there are $l^3mk'/(2\pi)^3\hbar^2$ states per unit energy per unit solid angle. Therefore

$$\sum_{\mathbf{k}' \text{ in } d\Omega} \to d\Omega' \int_0^\infty \frac{l^3 mk'}{(2\pi)^3\hbar^2} \, d\varepsilon_{k'} \tag{12-29}$$

so that from (12-26) and (12-27) we find

$$d\Gamma = \frac{d\Omega'}{l^3} \, \frac{mk}{4\pi^2\hbar^3} \, |V_{\mathbf{k}'-\mathbf{k}}|^2, \tag{12-30}$$

where in this formula $\mathbf{k}'$ is a vector in $d\Omega$ of length k.

The example we just considered is very much like a scattering experiment; in a sense, the potential is "turned on" in a scattering experiment, since when the incident particle is far from the target, it sees no potential. In a real scattering experiment one uses a beam of incident particles. Then (12-30) can be interpreted as the rate at which particles scatter into $d\Omega$ per incident particle in a volume l^3. The flux of particles per incident particle of momentum $\hbar k$ in a volume l^3 is $\hbar k/ml^3$. Dividing $d\Gamma$ by this flux and by $d\Omega$ we find the differential cross section

$$\frac{d\sigma}{d\Omega} = \frac{m^2}{4\pi^2\hbar^4} |V_{\mathbf{k}'-\mathbf{k}}|^2. \tag{12-31}$$

This is simply the Born approximation formula for the differential cross section.

As a second example let us consider the radioactive decay of a nucleus in which the nucleus emits a particle in momentum state $|\hbar\mathbf{k}\rangle$. Let $|i\rangle$ be the initial state of the nucleus, and $|f\rangle$ the final state of the residual nucleus; let us neglect the recoil of the nucleus, and assume that we know the matrix element $\langle f, \hbar\mathbf{k}|V|i\rangle$ of the interaction responsible for the decay. The initial energy of the nucleus is E_i and the final energy of the nucleus and emitted particle is $E_f + \varepsilon_k$. Then the rate at which the particle is emitted into solid angle $d\Omega$ is

$$\begin{aligned} d\Gamma &= d\Omega \int_0^\infty \frac{mkl^3}{(2\pi)^2\hbar^3} d\varepsilon_k |\langle f, \hbar\mathbf{k}|V|i\rangle|^2 \delta(E_i - E_f - \varepsilon_k) \\ &= d\Omega \frac{mkl^3}{(2\pi)^2\hbar^3} |\langle f, \hbar\mathbf{k}|V|i\rangle|^2 \end{aligned} \tag{12-32}$$

where in this last expression $\mathbf{k}$ is a vector in the solid angle $d\Omega$ and of length $[2m(E_i - E_f)/\hbar^2]^{1/2}$; $\hbar\mathbf{k}$ is the momentum of the emitted particle. The total transition rate is then found by integrating (12-32) over all angles

$$\Gamma = \frac{mkl^3}{(2\pi)^2\hbar^3} \int d\Omega |\langle f, \hbar\mathbf{k}|V|i\rangle|^2. \tag{12-33}$$

[The matrix element will be proportional to $l^{-3/2}$, from the normalization of the emitted particle's wave function, so that Γ is actually independent of the volume of the box.] This Γ is the rate at which the nucleus decays. Thus if we start out with N_0 nuclei at $t = 0$, the perturbation theory tells us that we would have $\Delta N = N_0 \Gamma t$ fewer undecayed nuclei at a time t later. This expression is only valid for short times, since the original N_0 nuclei become depleted. Correctly, $N(t) = e^{-\Gamma t} N_0$. Thus in using the golden rule expression for a constant transition rate, one must be careful to allow for the possibility of depletion of the initial state.

We assumed in deriving (12-14) that the interaction was turned on sharply at t_0. In many problems, it is more realistic to think of the potential as being turned on slowly. For example, if we begin to shine light on an atom, the wave front of the light as it strikes the atom is not perfectly sharp; generally the amplitude striking the atom takes some time to build up to its steady state value. During this time the electrons in the atom can make many, many orbits, so that the light, from the point of view of the atom, is turned on slowly.

One useful way of representing the situation in which the perturbation is turned on slowly to a final value V is to write

$$V_t = e^{\eta t} V \tag{12-34}$$

where $\eta > 0$, and then to let η approach zero at the end of the calculation. Let us take V independent of time and again calculate the transition amplitude from $|0\rangle$ to $|n\rangle$, this time using the "slow turn on" representation (12-34). From (12-7) we have to first order

$$\langle n|\psi(t)\rangle = \frac{1}{i\hbar}\int_{t_0}^{t} dt'\ e^{\eta t'}\ e^{i(\varepsilon_n - \varepsilon_0)t'/\hbar}\langle n|V|0\rangle \tag{12-35}$$

since $|\psi(t_0)\rangle = |0\rangle$. In the far distant past $e^{\eta t'}$ is vanishingly small, so that we can replace the t_0 in (12-35) by $-\infty$. Thus

$$\langle n|\psi(t)\rangle = \frac{e^{\eta t + i(\varepsilon_n - \varepsilon_0)t/\hbar}}{\varepsilon_0 - \varepsilon_n + i\eta\hbar}\langle n|V|0\rangle . \tag{12-36}$$

The probability of a transition to $|n\rangle$ is therefore

$$|\langle n|\psi(t)\rangle|^2 = \frac{e^{2\eta t}}{(\varepsilon_0 - \varepsilon_n)^2 + (\eta\hbar)^2}|\langle n|V|0\rangle|^2. \tag{12-37}$$

This form, shown as a function of ε_n in Fig. 12-2, is very similar to (12-14). The range of likely final energies is

$$\Delta\varepsilon \lesssim \eta\hbar;$$

η^{-1} is effectively the length of time the potential has been on.

Again suppose that $|n\rangle$ is part of a continuum of states. Then the time rate of change of (12-37) is

$$e^{2\eta t}\frac{2\eta}{(\varepsilon_0 - \varepsilon_n)^2 + (\eta\hbar)^2}|\langle n|V|0\rangle|^2.$$

Now as we take the limit $\eta \to 0$ (arbitarily slow turn on), $e^{2\eta t} \to 1$. Also $2\eta/[(\varepsilon_0 - \varepsilon_n)^2 + (\eta\hbar)^2]$ has total area $2\pi/\hbar$ under it, considered

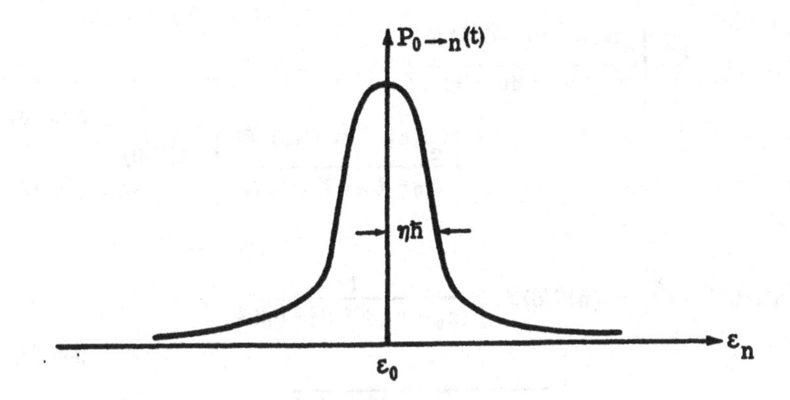

Fig. 12-2

Probability of a first-order transition from $|0\rangle$ to $|n\rangle$ for a slowly turned on potential.

as a function of ε_n, and becomes more and more sharply peaked about $\varepsilon_n = \varepsilon_0$; thus it approaches $(2\pi/\hbar)\delta(\varepsilon_0 - \varepsilon_n)$ as $\eta \to 0$. Hence the time derivative of (12-37) becomes

$$\Gamma_{0\to n} = \frac{2\pi}{\hbar} |\langle n|V|0\rangle|^2 \delta(\varepsilon_0 - \varepsilon_n), \tag{12-38}$$

which is the same result as we found for the transition rate by turning on the potential rapidly. Actually this expression for the transition rate is fairly insensitive to the details of how the potential is turned on.

HARMONIC PERTURBATIONS

So far we have only considered the effects of time-independent perturbations which have been turned on in various ways. Next, let us study the effects of a perturbation that varies harmonically in time

$$V_t = V \cos \omega t = \frac{V}{2}(e^{-i\omega t} + e^{i\omega t}) \tag{12-39}$$

and is turned on slowly with an $e^{\eta t}$ factor. Then from (12-11), taking $|\psi_{t_0}\rangle = |0\rangle$ and letting $t_0 \to -\infty$, we find

$$\langle n|\psi(t)\rangle = \frac{e^{\eta t}}{2}\left[\frac{e^{i(\varepsilon_n-\varepsilon_0-\hbar\omega)t/\hbar}}{\varepsilon_0-\varepsilon_n+\hbar\omega+i\eta\hbar} + \frac{e^{i(\varepsilon_n-\varepsilon_0+\hbar\omega)t/\hbar}}{\varepsilon_0-\varepsilon_n-\hbar\omega+i\eta\hbar}\right]\langle n|V|0\rangle \tag{12-40}$$

and

$$|\langle n|\psi(t)\rangle|^2 = \frac{e^{2\eta t}}{4}|\langle n|V|0\rangle|^2\left\{\frac{1}{(\varepsilon_0-\varepsilon_n+\hbar\omega)^2+(\eta\hbar)^2} + \frac{1}{(\varepsilon_0-\varepsilon_n-\hbar\omega)^2+(\eta\hbar)^2} + 2\,\mathrm{Re}\,\frac{e^{-2i\omega t}}{(\varepsilon_0-\varepsilon_n+\hbar\omega+i\eta\hbar)(\varepsilon_0-\varepsilon_n-\hbar\omega-i\eta\hbar)}\right\}. \tag{12-41}$$

The first term in (12-41) is the effect of the positive frequency $[e^{-i\omega t}]$ part of V_t, the second term is the effect of the negative frequency $[e^{i\omega t}]$ part of V_t, while the last term is due to the interference between the positive and negative frequency parts of (12-40).

Differentiating (12-41) with respect to t we find

$$\frac{d}{dt}P_{0\to n}(t) = \frac{e^{2\eta t}}{4}|\langle n|V|0\rangle|^2\left\{\left[\frac{2\eta}{(\varepsilon_0-\varepsilon_n+\hbar\omega)^2+(\eta\hbar)^2} + \frac{2\eta}{(\varepsilon_0-\varepsilon_n-\hbar\omega)^2+(\eta\hbar)^2}\right](1-\cos 2\omega t) + 2(\sin 2\omega t)\left[\frac{\varepsilon_0-\varepsilon_n+\hbar\omega}{(\varepsilon_0-\varepsilon_n+\hbar\omega)^2+(\eta\hbar)^2} - \frac{\varepsilon_0-\varepsilon_n-\hbar\omega}{(\varepsilon_0-\varepsilon_n-\hbar\omega)^2+(\eta\hbar)^2}\right]\right\}. \tag{12-42}$$

The cos and sin terms arise from the interference term. Again let us assume that $|n\rangle$ is part of a continuum of states and take the limit $\eta \to 0$. The first two terms in (12-42) are nonzero only if the final state energy differs from ε_0 by $\pm\hbar\omega$; the last term imposes no such restriction. However, if we average $dP_{0\to n}/dt$ over a few cycles of V_t, and call the result $\Gamma_{0\to n}$, then the sin ωt and cos ωt terms average to zero, and only the first two terms remain. They give

$$\Gamma_{0\to n} = \frac{2\pi}{\hbar}\frac{|\langle n|V|0\rangle|^2}{4}[\delta(\varepsilon_n-\varepsilon_0-\hbar\omega)+\delta(\varepsilon_n-\varepsilon_0+\hbar\omega)]. \tag{12-43}$$

[This formula is also valid if the states are discrete but ω is part of a continuum of frequencies applied to the system.] Thus the positive and negative frequency parts act independently – the interference effects average to zero. The positive frequency part, $e^{-i\omega t}$, of the interaction causes in first order an increase in the energy of the system by $\hbar\omega$, the negative frequency part, $e^{i\omega t}$, leads to decrease in energy by $\hbar\omega$.

SECOND-ORDER TRANSITIONS

It can occasionally happen that the matrix element $\langle n|V_t|0\rangle$ vanishes for all time. Then the perturbation can't cause any transitions from $|0\rangle$ to $|n\rangle$ in first order. It is possible though that transitions can occur in second or higher order. Let us study second order in detail in the case of a constant potential V turned on slowly.

From (12-8) we find that the second-order contribution to $\langle n|\psi(t)\rangle$ is

$$\langle n|\psi(t)\rangle^{(2)} = \frac{1}{(i\hbar)^2}\int_{t_0}^{t} dt' \int_{t_0}^{t'} dt'' \sum_m \langle n|V|m\rangle e^{i(\varepsilon_n-\varepsilon_m)t'/\hbar+\eta t'} \times \langle m|V|0\rangle e^{i(\varepsilon_m-\varepsilon_0)t''/\hbar+\eta t''}, \tag{12-44}$$

where the sum is over a complete set of eigenstates of H_0. Doing the t'' and t' integrals, and letting $t_0 \to -\infty$ we find

$$\langle n|\psi(t)\rangle^{(2)} = e^{-i(\varepsilon_0-\varepsilon_n)t/\hbar}\,\frac{e^{2\eta t}}{\varepsilon_0-\varepsilon_n+2i\eta\hbar}\sum_m \frac{\langle n|V|m\rangle\langle m|V|0\rangle}{\varepsilon_0-\varepsilon_m+i\eta\hbar}. \tag{12-45}$$

If the matrix element $\langle n|V|0\rangle = 0$, then the first-order transition amplitude is zero. Squaring (12-45), taking its time derivative, and letting $\eta \to 0$ we then find, in the case that $|n\rangle$ is part of a continuum, a second-order version of the golden rule

$$\Gamma_{0\to n} = \frac{2\pi}{\hbar}\left|\sum_m \frac{\langle n|V|m\rangle\langle m|V|0\rangle}{\varepsilon_0-\varepsilon_m+i\eta\hbar}\right|^2 \delta(\varepsilon_n-\varepsilon_0), \tag{12-46}$$

with the limit $\eta \to 0$ understood. One can interpret this formula by saying that the system makes the "forbidden" transition from $|0\rangle$ to

$|n\rangle$ in two steps. First it goes from $|0\rangle$ to an *intermediate state* $|m\rangle$ [which needn't have the same energy as $|0\rangle$] and then from $|m\rangle$ to $|n\rangle$. The amplitude for this double step is proportional to $\langle n|V|m\rangle\langle m|V|0\rangle/(\varepsilon_0 - \varepsilon_m + i\eta\hbar)$. To calculate the transition rate we must sum over all intermediate states *before* squaring; the contributions of the different intermediate states often interfere with each other. One speaks of the amplitude

$$\sum_m \frac{\langle n|V|m\rangle\langle m|V|0\rangle}{\varepsilon_0 - \varepsilon_m + i\eta\hbar} \tag{12-47}$$

as the "second-order matrix element," since it plays the same role in the golden rule as $\langle n|V|0\rangle$ did in first order. The $i\eta\hbar$ is very important in evaluating this matrix element when $|m\rangle$ is part of a continuum in the region $\varepsilon_m \approx \varepsilon_0$.

In case the first-order matrix element is nonvanishing, then it is easy to verify that the correct transition rate is formed by adding the second-order term (12-47) (and higher-order terms) to $\langle n|V|0\rangle$, squaring and multiplying by $2\pi\delta(\varepsilon_0 - \varepsilon_n)/\hbar$.

FORWARD "SCATTERING" AMPLITUDE

Up to now we have concentrated entirely on the amplitude for the system to make transitions from the eigenstate $|0\rangle$ of H_0 to the eigenstate $|n\rangle$ of H_0. Let us look, for a moment, at the amplitude $\langle 0|\psi_t\rangle$ that the system remains in the state it starts out in. Clearly $\langle 0|\psi_t\rangle = e^{-i\varepsilon_0 t/\hbar}\langle 0|\psi(t)\rangle$. To find an expression for $\langle 0|\psi(t)\rangle$ we multiply (12-4) on the left by $\langle 0|$. Then

$$\begin{aligned} i\hbar\frac{\partial}{\partial t}\langle 0|\psi(t)\rangle &= \langle 0|V(t)|\psi(t)\rangle = \sum_n \langle 0|V(t)|n\rangle\langle n|\psi(t)\rangle \\ &= \langle 0|V(t)|0\rangle\langle 0|\psi(t)\rangle + \sum_n{}' \langle 0|V(t)|n\rangle\langle n|\psi(t)\rangle \end{aligned} \tag{12-48}$$

where the prime means that the state $|0\rangle$ is not included in the sum over states. Dividing both sides of (12-48) by $\langle 0|\psi(t)\rangle$ we find

$$i\hbar\frac{\partial}{\partial t}\ln\langle 0|\psi(t)\rangle = \langle 0|V(t)|0\rangle + \sum_n{}' \langle 0|V(t)|n\rangle\frac{\langle n|\psi(t)\rangle}{\langle 0|\psi(t)\rangle}. \tag{12-49}$$

Let us only keep terms to order V^2 on the right side of this equation; since $\langle n|\psi(t)\rangle$ to lowest order is given by (12-7) which is first order in V, we can replace $\langle 0|\psi(t)\rangle$ in the denominator of the last term on the right by $\langle 0|0\rangle = 1$. Thus to second order we have

$$i\hbar \frac{\partial}{\partial t} \ln \langle 0|\psi(t)\rangle = \langle 0|V(t)|0\rangle + {\sum_n}' \frac{1}{i\hbar} \int_{t_0}^{t} dt' \langle 0|V(t)|n\rangle \langle n|V(t')|0\rangle . \quad (12\text{-}50)$$

Let us look at the case that V(t) is a constant potential V that is turned on slowly with an $e^{\eta t}$. Then in the limit $\eta \to 0$

$$i\hbar \frac{\partial}{\partial t} \ln \langle 0|\psi(t)\rangle = \langle 0|V|0\rangle + {\sum_n}' \frac{|\langle 0|V|n\rangle|^2}{\varepsilon_0 - \varepsilon_n + i\eta\hbar} . \quad (12\text{-}51)$$

First of all, in the case that the levels are discrete and ε_0 is nondegenerate, the $i\eta\hbar$ has no effect in the limit $\eta \to 0$. Then we recognize that the right side of (12-51) is just the expression for the second-order shift in the energy level ε_0 produced by a stationary perturbation V. Thus we find

$$\langle 0|\psi_t\rangle \sim e^{-iE_0 t/\hbar} \quad (12\text{-}52)$$

where E_0 is the energy eigenvalue of $H_0 + V$ corresponding to the eigenvalue ε_0 of H_0. We can interpret this result by noting that when we turn on the potential V very slowly its effect, when the levels of H_0 are discrete, is to adiabatically transform the eigenstates of H_0 into the eigenstates of $H_0 + V$. The state $|\psi_{t_0}\rangle$ is thus turned into an eigenstate $|\psi_t\rangle$ of $H_0 + V$, which, because it is an energy eigenstate, varies in time as $e^{-iE_0 t/\hbar}$. Stationary state perturbation theory can thus be looked upon as a special case of time-dependent perturbation theory.

Next, let us suppose that there is a continuum of energy levels $|n\rangle$ having energy in the neighborhood of ε_0. Let us write

$$\frac{1}{\varepsilon_0 - \varepsilon_n + i\eta\hbar} = \frac{\varepsilon_0 - \varepsilon_n}{(\varepsilon_0 - \varepsilon_n)^2 + (\eta\hbar)^2} - \frac{i\eta\hbar}{(\varepsilon_0 - \varepsilon_n)^2 + (\eta\hbar)^2}.$$

The second term on the right becomes $-\pi i \delta(\varepsilon_0 - \varepsilon_n)$ as $\eta \to 0$. The first term on the right occurring inside an integral gives just the principal value integral as $\eta \to 0$:

$$\lim_{\eta \to 0} \int d\varepsilon_n \frac{\varepsilon_0 - \varepsilon_n}{(\varepsilon_0 - \varepsilon_n)^2 + (\eta\hbar)^2} f(\varepsilon_n) = P \int d\varepsilon_n \frac{f(\varepsilon_n)}{\varepsilon_0 - \varepsilon_n}$$

Thus we can write, as $\eta \to 0$

$$\frac{1}{\varepsilon_0 - \varepsilon_n + i\eta\hbar} = \frac{P}{\varepsilon_0 - \varepsilon_n} - \pi i \delta(\varepsilon_0 - \varepsilon_n), \tag{12-53}$$

and hence

$$\sum_n{}' \frac{|\langle 0|V|n\rangle|^2}{\varepsilon_0 - \varepsilon_n + i\eta\hbar} = P\sum_n{}' \frac{|\langle 0|V|n\rangle|^2}{\varepsilon_0 - \varepsilon_n} - \frac{i\hbar}{2}\Gamma \tag{12-54}$$

where Γ is the total rate of transitions out of $|0\rangle$, in lowest order, Thus the solution to (12-51) can be written as

$$\langle 0|\psi_t\rangle \sim e^{-\Gamma t/2} e^{-iE_0 t/\hbar}, \tag{12-55}$$

where

$$E_0 = \varepsilon_0 + \langle 0|V|0\rangle + P\sum_n \frac{|\langle 0|V|n\rangle|^2}{\varepsilon_0 - \varepsilon_n}. \tag{12-56}$$

We can interpret this result by saying that the effect of the perturbation is twofold. First, it shifts the original energy level ε_0 to E_0, as in the case of discrete levels. In evaluating this shift the sum in the second term becomes the principal value integral $P\int d\varepsilon_n \rho(\varepsilon_n)\ldots$ where $\rho(\varepsilon_n)$ is the density of $|n\rangle$ states. Furthermore the potential induces transitions at a rate Γ from the state $|0\rangle$ to the states $|n\rangle$. We see from squaring (12-55) that

$$|\langle 0|\psi_t\rangle|^2 \sim e^{-\Gamma t};$$

the amplitude for remaining in the initial state decays exponentially when there is a constant rate of transitions to other states. One occasionally says, in this case, that $|\psi_t\rangle$ behaves as if it had a complex energy $E_0 - i\hbar\Gamma/2$.

The left side of (12-54) is the *second-order* amplitude to go from $|0\rangle$ to $|0\rangle$, and its imaginary part is proportional to the total transition rate; this is a variation on the optical theorem of scattering theory.

PROBLEMS

1. An electron is in the ground state of a three-dimensional square well potential of depth V_0 and radius a. A uniform electric field, $\mathscr{E}\cos\omega t$, is applied to the electron. Calculate the rate

of ionization of the electron as a function of the frequency ω.

2. A uniform electric field $\mathcal{E} \cos \omega t$ is applied to a hydrogen atom in its ground state. Calculate the rate of ionization of the atom as a function of ω. Do not assume that the proton is fixed in space. However, take plane wave functions for the continuum states.
3. Using Eqs. (12-54) and (12-31) rederive the optical theorem [Eq. (9-40)] for the Born approximation.
4. (a) Suppose that a system in an energy eigenstate $|0\rangle$ at $t = 0$ is acted on by an external perturbation, $H'(t)$. Let θ be some observable of the system. Show that the expectation value of θ at time t is given to first order in H' by

$$\frac{1}{i\hbar}\int_0^t \langle 0|[\theta(t), H'(t')]|0\rangle dt'$$

plus the zeroth-order term. The operators in this expression are in the interaction representation.

(b) Calculate the time-dependent dipole moment induced in a one-dimensional harmonic oscillator by an electric field $\mathcal{E} \cos \omega t$ turned on at $t = 0$; repeat for the case that the electric field is turned on very slowly.

Chapter 13
INTERACTION OF RADIATION WITH MATTER

First let us review the *classical* description of radiation. It is most convenient to work in the *transverse* gauge in which the scalar potential $\varphi(\mathbf{r}t)$ vanishes and the vector potential $\mathbf{A}(\mathbf{r}t)$ is divergence-free

$$\nabla \cdot \mathbf{A}(\mathbf{r}t) = 0. \tag{13-1}$$

The electric and magnetic fields are given in terms of $\mathbf{A}(\mathbf{r}t)$ by

$$\mathcal{E}(\mathbf{r}t) = -\frac{1}{c}\frac{\partial \mathbf{A}(\mathbf{r}t)}{\partial t} \tag{13-2}$$

$$\mathcal{H}(\mathbf{r}t) = \nabla \times \mathbf{A}(\mathbf{r}t). \tag{13-3}$$

The energy of the radiation is

$$E = \int d^3r\, \frac{\mathcal{E}^2(\mathbf{r}t) + \mathcal{H}^2(\mathbf{r}t)}{8\pi}, \tag{13-4}$$

while the rate (ergs/cm^2/sec) and direction of energy transport are given by the Poynting vector

$$\mathcal{P}(\mathbf{r}t) = \frac{c}{4\pi}\, \mathcal{E}(\mathbf{r}t) \times \mathcal{H}(\mathbf{r}t) \tag{13-5}$$

The radiation generated by a classical current $\mathcal{J}(\mathbf{r}t)$ is found by solving

$$\Box^2 \mathbf{A}(\mathbf{r}t) = \left(\nabla^2 - \frac{1}{c^2}\frac{\partial^2}{\partial t^2}\right)\mathbf{A}(\mathbf{r}t) = -\frac{4\pi}{c}\,\mathcal{J}_\perp(\mathbf{r}t) \tag{13-6}$$

where $\perp$ denotes the transverse, or divergence-free part of $\mathcal{J}(\mathbf{r}t)$.

Consider, for example, a plane wave solution to (13-6) in a region where $\mathcal{J} = 0$.

$$\mathbf{A}(\mathbf{r}t) = \alpha\boldsymbol{\lambda} e^{i\mathbf{k}\cdot\mathbf{r} - i\omega t} + \alpha^*\boldsymbol{\lambda}^* e^{-i\mathbf{k}\cdot\mathbf{r} + i\omega t}. \tag{13-7}$$

To solve $\Box^2 A = 0$ we need

$$\omega = ck, \tag{13-8}$$

while the polarization vector $\boldsymbol{\lambda}$, to satisfy (13-1) must obey

$$\boldsymbol{\lambda}\cdot\mathbf{k} = 0. \tag{13-9}$$

We can choose $|\boldsymbol{\lambda}|^2 = 1$; $\boldsymbol{\lambda}$ is real for plane polarized radiation.[1] The amplitude α is a constant. Since the vector potential must be real we must add to the first term in (13-7) its complex conjugate.

The energy per unit volume in the wave (13-7) is

$$\frac{\mathcal{E}^2 + \mathcal{H}^2}{8\pi} = \frac{\omega^2}{2\pi c^2}\,[|\alpha|^2 - \mathrm{Re}\,(\alpha^2\lambda^2 e^{2i\mathbf{k}\cdot\mathbf{r} - 2i\omega t})]. \tag{13-10}$$

The last term oscillates in time and averages to zero over one period. Thus the average energy density in the wave, per unit volume, is simply

$$\frac{E}{\mathrm{vol}} = \frac{\omega^2}{2\pi c^2}\,|\alpha|^2. \tag{13-11}$$

Since the wave travels in the direction $\mathbf{k}$ with velocity c, the time average over a few cycles of the Poynting vector is

$$\hat{\mathbf{k}}\,\frac{\omega^2}{2\pi c}\,|\alpha|^2. \tag{13-12}$$

This result can also be obtained directly from (13-5).

Any wave $\mathbf{A}(\mathbf{r}t)$ in free space can be written as a linear superposition of plane waves of the form (13-7)

$$\mathbf{A}(\mathbf{r}t) = \sum_{\mathbf{k}\lambda}\left[A_{\mathbf{k}\lambda}\,\boldsymbol{\lambda}\,\frac{e^{i\mathbf{k}\cdot\mathbf{r} - i\omega t}}{\sqrt{V}} + A_{\mathbf{k}\lambda}^*\,\boldsymbol{\lambda}^*\,\frac{e^{-i\mathbf{k}\cdot\mathbf{r} + i\omega t}}{\sqrt{V}}\right] \tag{13-13}$$

[1] The two components of the polarization vector perpendicular to **k** are just the components of the photon polarization state considered in Chapter 1.

where $\omega = ck$ and λ is summed over two orthogonal polarizations for each $\mathbf{k}$; these must also be orthogonal to $\mathbf{k}$. [We shall always use this meaning of $\sum_\lambda$.] We shall work in a huge box of volume V, letting the waves $e^{i\mathbf{k}\cdot\mathbf{r}}$ obey periodic boundary conditions; this is merely a convenience and involves no real restrictions. From (13-4), we find that the total energy of the wave (13-13), averaged over a few cycles, is

$$E = \sum_{\mathbf{k}\lambda} \frac{\omega^2}{2\pi c^2} |A_{\mathbf{k}\lambda}|^2. \tag{13-14}$$

INTERACTION HAMILTONIAN

Next let us turn to the problem of describing how classical electromagnetic radiation interacts with a quantum mechanical particle. We shall temporarily drop the restriction to the transverse gauge. Classically, the Hamiltonian for a particle of charge e in an electromagnetic field is

$$H = \frac{\left[\mathbf{p} - \frac{e}{c}\mathbf{A}(\mathbf{r}t)\right]^2}{2m} + e\varphi(\mathbf{r}t) + V(\mathbf{r}t) \tag{13-15}$$

where $V(\mathbf{r}t)$ describes all other potentials seen by the particle. As we've already mentioned in Chapter 3, the quantum mechanical Hamiltonian for the particle is the same as (13-15), where $\mathbf{p}$ and $\mathbf{r}$ are the momentum and position operators for the particle. The Schrödinger equation, using the Hamiltonian (13-15), is

$$i\hbar \frac{\partial\psi(\mathbf{r}t)}{\partial t} = \left[\frac{1}{2m}\left(\frac{\hbar}{i}\nabla - \frac{e}{c}\mathbf{A}(\mathbf{r}t)\right)^2 + e\varphi(\mathbf{r}t) + V(\mathbf{r}t)\right]\psi(\mathbf{r}t). \tag{13-16}$$

Since the wave function depends on the potentials, it must change if we change the potentials by a gauge transformation. In such a transformation we let

$$\mathbf{A}'(\mathbf{r}t) = \mathbf{A}(\mathbf{r}t) + \nabla\chi(\mathbf{r}t)$$

$$\varphi'(\mathbf{r}t) = \varphi(\mathbf{r}t) - \frac{1}{c}\frac{\partial\chi(\mathbf{r}t)}{\partial t}; \tag{13-17}$$

where $\chi(\mathbf{r}t)$ is a scalar function of $\mathbf{r}$ and t. The fields $\mathcal{E}$ and $\mathcal{H}$

derived from $\mathbf{A}'$ and φ' are the same as those derived from $\mathbf{A}$ and φ. The wave function ψ' in this new gauge obeys

$$i\hbar \frac{\partial \psi'(\mathbf{r}t)}{\partial t} = \left[\frac{1}{2m}\left(\frac{\hbar}{i}\nabla - \frac{e}{c}\mathbf{A}'(\mathbf{r}t)\right)^2 + e\varphi'(\mathbf{r}t) + V(\mathbf{r}t)\right]\psi'(\mathbf{r}t); \qquad (13\text{-}18)$$

it is easy to verify, using (13-16), (13-17), and (13-18) that the wave function in the new gauge is related to the wave function in the original gauge by

$$\psi'(\mathbf{r}t) = e^{ie\chi(\mathbf{r}t)/\hbar c}\psi(\mathbf{r}t). \qquad (13\text{-}19)$$

The momentum $\mathbf{p}$ is not a gauge invariant quantity; any matrix element of $\mathbf{p}$ between states of the system, e.g.,

$$\int d^3r\, \psi_1^*(\mathbf{r}t)\frac{\hbar}{i}\nabla\psi_2(\mathbf{r}t) \qquad (13\text{-}20)$$

will change if we make a gauge transformation. This is because in the new gauge, (13-20) becomes

$$\int d^3r \left[e^{ie\chi(\mathbf{r}t)/\hbar c}\psi_1(\mathbf{r}t)\right]^* \frac{\hbar}{i}\nabla\left[e^{ie\chi(\mathbf{r}t)/\hbar c}\psi_2(\mathbf{r}t)\right] \qquad (13\text{-}21)$$

and the gradient acting on the exponential adds an extra term

$$\int \psi_1^*(\mathbf{r}t)\left[\frac{e}{c}\nabla\chi(\mathbf{r}t)\right]\psi_2(\mathbf{r}t)\, d^3r \qquad (13\text{-}22)$$

onto (13-20). However, the combination $\mathbf{p} - (e/c)\mathbf{A}(\mathbf{r}t)$ has gauge invariant matrix elements, since the extra term (13-22) is exactly canceled by the change in $\mathbf{A}$ on going to the new gauge. This combination in classical mechanics is m times the velocity of the particle; similarly in quantum mechanics we can write the operator relation, in the Heisenberg representation,

$$m\frac{d\mathbf{r}(t)}{dt} = \mathbf{p}(t) - \frac{e}{c}\mathbf{A}(\mathbf{r}(t), t). \qquad (13\text{-}23)$$

This is easily verified by commuting $\mathbf{r}$ with the Hamiltonian (13-15). The general rule is that, even though the wave functions and the electromagnetic potentials change with the gauge, all *physically observable* quantities are independent of the gauge.

We can write the Hamiltonian (13-15) in the form

$$H = H_0 + H_{\text{int}} \qquad (13\text{-}24)$$

where $H_0 = p^2/2m + V(\mathbf{r}, t)$ is the Hamiltonian in the absence of the electromagnetic potentials, and

$$H_{int} = -\frac{e}{2mc}(\mathbf{p}\cdot\mathbf{A}(\mathbf{r}t) + \mathbf{A}(\mathbf{r}t)\cdot\mathbf{p}) + \frac{e^2}{2mc^2}A^2(\mathbf{r}t) + e\varphi(\mathbf{r}t) \tag{13-25}$$

is the operator for the interaction of the system with the radiation. In general $\mathbf{p}\cdot\mathbf{A} \neq \mathbf{A}\cdot\mathbf{p}$ since $\mathbf{p}\cdot\mathbf{A} - \mathbf{A}\cdot\mathbf{p} = -i\hbar\nabla\cdot\mathbf{A}$; only in the transverse gauge does $\mathbf{p}\cdot\mathbf{A} = \mathbf{A}\cdot\mathbf{p}$.

For a system of many particles, say N, interacting with radiation, the Hamiltonian becomes

$$H = \sum_{i=1}^{N} \frac{[\mathbf{p}_i - (e_i/c)\mathbf{A}(\mathbf{r}_i, t)]^2}{2m_i} + \sum_{i=1}^{N} e_i\varphi(\mathbf{r}_i, t) + V \tag{13-26}$$

where e_i and m_i are the charge and mass of the ith particle, $\mathbf{r}_i$ and $\mathbf{p}_i$ are the coordinate and momentum operators for the ith particle, and V is all the other interaction terms in the Hamiltonian. Thus, for N particles

$$H_{int} = \sum_{i=1}^{N} \left\{ -\frac{e_i}{2m_ic}(\mathbf{p}_i\cdot\mathbf{A}(\mathbf{r}_i, t) + \mathbf{A}(\mathbf{r}_i, t)\cdot\mathbf{p}_i) + \frac{e_i^2}{2m_ic^2}A^2(\mathbf{r}_i, t) + e_i\varphi(\mathbf{r}_i, t) \right\}. \tag{13-27}$$

For simplicity, let us assume that all the particles have charge e and mass m. Then we can write, for instance,

$$\sum_i e\varphi(\mathbf{r}_i, t) = \sum_i \int d^3r\; e\delta(\mathbf{r} - \mathbf{r}_i)\varphi(\mathbf{r}, t) = \int d^3r\; e\rho(\mathbf{r})\varphi(\mathbf{r}, t) \tag{13-28}$$

where

$$\rho(\mathbf{r}) = \sum_i \delta(\mathbf{r} - \mathbf{r}_i). \tag{13-29}$$

In (13-28), $\varphi(\mathbf{r}, t)$ is no longer an operator since all the position operators, $\mathbf{r}_i$, are in $\rho(\mathbf{r})$. This operator, $\rho(\mathbf{r})$, may be interpreted as the operator for the density of particles; the integral of $\rho(\mathbf{r})$ over all space is just N, the total number of particles. Also, if we integrate $\rho(\mathbf{r})$ over a finite volume of space, each $\delta(\mathbf{r} - \mathbf{r}_i)$ will give either one or zero depending on whether the ith particle is in the volume or not, and thus the integral of $\rho(\mathbf{r})$ over the volume tells us the number of particles in the volume.

Similarly, we can write the first term on the right side of (13-27) as

$$-\frac{e}{c}\int d^3r\, \mathbf{j}(\mathbf{r})\cdot\mathbf{A}(\mathbf{r}t), \tag{13-30}$$

where

$$\mathbf{j}(\mathbf{r}) = \frac{1}{2}\sum_i\left(\frac{\mathbf{p}_i}{m}\delta(\mathbf{r}-\mathbf{r}_i)+\delta(\mathbf{r}-\mathbf{r}_i)\frac{\mathbf{p}_i}{m}\right) \tag{13-31}$$

may be interpreted as the operator for the particle current in the system. It is a sum of terms of the form of the velocity of the ith particle times $\delta(\mathbf{r}-\mathbf{r}_i)$, the density operator for the ith particle. The symmetric combination in (13-31) makes $\mathbf{j}(\mathbf{r})$ a Hermitian operator. Actually, as we have discussed, $\mathbf{p}_i/m$ is not the velocity of the ith particle in the presence of an electromagnetic field; rather the velocity is $(\mathbf{p}_i/m)-(e/mc)\mathbf{A}(\mathbf{r}_i, t)$. This means that the operator

$$\mathbf{J}(\mathbf{r}) = \mathbf{j}(\mathbf{r}) - \frac{e}{mc}\mathbf{A}(\mathbf{r}t)\rho(\mathbf{r}) \tag{13-32}$$

is the true operator for the particle current. The first term $\mathbf{j}(\mathbf{r})$ is called the *paramagnetic* current while the second term is called the *diamagnetic* current. In terms of $\rho(\mathbf{r})$ and $\mathbf{j}(\mathbf{r})$ the electromagnetic interaction (13-27) becomes

$$H_{int} = \int d^3r\left[-\frac{e}{c}\mathbf{j}(\mathbf{r})\cdot\mathbf{A}(\mathbf{r}t)+\frac{e^2}{2mc^2}\rho(\mathbf{r})A^2(\mathbf{r}t)+e\rho(\mathbf{r})\varphi(\mathbf{r}t)\right]. \tag{13-33}$$

ABSORPTION OF LIGHT

We have now developed sufficient machinery to begin to calculate the effects of radiation on matter. For radiation in the transverse gauge the last term in (13-33) is zero. Generally, if the electric fields described by $\mathbf{A}$ are small compared with "atomic" fields, e/a_0^2, the effects of the ρA^2 term, second order in $\mathbf{A}$, are small compared with the effects of the $\mathbf{j}\cdot\mathbf{A}$ term. In particular, we may neglect the ρA^2 term in calculating the rate of absorption of light by a system of charged particles, such as an atom. In such a case we may take simply

$$H_{int} = -\frac{e}{c}\int d^3r\, \mathbf{j}(\mathbf{r})\cdot\mathbf{A}(\mathbf{r}t); \tag{13-34}$$

if we represent $\mathbf{A}(\mathbf{r}t)$ by a sum of plane waves, as in (13-13), then (13-34) becomes

$$H_{int} = -\frac{e}{c}\sum_{\mathbf{k}\lambda}\left[A_{\mathbf{k}\lambda}\mathbf{j}_{-\mathbf{k}}\cdot\lambda\,\frac{e^{-i\omega t}}{\sqrt{V}} + A_{\mathbf{k}\lambda}{}^{*}\mathbf{j}_{\mathbf{k}}\cdot\lambda^{*}\,\frac{e^{i\omega t}}{\sqrt{V}}\right] \qquad (13\text{-}35)$$

where

$$\mathbf{j}_{\mathbf{k}} = \int d^3r\, e^{-i\mathbf{k}\cdot\mathbf{r}}\mathbf{j}(\mathbf{r}) = \frac{1}{2}\sum_i\left[\frac{\mathbf{p}_i}{m}e^{-i\mathbf{k}\cdot\mathbf{r}_i} + e^{-i\mathbf{k}\cdot\mathbf{r}_i}\frac{\mathbf{p}_i}{m}\right]. \qquad (13\text{-}36)$$

Let us calculate the rate of absorption, by an atom in an arbitrary state $|0\rangle$, of a beam of light represented by (13-13). Let us assume that the beam is incoherent in the sense that there are no phase correlations between the different Fourier components of the beam. For example, the light emitted by a hot gas, such as mercury vapor, is the sum of contributions from all the different atoms in the gas, and these different contributions have random phase relationships. If there are no phase correlations between the different Fourier components of the incident beam, we may calculate the effect of each component separately, and then sum over the different components; in other words, we shall assume that there are no interference effects, in the interaction of the light with the atom, between different Fourier components of the light.

To calculate the rate of absorption of the light we may use the golden rule – even though the final state of the atom may not be part of a continuum, we will get time proportional transition probabilities if there is a continuum of frequencies present in the incident light. As we saw in the last section a given Fourier component, $\mathbf{k}\lambda$, of the incident beam will in first order induce both upward and downward transitions of the atom by energy $\hbar\omega = \hbar ck$; the upward transitions are caused by the positive frequency component of the perturbation, while the downward transitions are caused by the negative frequency component. The rate of upward transitions caused by the $A_{\mathbf{k}\lambda}$ component of (13-35) is, from the golden rule,

$$\Gamma^{abs}_{0\to n;\mathbf{k}\lambda} = \frac{2\pi}{\hbar}\,\delta(\varepsilon_n - \varepsilon_0 - \hbar\omega)\frac{e^2}{Vc^2}\,|A_{\mathbf{k}\lambda}|^2\,|\langle n|\mathbf{j}_{-\mathbf{k}}\cdot\lambda|0\rangle|^2. \qquad (13\text{-}37)$$

Summing this over all $\mathbf{k}$ and λ (two orthogonal polarizations for each $\mathbf{k}$), we find the total rate of transitions upward from $|0\rangle$ to $|n\rangle$ is

$$\Gamma^{abs}_{0\to n} = \frac{1}{V}\sum_{\mathbf{k}\lambda}\frac{2\pi}{\hbar}\,\delta(\varepsilon_n - \varepsilon_0 - \hbar\omega)\,\frac{e^2}{c^2}\,|A_{\mathbf{k}\lambda}|^2\,|\langle n|\mathbf{j}_{-\mathbf{k}}\cdot\lambda|0\rangle|^2. \qquad (13\text{-}38)$$

Converting the sum over **k** into an integral:

$$\frac{1}{V}\sum_{k} \to \int \frac{k^2\,dk\,d\Omega}{(2\pi)^3} = \int \frac{\omega^2\,d\omega\,d\Omega}{(2\pi c)^3} \tag{13-39}$$

we find

$$\Gamma_{0\to n}^{abs} = \frac{2\pi e^2}{\hbar^2 c^2}\frac{\omega^2}{(2\pi c)^3}\int d\Omega \sum_{\lambda} |\langle n|\mathbf{j}_{-\mathbf{k}}\cdot\boldsymbol{\lambda}|0\rangle|^2 |A_{\mathbf{k}\lambda}|^2. \tag{13-40}$$

where $\omega = (\varepsilon_n - \varepsilon_0)/\hbar$.

Suppose that the incident beam subtends a solid angle $d\Omega$ and that it is polarized, with polarization vector $\boldsymbol{\lambda}$. Then from (13-12) we find that the total rate of energy transport in the beam is

$$\frac{1}{V}\sum_{k} \frac{\omega^2}{2\pi c}|A_{\mathbf{k}\lambda}|^2 = d\Omega \int d\omega \frac{\omega^4}{(2\pi c)^4}|A_{\mathbf{k}\lambda}|^2. \tag{13-41}$$

Thus

$$I(\omega) = \frac{d\Omega\,\omega^4 |A_{\mathbf{k}\lambda}|^2}{(2\pi c)^4} \tag{13-42}$$

is the intensity (ergs/cm^2-rad) of the incident beam per unit frequency. In terms of $I(\omega)$ we may write the absorption rate as

$$\Gamma_{0\to n}^{abs} = \frac{4\pi^2 e^2}{\hbar^2 c\omega^2} I(\omega)|\langle n|\mathbf{j}_{-\mathbf{k}}\cdot\boldsymbol{\lambda}|0\rangle|^2. \tag{13-43}$$

We similarly find that the rate of downward transitions induced by the incident beam from *initial* state $|n\rangle$ to *final* state $|0\rangle$ is given by

$$\Gamma_{n\to 0}^{ind.\,emm.} = \frac{1}{V}\sum_{k\lambda}\frac{2\pi}{\hbar}\delta(\varepsilon_n - \varepsilon_0 - \hbar\omega)\frac{e^2}{c^2}|A_{\mathbf{k}\lambda}|^2|\langle 0|\mathbf{j}_{\mathbf{k}}\cdot\boldsymbol{\lambda}^*|n\rangle|^2 \tag{13-44}$$

$$= \frac{4\pi^2 e^2}{\hbar^2 c\omega^2} I(\omega)|\langle 0|\mathbf{j}_{\mathbf{k}}\cdot\boldsymbol{\lambda}^*|n\rangle|^2, \tag{13-45}$$

where in (13-45), $\omega = (\varepsilon_n - \varepsilon_0)/\hbar$. Notice that

$$\Gamma_{0\to n}^{abs} = \Gamma_{n\to 0}^{ind.\,emm.} \tag{13-46}$$

since

$$\langle 0|\mathbf{j}_{\mathbf{k}}\cdot\boldsymbol{\lambda}^*|n\rangle = \langle n|\mathbf{j}_{-\mathbf{k}}\cdot\boldsymbol{\lambda}|0\rangle^*.$$

The reason for describing the upward transition rate as the absorption rate is that one may interpret the upward transition process as the absorption of one *photon* of energy $\hbar\omega = \varepsilon_n - \varepsilon_0$; then the energy gain of the electron is compensated by a corresponding energy loss of the electromagnetic field. Similarly the downward transition process may be thought of as the emission of one photon of energy $\hbar\omega = \varepsilon_n - \varepsilon_0$. This process is called *induced* (or stimulated) *emission* since the rate at which it occurs is proportional to the intensity of the applied radiation.

Each photon of frequency ω in the incident beam has energy $\hbar\omega$. Thus the total energy in the incident beam is

$$E = \sum_{\mathbf{k}\lambda} \hbar\omega N_{\mathbf{k}\lambda} \tag{13-47}$$

where $N_{\mathbf{k}\lambda}$ is the number of photons in the mode $\mathbf{k}\lambda$ in the incident beam. Comparing (13-47) with (13-14) we see that the amplitude $A_{\mathbf{k}\lambda}$ is related to the number of photons in the mode $\mathbf{k}\lambda$ by $\hbar\omega N_{\mathbf{k}\lambda} = \omega^2 |A_{\mathbf{k}\lambda}|^2/2\pi c^2$ or

$$|A_{\mathbf{k}\lambda}|^2 = \frac{2\pi\hbar c^2}{\omega} N_{\mathbf{k}\lambda}. \tag{13-48}$$

We may write the absorption and emission rates in terms of the numbers of photons as

$$\Gamma_{0\to n}^{\text{abs}} = \sum_{\mathbf{k}\lambda} \frac{4\pi^2 e^2}{\omega V} \delta(\varepsilon_n - \varepsilon_0 - \hbar\omega) |\langle n|\mathbf{j}_{-\mathbf{k}} \cdot \boldsymbol{\lambda}|0\rangle|^2 N_{\mathbf{k}\lambda} = \Gamma_{n\to 0}^{\text{ind.emm.}} \tag{13-49}$$

We can only express the magnitude of $A_{\mathbf{k}\lambda}$, not its phase, in terms of $N_{\mathbf{k}\lambda}$; however, the assumption that the incident beam is incoherent is equivalent to having no information about the relative phases of the different $A_{\mathbf{k}\lambda}$. The procedure of calculating the effects of each Fourier component of the beam individually, and then summing the result over the different components is equivalent to averaging over all possible phases of the $A_{\mathbf{k}\lambda}$'s. Thus an incoherent beam is sufficiently specified by giving the number of photons present in each mode. In other words, if all we know about a beam are the photon numbers $N_{\mathbf{k}\lambda}$, then we have no information about the phase relationships between the different Fourier components of the beam. In the complete quantum mechanical description of radiation, photon numbers and phases are complimentary quantities, like momentum and position — the more accurately one is specified, the more uncertain is the other. [See Problem 10, Chapter 1.]

The nature of the assumption that the incident beam is incoherent may be seen from another point of view if we note that an alternative method of arriving at the result (13-43) is the following: assume that we shine a pulse of light on the atom, and then calculate the transition probability $P_{0\to n}$ induced by the pulse. Then if we shine a sequence of such pulses on the atom, with no phase relationships between different pulses, the rate of transitions induced in the atom is $\overline{P}_{0\to n}$, the average of $P_{0\to n}$ over the phases (which are random from pulse to pulse) times the number of pulses per second; carrying through this calculation one finds just (13-43), where $I(\omega)$ is the average intensity of light of frequency ω seen by the atom.

The total rate of absorption of a beam of $N_{\mathbf{k}\lambda}$ incident photons in the mode $\mathbf{k}\lambda$ is, from (13-49)

$$\Gamma^{\text{abs}}(\omega) = \left[\frac{N_{\mathbf{k}\lambda}c}{V}\right]\frac{4\pi^2e^2}{\omega c}\sum_n |\langle n|\mathbf{j}_{-\mathbf{k}}\cdot\lambda|0\rangle|^2\,\delta(\varepsilon_n-\varepsilon_0-\hbar\omega). \qquad (13\text{-}50)$$

Now $N_{\mathbf{k}\lambda}/V$ is the density of photons per unit volume and $N_{\mathbf{k}\lambda}c/V$ is therefore the incident photon flux. The ratio of Γ^{abs} to this flux defines the total *absorption cross section* of the matter system

$$\sigma_{\text{abs}}(\omega) = \frac{4\pi^2e^2}{\omega c}\sum_n |\langle n|\mathbf{j}_{-\mathbf{k}}\cdot\lambda|0\rangle|^2\,\delta(\varepsilon_n-\varepsilon_0-\hbar\omega). \qquad (13\text{-}51)$$

The absorption cross section depends on the direction and polarization of the radiation, in addition to its frequency.

QUANTIZED RADIATION FIELD

So far we have described the process of the absorption of light by a system in terms of the effect of an external classical electromagnetic field on the system. An alternative way of discussing this process is also to describe the electromagnetic field in terms of states, and then to say that in the absorption process, the system makes a transition from $|0\rangle$ to $|n\rangle$ while the electromagnetic field makes a transition from its initial state to a state with one fewer photon (the absorbed photon). As we have seen, an incoherent beam of light is specified by saying how many photons there are in each mode $\mathbf{k}, \lambda$. Thus we may write the (normalized) initial state of the electromagnetic field as

$$|N_{\mathbf{k}_1\lambda_1}, N_{\mathbf{k}_2\lambda_2}, \ldots, N_{\mathbf{k}\lambda}, \ldots\rangle, \qquad (13\text{-}52)$$

where $N_{\mathbf{k}_1\lambda_1}$ is the number of photons in the mode $\mathbf{k}_1, \lambda_1$, etc. Two such states are orthogonal unless they have the same number of photons in each mode. The final state of the electromagnetic field, after absorption by the system of one photon in the mode $\mathbf{k}\lambda$ is

$$|N_{\mathbf{k}_1\lambda_1}, N_{\mathbf{k}_2\lambda_2}, \ldots, N_{\mathbf{k}\lambda} - 1, \ldots\rangle. \tag{13-53}$$

We can then say that the transition from the state of the system $|0\rangle$ and the state of the radiation (13-52) to the state of the system $|n\rangle$ and the state of the radiation (13-53) is caused by some operator $\hat{H}_{int}$ that couples the radiation to the matter. The form of this operator is determined by the requirement that this alternative "quantum mechanical" description of the absorption of light by matter, in terms of states of the radiation field, gives the same result, (13-37), (13-49), as the description in terms of a classical external field impinging on the matter. Let us see how this works.

What we want then is the transition rate from an initial state of matter and radiation

$$|0; N_{\mathbf{k}_1\lambda_1}, N_{\mathbf{k}_2\lambda_2}, \ldots, N_{\mathbf{k}\lambda}, \ldots\rangle \tag{13-54}$$

to a final state of matter and radiation

$$|n; N_{\mathbf{k}_1\lambda_1}, N_{\mathbf{k}_2\lambda_2}, \ldots, N_{\mathbf{k}\lambda} - 1, \ldots\rangle, \tag{13-55}$$

caused by a coupling $\hat{H}_{int}$ between the matter and the radiation. The state (13-54) stands for the matter in $|0\rangle$ and the radiation in (13-52), with no coupling between them; similarly the state (13-55) stands for the matter in $|n\rangle$ and the radiation in (13-53), with no coupling between them. The energy of the state (13-54) is

$$\varepsilon_0 + \sum_{\mathbf{k}'\lambda'} \hbar c k' N_{\mathbf{k}'\lambda'}$$

while the energy of the state (13-55) is

$$\varepsilon_n + \sum_{\mathbf{k}'\lambda'} \hbar c k' N_{\mathbf{k}'\lambda'} - \hbar c k.$$

Thus according to the golden rule the rate of transition from (13-54) to (13-55) is

$$\frac{2\pi}{\hbar}\delta(\varepsilon_n - \hbar c k - \varepsilon_0) \times |\langle n; N_{\mathbf{k}_1\lambda_1}, \ldots, N_{\mathbf{k}\lambda} - 1, \ldots|\hat{H}_{int}|0; N_{\mathbf{k}_1\lambda_1}, \ldots, N_{\mathbf{k}\lambda}, \ldots\rangle|^2; \tag{13-56}$$

in order that this be the same as (13-37) we see that the matrix element in (13-56) must obey

$$|\langle n; N_{k_1\lambda_1}, \ldots, N_{k\lambda} - 1, \ldots |\hat{H}_{int}| 0; N_{k_1\lambda_1}, \ldots, N_{k\lambda}, \ldots\rangle|^2$$

$$= \frac{e^2}{Vc^2} |A_{k\lambda}|^2 |\langle n|j_{-k} \cdot \lambda|0\rangle|^2 = \frac{e^2}{Vc^2} \frac{2\pi\hbar c^2}{\omega} N_{k\lambda} |\langle n|j_{-k} \cdot \lambda|0\rangle|^2. \tag{13-57}$$

Hence $\hat{H}_{int}$ must include a part $j_{-k} \cdot \lambda$ that acts on the matter part of the states, times another part that decreases the number of photons in k, λ by one; in fact [cf. (13-35)] $\hat{H}_{int}$ must be of the form

$$\hat{H}_{int} = -\frac{e}{cV^{1/2}} \sum_{k'\lambda'} (j_{-k'} \cdot \lambda' A_{k'\lambda'}^{(op)} + j_{k'} \cdot \lambda'^* A_{k'\lambda'}^{(op)\dagger}) \tag{13-58}$$

where $A_{k'\lambda'}^{(op)}$ is an operator that acting on a photon state reduces the number of photons in $k'\lambda'$ by one. The final term in (13-58) makes $\hat{H}_{int}$ Hermitian. The matrix element of $\hat{H}_{int}$ between the states (13-55) on the left, and (13-54) on the right, picks out only the $k\lambda$ term in (13-58) since two photon states, with different sets of photon occupation numbers $N_{k_i\lambda_i}$, are orthogonal. For example, if k_1, $\lambda_1 \neq k$, λ, then

$$A_{k_1\lambda_1}|N_{k_1\lambda_1}, \ldots, N_{k\lambda}, \ldots\rangle \sim |N_{k_1\lambda_1} - 1, \ldots, N_{k\lambda}, \ldots\rangle$$

is orthogonal to a state with $N_{k\lambda} - 1$ photons. Similarly, the $A_{k'\lambda'}^{(op)\dagger}$ part of (13-58), as we shall see, must increase the number of photons when acting on a state, and therefore it doesn't contribute to the matrix element of $\hat{H}_{int}$ in (13-57). Thus

$$\langle n; N_{k_1\lambda_1}, \ldots, N_{k\lambda} - 1, \ldots |\hat{H}_{int}| 0; N_{k_1\lambda_1}, \ldots, N_{k\lambda}, \ldots\rangle$$

$$= -\frac{e}{c\sqrt{V}} \langle n|j_{-k} \cdot \lambda|0\rangle \langle \ldots, N_{k\lambda} - 1, \ldots |A_{k\lambda}^{(op)}| \ldots, N_{k\lambda}, \ldots\rangle \tag{13-59}$$

Comparing (13-59) with (13-57) we see that

$$\langle \ldots, N_{k\lambda} - 1, \ldots |A_{k\lambda}^{(op)}| \ldots, N_{k\lambda}, \ldots\rangle = \sqrt{\frac{2\pi\hbar c^2}{\omega}} \sqrt{N_{k\lambda}}, \tag{13-60}$$

to within a phase factor. We are free, however, to choose the relative phases of the states (13-50) such that this phase factor equals one. Notice that this matrix element in this new picture plays exactly the same role as $A_{k\lambda}$ in the old procedure.

Taking the complex conjugate of (13-60) gives

$$\langle \ldots, N_{k\lambda} - 1, \ldots | A_{k\lambda}^{(op)} | \ldots, N_{k\lambda}, \ldots \rangle^*$$

$$= \langle \ldots, N_{k\lambda}, \ldots | A_{k\lambda}^{(op)\dagger} | \ldots, N_{k\lambda} - 1, \ldots \rangle \tag{13-61}$$

$$= \sqrt{\frac{2\pi\hbar c^2}{\omega}} \sqrt{N_{k\lambda}};$$

we see explicitly that $A_{k\lambda}{}^{(op)\dagger}$ is an operator that increases the number of photons in the mode $k\lambda$ by one. The operators $A_{k\lambda}{}^{(op)}$ and $A_{k\lambda}{}^{(op)\dagger}$ thus obey

$$A_{k\lambda}^{(op)} |N_{k_1\lambda_1}, \ldots, N_{k\lambda}, \ldots\rangle = \sqrt{\frac{2\pi\hbar c^2}{\omega}} \sqrt{N_{k\lambda}} |N_{k_1\lambda_1}, \ldots, N_{k\lambda} - 1, \ldots\rangle$$

$$A_{k\lambda}^{(op)\dagger} |N_{k_1\lambda_1}, \ldots, N_{k\lambda}, \ldots\rangle \tag{13-62}$$

$$= \sqrt{\frac{2\pi\hbar c^2}{\omega}} \sqrt{N_{k\lambda} + 1} |N_{k_1\lambda_1}, \ldots, N_{k\lambda} + 1, \ldots\rangle$$

Aside from the factor $\sqrt{2\pi\hbar c^2/\omega}$, these operators are identical to the a and $a^\dagger$ of the harmonic oscillator. The quantum mechanical description of the radiation field is simply that it is an infinite number of "harmonic oscillators," one per mode; the quanta of these oscillators are the photons.

It is convenient to define a (Hermitian) electromagnetic field operator

$$\mathbf{A}^{(op)}(\mathbf{r}) = \sum_{k\lambda} \left(A_{k\lambda}^{(op)} \lambda \frac{e^{i\mathbf{k}\cdot\mathbf{r}}}{\sqrt{V}} + A_{k\lambda}^{(op)\dagger} \lambda^* \frac{e^{-i\mathbf{k}\cdot\mathbf{r}}}{\sqrt{V}} \right) \tag{13-63}$$

in terms of which we can write $\hat{H}_{int}$, the operator for the interaction between the matter and the quantum mechanically described radiation, now including the quantum mechanical generalization of the ρA^2 term,

$$\hat{H}_{int} = \int d^3r \left[-\frac{e}{c} \mathbf{j}(\mathbf{r}) \cdot \mathbf{A}^{(op)}(\mathbf{r}) + \frac{e^2}{2mc^2} \rho(\mathbf{r}) [\mathbf{A}^{(op)}(\mathbf{r})]^2 \right] \tag{13-64}$$

[cf. (13-33)]. In the interaction representation, $\mathbf{A}^{(op)}(\mathbf{r}, t)$ develops in time by

$$\mathbf{A}^{(op)}(\mathbf{r}, t) = e^{iH_{em}t/\hbar} \mathbf{A}^{(op)}(\mathbf{r}) e^{-iH_{em}t/\hbar} \tag{13-65}$$

where H_{em} is the Hamiltonian operator for free radiation. It is easy to see that

$$e^{iH_{em}t/\hbar} A_{k\lambda}^{(op)} e^{-iH_{em}t/\hbar} = e^{-ickt} A_{k\lambda}^{(op)} \tag{13-66}$$

since $A_{k\lambda}^{(op)}$ acting on an energy eigenstate removes a photon in the mode $k\lambda$ and thus reduces the energy of the state by $\hbar ck$. Thus from (13-66) and its Hermitian adjoint we find that we can write $\mathbf{A}^{(op)}(\mathbf{r}t)$ as

$$\mathbf{A}^{(op)}(\mathbf{r}t) = \sum_{k\lambda} \left[A_{k\lambda}^{(op)} \lambda \frac{e^{i\mathbf{k}\cdot\mathbf{r} - i\omega t}}{\sqrt{V}} + A_{k\lambda}^{(op)\dagger} \lambda^* \frac{e^{-i\mathbf{k}\cdot\mathbf{r} + i\omega t}}{\sqrt{V}} \right] \tag{13-67}$$

By construction, the description of the absorption of fully quantum mechanical radiation is, to lowest order, identical to the description of the absorption of classical radiation. Let us now carry out this completely quantum mechanical description of the emission process. We ask for the transition rate from a state (13-54) with energy $\varepsilon_n + \sum_{k'\lambda'} \hbar ck' N_{k'\lambda'}$, to a new state

$$|0; N_{k_1\lambda_1}, \ldots, N_{k\lambda}+1, \ldots\rangle \tag{13-68}$$

with energy $\varepsilon_0 + \sum_{k'\lambda'} \hbar ck' N_{k'\lambda'} + \hbar ck$; from the golden rule, this rate is

$$\frac{2\pi}{\hbar} \delta(\varepsilon_n - \hbar ck - \varepsilon_0) |\langle 0; \ldots, N_{k\lambda}+1, \ldots | \hat{H}_{int} | n; \ldots, N_{k\lambda}, \ldots\rangle|^2. \tag{13-69}$$

Now the matrix element in (13-69) is

$$\langle 0; \ldots, N_{k\lambda}+1, \ldots | \hat{H}_{int} | n; \ldots, N_{k\lambda}, \ldots\rangle$$

$$= -\frac{e}{c\sqrt{V}} \langle 0 | \mathbf{j}_k \cdot \lambda^* | n\rangle \langle \ldots, N_{k\lambda}+1, \ldots | A_{k\lambda}^{(op)\dagger} | \ldots, N_{k\lambda}, \ldots\rangle$$

$$= -\frac{e}{c}\sqrt{\frac{2\pi\hbar c^2}{\omega V}} \langle 0 | \mathbf{j}_k \cdot \lambda^* | n\rangle \sqrt{N_{k\lambda}+1}, \tag{13-70}$$

using (13-61) to determine the matrix element of $A_{k\lambda}^{(op)\dagger}$. Substituting (13-70) in (13-69) we find that the emission rate to $k\lambda$ is given by

$$\Gamma^{\text{emm.}}_{n\to 0;\mathbf{k}\lambda} = \frac{4\pi^2 e^2}{\omega V}\delta(\varepsilon_n - \varepsilon_0 - \hbar\omega)|\langle 0|\mathbf{j}_{\mathbf{k}}\cdot\boldsymbol{\lambda}^*|n\rangle|^2(N_{\mathbf{k}\lambda}+1). \tag{13-71}$$

We don't get quite the same result as in (13-49), the $N_{\mathbf{k}\lambda}$ part is just our old result for the rate of induced emission. But we now find an additional term, the +1 in $N_{\mathbf{k}\lambda} + 1$, which represents emission that can take place even if there is no external field present. This additional emission process is called *spontaneous emission.* The total emission rate is the sum of the induced and the spontaneous rates. Spontaneous emission is just the quantum mechanical version of the classical phenomenon of radiation from an accelerating charge.

A very important consequence of the fact that the electromagnetic field is quantized is that there are *vacuum fluctuations* of the field, analogous to the zero point motion of a harmonic oscillator. Thus in the vacuum state $|0, 0, 0, \ldots\rangle$, $\langle \mathbf{A}^{(op)}(\mathbf{r}, t)\rangle = 0$ but $\langle \mathbf{A}^{(op)}(\mathbf{r}, t)\mathbf{A}^{(op)}(\mathbf{r}', t)\rangle \neq 0$, and so, for example, $\langle \mathbf{E}^{(op)}(\mathbf{r}, t)\mathbf{E}^{(op)}(\mathbf{r}', t)\rangle$, the expectation value of the product of the (transverse) electric field operator $[= -c^{-1}\partial\mathbf{A}^{(op)}(\mathbf{r}, t)/\partial t]$ at two different points is nonzero. [See Problem 5, this chapter.] In a sense, one can regard spontaneous emission as induced emission due to the vacuum fluctuations of the electromagnetic field.

EINSTEIN'S A AND B COEFFICIENTS

Einstein gave a simple statistical argument to determine the rate at which spontaneous emission takes place. According to the Boltzmann distribution law of statistical mechanics, the probability of finding a system, which is at temperature T, having energy E, is proportional to $e^{-E/KT}$, where K is Boltzmann's constant. If we apply this law to photons in a cavity whose walls are at temperature T, we find that the relative probability of having N photons in a mode $\mathbf{k}\lambda$ of the cavity is proportional to $e^{-N\hbar ck/KT}$, since N such photons have energy $N\hbar ck$. Thus the average number of photons in the mode $\mathbf{k}\lambda$ is

$$\bar{N}_{\mathbf{k}\lambda} = \frac{\sum_{N=0}^{\infty} N e^{-N\hbar ck/KT}}{\sum_{N=0}^{\infty} e^{-N\hbar ck/KT}} = \frac{1}{e^{\hbar ck/KT} - 1}. \tag{13-72}$$

This relation is Planck's distribution law; from it we see that the average energy per mode is

$$E_{k\lambda} = \hbar ck N_{k\lambda} = \frac{\hbar ck}{e^{\hbar ck/KT} - 1}. \tag{13-73}$$

Now Einstein asked the question: in a cavity, photons are constantly being absorbed and emitted by the walls; what relations must there be between the absorption and emission rates to guarantee that $N_{k\lambda}$ on the average is given by (13-72)? Suppose, for simplicity, that the walls of the cavity are made of atoms having two levels, ε_0 and ε_n, with $\varepsilon_n > \varepsilon_0$. According to the Boltzmann law, the probability, P_n, of an atom being in the upper state is proportional to $e^{-\varepsilon_n/KT}$. Thus

$$\frac{P_n}{P_0} = e^{-(\varepsilon_n - \varepsilon_0)/KT}. \tag{13-74}$$

Consider the equilibrium between these atoms and radiation of frequency $\omega = (\varepsilon_n - \varepsilon_0)/\hbar$ in the cavity. Photons of this frequency are absorbed at a rate proportional to the number, N, of such photons present, and proportional to the probability of the atoms being in their lower level

$$\left(\frac{dN}{dt}\right)_{abs} = -BNP_0. \tag{13-75}$$

Emission of such photons is induced at a rate

$$\left(\frac{dN}{dt}\right)_{ind.emm.} = BNP_n; \tag{13-76}$$

the coefficient is also B because of (13-46). If these were the only absorption and emission processes, all the photons would soon disappear, because $P_n < P_0$. Thus there must be an additional emission process, spontaneous emission, at a rate

$$\left(\frac{dN}{dt}\right)_{\substack{\text{spontaneous}\\ \text{emission}}} = AP_n. \tag{13-77}$$

The problem is to determine the coefficient A.

In equilibrium, the net dN/dt is zero and $N = \bar{N}$. Adding (13-75), (13-76), and (13-77), we find $B\bar{N}(P_0 - P_n) + AP_n = 0$. (13-74) implies that

$$A = B\bar{N}\left(e^{(\varepsilon_n - \varepsilon_0)/KT} - 1\right),$$

or from (13-72)

$$A = B. \tag{13-78}$$

Thus the coefficient in spontaneous emission rate is exactly the same as in the induced emission rate; the total emission rate is

$$\left(\frac{dN}{dt}\right)_{\text{emm.}} = BP_n(N+1). \tag{13-79}$$

This agrees with our result (13-71) in its dependence on the number of photons N. Notice that the form of the emission rate (13-79) was deduced without knowing the actual value of B; only *detailed balancing* arguments were used.

DETAILS OF SPONTANEOUS EMISSION

Let us now turn to examining the details of the spontaneous emission process. Equation (13-71) with $N_{\mathbf{k}\lambda} = 0$ is the rate of spontaneous emission into the mode $\mathbf{k}\lambda$. The power dP of light of polarization λ, radiated into a small solid angle $d\Omega$ in the direction $\hat{\mathbf{k}}$, due to spontaneous transitions from $|n\rangle$ to $|0\rangle$, is just

$$\begin{aligned} dP &= \sum_{\mathbf{k}\text{ in }d\Omega} \hbar\omega\, \Gamma^{\text{emm.}}_{n\to 0;\mathbf{k}\lambda} \\ &= d\Omega \int \frac{\omega^2\, d\omega}{(2\pi c)^3}\, \hbar\omega\, \frac{4\pi^2 e^2}{\omega}\, |\langle 0|\mathbf{j}_{\mathbf{k}}\cdot\boldsymbol{\lambda}^*|n\rangle|^2 \delta(\varepsilon_n - \varepsilon_0 - \hbar\omega). \end{aligned} \tag{13-80}$$

Thus the power per unit solid angle in a given direction and for a given polarization is

$$\frac{dP}{d\Omega} = \frac{\omega^2 e^2}{2\pi c^3}\, |\langle 0|\mathbf{j}_{\mathbf{k}}\cdot\boldsymbol{\lambda}^*|n\rangle|^2. \tag{13-81}$$

We can get some feeling for this result by comparing it with the formula for the power radiated by a transverse classical current of frequency ω:

$$\mathcal{J}(\mathbf{r}, t) = \mathcal{J}(\mathbf{r})e^{-i\omega t} + \mathcal{J}(\mathbf{r})^* e^{i\omega t}. \tag{13-82}$$

The solution of Eq. (13-6) for $\mathbf{A}(\mathbf{r}t)$ is

$$\mathbf{A}(\mathbf{r}t) = \int d^3r'\, \frac{\mathcal{J}(\mathbf{r}', t - |\mathbf{r}-\mathbf{r}'|/c)}{c|\mathbf{r}-\mathbf{r}'|},$$

and in the far field, far from the current, this becomes on using (13-82)

$$\mathbf{A}(\mathbf{r}t) = \frac{1}{rc}\int d^3r'\, \mathbf{j}(\mathbf{r}')e^{-i\omega(t - r/c + \hat{\mathbf{r}}\cdot\mathbf{r}'/c)} + \text{c.c.}$$
$$= \frac{e^{-i\omega(t-r/c)}}{rc}\,\mathbf{j}_{\mathbf{k}} + \text{c.c.} \qquad (13\text{-}83)$$

where

$$\mathbf{k} = \frac{\omega\hat{\mathbf{r}}}{c} \qquad (13\text{-}84)$$

is the wave vector of the light as seen in the far field. The component of $\mathbf{A}$ of polarization $\boldsymbol{\lambda}$ is $\mathbf{A}(\mathbf{r}t)\cdot\boldsymbol{\lambda}^*$, and the radiated power per unit area carried by this component is

$$\frac{\omega^2}{2\pi c}\,\frac{|\mathbf{j}_{\mathbf{k}}\cdot\boldsymbol{\lambda}^*|^2}{(rc)^2};$$

the power per unit solid angle is therefore

$$\frac{dP}{d\Omega} = \frac{\omega^2}{2\pi c^3}|\mathbf{j}_{\mathbf{k}}\cdot\boldsymbol{\lambda}^*|^2. \qquad (13\text{-}85)$$

If we compare this with the formula for spontaneous emission, we see that the matrix element $\langle n|e\mathbf{j}_{\mathbf{k}}|0\rangle$ plays the same role in spontaneous emission as does $\mathbf{j}_{\mathbf{k}}$ in the case of radiation from a classical current. Thus we may regard spontaneous emission as being the radiation from an effective current $\langle 0|e\mathbf{j}(\mathbf{r}t)|n\rangle$, with spatial Fourier components:

$$\mathbf{j}_{\mathbf{k}}^{\text{eff}}(t) = \langle 0|e\mathbf{j}_{\mathbf{k}}(t)|n\rangle; \qquad (13\text{-}86)$$

this effective current contains only a positive frequency component when ε_0 is less than ε_n.

We can learn the details of the angular distribution of emitted radiation from (13-81). First, let us consider the case that the initial state of the matter is a momentum eigenstate, with total momentum $\hbar\mathbf{q}_n$, and that we look for the transition rate to a final state of the matter with total momentum $\hbar\mathbf{q}_0$. We expect in this case that in order to conserve momentum the emitted photon must have momentum $\hbar\mathbf{q}_n - \hbar\mathbf{q}_0$ and hence wave vector $\mathbf{q}_n - \mathbf{q}_0$. First of all, if the matter consists of a single charged free particle of mass m, then the current matrix element $\langle \mathbf{q}_0|\mathbf{j}_{\mathbf{k}}|\mathbf{q}_n\rangle$, using (3-36), is given by

$$\langle q_0|j_k|q_n\rangle = \int d^3r\ \frac{e^{-iq_0\cdot r}}{\sqrt{V}}\left[\frac{\hbar}{2im}\nabla e^{-ik\cdot r} + e^{-ik\cdot r}\frac{\hbar}{2im}\nabla\right]\frac{e^{iq_n\cdot r}}{\sqrt{V}}$$
$$= \frac{\hbar}{2m}(q_0 + q_n)\delta_{q_n - q_0, k} \tag{13-87}$$

which is nonvanishing only if $k = q_n - q_0$. Notice, though, that for a single free particle it is impossible to satisfy the momentum conservation condition

$$\hbar k = \hbar q_n - \hbar q_0$$

and the energy conservation condition

$$\hbar ck = \frac{\hbar^2 q_n^2}{2m} - \frac{\hbar^2 q_0^2}{2m}$$

simultaneously; a free particle can't radiate.

To show in general that the radiation can only come off with $k = q_n - q_0$, let us notice that [Eq. (6-23)] for the ith particle

$$e^{ip_i\cdot a/\hbar}\, r_i e^{-ip_i\cdot a/\hbar} = r_i + a,$$

where a is a c-number vector; since p_j commutes with r_i if $i \neq j$

$$e^{iP\cdot a/\hbar}\, r_i e^{-iP\cdot a/\hbar} = r_i + a \tag{13-88}$$

where

$$P = \sum_j p_j \tag{13-89}$$

is the total momentum operator of the system. Equation (13-88) says that the unitary operator $e^{iP\cdot a/\hbar}$ has the effect of translating any coordinate operator by a. Furthermore, since P commutes with all the p_j,

$$e^{iP\cdot a/\hbar} f(r_1, \ldots, r_N, p_1, \ldots, p_N) e^{-iP\cdot a/\hbar}$$
$$= f(r_1 + a, \ldots, r_N + a, p_1, \ldots, p_N) \tag{13-90}$$

where f is any function of the momenta and coordinates of the particles. In particular,

$$e^{i\mathbf{P}\cdot\mathbf{a}/\hbar}\mathbf{j}(\mathbf{r})e^{-i\mathbf{P}\cdot\mathbf{a}/\hbar}$$
$$= \sum_i \frac{1}{2m}[\mathbf{p}_i\delta(\mathbf{r}_i+\mathbf{a}-\mathbf{r}) + \delta(\mathbf{r}_i+\mathbf{a}-\mathbf{r})\mathbf{p}_i] = \mathbf{j}(\mathbf{r}-\mathbf{a}), \tag{13-91}$$

so that we can write

$$\mathbf{j}(\mathbf{r}) = e^{-i\mathbf{P}\cdot\mathbf{r}/\hbar}\mathbf{j}(\mathbf{r}=0)e^{i\mathbf{P}\cdot\mathbf{r}/\hbar}. \tag{13-92}$$

This equation expresses the current operator at point $\mathbf{r}$ in terms of the current operator at the origin and the translation operator. [$\mathbf{r}$ in Eqs. (13-91) and (13-92) is *not* an operator — it is the point in space where one observes the current.] We can use (13-92) to write the matrix element $\langle 0|\mathbf{j}_\mathbf{k}|n\rangle$ as

$$\int d^3r\, e^{-i\mathbf{k}\cdot\mathbf{r}}\langle 0|e^{-i\mathbf{P}\cdot\mathbf{r}/\hbar}\mathbf{j}(0)\, e^{i\mathbf{P}\cdot\mathbf{r}/\hbar}|n\rangle.$$

If $|0\rangle$ and $|n\rangle$ are eigenstates of total momentum with eigenvalues $\hbar\mathbf{q}_0$ and $\hbar\mathbf{q}_n$ then this matrix element becomes

$$\langle 0|\mathbf{j}_\mathbf{k}|n\rangle = \int d^3r\, e^{-i\mathbf{k}\cdot\mathbf{r}}\langle 0|e^{-i\mathbf{q}_0\cdot\mathbf{r}}\mathbf{j}(0)e^{i\mathbf{q}_n\cdot\mathbf{r}}|n\rangle$$
$$= V\langle 0|\mathbf{j}(0)|n\rangle\delta_{\mathbf{k},\mathbf{q}_n-\mathbf{q}_0}, \tag{13-93}$$

thus showing that the matrix element is nonzero only for those final photons that conserve momentum.

Thus if we begin with a system in a momentum eigenstate and then observe it to be in a new momentum eigenstate, due to spontaneous emission, the direction at which we can observe the emitted photon is completely determined. Conversely, if we observe the direction and frequency of the emitted photon then the final state of the matter must be a momentum eigenstate, with the value of momentum necessary to conserve total momentum. In other words, if the initial state has momentum $\hbar\mathbf{q}_n$, the final state of the matter and light is a linear superposition of states for all possible $\mathbf{q}_0$ in which the matter has momentum $\hbar\mathbf{q}_0$ and the photon has momentum $\hbar(\mathbf{q}_n - \mathbf{q}_0)$. If the measurement of $\mathbf{q}_0$ has some uncertainty, this implies a spread in the observable photon momenta.

ELECTRIC DIPOLE TRANSITIONS

Next let us consider the radiation from an atom whose nucleus is well localized in space. Typically, the wave lengths of light emitted by an atom are several orders of magnitude greater than the spatial extent of the electron orbits about the nucleus. This means that in evaluating the matrix element

$$\langle 0|\mathbf{j}_{\mathbf{k}}|n\rangle = \int e^{-i\mathbf{k}\cdot\mathbf{r}}\langle 0|\mathbf{j}(\mathbf{r})|n\rangle\, d^3r \tag{13-94}$$

it is a good approximation to expand $e^{-i\mathbf{k}\cdot\mathbf{r}}$ about $\mathbf{r}_0$, the position of the nucleus. For convenience, let us choose $\mathbf{r}_0 = 0$. Then (13-94) becomes

$$\begin{aligned}\langle 0|\mathbf{j}_{\mathbf{k}}|n\rangle &= \int d^3r\,[1 - i\mathbf{k}\cdot\mathbf{r} + \cdots]\langle 0|\mathbf{j}(\mathbf{r})|n\rangle \\ &= \langle 0|\mathbf{j}_0|n\rangle - i\langle 0|\int d^3r\,\mathbf{k}\cdot\mathbf{r}\,\mathbf{j}(\mathbf{r})|n\rangle + \cdots\end{aligned} \tag{13-95}$$

Now

$$\mathbf{j}_0 = \frac{\mathbf{P}}{m} = \frac{1}{i\hbar}[\mathbf{R}, H_0] \tag{13-96}$$

where $\mathbf{P}$ is the total momentum operator,

$$\mathbf{R} = \sum_i \mathbf{r}_i \tag{13-97}$$

is the dipole moment operator, and H_0 is the Hamiltonian of the atom, without radiation. The first term in (13-95) becomes

$$\frac{1}{i\hbar}\langle 0|(\mathbf{R}H_0 - H_0\mathbf{R})|n\rangle = \frac{\varepsilon_n - \varepsilon_0}{i\hbar}\langle 0|\mathbf{R}|n\rangle;$$

$(\varepsilon_n - \varepsilon_0)/\hbar$ is the frequency of the emitted radiation. The matrix element, to lowest order in $\mathbf{k}\cdot\mathbf{r}$ is

$$\langle 0|\mathbf{j}_{\mathbf{k}}|n\rangle = -i\omega\mathbf{d}_{0n} \tag{13-98}$$

where the vector

$$\mathbf{d}_{0n} = \langle 0|\mathbf{R}|n\rangle \tag{13-99}$$

is the off-diagonal matrix element of the dipole moment operator. The transition described by the matrix element (13-98) is called an *electric dipole transition;* $e\mathbf{R}$ is the electric dipole operator. To this order the formula for the spontaneously emitted power, of a given polarization $\boldsymbol{\lambda}$, becomes

$$\frac{dP_\lambda}{d\Omega} = \frac{\omega^4 e^2}{2\pi c^3}|\mathbf{d}_{0n}\cdot\boldsymbol{\lambda}^*|^2. \tag{13-100}$$

This equation says that the amplitude of emitted light of polarization

$\boldsymbol{\lambda}$ is proportional to the component of $\mathbf{d}_{0n}$ along $\boldsymbol{\lambda}$; the power depends on the absolute square of this component. The polarization vector of the emitted light is essentially proportional to $\mathbf{d}_{0n}$; more precisely, since the polarization vector must be orthogonal to the wave vector $\mathbf{k}$, the polarization of the emitted light is proportional to $(\mathbf{d}_{0n})_\perp$, the component of $\mathbf{d}_{0n}$ orthogonal to $\mathbf{k}$:

$$(\mathbf{d}_{0n})_\perp = \mathbf{d}_{0n} - (\mathbf{d}_{0n} \cdot \hat{\mathbf{k}})\hat{\mathbf{k}}. \qquad (13\text{–}101)$$

[For $\boldsymbol{\lambda} \cdot \mathbf{k} = 0$ then $(\mathbf{d}_{0n})_\perp \cdot \boldsymbol{\lambda}^* = \mathbf{d}_{0n} \cdot \boldsymbol{\lambda}^*$].

Suppose that the initial and final states are eigenstates of the square and z component of the total orbital angular momentum

$$\mathbf{L} = \sum_i \mathbf{r}_i \times \mathbf{p}_i. \qquad (13\text{–}102)$$

What can we then say about the polarization of the emitted radiation? First, we notice that

$$[L_z, R_z] = 0. \qquad (13\text{–}103)$$

Therefore, if the initial and final eigenstates have L^2, L_z eigenvalues $\hbar^2 l(l+1)$, $\hbar m$, and $\hbar^2 l'(l'+1)$, $\hbar m'$ respectively, we have

$$\langle l'm'|[L_z, R_z]|lm\rangle = \hbar(m' - m)\langle l'm'|R_z|lm\rangle = 0. \qquad (13\text{–}104)$$

This implies the selection rule

$$\langle l'm'|R_z|lm\rangle = 0 \quad \text{for } m \neq m'. \qquad (13\text{–}105)$$

Furthermore

$$[L_z, R_x] = i\hbar R_y$$

$$[L_z, R_y] = -i\hbar R_x; \qquad (13\text{–}106)$$

Thus

$$\langle l'm'|[L_z, R_x]|lm\rangle = \hbar(m' - m)\langle l'm'|R_x|lm\rangle = i\hbar\langle l'm'|R_y|lm\rangle$$

$$\langle l'm'|[L_z, R_y]|lm\rangle = \hbar(m' - m)\langle l'm'|R_y|lm\rangle = -i\hbar\langle l'm'|R_x|lm\rangle. \qquad (13\text{–}107)$$

Substituting the first equation into the second we find

$$(m' - m)^2\langle l'm'|R_x|lm\rangle = \langle l'm'|R_x|lm\rangle. \qquad (13\text{–}108)$$

Therefore

$$\langle l'm'|R_x|lm\rangle = 0 \quad \text{for } m' \neq m \pm 1$$

$$\langle l'm'|R_y|lm\rangle = 0 \quad \text{for } m' \neq m \pm 1. \tag{13-109}$$

This equation and (13-105) imply that electric dipole radiation occurs only if

$$m' = m \quad \text{or} \quad m \pm 1. \tag{13-110}$$

In the case m' = m, the dipole matrix element, $\mathbf{d}_{on}$, is along the z axis. The emitted photon is always plane polarized in the plane defined by **k** and the z axis, as in Fig. 13-1. The emission rate, proportional to $(\mathbf{d}_{on})_\perp$ squared, thus goes as $\sin^2\theta$. In particular, there is never any radiation along the z axis when m' = m.

Next in the case m' = m − 1, the vector $\mathbf{d}_{on}$ lies in the x, y plane; from Eq. (13-107)

$$\langle l'm'|R_y|lm\rangle = i\langle l'm'|R_x|lm\rangle \tag{13-111}$$

and therefore

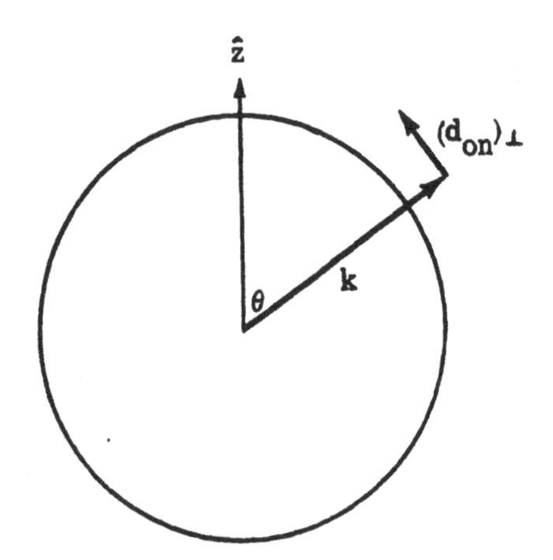

Fig. 13-1

Direction of the component of the dipole matrix element orthogonal to k in a $\Delta L_z = 0$ transition.

$$d_{0n} \sim \begin{pmatrix} 1 \\ i \\ 0 \end{pmatrix} \tag{13-112}$$

in the x, y, z coordinate system. Thus if **k** is along the positive z axis, the photon is right circularly polarized. This is the only possibility consistent with the conservation of L_z; the L_z component of the system changes by $-\hbar$, and as we know, a right circularly polarized photon carries angular momentum $+\hbar$ along its direction of motion. Similarly, if **k** is along the negative z axis, the photon whose polarization is (13-112) is left circularly polarized, has angular momentum $-\hbar$ along its direction of travel and consequently angular momentum $+\hbar$ along the positive z axis. If **k** lies in the equatorial plane, the emitted photon has plane polarization in the equatorial plane and perpendicular to **k**. A photon emitted in an arbitrary direction is elliptically polarized. In the case $m' = m + 1$,

$$d_{0n} \sim \begin{pmatrix} 1 \\ -i \\ 0 \end{pmatrix} \tag{13-113}$$

and a photon emitted along the +z axis is left circularly polarized, etc.

The selection rules for l are

$$l' = l \pm 1. \tag{13-114}$$

We shall understand why this must be so when we study the theory of the addition of quantum mechanical angular momenta. For now we can show (13-114) by a simple argument given by Dirac. By a direct dull calculation

$$[L^2, [L^2, \mathbf{R}]] = 2\hbar^2(\mathbf{R}L^2 + L^2\mathbf{R}). \tag{13-115}$$

Taking matrix elements of this relation we find

$$[l'(l'+1) - l(l+1)]^2 \langle l'm' | \mathbf{R} | lm \rangle = 2[l(l+1) + l'(l'+1)] \langle l'm' | \mathbf{R} | lm \rangle,$$

or

$$(l + l')(l + l' + 2)[(l - l')^2 - 1] \langle l'm' | \mathbf{R} | lm \rangle = 0. \tag{13-116}$$

Since l and l' are nonnegative the dipole matrix element can be nonzero only if $l + l' = 0$ or $l' = l \pm 1$. However, $l + l'$ can be zero only if $l = l' = 0$. In this case $\langle 00 | \mathbf{R} | 00 \rangle = 0$ anyhow since the states with $l = 0$ are spherically symmetric, leaving the vector $\langle 00 | \mathbf{R} | 00 \rangle$ with no direction to point. This leaves only the possibility (13-114).

For a single electron making a transition from $|n, l, m\rangle$ to $|n', l', m'\rangle$ the dipole matrix element is

$$\langle n', l', m' | \mathbf{r} | n, l, m \rangle = \int d^3r \, [R_{n'}(r) Y_{l'm'}(\theta, \varphi)]^* \mathbf{r} \, [R_n(r) Y_{lm}(\theta, \varphi)]$$
$$= \int_0^\infty r^3 \, dr \, R_{n'}^*(r) R_n(r) \int d\Omega \, Y_{l'm'}^*(\theta, \varphi) \hat{\mathbf{r}} \, Y_{lm}(\theta, \varphi) \tag{13-117}$$

where R_n is the initial radial wave function and $R_{n'}$ is the final state radial wave function (not necessarily hydrogenic). One can check the selection rules (13-105), (13-109), and (13-114) by direct calculation of the angular integrals in (13-117). For example, the x component plus i times the y component of (13-117) is

$$\int d\Omega \, Y_{l'm'}^*(\theta, \varphi) \sin\theta \, e^{i\varphi} Y_{lm}(\theta, \varphi)$$

which is nonzero only if $m' = m + 1$ and $l' = l \pm 1$.

In actual atoms the energy eigenstates are not simply eigenstates of orbital L^2 and L_z, because of spin-orbit coupling. We shall study the general form of the selection rules in Chapter 17.

The total absorption cross section (13-51) assumes a particularly simple form for electric dipole transitions; on using (13-98) we find:

$$\sigma_{\text{dip.abs.}} = \frac{4\pi^2 e^2 \omega}{c} \sum_n |\mathbf{d}_{n0} \cdot \boldsymbol{\lambda}|^2 \delta(\varepsilon_n - \varepsilon_0 - \hbar\omega). \tag{13-118}$$

The dipole absorption cross section for a system in its ground state obeys a particularly simple *sum rule* when the incident light is plane polarized: namely:

$$\int_0^\infty d\omega \, \sigma_{\text{dip.abs.}}(\omega) = \frac{2\pi^2 e^2 N}{mc}, \tag{13-119}$$

where N is the total number of particles (all assumed of mass m and charge e) in the system. To derive this sum rule we note that if $\hat{\mathbf{u}}$ is a real unit vector, then

$$[\mathbf{P} \cdot \hat{\mathbf{u}}, \mathbf{R} \cdot \hat{\mathbf{u}}] = \left[\sum_{i=1}^N \mathbf{p}_i \cdot \hat{\mathbf{u}}, \sum_{j=1}^N \mathbf{r}_j \cdot \hat{\mathbf{u}} \right] = -i\hbar \sum_{ij} \delta_{ij} = -i\hbar N, \tag{13-120}$$

where N is the number of electrons in the system. Taking the expectation value of (13-120) in the initial state $|0\rangle$ (not necessarily the ground state) we have

$$\langle 0 | [(\mathbf{P} \cdot \hat{\mathbf{u}})(\mathbf{R} \cdot \hat{\mathbf{u}}) - (\mathbf{R} \cdot \hat{\mathbf{u}})(\mathbf{P} \cdot \hat{\mathbf{u}})] | 0 \rangle = -i\hbar N.$$

Inserting a complete set of states between the $\mathbf{P}$ and the $\mathbf{R}$ we find

$$\sum_n \langle 0|\mathbf{P}\cdot\hat{\mathbf{u}}|n\rangle\langle n|\mathbf{R}\cdot\hat{\mathbf{u}}|0\rangle - \sum_n \langle 0|\mathbf{R}\cdot\hat{\mathbf{u}}|n\rangle\langle n|\mathbf{P}\cdot\hat{\mathbf{u}}|0\rangle = -i\hbar N.$$

Now (in the absence of velocity dependent forces)

$$\mathbf{P} = \frac{m}{i\hbar}[\mathbf{R}, H_0]$$

so that (13–119) becomes

$$\sum_n (\varepsilon_n - \varepsilon_0)|\langle n|\mathbf{R}|0\rangle\cdot\hat{\mathbf{u}}|^2 = \frac{\hbar^2 N}{2m}. \qquad (13\text{–}121)$$

This relation is called the *dipole sum rule;* it is a consequence of the commutation relation between r and p. The sum is over all states.[2] It tells us, for example, that the dipole matrix elements must certainly decrease in energy faster than $(\varepsilon_n - \varepsilon_0)^{1/2}$, in order that the left side of (13–121) converge.

Now integrating (13–118) over all frequencies from 0 to ∞ we find

$$\int_0^\infty d\omega\, \sigma_{\text{dip.abs.}}(\omega) = \frac{4\pi^2 e^2}{\hbar^2 c}\sum_n (\varepsilon_n - \varepsilon_0)|\mathbf{d}_{n0}\cdot\boldsymbol{\lambda}|^2,$$

where the sum is only over states with energy $\geq \varepsilon_0$. If $|0\rangle$ is the ground state then the sum is over all $|n\rangle$; then if $\boldsymbol{\lambda}$ is a real vector, as for plane polarization, (13–121) gives us the sum rule (13–119). Note that (13–119) is also valid for unpolarized incident light.

MAGNETIC DIPOLE AND ELECTRIC QUADRUPOLE TRANSITIONS

It may happen that because of a selection rule, or accidently, the dipole matrix element between an initial state $|n\rangle$ and a final state $|0\rangle$ vanishes. In that case the transition from n to 0 is called *forbidden.* There still is the possibility that the transition can occur through higher-order terms in (13–95). Let us examine the next order

[2]The N on the right side is the number of electrons counted in the dipole operator R. If, for example, one is considering only processes in a heavy atom in which the inner core electrons never participate, then if one only includes excited states of the *outer* electrons in the sum, the N on the right should be taken to be the number of such outer electrons.

term in $\langle 0 | \mathbf{j}_{\mathbf{k}} \cdot \boldsymbol{\lambda}^* | n \rangle$ beyond the dipole:

$$\langle 0 | \int d^3r \; (\mathbf{k} \cdot \mathbf{r})(\mathbf{j}(\mathbf{r}) \cdot \boldsymbol{\lambda}^*) | n \rangle . \tag{13-122}$$

times $-i$. We may write

$$(\mathbf{k} \cdot \mathbf{r})(\mathbf{j}(\mathbf{r}) \cdot \boldsymbol{\lambda}^*) = \frac{1}{2} [(\mathbf{k} \cdot \mathbf{r})(\mathbf{j}(\mathbf{r}) \cdot \boldsymbol{\lambda}^*) - (\boldsymbol{\lambda}^* \cdot \mathbf{r})(\mathbf{j}(\mathbf{r}) \cdot \mathbf{k})]$$
$$+ \frac{1}{2} [(\mathbf{k} \cdot \mathbf{r})(\mathbf{j}(\mathbf{r}) \cdot \boldsymbol{\lambda}^*) + (\boldsymbol{\lambda}^* \cdot \mathbf{r})(\mathbf{j}(\mathbf{r}) \cdot \mathbf{k}))]$$

whereupon (13-122) becomes

$$\frac{1}{2} (\mathbf{k} \times \boldsymbol{\lambda}^*) \cdot \langle 0 | \int d^3r \; \mathbf{r} \times \mathbf{j}(\mathbf{r}) | n \rangle + \sum_{a,b=x,y,z} \frac{1}{2} \lambda_a^* k_b \langle 0 | \int d^3r \, [j_a(\mathbf{r}) r_b + r_a \, j_b(\mathbf{r})] | n \rangle . \tag{13-123}$$

Now

$$\int d^3r \, \mathbf{r} \times \mathbf{j}(\mathbf{r}) = \sum_i \left[-\frac{\mathbf{p}_i}{2m} \times \mathbf{r}_i + \mathbf{r}_i \times \frac{\mathbf{p}_i}{2m} \right] = \frac{1}{m} \sum_i \mathbf{r}_i \times \mathbf{p}_i = \frac{\mathbf{L}}{m} . \tag{13-124}$$

The first term in (13-123) is thus proportional to the matrix element of the orbital angular momentum which in turn is proportional to the matrix element of the orbital magnetic dipole operator

$$\mathbf{M}_{\text{orb}} = \frac{e}{2mc} \mathbf{L} \tag{13-125}$$

Transitions arising through this term are *magnetic dipole transitions.* [Cf. Problem 11, this chapter.] Such transitions are much weaker than electric dipole transitions. The net matrix element

$$\frac{\mathbf{k} \times \boldsymbol{\lambda}^*}{2m} \cdot \langle 0 | \mathbf{L} | n \rangle$$

is of order $\hbar k/m$, since $\langle 0 | \mathbf{L} | n \rangle$ is of order $\hbar$. Now electric dipole terms, (13-98), are of order ωa_0, where a_0 is the Bohr radius, since $\langle 0 | \mathbf{R} | n \rangle$, if nonvanishing, is on the order of atomic dimensions. Thus magnetic dipole transition amplitudes are of order

$$\frac{\hbar k}{m \omega a_0} = \frac{e^2}{\hbar c} \equiv \alpha = \frac{1}{137}$$

smaller than electric dipole transition amplitudes. Equivalently the ratio of the amplitudes is of order v/c where v is a typical electron velocity in an atom ($\sim\alpha c$).

The integral in the second term equals

$$\frac{1}{2m}\sum_i (p_{i,a}r_{i,b}+r_{i,b}p_{i,a}+p_{i,b}r_{i,a}+r_{i,a}p_{i,b}).$$

Using the facts that

$$[r_{i,a}r_{i,b}, H_0] = \frac{i\hbar}{m}(r_{i,a}p_{i,b}+p_{i,a}r_{i,b})$$

and $\mathbf{k}\cdot\boldsymbol{\lambda}^* = 0$ we find that the second term in (13-123) becomes

$$-\frac{1}{2}\omega\sum_{ab} k_a\lambda_b^*\langle 0|\sum_i(r_{i,a}r_{i,b}-\frac{1}{3}\delta_{ab}r_i^2)|n\rangle; \tag{13-126}$$

the operator $e\sum_i(r_{i,a}r_{i,b}-\frac{1}{3}\delta_{ab}r_i^2)$ is the electric quadrupole moment operator and transitions occurring through the amplitude (13-126) are called *electric quadrupole transitions.* Comparing (13-126) with (13-98) we see that electric quadrupole amplitudes are of order ka_0 smaller than electric dipole amplitudes. In Chapter 17, we shall examine the general structure of multipole radiation.

Now it may happen that for a given transition the entire matrix element $\langle 0|\mathbf{j_k}|n\rangle$ vanishes. Then the transition is called *strictly forbidden.* But one need not despair — it is still possible for the transition to occur in a higher order in perturbation theory than the first. Such higher-order processes are associated with the emission or absorption of more than one photon.

SCATTERING OF LIGHT

Let us now study the scattering of light by a charged system. We shall use the quantum mechanical description of light to discuss the scattering. From this point of view, we have an initial state $|0; N_{\mathbf{k}\lambda}, N_{\mathbf{k}'\lambda'} = 0\rangle$ with $N_{\mathbf{k}\lambda}$ photons in the mode $\mathbf{k}$, $\boldsymbol{\lambda}$ and none in $\mathbf{k}'$, $\boldsymbol{\lambda}'$, and we ask for the transition amplitude to a state $|n; N_{\mathbf{k}\lambda}-1, N_{\mathbf{k}'\lambda'} = 1\rangle$ with one photon in the mode $\mathbf{k}'\boldsymbol{\lambda}'$ and one fewer $\mathbf{k}\boldsymbol{\lambda}$ photon; this corresponds to the scattering of a $\mathbf{k}$, $\boldsymbol{\lambda}$ photon to $\mathbf{k}'$, $\boldsymbol{\lambda}'$. This process can occur either through the $(e^2/2mc^2)\int d^3r\ \rho(\mathbf{r})A^2(\mathbf{r})$ term in first order, or through the $-(e/c)\int d^3r\ \mathbf{j}(\mathbf{r})\cdot\mathbf{A}(\mathbf{r})$ term in second order. First we look at the effects of the ρA^2 term.

In the fully quantum mechanical description of the interaction of light with matter, the semiclassical ρA^2 interaction term is replaced by

$$\frac{e^2}{2mc^2}\int d^3r\ \rho(\mathbf{r})A^{(op)}(\mathbf{r})^2. \tag{13-127}$$

The matrix element of (13-127) is

$$\begin{aligned}
&\langle n;\ N_{\mathbf{k}\lambda}-1, N_{\mathbf{k}'\lambda'}=1|\frac{e^2}{2mc^2}\int d^3r\ \rho(\mathbf{r})A^{(op)}(\mathbf{r})^2|0;\ N_{\mathbf{k}\lambda}, N_{\mathbf{k}'\lambda'}=0\rangle\\
&=\frac{e^2}{2mc^2}\int d^3r\ \langle n|\rho(\mathbf{r})|0\rangle\\
&\quad\times\langle N_{\mathbf{k}\lambda}-1, N_{\mathbf{k}'\lambda'}=1|\mathbf{A}^{(op)}(\mathbf{r})\cdot\mathbf{A}^{(op)}(\mathbf{r})|N_{\mathbf{k}\lambda}, N_{\mathbf{k}'\lambda'}=0\rangle.
\end{aligned} \tag{13-128}$$

The operator $\mathbf{A}^{(op)}(\mathbf{r})$ creates and annihilates photons one at a time. The only way the photon matrix element in (13-128) can be nonzero is if the $\mathbf{A}^{(op)}$ on the right annihilates a $\mathbf{k},\lambda$ photon and then the $\mathbf{A}^{(op)}$ on the left creates a $\mathbf{k}',\lambda'$ photon, or vice versa. Thus

$$\begin{aligned}
&\langle N_{\mathbf{k}\lambda}-1, N_{\mathbf{k}'\lambda'}=1|A^{(op)}(\mathbf{r})^2|N_{\mathbf{k}\lambda}, N_{\mathbf{k}'\lambda'}=0\rangle\\
&\quad=\langle N_{\mathbf{k}\lambda}-1, N_{\mathbf{k}'\lambda'}=1|\mathbf{A}^{(op)}(\mathbf{r})|N_{\mathbf{k}\lambda}-1, N_{\mathbf{k}'\lambda'}=0\rangle\\
&\qquad\cdot\langle N_{\mathbf{k}\lambda}-1, N_{\mathbf{k}'\lambda'}=0|\mathbf{A}^{(op)}(\mathbf{r})|N_{\mathbf{k}\lambda}, N_{\mathbf{k}'\lambda'}=0\rangle\\
&\quad+\langle N_{\mathbf{k}\lambda}-1, N_{\mathbf{k}'\lambda'}=1|\mathbf{A}^{(op)}(\mathbf{r})|N_{\mathbf{k}\lambda}, N_{\mathbf{k}'\lambda'}=1\rangle\\
&\qquad\cdot\langle N_{\mathbf{k}\lambda}, N_{\mathbf{k}'\lambda'}=1|\mathbf{A}^{(op)}(\mathbf{r})|N_{\mathbf{k}\lambda}, N_{\mathbf{k}'\lambda'}=0\rangle\\
&\quad=2\sqrt{\frac{2\pi\hbar c^2}{\omega V}}\sqrt{\frac{2\pi\hbar c^2}{\omega' V}}\sqrt{N_{\mathbf{k}\lambda}}\ \boldsymbol{\lambda}\cdot\boldsymbol{\lambda}'^{*}\,e^{i\mathbf{k}\cdot\mathbf{r}}\,e^{-i\mathbf{k}'\cdot\mathbf{r}},
\end{aligned}$$

using (13-63), (13-60), and (13-61). The matrix element (13-128) is thus

$$r_0\,\frac{2\pi\hbar c^2}{\sqrt{\omega\omega'}}\,\frac{\sqrt{N_{\mathbf{k}\lambda}}}{V}\,\langle n|\int d^3r\,\rho(\mathbf{r})e^{i(\mathbf{k}-\mathbf{k}')\cdot\mathbf{r}}|0\rangle\,\boldsymbol{\lambda}\cdot\boldsymbol{\lambda}'^{*} \tag{13-129}$$

Where $r_0 = e^2/mc^2 = 2.8\times 10^{-13}$ cm is the *classical radius of the electron.*

To calculate the scattering rate arising from this matrix element we use the golden rule and find

$$\Gamma_{\substack{0\to n\\ \mathbf{k}\lambda\to\mathbf{k}'\lambda'}} = \frac{2\pi}{\hbar}\,\delta(\varepsilon_0+\hbar\omega-\varepsilon_n-\hbar\omega')\left(\frac{2\pi\hbar r_0 c^2}{V}\right)^2 \times \frac{N_{\mathbf{k}\lambda}}{\omega\omega'}\,|\langle n|\rho_{\mathbf{k}'-\mathbf{k}}|0\rangle|^2\,|\boldsymbol{\lambda}\cdot\boldsymbol{\lambda}'^*|^2. \tag{13-130}$$

Summing over all $\mathbf{k}'$ in a small solid angle $d\Omega$ we find

$$\frac{d\Gamma_{0\to n}}{d\Omega'} = c\,\frac{N_{\mathbf{k}\lambda}}{V}\,r_0^2\,\frac{\omega'}{\omega}\,|\boldsymbol{\lambda}\cdot\boldsymbol{\lambda}'^*|^2\,|\langle n|\rho_{\mathbf{k}'-\mathbf{k}}|0\rangle|^2. \tag{13-131}$$

Now $N_{\mathbf{k}\lambda}/V$ is the density of incident photons and therefore $cN_{\mathbf{k}\lambda}/V$ is the incident photon flux. Dividing by this flux we find the differential cross section for the scattering of $\mathbf{k}$, $\boldsymbol{\lambda}$ photons into the solid angle $d\Omega'$ with polarization $\boldsymbol{\lambda}'^*$, the matter going from $|0\rangle$ to $|n\rangle$:

$$\frac{d\sigma_{0\to n}}{d\Omega'} = r_0^2\,\frac{\omega'}{\omega}\,|\boldsymbol{\lambda}\cdot\boldsymbol{\lambda}'^*|^2\,|\langle n|\rho_{\mathbf{k}'-\mathbf{k}}|0\rangle|^2. \tag{13-132}$$

Suppose that the initial state $|0\rangle$ is a momentum eigenstate with momentum $\hbar\mathbf{q}_0$, and that $|n\rangle$ is a momentum eigenstate with momentum $\hbar\mathbf{q}_n$. Then by the same argument that led to Eq. (13-93) we find

$$\langle n|\rho_{\mathbf{k}'-\mathbf{k}}|0\rangle \sim \delta_{\mathbf{k}+\mathbf{q}_0,\,\mathbf{k}'+\mathbf{q}_n} \tag{13-133}$$

showing the conservation of momentum in the scattering. In particular, for a single electron

$$\langle n|\rho_{\mathbf{k}'-\mathbf{k}}|0\rangle = \delta_{\mathbf{k}+\mathbf{q}_0,\,\mathbf{k}'+\mathbf{q}_n}. \tag{13-134}$$

Then

$$\frac{d\sigma_{0\to n}}{d\Omega'} = r_0^2\,\frac{\omega'}{\omega}\,|\boldsymbol{\lambda}\cdot\boldsymbol{\lambda}'^*|^2\,\delta_{\mathbf{q}_0+\mathbf{k},\,\mathbf{q}_n+\mathbf{k}'}; \tag{13-135}$$

the total scattering cross section is found by summing over all electron final states, $\mathbf{q}_n$:

$$\frac{d\sigma}{d\Omega'} = r_0^2\,\frac{\omega'}{\omega}\,|\boldsymbol{\lambda}\cdot\boldsymbol{\lambda}'^*|^2. \tag{13-136}$$

This result is the *Thomson cross section;* it is the nonrelativistic limit of the cross section for Compton scattering.

In the scattering of long wavelength light from an atom or molecule, we can, to a first approximation, neglect the possibility of excitation of internal degrees of freedom of the matter. The matter then behaves as a particle of charge Ne, where N is the number of electrons in the atom or molecule. We don't have to worry about the scattering of light from the nucleus, since this scattering leads to a term of order $(e^2/Mc^2)^2$ in the cross section, where M is the mass of the proton; the scattering from the protons is six orders of magnitude weaker than the scattering from the electrons. For long wavelengths then, the matrix element $|\langle n|\rho_{\mathbf{k}'-\mathbf{k}}|0\rangle|$ becomes simply[3] $N\delta_{\mathbf{k}+\mathbf{q}_0,\mathbf{k}'+\mathbf{q}_n}$, and the total cross section becomes

$$\frac{d\sigma}{d\Omega'} = N^2 r_0^2 \frac{\omega'}{\omega} |\boldsymbol{\lambda} \cdot \boldsymbol{\lambda}'^*|^2. \qquad (13\text{-}137)$$

The scattering is proportional to the square of the number of electrons since the scattering of long wavelength light by the N electrons is coherent; this means that the amplitude is N times the amplitude from a single electron, and the intensity is N^2 as great. [This effect explains why we can see clouds, which are composed of dense coherently scattering droplets, even though we cannot see water vapor.]

To use Eqs. (13-136) and (13-137) it is necessary to find ω' as a function of the scattering angle. To do so one must solve the conservation equations

$$\mathbf{k} + \mathbf{q}_0 = \mathbf{k}' + \mathbf{q}_n, \qquad \hbar\omega + \frac{\hbar^2 q_0^2}{2m} = \hbar\omega' + \frac{\hbar^2 q_n^2}{2m} \qquad (13\text{-}138)$$

for ω' as a function of $\mathbf{k}$, $\mathbf{q}_0$, and the direction of $\mathbf{k}'$. The frequency shift is small, however, in the nonrelativistic limit ($\hbar q_0/m \ll c$, $\hbar q_n/m \ll c$) where our calculations are valid. To be in this limit it is necessary that $\hbar\omega$ be $\ll mc^2$.

The factor $|\boldsymbol{\lambda} \cdot \boldsymbol{\lambda}'^*|^2$ in the transition rate arises because the scattered photon has essentially the same polarization as the incident photon — the change in polarization is due only to the fact that the polarization vector must be orthogonal to the direction of travel.

As we mentioned, the scattering of light by a system can also take place through the $\mathbf{j} \cdot \mathbf{A}$ term in second order. There are two possible ways this can happen. First, the $\mathbf{k}$, $\boldsymbol{\lambda}$ photon can be absorbed leaving the system in an intermediate state $|m\rangle$, and then the system radiates a $\mathbf{k}'$, $\boldsymbol{\lambda}'$ photon, as in Fig. 13-2(a). Or else, the system can first radiate a $\mathbf{k}'$, $\boldsymbol{\lambda}'$ photon and then absorb the $\mathbf{k}$, $\boldsymbol{\lambda}$ photon, as in Fig. 13-2(b). In the first possibility the intermediate state is $|m; N_{\mathbf{k}\lambda} - 1, N_{\mathbf{k}'\lambda'} = 0\rangle$. The energy of this state is

[3]This form is always valid for a small angle elastic scattering; $ka \sin(\theta/2) \ll 1$, where a is the size of the system and θ is the photon scattering angle.

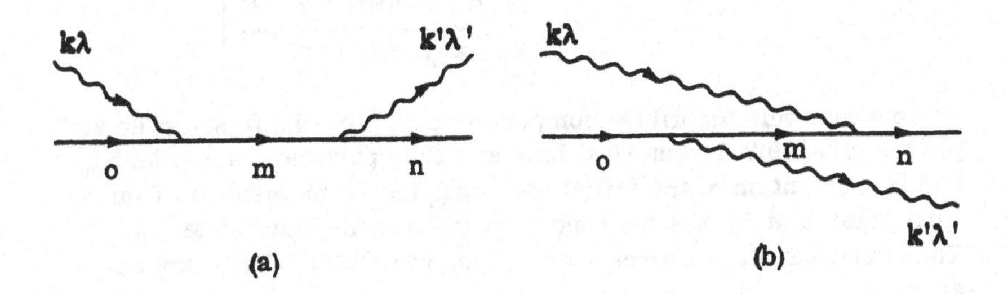

Fig. 13-2

The two processes leading to scattering of light in second order.

$\varepsilon_m + (N_{k\lambda} - 1)\hbar ck$, and thus the second-order transition amplitude is [Eq. (12-47)]

$$\sum_m \frac{\langle n; N_{k\lambda} - 1, N_{k'\lambda'} = 1|\hat{H}_{int}|m; N_{k\lambda} - 1, N_{k'\lambda'} = 0\rangle}{(\varepsilon_0 + N_{k\lambda}\hbar ck) - (\varepsilon_m + (N_{k\lambda} - 1)\hbar ck) + i\eta} \tag{13-139}$$

$$\times \langle m; N_{k\lambda} - 1, N_{k'\lambda'} = 0|\hat{H}_{int}|0; N_{k\lambda}, N_{k'\lambda'} = 0\rangle ,$$

where $\hat{H}_{int}$ is given by (13-64). Evaluating the matrix elements, we find that this contribution to the second-order transition amplitude is

$$\left(\frac{e}{c}\right)^2 \sqrt{\frac{2\pi\hbar c^2 N_{k\lambda}}{\omega V}} \sqrt{\frac{2\pi\hbar c^2}{\omega' V}} \sum_m \frac{\langle n|j_{k'} \cdot \lambda'^*|m\rangle\langle m|j_{-k} \cdot \lambda|0\rangle}{\hbar\omega + \varepsilon_0 - \varepsilon_m + i\eta} . \tag{13-140}$$

Similarly, the intermediate state in the second process is $|m; N_{k\lambda}, N_{k'\lambda'} = 1\rangle$, with energy $\varepsilon_m + N_{k\lambda}\hbar ck + \hbar ck'$, and the contribution of this process to the transition amplitude is

$$\left(\frac{e}{c}\right)^2 \sqrt{\frac{2\pi\hbar c^2 N_{k\lambda}}{\omega V}} \sqrt{\frac{2\pi\hbar c^2}{\omega' V}} \sum_m \frac{\langle n|j_{-k} \cdot \lambda|m\rangle\langle m|j_{k'} \cdot \lambda'^*|0\rangle}{\varepsilon_0 - \varepsilon_m - \hbar\omega' + i\eta} . \tag{13-141}$$

Since these two processes lead to the same final state from the same initial state their amplitudes add coherently. The total second-order amplitude is then the sum of (13-140) and (13-141):

$$\frac{2\pi\hbar c^2}{\sqrt{\omega\omega'}} \frac{\sqrt{N_{k\lambda}}}{V} \sum_m \left[\frac{\langle n|j_{k'}\cdot\lambda'^*|m\rangle\langle m|j_{-k}\cdot\lambda|0\rangle}{\hbar\omega+\varepsilon_0-\varepsilon_m+i\eta} + \frac{\langle n|j_{-k}\cdot\lambda|m\rangle\langle m|j_{k'}\cdot\lambda'^*|0\rangle}{\varepsilon_0-\varepsilon_m-\hbar\omega'+i\eta} \right]. \tag{13-142}$$

How does this amplitude compare in size with the first-order amplitude (13-129)? Let us first look at a free particle. Again let $\hbar q_0$ be the momentum of the initial state, $\hbar q_n$ be the momentum of the final state, and $\hbar q'$ the momentum of the intermediate state $|m\rangle$. The current matrix elements are given by (13-87). Thus, for example

$$\langle m|j_{-k}\cdot\lambda|0\rangle = \frac{\hbar}{2m}(q_0+q')\cdot\lambda\,\delta_{q_0+k,q'} = \frac{\hbar}{m} q_0\cdot\lambda\,\delta_{q_0+k,q'} \tag{13-143}$$

since $k\cdot\lambda = 0$. Performing the sum over intermediate states we find that (13-142) becomes

$$r_0 \frac{2\pi\hbar c^2}{\sqrt{\omega\omega'}} \frac{\sqrt{N_{k\lambda}}}{V} \left[\frac{\hbar^2(q\cdot\lambda)(q'\cdot\lambda'^*)/m}{\hbar\omega+\varepsilon_q-\varepsilon_{q+k}+i\eta} + \frac{\hbar^2(q'\cdot\lambda)(q\cdot\lambda'^*)/m}{\varepsilon_q-\varepsilon_{q-k'}-\hbar\omega+i\eta} \right]. \tag{13-144}$$

In the first-order amplitude the quantity in brackets is replaced by 1. The denominator of the first term in the brackets is

$$\hbar ck + \frac{\hbar^2}{2m}[q^2-(q+k)^2] = \hbar k \left[c + \frac{\hbar k}{2m} - \frac{\hbar}{m} q\cdot k \right];$$

the electron energy terms are negligible for $v = \hbar q/m \ll c$ and $\hbar\omega \ll mc^2$; the denominator is thus essentially $\hbar\omega$. The numerator, on the other hand, is of order of magnitude the initial energy of the electron, $\hbar^2 q^2/2m$, which for $q \lesssim k$ is of order v/c smaller than the denominator. A similar argument holds for the second term. Furthermore, the second term has opposite sign to the first, since it has $-\hbar\omega'$ in the denominator. Thus the two terms in (13-144) tend to cancel each other. We conclude then that for a free particle, the second-order amplitude is at least of order v/c smaller than the first-order amplitude; it is quite legal to ignore the second-order amplitude in calculating the nonrelativistic Thomson scattering cross section.

RAMAN SCATTERING

While the second-order amplitude can be neglected in the scattering of light by a free charged particle, it is very important in determining the scattering of light by an atom (or molecule) when internal degrees of freedom of the atom are excited. Such scattering is frequently called *Raman scattering*. We shall assume that the atom is anchored in space at $\mathbf{r} = 0$; then we may directly use the dipole approximation to evaluate the matrix elements. Let us first calculate the second-order term (13-142). The expression in the brackets, in the dipole approximation (13-98) becomes

$$\frac{1}{i\hbar m}\left[\frac{\langle n|\mathbf{P}\cdot\boldsymbol{\lambda}'^{*}|m\rangle\langle m|\mathbf{R}\cdot\boldsymbol{\lambda}|0\rangle(\varepsilon_0-\varepsilon_m)}{\varepsilon_0-\varepsilon_m+\hbar\omega+i\eta} + \frac{\langle n|\mathbf{R}\cdot\boldsymbol{\lambda}|m\rangle\langle m|\mathbf{P}\cdot\boldsymbol{\lambda}'^{*}|0\rangle(\varepsilon_m-\varepsilon_n)}{\varepsilon_0-\varepsilon_m-\hbar\omega'+i\eta}\right]. \tag{13-145}$$

By energy conservation in the golden rule

$$\varepsilon_0+\hbar\omega = \varepsilon_n+\hbar\omega' \tag{13-146}$$

and thus we can write the denominator of the second term as $\varepsilon_n - \varepsilon_m - \hbar\omega + i\eta$. Then by simple algebra (13-145) becomes

$$\frac{1}{i\hbar m}\left(\langle n|\mathbf{P}\cdot\boldsymbol{\lambda}'^{*}|m\rangle\langle m|\mathbf{R}\cdot\boldsymbol{\lambda}|0\rangle - \langle n|\mathbf{R}\cdot\boldsymbol{\lambda}|m\rangle\langle m|\mathbf{P}\cdot\boldsymbol{\lambda}'^{*}|0\rangle\right)$$
$$+\,i\omega\left[\frac{\langle n|\mathbf{P}\cdot\boldsymbol{\lambda}'^{*}/m|m\rangle\langle m|\mathbf{R}\cdot\boldsymbol{\lambda}|0\rangle}{\varepsilon_n-\varepsilon_m+\hbar\omega'+i\eta} + \frac{\langle n|\mathbf{R}\cdot\boldsymbol{\lambda}|m\rangle\langle m|\mathbf{P}\cdot\boldsymbol{\lambda}'^{*}/m|0\rangle}{\varepsilon_0-\varepsilon_m-\hbar\omega'+i\eta}\right]. \tag{13-147}$$

Again let us use $\mathbf{P}/m = [\mathbf{R}, H_0]/i\hbar$. Then after a little more algebra, (13-147) becomes

$$\frac{1}{i\hbar m}\left(\langle n|\mathbf{P}\cdot\boldsymbol{\lambda}'^{*}|m\rangle\langle m|\mathbf{R}\cdot\boldsymbol{\lambda}|0\rangle - \langle n|\mathbf{R}\cdot\boldsymbol{\lambda}|m\rangle\langle m|\mathbf{P}\cdot\boldsymbol{\lambda}'^{*}|0\rangle\right)$$
$$-\frac{\omega}{\hbar}\left(\langle n|\mathbf{R}\cdot\boldsymbol{\lambda}'^{*}|m\rangle\langle m|\mathbf{R}\cdot\boldsymbol{\lambda}|0\rangle - \langle n|\mathbf{R}\cdot\boldsymbol{\lambda}|m\rangle\langle m|\mathbf{R}\cdot\boldsymbol{\lambda}'^{*}|0\rangle\right) \tag{13-148}$$
$$+\,\omega\omega'\left[\frac{\langle n|\mathbf{R}\cdot\boldsymbol{\lambda}'^{*}|m\rangle\langle m|\mathbf{R}\cdot\boldsymbol{\lambda}|0\rangle}{\varepsilon_n-\varepsilon_m+\hbar\omega'+i\eta} + \frac{\langle n|\mathbf{R}\cdot\boldsymbol{\lambda}|m\rangle\langle m|\mathbf{R}\cdot\boldsymbol{\lambda}'^{*}|0\rangle}{\varepsilon_0-\varepsilon_m-\hbar\omega'+i\eta}\right].$$

Now let us sum over all intermediate states $|m\rangle$. The first term in (13-148) becomes

$$\frac{1}{i\hbar m}\langle n|[(\mathbf{P}\cdot\boldsymbol{\lambda}'^*)(\mathbf{R}\cdot\boldsymbol{\lambda}) - (\mathbf{R}\cdot\boldsymbol{\lambda})(\mathbf{P}\cdot\boldsymbol{\lambda}'^*)]|0\rangle$$

$$= \frac{1}{i\hbar m}\langle n|[\mathbf{P}\cdot\boldsymbol{\lambda}'^*, \mathbf{R}\cdot\boldsymbol{\lambda}]|0\rangle = \frac{-\boldsymbol{\lambda}\cdot\boldsymbol{\lambda}'^*}{m}\langle n|N|0\rangle. \tag{13-149}$$

The second term becomes

$$-\frac{\omega}{\hbar}\langle n|[(\mathbf{R}\cdot\boldsymbol{\lambda}'^*)(\mathbf{R}\cdot\boldsymbol{\lambda}) - (\mathbf{R}\cdot\boldsymbol{\lambda})(\mathbf{R}\cdot\boldsymbol{\lambda}'^*)]|0\rangle = 0. \tag{13-150}$$

Hence the entire second-order matrix element equals

$$\frac{2\pi\hbar e^2}{\sqrt{\omega\omega'}}\frac{\sqrt{N_{k\lambda}}}{V}\left\{-\frac{\boldsymbol{\lambda}\cdot\boldsymbol{\lambda}'^*}{m}\langle n|N|0\rangle \right. \tag{13-151}$$

$$\left. +\omega\omega'\sum_m\left[\frac{\langle n|\mathbf{R}\cdot\boldsymbol{\lambda}'^*|m\rangle\langle m|\mathbf{R}\cdot\boldsymbol{\lambda}|0\rangle}{\varepsilon_0-\varepsilon_m+\hbar\omega+i\eta} + \frac{\langle n|\mathbf{R}\cdot\boldsymbol{\lambda}|m\rangle\langle m|\mathbf{R}\cdot\boldsymbol{\lambda}'^*|0\rangle}{\varepsilon_0-\varepsilon_m-\hbar\omega'+i\eta}\right]\right\}.$$

In the dipole approximation $\rho_{k'-k}$ to lowest order equals ρ_0, which is just N. Thus the first-order transition amplitude (13-129), which must be added to (13-151), exactly cancels the first term in (13-151). The net amplitude, including both the first-order ρA^2 and the second order $\mathbf{j}\cdot\mathbf{A}$ processes, is therefore

$$2\pi\hbar e^2\frac{\sqrt{N_{k\lambda}}}{V}\sqrt{\omega\omega'}\sum_m\left[\frac{\langle n|\mathbf{R}\cdot\boldsymbol{\lambda}'^*|m\rangle\langle m|\mathbf{R}\cdot\boldsymbol{\lambda}|0\rangle}{\varepsilon_0-\varepsilon_m+\hbar\omega+i\eta}\right. \tag{13-152}$$

$$\left. + \frac{\langle n|\mathbf{R}\cdot\boldsymbol{\lambda}|m\rangle\langle m|\mathbf{R}\cdot\boldsymbol{\lambda}'^*|0\rangle}{\varepsilon_0-\varepsilon_m-\hbar\omega'+i\eta}\right].$$

The frequency change of the light in Raman scattering is given by $\omega' - \omega = (\varepsilon_n - \varepsilon_0)/\hbar$. Thus one expects to see only discrete frequency shifts, in the case of a stationary atom, corresponding to transitions between bound states. Such scattering is good evidence that the bound state energy levels of atoms and molecules are quantized, and not continuous as they would be classically. One speaks of scattering that raises the energy of the atom and therefore decreases the frequency of the light as *Stokes scattering;* when the light gains energy from the atom it is called *anti-Stokes scattering.*

We see from (13-152) that in the situation that the states have definite parities, the parity of $|m\rangle$ must be opposite that of $|0\rangle$, else

$\langle m|\mathbf{R}|0\rangle = 0$. Similarly, the parity of $|n\rangle$ must also be opposite that of $|m\rangle$ and hence equal to the parity of $|0\rangle$. Thus dipole Raman scattering does not change the parity of the state of the atom. On the other hand, in an absorptive dipole transition, the parity of the atomic state changes sign. We therefore reach the conclusion that transitions that give strong absorption give only weak, higher-order, Raman scattering, and vice versa. For example, in hydrogen, the transition $|1S\rangle \to |2P\rangle$ is strongly absorptive but leads to only weak Raman scattering. On the other hand the transition $|1S\rangle \to |3S\rangle$ gives relatively strong Raman scattering but only weak absorption.

To calculate the total scattering cross section, we put the amplitude (13-152) into the golden rule, sum over all final photon frequencies (but not directions) and divide by the incident photon flux $N_{k\lambda}c/V$. This yields

$$\frac{d\sigma_{0\to n}}{d\Omega'} = \frac{e^4\omega\omega'^3}{c^4}\left|\sum_m\right|^2 \qquad (13\text{-}153)$$

where the sum over m is that in (13-152) and $\hbar\omega' = \hbar\omega + \varepsilon_0 - \varepsilon_n$. For small ω the incident photons are not energetic enough to excite the system, if it is in its ground state. In this case $|n\rangle = |0\rangle$ and we find that the low-frequency cross section is

$$\frac{d\sigma}{d\Omega'} = \left(\frac{\omega a_0}{c}\right)^4 a_0^2|\lambda'^*\cdot\mathbf{M}\cdot\lambda|^2 = \frac{r_0^2}{16}\left(\frac{\hbar\omega}{Ry}\right)^4|\lambda'^*\cdot\mathbf{M}\cdot\lambda|^2 \qquad (13\text{-}154)$$

where Ry is the Rydberg and

$$\lambda'^*\cdot\mathbf{M}\cdot\lambda = \frac{e^2}{a_0^3}\sum_m \frac{\langle 0|\mathbf{R}\cdot\lambda'^*|m\rangle\langle m|\mathbf{R}\cdot\lambda|0\rangle + \langle 0|\mathbf{R}\cdot\lambda|m\rangle\langle m|\mathbf{R}\cdot\lambda'^*|0\rangle}{\varepsilon_0 - \varepsilon_m + i\eta} \qquad (13\text{-}155)$$

is dimensionless and ~ 1, give or take an order of magnitude. Equation (13-154) is the cross section for Rayleigh scattering; it behaves as ω^4.

Notice that the amplitude (13-152) is exactly what we would have found had we taken the interaction between the electromagnetic field and the atom to be

$$H_{dipole} = -e\mathbf{R}\cdot\mathbf{E}^{(op)}(0) \qquad (13\text{-}156)$$

where $\mathbf{E}^{(op)}(\mathbf{r})$ is the quantum mechanical (transverse) electric field operator:

$$\mathbf{E}^{(\mathrm{op})}(\mathbf{r}) = \sum_{\mathbf{k}\lambda} \left[\frac{i\omega}{c} A_{\mathbf{k}\lambda}^{(\mathrm{op})}\lambda \frac{e^{i\mathbf{k}\cdot\mathbf{r}}}{\sqrt{V}} - \frac{i\omega}{c} A_{\mathbf{k}\lambda}^{(\mathrm{op})\dagger}\lambda^* \frac{e^{-i\mathbf{k}\cdot\mathbf{r}}}{\sqrt{V}} \right], \qquad (13\text{-}157)$$

$$\mathbf{E}^{(\mathrm{op})}(\mathbf{r}, t) = -\frac{1}{c} \frac{\partial \mathbf{A}^{(\mathrm{op})}(\mathbf{r}, t)}{\partial t}. \qquad (13\text{-}158)$$

Semiclassically the interaction (13-156) is just the coupling of a slowly varying electric field to a system of charged particles.

We can, by means of a simple gauge transformation, derive the fact that (13-156) is the effective interaction for the scattering of light by a localized system. Let us do this, treating the radiation semiclassically; to get the fully quantum mechanical interaction we need only put "op's" on the electromagnetic fields. Let us choose as a gauge function

$$\chi(\mathbf{r}, t) = -\mathbf{r}\cdot\mathbf{A}(\mathbf{r}, t) \qquad (13\text{-}159)$$

where $\mathbf{A}$ is the vector potential itself. Then $(a, b = x, y, z)$

$$A'_a(\mathbf{r}, t) = (\mathbf{A} + \nabla\chi)_a = -\sum_b r_b \nabla_a A_b(\mathbf{r}, t)$$

$$\varphi'(\mathbf{r}, t) = -\frac{1}{c}\frac{\partial\chi}{\partial t} = \frac{1}{c}\mathbf{r}\cdot\frac{\partial \mathbf{A}}{\partial t} = -\mathbf{r}\cdot\mathbf{E}(\mathbf{r}, t). \qquad (13\text{-}160)$$

In this new gauge, the electromagnetic interaction (13-33) becomes

$$H_{\mathrm{int}} = \int d^3r \left[-\frac{e}{c} j_a(\mathbf{r}) r_b \nabla_a A_b(\mathbf{r}, t) \right] - \int d^3r \; e\mathbf{r}\cdot\mathbf{E}(\mathbf{r}, t)\rho(\mathbf{r})$$
$$+ \frac{e^2}{2mc^2} \int d^3r \; \rho(\mathbf{r}) [r_b \nabla_a A_b(\mathbf{r}, t)]^2 , \qquad (13\text{-}161)$$

using a summation convention for the spatial indices. Let us consider a localized system. Then the first term in (13-161) which is proportional to the velocity of the particles, is of order v/c smaller than the second term and therefore can be neglected. In the last term, $\nabla\mathbf{A}$ is the same order of magnitude as $\mathbf{E}$ in the second term. The scattering produced by the second term, in second order, is much greater than that produced by the last term in first order, because of the mc^2 in the denominator of the last term. Thus the last term can also be neglected. Finally, for relatively long wavelength fields we can replace $\mathbf{E}(\mathbf{r}, t)$ by $\mathbf{E}(\mathbf{r} = 0, t)$, so that (13-161) reduces to

$$H_{int} = -e\mathbf{R}\cdot\mathbf{E}(0, t), \tag{13-162}$$

which is just the dipole interaction (13-156) in semiclassical form.

PROBLEMS

1. Consider an N-particle system in an external time-dependent electromagnetic field. Show that if one makes a gauge transformation of the electromagnetic potentials with a gauge function $\chi(\mathbf{r}, t)$, then the wave function, $\psi(\mathbf{r}_1, \mathbf{r}_2 \ldots \mathbf{r}_N, t)$ of the N-particle system is transformed by

$$\psi(\mathbf{r}_1, \ldots, \mathbf{r}_N, t) \rightarrow \left\{ \exp\left[\frac{ie}{\hbar c} \sum_{i=1}^{N} \chi(\mathbf{r}_i, t)\right]\right\} \psi(\mathbf{r}_1, \ldots, \mathbf{r}_N, t).$$

2. (a) Show that all matrix elements of the current operator

$$\mathbf{J}(\mathbf{r}) = \mathbf{j}(\mathbf{r}) - \frac{e}{mc}\mathbf{A}(\mathbf{r})\rho(\mathbf{r})$$

between states of the system are gauge invariant.
(b) Show that in the Heisenberg representation

$$\frac{\partial\rho(\mathbf{r}t)}{\partial t} + \nabla\cdot\mathbf{J}(\mathbf{r}, t) = 0.$$

3. (a) Show that the quantum mechanical electromagnetic field operator has the following properties

(i) $$\left[A_{\mathbf{k}\lambda}^{(op)}, A_{\mathbf{k}'\lambda'}^{(op)}\right] = 0$$

(ii) $$\left[A_{\mathbf{k}\lambda}^{(op)\dagger}, A_{\mathbf{k}'\lambda'}^{(op)\dagger}\right] = 0$$

(iii) $$\left[A_{\mathbf{k}\lambda}^{(op)}, A_{\mathbf{k}'\lambda'}^{(op)\dagger}\right] = \frac{2\pi\hbar c}{k}\,\delta_{\mathbf{k}\mathbf{k}'}\,\lambda\cdot\lambda'^{*}$$

(b) What is the Hamiltonian, H_{em}, of the electromagnetic field, in terms of the $A_{\mathbf{k}\lambda}^{(op)}$ and $A_{\mathbf{k}\lambda}^{(op)\dagger}$.

4. Let $\mathcal{H}^{(op)}(\mathbf{r}, t) = \nabla\times\mathbf{A}^{(op)}(\mathbf{r}, t)$ denote the *magnetic field operator*, and $\mathbf{E}^{(op)}(\mathbf{r}, t)$ be the electric field operator, Eq. (13-157). Calculate $[\mathcal{H}^{(op)}(\mathbf{r}, t), \mathbf{E}^{(op)}(\mathbf{r}', t')]$. What does this say about

the simultaneous measurability of $\mathcal{H}$ and $\mathbf{E}$?

5. (a) Calculate $\langle \mathbf{E}^{(op)}(\mathbf{r})\ \mathbf{E}^{(op)}(\mathbf{r}')\rangle$, the expectation value of the electron field at two points in the vacuum (no photons).
   (b) Calculate $\langle(\bar{\mathbf{E}}^{(op)})^2\rangle$ where $\bar{\mathbf{E}}^{op}$ is the average of the electric field over a small volume Ω. What happens as $\Omega \to 0$?
6. A right circularly polarized beam of light with wavelength $\lambda = 7 \times 10^{-5}$ cm and intensity = 1 watt/cm^2 illuminates a (low density) gas of 10^{23} electrons. Assume that the velocity of the electrons is negligibly small. Calculate the frequency and intensity of left circularly polarized light at angles of 0°, 90°, and 180° with respect to the direction of the incident beam.
7. A hydrogen atom in its $2P_1$ state is at rest, i.e., has zero momentum, in a large box. Calculate the details of the spontaneous emission from the atom. In your calculation, devise the appropriate form of the dipole approximation for this situation. Compare the lifetime of the $2P_1$ state here with the lifetime when the proton is anchored in space.
8. A hydrogen atom is in its ground state and *at rest* in a large box. Light of wave vector $\mathbf{k}$ falls on the atom. Write down an explicit expression in terms of dipole matrix elements for the Raman scattering cross section for the transition to a D state. Estimate the transition rate.
9. Determine, for a one-electron atom, the parity, L_z and L^2 selection rules for electric quadrupole and magnetic dipole transitions.
10. In scattering x-rays from a solid one usually does not analyze the frequencies of the scattered rays (why?), but rather measures the total scattered intensity as a function of scattering angle.
    (a) Show that the differential scattering cross section for the frequency integrated intensity is given to within small corrections by

$$\frac{d\sigma}{d\Omega} = r_0^2\, \frac{1+\cos^2\theta}{2}\, S(\mathbf{k}_i - \mathbf{k}_f)$$

where $\mathbf{k}_i$ is the wave vector of the incident x-rays, $\mathbf{k}_f$ points in the direction of the detector from the solid, its magnitude equals k_i, θ is the angle between $\mathbf{k}_f$ and $\mathbf{k}_i$, and

$$S(\mathbf{q}) = \sum_n |\langle n|\rho_{-\mathbf{q}}|0\rangle|^2$$

where ρ is the electron density operator. Assume the solid to be in its ground state initially.

(b) Estimate the importance of the second-order $\mathbf{j} \cdot \mathbf{A}$ scattering process.

(c) The term in $S(q)$ where $|n\rangle = |0\rangle$ gives the elastic scattering. Show that because of the periodicity of the electron distribution in the state $|0\rangle$ this term gives rise to peaks in the scattering cross section in well-defined directions (Laue spots). From these peaks one can determine the crystal structure of the solid.

11. Show from (13-13) that the interaction of a weak uniform magnetic field $\mathcal{H}$ with a charged (spinless) system is given by $H_{int} = -\mathbf{M}_{orb} \cdot \mathcal{H}$, where the orbital magnetic moment is given by (13-125).

Chapter 14
SPIN 1/2

The electron, proton, and neutron, as well as the mu meson, neutrino, Λ, Σ, and Ξ particles, all have an internal angular momentum, like a ball spinning on its own axis, with the property that if one measures the component of this angular momentum along any direction one finds either the value $+\hbar/2$ or $-\hbar/2$. Such an internal angular momentum, corresponding to a total angular momentum quantum number l equal to 1/2, is called *spin ½.*

Spin angular momentum is described by a vector operator $\mathbf{S} = (S_x, S_y, S_z)$, whose components obey the angular momentum commutation relations (6–10).

$$[S_i, S_j] = i\hbar\varepsilon_{ijk}S_k. \tag{14–1}$$

The statement that the component of the spin along any direction, say $\hat{\mathbf{m}}$, can only take on the values $\pm\hbar/2$ says that the operator $\mathbf{S}\cdot\hat{\mathbf{m}}$, which is the component of $\mathbf{S}$ along $\hat{\mathbf{m}}$, has just two eigenvalues, $\pm\hbar/2$. Denoting the two eigenstates corresponding to these eigenvalues by $|\hat{\mathbf{m}}\uparrow\rangle$ and $|\hat{\mathbf{m}}\downarrow\rangle$ we have

$$\mathbf{S}\cdot\hat{\mathbf{m}}|\hat{\mathbf{m}}\uparrow\rangle = \frac{\hbar}{2}|\hat{\mathbf{m}}\uparrow\rangle$$

$$\mathbf{S}\cdot\hat{\mathbf{m}}|\hat{\mathbf{m}}\downarrow\rangle = -\frac{\hbar}{2}|\hat{\mathbf{m}}\downarrow\rangle. \tag{14–2}$$

If the spin of a particle is in the state $|\hat{\mathbf{m}}\downarrow\rangle$, for example, then a measurement of the component of the spin along $\hat{\mathbf{m}}$ will always yield the value $-\hbar/2$.

The two spin eigenstates $|\hat{\mathbf{m}}\uparrow\rangle$ and $|\hat{\mathbf{m}}\downarrow\rangle$ for a given direction $\hat{\mathbf{m}}$ form a basis; any state $|\Psi\rangle$ must be expressible as a linear combination of these two states. The component $\langle\hat{\mathbf{m}}\uparrow|\Psi\rangle$ is the amplitude for finding the spin "pointed along $\hat{m}$," while $\langle \hat{m}\downarrow|\Psi\rangle$ is the amplitude for finding the spin "pointed opposite to $\hat{m}$." Since there are no other possible results for a measurement of the component of spin along $\hat{m}$ these two amplitudes completely specify the spin state $|\Psi\rangle$.

The most commonly used basis is formed of the eigenstates of S_z. In this basis any state $|\Psi\rangle$ is written as a two component vector

$$\begin{pmatrix} \langle \hat{z}\uparrow|\Psi\rangle \\ \langle \hat{z}\downarrow|\Psi\rangle \end{pmatrix}. \tag{14-3}$$

In particular, the state with spin "up" $|\hat{z}\uparrow\rangle$ becomes the vector $\begin{pmatrix} 1 \\ 0 \end{pmatrix}$ while $|\hat{z}\downarrow\rangle$, the state with spin "down" becomes the vector $\begin{pmatrix} 0 \\ 1 \end{pmatrix}$. The operator S_z in this basis becomes the matrix

$$\begin{pmatrix} \langle \hat{z}\uparrow|S_z|\hat{z}\uparrow\rangle & \langle \hat{z}\uparrow|S_z|\hat{z}\downarrow\rangle \\ \langle \hat{z}\downarrow|S_z|\hat{z}\uparrow\rangle & \langle \hat{z}\downarrow|S_z|\hat{z}\downarrow\rangle \end{pmatrix} = \frac{\hbar}{2}\begin{pmatrix} 1 & 0 \\ 0 & -1 \end{pmatrix}. \tag{14-4}$$

What are the matrices that represent S_x and S_y? It is easy to verify that the only form for these Hermitian matrices consistent with the commutation relation (14-1) is

$$S_x = \frac{\hbar}{2}\begin{pmatrix} 0 & \delta \\ \delta^* & 0 \end{pmatrix}, \quad S_y = \frac{\hbar}{2}\begin{pmatrix} 0 & -i\delta \\ i\delta^* & 0 \end{pmatrix} \tag{14-5}$$

where $|\delta| = 1$. For example, from the commutation relation $S_xS_z - S_zS_x = -i\hbar S_y$ we conclude

$$\langle \hat{z}\uparrow|(S_xS_z - S_zS_x)|\hat{z}\downarrow\rangle = -\hbar\langle \hat{z}\uparrow|S_x|\hat{z}\downarrow\rangle = -i\hbar\langle \hat{z}\uparrow|S_y|\hat{z}\downarrow\rangle$$

or $\langle \hat{z}\uparrow|S_y|\hat{z}\downarrow\rangle = -i\langle \hat{z}\uparrow|S_x|\hat{z}\downarrow\rangle$, etc. The phase of δ depends on the relative phase of the states $|\hat{z}\uparrow\rangle$ and $|\hat{z}\downarrow\rangle$. It is customary to choose this phase to make $\delta = 1$ so that the spin operator is represented by

$$\mathbf{S} = \frac{\hbar}{2}\boldsymbol{\sigma}, \tag{14-6}$$

where the *Pauli spin matrices* $\boldsymbol{\sigma} = (\sigma_x, \sigma_y, \sigma_z)$ are given by

$$\sigma_x = \begin{pmatrix} 0 & 1 \\ 1 & 0 \end{pmatrix}, \quad \sigma_y = \begin{pmatrix} 0 & -i \\ i & 0 \end{pmatrix}, \quad \sigma_z = \begin{pmatrix} 1 & 0 \\ 0 & -1 \end{pmatrix}. \tag{14-7}$$

The Pauli spin matrices have many useful properties. First, they each have the two eigenvalues ± 1. From (14-1) and (14-6) we see that they obey the commutation relation

$$\sigma_i\sigma_j - \sigma_j\sigma_i = 2i\varepsilon_{ijk}\sigma_k \,. \tag{14-8}$$

By direct calculation one finds $\sigma_x\sigma_y = i\sigma_z$, $\sigma_y\sigma_x = -i\sigma_z$, and in general, for $i \neq j$.

$$\sigma_i\sigma_j = i\varepsilon_{ijk}\sigma_k, \quad (i \neq j).$$

Thus, for $i \neq j$, the spin matrices obey an *anticommutation* relation

$$\sigma_i\sigma_j + \sigma_j\sigma_i = 0, \ (i \neq j). \tag{14-9}$$

Furthermore

$$\sigma_x^2 = \sigma_y^2 = \sigma_z^2 = 1 = \begin{pmatrix} 1 & 0 \\ 0 & 1 \end{pmatrix}, \tag{14-10}$$

so that the general anticommutation relation can be written as

$$\sigma_i\sigma_j + \sigma_j\sigma_i = 2\delta_{ij}. \tag{14-11}$$

All these algebraic properties can be subsumed under the relation we get by adding (14-8) to (14-11)

$$\sigma_i\sigma_j = \delta_{ij} + i\varepsilon_{ijk}\sigma_k; \tag{14-12}$$

equivalently

$$S_iS_j = \frac{\hbar^2}{4}\delta_{ij} + i\varepsilon_{ijk}\frac{\hbar}{2}S_k. \tag{14-13}$$

We see explicity from (14-13) that

$$S^2 = \frac{\hbar^2}{4}(\sigma_x^2 + \sigma_y^2 + \sigma_z^2) = \frac{3\hbar^2}{4} = \hbar^2\frac{1}{2}\frac{1}{2} + 1 \ .$$

As an example, let $\mathbf{a}$ and $\mathbf{b}$ be two c-number vectors. Then from (14-12) (with the summation convention as on p. 149)

$$(\mathbf{a}\cdot\boldsymbol{\sigma})(\mathbf{b}\cdot\boldsymbol{\sigma}) = a_ib_j\sigma_i\sigma_j = a_ib_j(\delta_{ij} + i\varepsilon_{ijk}\sigma_k) = \mathbf{a}\cdot\mathbf{b} + i(\mathbf{a}\times\mathbf{b})\cdot\boldsymbol{\sigma}. \tag{14-14}$$

This relation is valid even for vector operators $\mathbf{a}$ and $\mathbf{b}$ that commute with the spin. Another useful property of the σ's is that any

2×2 matrix can be written in the form $\alpha 1 + \boldsymbol{\beta}\cdot\boldsymbol{\sigma}$, where 1 is the unit 2×2 matrix, and α, β_x, β_y, and β_z are numbers.

ROTATIONS IN SPIN SPACE

How can we express the eigenstates of the operator $\mathbf{S}\cdot\hat{\mathbf{n}}$ in the $\hat{\mathbf{z}}$ basis? To answer this, let us first consider a somewhat more general question. Suppose $\hat{\mathbf{n}}$ is a unit vector in a direction gotten by rotating the unit vector $\hat{\mathbf{m}}$ about an axis $\hat{\boldsymbol{\alpha}}$ in a right-handed sense by an angle $|\alpha|$, as in Fig. 14-1. How are the eigenstates $|\hat{\mathbf{n}}\uparrow\rangle$ and $|\hat{\mathbf{n}}\downarrow\rangle$ of $\mathbf{S}\cdot\hat{\mathbf{n}}$ related to the eigenstates of $\mathbf{S}\cdot\hat{\mathbf{m}}$? In the case that $|\alpha|$ is an infinitesimal angle, we can write

$$\hat{\mathbf{n}} = \hat{\mathbf{m}} + \boldsymbol{\alpha}\times\hat{\mathbf{m}} \tag{14-15}$$

where $\boldsymbol{\alpha} = |\alpha|\hat{\boldsymbol{\alpha}}$; then we have

$$\mathbf{S}\cdot\hat{\mathbf{n}} = \mathbf{S}\cdot\hat{\mathbf{m}} + \mathbf{S}\cdot(\boldsymbol{\alpha}\times\hat{\mathbf{m}}) = \mathbf{S}\cdot\hat{\mathbf{m}} + \varepsilon_{ijk}\alpha_i\hat{m}_j S_k. \tag{14-16}$$

Substituting for $\varepsilon_{ijk}S_k$ from (14-1) this equation becomes

$$\mathbf{S}\cdot\hat{\mathbf{n}} = \mathbf{S}\cdot\hat{\mathbf{m}} + \frac{1}{i\hbar}[S_i, S_j]\,\alpha_i\hat{m}_j = \mathbf{S}\cdot\hat{\mathbf{m}} + \frac{i}{\hbar}[\mathbf{S}\cdot\hat{\mathbf{m}}, \mathbf{S}\cdot\boldsymbol{\alpha}] \tag{14-17}$$

To first order in $\boldsymbol{\alpha}$, this equation is equivalent to

$$\mathbf{S}\cdot\hat{\mathbf{n}} = e^{-i\mathbf{S}\cdot\boldsymbol{\alpha}/\hbar}\,\mathbf{S}\cdot\hat{\mathbf{m}}\,e^{i\mathbf{S}\cdot\boldsymbol{\alpha}/\hbar}. \tag{14-18}$$

In fact, (14-18) is valid even if $\boldsymbol{\alpha}$ is a finite rotation. The proof is

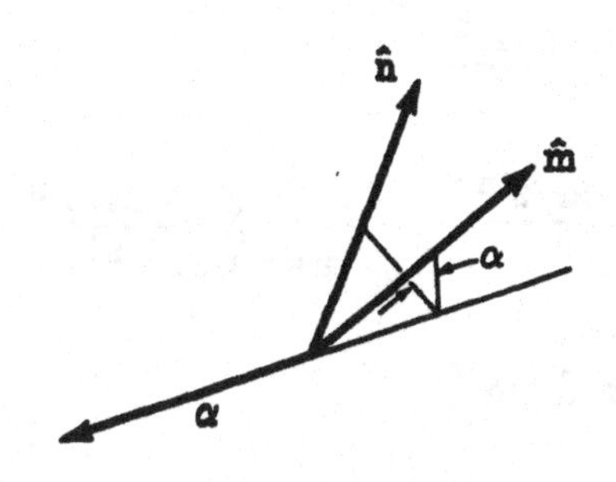

Fig. 14-1

equivalent to that given on p. 151. Thus (14-10) enables us to express the component of the spin operator along any direction $\hat{n}$ in terms of the components of the spin operator along any other direction $\hat{m}$ and along the direction $\hat{\alpha}$ of the rotation that turns $\hat{m}$ into $\hat{n}$.

Let us act with both sides of (14-18) on the state $e^{-i\mathbf{S}\cdot\boldsymbol{\alpha}/\hbar}|\hat{m}\uparrow\rangle$. This gives

$$\mathbf{S}\cdot\hat{n}(e^{-i\mathbf{S}\cdot\boldsymbol{\alpha}/\hbar}|\hat{m}\uparrow\rangle) = e^{-i\mathbf{S}\cdot\boldsymbol{\alpha}/\hbar}\,\mathbf{S}\cdot\hat{m}|\hat{m}\uparrow\rangle = \frac{\hbar}{2}(e^{-i\mathbf{S}\cdot\boldsymbol{\alpha}/\hbar}|\hat{m}\uparrow\rangle); \tag{14-19}$$

we therefore conclude that

$$|\hat{n}\uparrow\rangle = e^{-i\mathbf{S}\cdot\boldsymbol{\alpha}/\hbar}|\hat{m}\uparrow\rangle \tag{14-20}$$

and similarly

$$|\hat{n}\downarrow\rangle = e^{-i\mathbf{S}\cdot\boldsymbol{\alpha}/\hbar}|\hat{m}\downarrow\rangle. \tag{14-21}$$

[The rotation that takes $\hat{m}$ to $\hat{n}$ is not unique, since one is always free to rotate by an arbitrary amount about $\hat{n}$ after rotating $\hat{m}$ into $\hat{n}$. This freedom corresponds simply to multiplying (14-20) and (14-21) by a phase factor.] Thus the unitary operator $e^{-i\mathbf{S}\cdot\boldsymbol{\alpha}/\hbar}$ has the effect of "rotating" the eigenstates of $\mathbf{S}\cdot\hat{m}$ into the eigenstates of $\mathbf{S}\cdot\hat{n}$. This operator performs rotations on the spin degrees of freedom completely analogously to the way $e^{-i\mathbf{L}\cdot\boldsymbol{\alpha}/\hbar}$ performs rotations on the spatial degrees of freedom, as in Eq. (6-20) for example. This ability of the spin operator to generate rotations is due entirely to the commutation relations (14-1).

In order to calculate explicitly we note that in the $\hat{z}$ basis the operator $e^{-i\mathbf{S}\cdot\boldsymbol{\alpha}/\hbar}$ becomes the matrix $e^{-i\boldsymbol{\sigma}\cdot\boldsymbol{\alpha}/2}$. To express such a matrix as a linear combination of σ's we expand it in a power series, noting that $(\boldsymbol{\sigma}\cdot\hat{\alpha})^2 = 1$:

$$\begin{aligned} e^{-i\boldsymbol{\sigma}\cdot\boldsymbol{\alpha}/2} &= \sum_{n=0}^{\infty}\frac{(-i\boldsymbol{\sigma}\cdot\boldsymbol{\alpha}/2)^n}{n!} \\ &= \sum_{n=0,2,4,\ldots}\frac{(-i\alpha/2)^n}{n!} + \sum_{n=1,3,5,\ldots}\frac{(-i\alpha/2)^n}{n!}\,\boldsymbol{\sigma}\cdot\hat{\alpha} \\ &= \cos\frac{\alpha}{2} - i\boldsymbol{\sigma}\cdot\hat{\alpha}\sin\frac{\alpha}{2}. \end{aligned} \tag{14-22}$$

Thus to write the eigenstate $|\hat{n}\uparrow\rangle$, for example, in the z basis, we let $\hat{m} = \hat{z}$. Then $|\hat{n}\uparrow\rangle$ in the z basis is, according to (14-20),

$$e^{-i\boldsymbol{\sigma}\cdot\boldsymbol{\alpha}/2}\begin{pmatrix}1\\0\end{pmatrix} = \left[\cos\frac{\alpha}{2} - i(\boldsymbol{\sigma}\cdot\hat{\boldsymbol{\alpha}})\sin\frac{\alpha}{2}\right]\begin{pmatrix}1\\0\end{pmatrix}$$

$$= \begin{pmatrix}\cos\frac{\alpha}{2} - i\hat{\alpha}_z\sin\frac{\alpha}{2}\\ [-i\hat{\alpha}_x + \hat{\alpha}_y]\sin\frac{\alpha}{2}\end{pmatrix}. \tag{14-23}$$

where $\boldsymbol{\alpha}$ is the rotation that takes $\hat{\mathbf{z}}$ into $\hat{\mathbf{n}}$.

It is important to notice that for any spin state $|\Psi\rangle$, there is some direction $\hat{\mathbf{n}}$, such that

$$\mathbf{S}\cdot\hat{\mathbf{n}}|\Psi\rangle = \frac{\hbar}{2}|\Psi\rangle. \tag{14-24}$$

Thus if the spin state is specified, one can always find a direction such that when the spin component along that direction is measured one will get the value $\hbar/2$. To prove (14-24) we need merely note that the most general normalized spin state can be written in the form (14-23).

As another example of the spin algebra, we let $\boldsymbol{\alpha}$ be a $-90°$ rotation about the x axis, $\boldsymbol{\alpha} = -(\pi/2)\hat{\mathbf{x}}$, and verify explicitly that the unitary transformation in (14-18) turns S_z into S_y, or equivalently σ_z into σ_y. To do this we note that from the anticommutation relation,

$$\sigma_z(\sigma_x)^n = (-\sigma_x)^n\sigma_z$$

or more generally, for functions of σ_x, σ_y, and σ_z,

$$\sigma_z f(\sigma_x) = f(-\sigma_x)\sigma_z,$$

and

$$\sigma_z f(\sigma_x, \sigma_y, \sigma_z) = f(-\sigma_x, -\sigma_y, \sigma_z)\sigma_z, \tag{14-25}$$

and so on. Thus in (14-18), we have

$$e^{i\pi\sigma_x/4}\sigma_z e^{-i\pi\sigma_x/4} = e^{i\pi\sigma_x/4}e^{i\pi\sigma_x/4}\sigma_z = e^{i\pi\sigma_x/2}\sigma_z$$

$$= \left[\cos\frac{\pi}{2} + i\sigma_x\sin\frac{\pi}{2}\right]\sigma_z = \sigma_y,$$

as we expect.

The unitary matrices

$$d(\boldsymbol{\alpha}) = e^{-i\boldsymbol{\sigma}\cdot\boldsymbol{\alpha}/2} \tag{14-26}$$

give one a useful way of representing rotations. Rotations have the property that two rotations performed in a row are equivalent to a single rotation. Thus if the rotation α rotates $\hat{m}$ into $\hat{n}$ and β rotates $\hat{n}$ into $\hat{o}$, we have, from (14–18)

$$\sigma \cdot \hat{n} = d(\alpha)\sigma \cdot \hat{m} d(\alpha)^{\dagger} \tag{14–27}$$

and

$$\sigma \cdot \hat{o} = d(\beta)\sigma \cdot \hat{n} d(\beta)^{\dagger} \tag{14–28}$$

where

$$d(\alpha)^{\dagger} = e^{i\sigma \cdot \alpha/2} = d(-\alpha). \tag{14–29}$$

Substituting (14–27) into (14–28) we find

$$\sigma \cdot \hat{o} = d(\beta)d(\alpha)\sigma \cdot \hat{m}[d(\beta)d(\alpha)]^{\dagger} \tag{14–30}$$

Thus the matrix that represents the single rotation γ, equivalent to α followed by β, is

$$d(\gamma) = d(\beta)d(\alpha). \tag{14–31}$$

This "spin representation" of rotations has one peculiar feature. Suppose that we rotate about some axis $\hat{\alpha}$ by 2π; this is equivalent to no rotation at all. However

$$d(2\pi\hat{\alpha}) = e^{-i\pi\sigma \cdot \hat{\alpha}} = \cos\pi - i\sigma \cdot \hat{\alpha} \sin\pi = -1.$$

Thus a 2π rotation is represented by -1. Since the rotations α and $\alpha + 2\pi\hat{\alpha}$ are physically equivalent, this representation of rotations is *double valued;* $d(\alpha)$ and $d(\alpha + 2\pi\hat{\alpha}) = -d(\alpha)$ both represent the same rotation.

INCLUDING SPATIAL DEGREES OF FREEDOM

The spin of a particle is a degree of freedom additional to the spatial degrees of freedom. It is independent of the spatial degrees of freedom in the sense that one can simultaneously specify the spin state and the spatial state of a particle; in other words, all operators referring to spatial degrees of freedom commute with all operators referring to spin degrees of freedom:

$$[\mathbf{S}, \mathbf{r}] = 0, \qquad [\mathbf{S}, \mathbf{p}] = 0, \qquad [\mathbf{S}, \mathbf{L}] = 0, \qquad \text{etc.} \tag{14–32}$$

Then to specify completely the state of a spinning particle one must give not only the amplitudes for finding it at various points in space, but the amplitudes for possible spin orientations as well. For example, if we choose $\hat{\mathbf{z}}$ as the spin "quantization" direction, then the state of the particle is described by giving, for all $\mathbf{r}$, both $\psi_\uparrow(\mathbf{r})$, the amplitude for finding the particle at $\mathbf{r}$ with its spin up ($S_z = \hbar/2$), and $\psi_\downarrow(\mathbf{r})$, the amplitude for finding the particle at $\mathbf{r}$ with spin down ($S_z = -\hbar/2$). One often writes the total wave function as a two-component vector

$$\Psi(\mathbf{r}) = \begin{pmatrix} \psi_\uparrow(\mathbf{r}) \\ \psi_\downarrow(\mathbf{r}) \end{pmatrix} \tag{14-33}$$

with spatially dependent components. This vector function is just the generalization of the vector (14-3) to include spatial degrees of freedom; the amplitudes for the different spin orientations can vary from point to point in space. If we write the state of the particle as $|\Psi\rangle$, then $\psi_\uparrow(\mathbf{r})$ is just the component of $|\Psi\rangle$ along the basis state $|\mathbf{r}, \hat{\mathbf{z}}\uparrow\rangle$, in which the particle is at $\mathbf{r}$ with its spin up:

$$\psi_\uparrow(\mathbf{r}) = \langle \mathbf{r}, \hat{\mathbf{z}}\uparrow|\Psi\rangle, \qquad \psi_\downarrow(\mathbf{r}) = \langle \mathbf{r}, \hat{\mathbf{z}}\downarrow|\Psi\rangle. \tag{14-34}$$

The total probability density for finding the particle at $\mathbf{r}$ is

$$|\psi_\uparrow(\mathbf{r})|^2 + |\psi_\downarrow(\mathbf{r})|^2,$$

while the total probability for finding the particle somewhere with its spin up is

$$\int d^3r\, |\psi_\uparrow(\mathbf{r})|^2.$$

The two-component wave function obeys the normalization condition

$$\langle\Psi|\Psi\rangle = \int d^3r\, (|\psi_\uparrow(\mathbf{r})|^2 + |\psi_\downarrow(\mathbf{r})|^2) = 1. \tag{14-35}$$

The total angular momentum $\mathbf{J}$ of a spinning particle is the sum of its orbital angular momentum and its spin angular momentum

$$\mathbf{J} = \mathbf{L} + \mathbf{S}. \tag{14-36}$$

The total angular momentum has the property that it generates rotations of both the spatial and spin degrees of freedom. If we act with $e^{-i\boldsymbol{\alpha}\cdot\mathbf{J}/\hbar}$ on a basis state $|\mathbf{r}_0, \hat{\mathbf{m}}\uparrow\rangle$ in which the particle is definitely at $\mathbf{r}_0$ with its spin component along $\hat{\mathbf{m}}$ equal to $\hbar/2$, we get the state $|\mathbf{r}_0', \hat{\mathbf{n}}\uparrow\rangle$, where $\mathbf{r}_0'$ is the vector $\mathbf{r}_0$ rotated by $\boldsymbol{\alpha}$ and $\hat{\mathbf{n}}$ is the vector $\hat{\mathbf{m}}$ rotated by $\boldsymbol{\alpha}$, as in Fig. 14-2. To see this we use the fact that

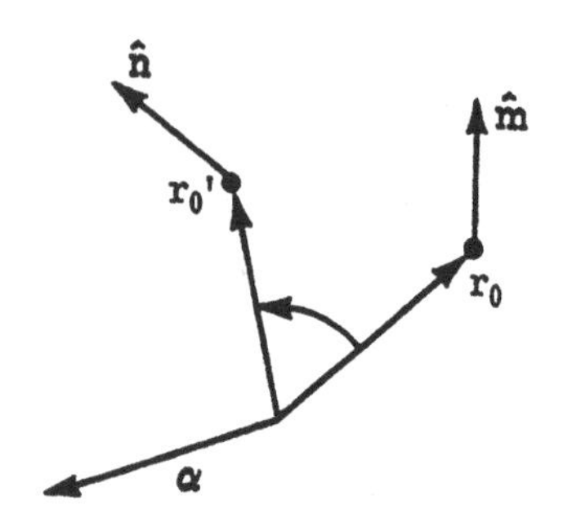

Fig. 14-2

$[\mathbf{L}, \mathbf{S}] = 0$ to write

$$e^{-i\alpha\cdot\mathbf{J}/\hbar}|\mathbf{r}_0, \hat{\mathbf{m}}\uparrow\rangle = e^{-i\alpha\cdot\mathbf{L}/\hbar}e^{-i\alpha\cdot\mathbf{S}/\hbar}|\mathbf{r}_0, \hat{\mathbf{m}}\uparrow\rangle.$$

Now the $e^{-i\alpha\cdot\mathbf{S}/\hbar}$ rotates the spin degrees of freedom according to (14-20) while the $e^{-i\alpha\cdot\mathbf{L}/\hbar}$ carries out the corresponding rotation of the spatial degrees of freedom, and therefore

$$e^{-i\alpha\cdot\mathbf{J}/\hbar}|\mathbf{r}_0, \hat{\mathbf{m}}\uparrow\rangle = |\mathbf{r}_0', \hat{\mathbf{n}}\uparrow\rangle. \tag{14-37}$$

Consider for example the state

$$|\Psi'\rangle = e^{i\alpha\cdot\mathbf{J}/\hbar}|\Psi\rangle. \tag{14-38}$$

The wave function $\langle\mathbf{r}_0, \mathbf{m}\uparrow|\Psi'\rangle$ of $|\Psi'\rangle$ is

$$\langle\mathbf{r}_0, \hat{\mathbf{m}}\uparrow|\Psi'\rangle = \langle\mathbf{r}_0, \hat{\mathbf{m}}\uparrow|e^{i\alpha\cdot\mathbf{J}/\hbar}|\Psi\rangle = \langle\mathbf{r}_0', \hat{\mathbf{n}}\uparrow|\Psi\rangle, \tag{14-39}$$

which is just the wave function of $|\Psi\rangle$ evaluated at the rotated point with rotated spin quantization direction.

SPIN MAGNETIC MOMENT

Let us turn now to the question of how one observes the spin of a particle. In high energy physics one can deduce directly the existence of the spin from the requirement that the total angular momentum be conserved in an elementary particle reaction. For example in the decay $\Lambda^0 \to p + \pi^-$, the relative orbital angular momentum of the proton and pion is often nonzero in the center of mass frame; however the orbital angular momentum of the Λ in the center of mass frame must be zero, since the Λ is at rest in that frame. Thus in order that the total angular momentum be conserved, one must conclude that the Λ (as well as the proton) must have internal spin degrees of freedom.

Nonrelativistically one must use more indirect methods of observing the spin. Classically, a spinning distribution of charge will have a magnetic moment; quantum mechanical spinning particles also have magnetic moments associated with their spins, and one tries to observe these magnetic moments in experiments.

A classical particle of charge[1] e and mass m, having orbital angular momentum $\mathbf{L}$ has a magnetic moment associated with its orbital motion given by

$$\mathbf{M}_{orb} = \frac{e}{2mc}\mathbf{L}. \tag{14-40}$$

This remains true quantum mechanically; however, because the different components of $\mathbf{L}$ cannot be simultaneously specified, neither can the different components of $\mathbf{M}_{orb}$ be simultaneously specified. The magnetic moment associated with the spin of a particle of charge e and mass m has the form

$$\mathbf{M}_{spin} = g\frac{e}{2mc}\mathbf{S}. \tag{14-41}$$

The constant g is called the *Landé g factor.* For an electron[2] $g = 2$, and since e is negative, the magnetic moment is opposite to the spin. The spin of the electron has twice as much magnetic moment associated with it, per unit of angular momentum, as has the orbital angular momentum. Notice that the total magnetic moment of an electron

$$\mathbf{M}_{tot} = \frac{e}{2mc}(\mathbf{L}+2\mathbf{S}) \tag{14-42}$$

is not parallel to the total angular momentum $\mathbf{J}$. For the proton $g = 5.59$; the spin magnetic moment of the proton is one thousand times smaller than that of the electron.

Neutral spinning particles, such as n, Λ, Σ^0, also have magnetic moments associated with their spins. One can think of these moments

[1] In this chapter e will always be a signed quantity, negative, for example, for electrons.

[2] Actually due to corrections from relativistic quantum electrodynamics, g isn't exactly two for electrons, but rather $g-2 = e^2/2\pi\hbar c$ plus terms of higher order in the fine structure constant $e^2/\hbar c$; this correction is the Schwinger term.

as arising from some internal charge distribution, in much the same way as a hydrogen atom in a $2P_1$ state, while neutral, has a magnetic moment. For the neutron $M_{spin} \approx -3.83\,(|e|/2Mc)\mathbf{S}$ where M is the neutron mass.

The energy of a magnetic moment in a magnetic field is $-\mathbf{M}\cdot\mathcal{H}$. Thus we have a term in the energy of a spinning particle in a magnetic field

$$H_{spin} = -\mathbf{M}_{spin}\cdot\mathcal{H}(\mathbf{r}, t). \tag{14-43}$$

It is primarily through this coupling that one observes spins nonrelativistically.

For example, let's look at the spectrum of a hydrogen atom in a uniform magnetic field $\mathcal{H}$, taking as the vector potential $\mathbf{A} = -{}^1/_2\,\mathbf{r}\times\mathcal{H}$. Assuming the proton to be fixed in space, we can write the Hamiltonian as $(e < 0)$

$$H = \frac{1}{2m}\left(p - \frac{e}{c}\mathbf{A}\right)^2 - \frac{e}{mc}\mathbf{S}\cdot\mathcal{H} - \frac{e^2}{r}, \tag{14-44}$$

which to first order in $\mathcal{H}$ is

$$\begin{aligned} H &= H_0 - \frac{e}{2mc}(\mathbf{p}\cdot\mathbf{A}+\mathbf{A}\cdot\mathbf{p}) - \frac{e}{mc}\mathbf{S}\cdot\mathcal{H} = H_0 - \frac{e\mathcal{H}}{2mc}\cdot(\mathbf{L}+2\mathbf{S}) \\ &= H_0 - \mathbf{M}_{tot}\cdot\mathcal{H}. \end{aligned} \tag{14-45}$$

Let us choose the z direction to be along the field. Then the perturbation is

$$H' = -\frac{e\mathcal{H}}{2mc}(L_z + 2S_z). \tag{14-46}$$

First, if the electron had no spin, the energy of the state $|nl\overline{m}\rangle$, where $\hbar\overline{m}$ is the component of $\mathbf{L}$ along the field, would be changed, to first order in H', by

$$(\Delta E)_{no\ spin} = \langle nl\overline{m}|H'|nl\overline{m}\rangle = -\frac{e\hbar}{2mc}\mathcal{H}\overline{m} = \mu_\beta\mathcal{H}\overline{m}, \tag{14-47}$$

where $\mu_\beta = |e\hbar/2mc| = 0.927\times10^{-20}$ ergs/gauss is the Bohr magneton. Thus the $2l + 1$-fold degeneracy due to rotational invariance is lifted, e.g., for $l = 1$ we have the level splitting in Fig. 14-3(a). This separation of the levels is an example of the *Zeeman effect.* Including the spin, and taking the spin quantization axis along the field, we find that

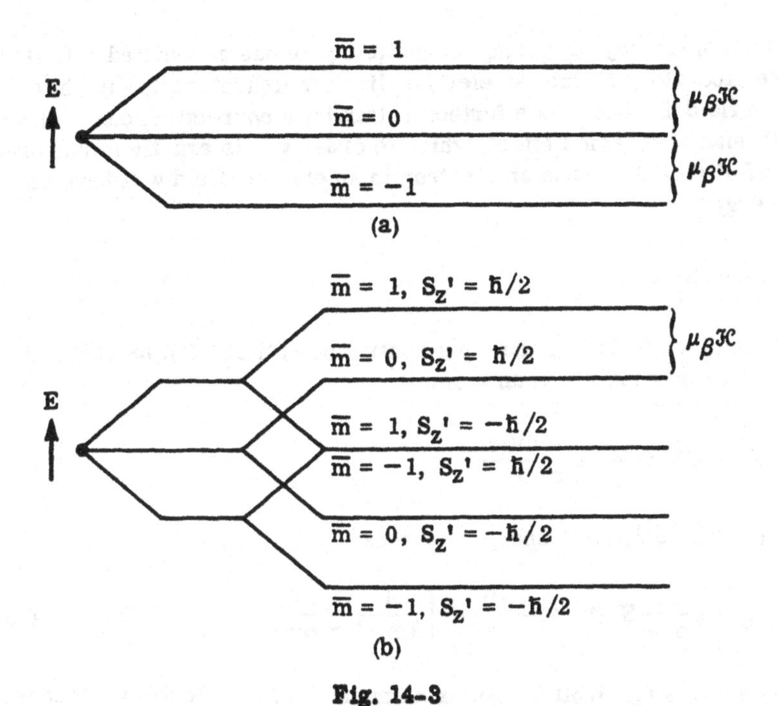

Fig. 14-3
Splitting of a hydrogenic P state in a strong magnetic field: (a) neglecting spin: (b) including spin, but neglecting spin orbit interaction.

$$\Delta E = \langle nl\overline{m}, S_z'|H'|nl\overline{m}, S_z'\rangle = \mu_\beta \mathcal{H}(\overline{m} + 2S_z'/\hbar), \tag{14-48}$$

where S_z' is the eigenvalue of S_z. Each m level is now doubly split depending on the direction of the electron spin. For $l = 1$ one now finds Fig. 14-3(b); there are five distinct levels, rather than the three one finds with no electron spin. Such additional level splitting was among the first evidence for the existence of the electron spin.

Actually, because of relativistic effects, this simple picture of the level splitting is valid only for relatively strong magnetic fields. For example, an electron traveling with velocity **v** through an electric field $\mathcal{E}$ will, according to the special theory of relativity, Eq. (23-3), feel a magnetic field equal to $-\mathbf{v}\times\mathcal{E}/c$, to lowest order in v/c. This magnetic field couples to the magnetic moment associated with the spin and leads to an additional interaction energy

$$-\frac{|e|}{mc}\mathbf{S}\cdot\left(\mathbf{v}\times\frac{\mathcal{E}}{c}\right). \tag{14-49}$$

Put another way, a moving magnetic dipole has associated with it, in the laboratory frame, an electric dipole which interacts with the electric field $\mathcal{E}$. There is a further relativistic correction, due to the Thomas precession effect, which to order v/c is exactly minus one half of (14-49). Thus an electron in an electric field will have an energy

$$-\frac{|e|}{2mc^2}\mathbf{S}\cdot(\mathbf{v}\times\mathcal{E}).$$

If the electric field is due to a central potential, $V(\mathbf{r})$, as occurs to a first approximation in an atom,

$$|e|\mathcal{E} = \nabla V(r) = \mathbf{r}\frac{1}{r}\frac{dV(r)}{dr}. \tag{14-50}$$

Thus this additional energy becomes

$$H_{s.o.} = \frac{1}{2mc^2}\mathbf{S}\cdot(\mathbf{r}\times\mathbf{v})\frac{1}{r}\frac{dV}{dr} = \left[\frac{1}{2m^2c^2}\frac{1}{r}\frac{dV}{dr}\right]\mathbf{L}\cdot\mathbf{S}, \tag{14-51}$$

where $\mathbf{L}$ is the orbital angular momentum of the electron. Because this interaction of an electric field with the spin depends on the velocity of the particle the coupling is known as the *spin-orbit interaction.* Thus for the level splitting (14-48) to be valid, the interaction with the applied field $\mathcal{H}$ must be much stronger than the spin-orbit interaction. We shall discuss the detailed effects of the spin-orbit interaction later when we study atoms.

The state of a spinning particle changes in time according to the usual Schrödinger equation

$$i\hbar\frac{d|\Psi(t)\rangle}{dt} = H|\Psi(t)\rangle. \tag{14-52}$$

In the position representation with the spin quantization axis along $\hat{\mathbf{z}}$, and neglecting the spin orbit force, this equation takes the form

$$i\hbar\frac{\partial}{\partial t}\begin{pmatrix}\psi_\uparrow(\mathbf{r}t)\\ \psi_\downarrow(\mathbf{r}t)\end{pmatrix} = \left[\frac{1}{2m}\left(\frac{\hbar}{i}\nabla - \frac{e}{c}\mathbf{A}(\mathbf{r}t)\right)^2\cdot 1 - \frac{g}{2}\frac{e\hbar}{2mc}\sigma\cdot\mathcal{H}(\mathbf{r}t) + V(\mathbf{r}t)\cdot 1\right]\begin{pmatrix}\psi_\uparrow(\mathbf{r}t)\\ \psi_\downarrow(\mathbf{r}t)\end{pmatrix} \tag{14-53}$$

where $V(\mathbf{r}t)$ is the potential energy operator, and 1 is the 2×2 unit matrix; also

$$\boldsymbol{\sigma} \cdot \mathcal{H} = \begin{pmatrix} \mathcal{H}_z & \mathcal{H}_x - i\mathcal{H}_y \\ \mathcal{H}_x + i\mathcal{H}_y & -\mathcal{H}_z \end{pmatrix}. \tag{14-54}$$

Equation (14-53) is known as the *Pauli equation.*

PRECESSION

Spins in magnetic fields have very interesting dynamical properties. Let us consider this motion neglecting at first the spatial motion of the particle in the field. The Hamiltonian of a spin in a (uniform) magnetic field, neglecting all orbital motion, is simply

$$H_{\text{spin}} = -\frac{g}{2}\frac{e}{mc}\mathbf{S} \cdot \mathcal{H}(t) = -\frac{g}{4}\frac{e\hbar}{mc}\boldsymbol{\sigma} \cdot \mathcal{H}(t). \tag{14-55}$$

The motion of the spin is most clearly seen in the Heisenberg representation where

$$i\hbar \frac{dS_i(t)}{dt} = [S_i(t), H_{\text{spin}}(t)] = -\frac{g}{2}\frac{e}{mc}[S_i(t), S_j(t)]\mathcal{H}_j(t)$$
$$= -i\frac{g}{2}\frac{e\hbar}{mc}\varepsilon_{ijk}S_k\mathcal{H}_j(t) \tag{14-56}$$

or

$$\frac{d\mathbf{S}(t)}{dt} = \frac{ge}{2mc}\mathbf{S}(t) \times \mathcal{H}(t) = \mathbf{M}_{\text{spin}}(t) \times \mathcal{H}(t). \tag{14-57}$$

The right side is just the torque exerted by a magnetic field on a magnetic moment. This equation says in operator language that the rate of change of the spin angular momentum is the applied torque. Spins of negatively charged particles thus precess in a positive sense about the field, as in Fig. 14-4, while spins of positively charged particles precess in a negative sense.

If $\mathcal{H}$ is independent of time and pointing in the z direction, $\mathcal{H} = \mathcal{H}_0\hat{\mathbf{z}}$, (14-57) becomes

$$\frac{dS_z(t)}{dt} = 0, \qquad \frac{dS_x(t)}{dt} = \frac{ge\mathcal{H}_0}{2mc}S_y(t), \qquad \frac{dS_y(t)}{dt} = -\frac{ge\mathcal{H}_0}{2mc}S_x(t) \tag{14-58}$$

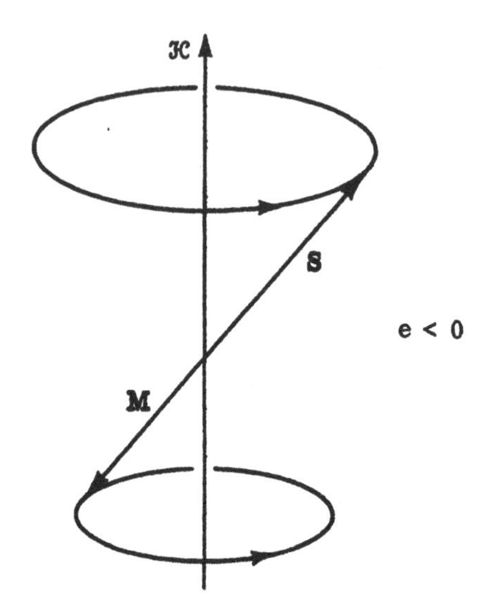

Fig. 14-4
Precession of the spin and magnetic moment about a constant magnetic field for e < 0.

and the solution is

$$S_x(t) = S_x(0) \cos \omega_0 t + S_y(0) \sin \omega_0 t$$

$$S_y(t) = -S_x(0) \sin \omega_0 t + S_y(0) \cos \omega_0 t$$

$$S_z(t) = S_z(0) \tag{14-59}$$

where

$$\omega_0 = \frac{ge\mathcal{H}_0}{2mc} \tag{14-60}$$

For electrons $\omega_0 < 0$ and $|\omega_0|/2\pi\mathcal{H}_0 = 2.8$ MHz/gauss. For example, if the spin at $t = 0$ is in the $\hat{x}$ direction, i.e., in an eigenstate of S_x with eigenvalue $\hbar/2$, then

$$\langle S_x(0)\rangle = \langle \hat{x}\uparrow | S_x(0) | \hat{x}\uparrow\rangle = \hbar/2 \qquad \langle S_y(0)\rangle = \langle S_z(0)\rangle = 0. \tag{14-61}$$

Thus taking the expectation value of (14-59) we find

$$\langle S_x(t)\rangle = \frac{\hbar}{2}\cos\omega_0 t, \quad \langle S_y(t)\rangle = -\frac{\hbar}{2}\sin\omega_0 t, \quad \langle S_z(t)\rangle = 0. \tag{14-62}$$

The expectation value of the spin thus rotates in a negative sense (for $e > 0$) in the x, y plane.

In the Schrödinger representation the state of the spin develops in time according to

$$|\Psi(t)\rangle = e^{-iHt/\hbar}|\Psi(0)\rangle = e^{i\omega_0 t S_z/\hbar}|\Psi(0)\rangle. \tag{14-63}$$

Comparing with (14-20) we see from this point of view that the direction of the spin rotates about $-\hat{\mathbf{z}}$ with angular velocity ω_0.

This precessional effect is employed in determining precisely the magnetic moment of the mu meson. The mu decays into an electron and two neutrinos

$$\mu \rightarrow e + \nu + \bar{\nu}.$$

This decay has the interesting property that the electron tends to come off, in the center of mass frame, in the direction opposite to the direction of the spin of the mu. One can measure the magnetic moment of the mu by preparing a beam of muons polarized opposite to their direction of motion, which is along $\hat{\mathbf{x}}$, say. The muons are then passed through a uniform magnetic field in the z direction for a time t, and the preferential direction for the electrons to come off (which is in the x, y plane) is then measured. Taking into account the Lorentz force, the angle with respect to the direction of motion of the μ is a direct measure of $g - 2$ for the muon.

SPIN RESONANCE

Since the spin precesses negatively about $\hat{\mathbf{z}}$ with angular velocity ω_0, if we observe the spin in a coordinate frame rotating negatively about $\hat{\mathbf{z}}$ with angular velocity ω_0, the spin will appear not to precess at all. One can think of there being an additional magnetic field in the rotating frame that exactly cancels the applied field $\mathcal{H}$. Generally, if the coordinate frame rotates with angular velocity $-\omega$ about $\hat{\mathbf{z}}$, the spin experiences an effective magnetic field in the z direction

$$\mathcal{H}_{\text{eff}} = \mathcal{H}_0 - \frac{2mc\omega}{ge}. \tag{14-64}$$

This suggests that if we could apply a weak field that remains stationary along the x axis of the frame rotating with angular velocity $-\omega_0$ about $\hat{\mathbf{z}}$, then the spin would, in this rotating frame, feel only this weak field, and would precess about it. For example, if the spin began pointing up, then after a 180° precession the spin would be pointing down; this would occur no matter how weak the field in the rotating x direction.

In the laboratory frame, such an experiment would require a rotating r.f. (radio frequency) magnetic field

$$\mathcal{H}_x = \frac{\mathcal{H}_1}{2}\cos\omega t, \quad \mathcal{H}_y = -\frac{\mathcal{H}_1}{2}\sin\omega t \tag{14-65}$$

with $\omega = \omega_0$ applied in addition to the static field in the z direction. Such a field is rather difficult to manufacture. Instead one uses an r.f. field

$$\mathcal{H}_x = \mathcal{H}_1 \cos\omega t, \qquad \mathcal{H}_y = 0; \tag{14-66}$$

for $\omega \approx \omega_0$ this field produces the same resonant effect. The reason for this is that the field (14-66) can be written as

$$\mathcal{H} = \frac{\mathcal{H}_1}{2}\begin{pmatrix}\cos\omega t\\ -\sin\omega t\\ 0\end{pmatrix} + \frac{\mathcal{H}_1}{2}\begin{pmatrix}\cos\omega t\\ \sin\omega t\\ 0\end{pmatrix}. \tag{14-67}$$

The first term is just the field (14-65) one needs. The second term rotates about $\hat{\mathbf{z}}$ with angular velocity $+\omega$, and thus rotates counter to the precessing spin. In the rotating frame, this counterrotating component tends to introduce only high-frequency wiggles into the precession of the spin about the z axis. It produces no net average effect on the spin motion, for $\omega \approx \omega_0$, and $\mathcal{H}_1 \ll \mathcal{H}_0$.

Now let's work out the mathematics of this process. In the Schrödinger representation the spin state obeys

$$i\hbar\frac{d|\Psi(t)\rangle}{dt} = -\frac{e\hbar}{2mc}\frac{g}{2}(\mathcal{H}_0\sigma_z + \mathcal{H}_1\cos\omega t\,\sigma_x)|\Psi(t)\rangle. \tag{14-68}$$

Rather than explicitly transforming to a rotating frame, let us write

$$|\Psi(t)\rangle = e^{i\omega t\sigma_z/2}|\Psi'(t)\rangle. \tag{14-69}$$

Substituting this into (14-68) and multiplying on the left by $e^{-i\omega\sigma_z t/2}$ we find

$$i\frac{d|\Psi'(t)\rangle}{dt}=\left[\frac{\omega-\omega_0}{2}\sigma_z-\omega_1\cos\omega t\left(e^{-i\omega\sigma_z t/2}\sigma_x e^{i\omega\sigma_z t/2}\right)\right]|\Psi'(t)\rangle \tag{14-70}$$

where $\omega_1 = ge\mathcal{H}_1/4mc$. Now

$$\cos\omega t\left(e^{-i\omega\sigma_z t/2}\sigma_x e^{i\omega\sigma_z t/2}\right) = \cos\omega t\ \sigma_x e^{i\omega\sigma_z t}$$

$$= \sigma_x(\cos^2\omega t + i\sigma_z\cos\omega t\sin\omega t) = \frac{\sigma_x}{2}+\frac{1}{2}(\sigma_x\cos 2\omega t+\sigma_y\sin 2\omega t).$$

The two high-frequency terms are due to the counterrotating component of the r.f. field; they produce high-frequency wiggles in $|\Psi'(t)\rangle$, and for studying the average motion of the spin they may be neglected. (14-70) then becomes

$$i\frac{d|\Psi'(t)\rangle}{dt}=\left[\frac{\omega-\omega_0}{2}\sigma_z-\frac{\omega_1}{2}\sigma_x\right]|\Psi'(t)\rangle. \tag{14-71}$$

This equation clearly has the solution

$$|\Psi'(t)\rangle = e^{-i\Omega t\hat{\sigma}/2}|\Psi'(0)\rangle$$

where

$$\Omega = [(\omega-\omega_0)^2+\omega_1^2]^{1/2} \tag{14-72}$$

and

$$\hat{\sigma}=\frac{\omega-\omega_0}{\Omega}\sigma_z-\frac{\omega_1}{\Omega}\sigma_x. \tag{14-73}$$

Note that $\hat{\sigma}^2 = 1$. Thus

$$|\Psi(t)\rangle = e^{i\omega t\sigma_z/2}e^{-i\Omega t\hat{\sigma}/2}|\Psi(0)\rangle \tag{14-74}$$

is the solution to (14-68) neglecting the counterrotating component of the r.f. field.

Suppose, for example, that the initial state of the spin is along $\hat{\mathbf{z}}$:

$$|\Psi(0)\rangle = |\hat{\mathbf{z}}\uparrow\rangle.$$

The amplitude for the spin having flipped over to the state $|\hat{\mathbf{z}}\downarrow\rangle$ at

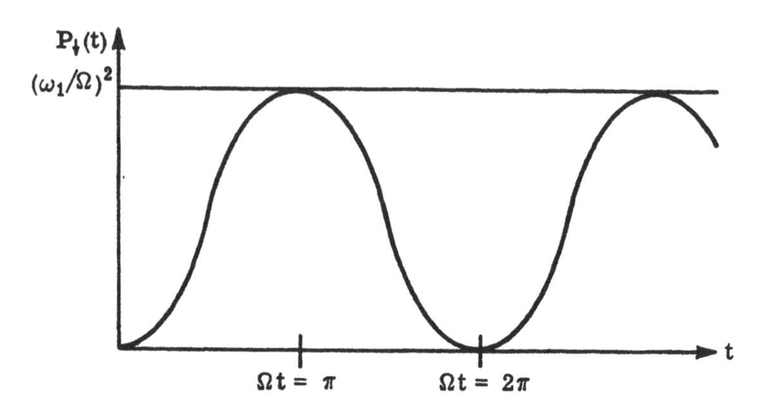

Fig. 14-5
Probability of spin flip as a function of time in a spin resonance experiment.

time t is thus

$$\begin{aligned}\langle \hat{\mathbf{z}}\downarrow|\Psi(t)\rangle &= e^{-i\omega t/2}\langle \hat{\mathbf{z}}\downarrow|\left(\cos\frac{\Omega t}{2} - i\hat{\sigma}\sin\frac{\Omega t}{2}\right)|\hat{\mathbf{z}}\uparrow\rangle \\ &= -ie^{-i\omega t/2}\langle \hat{\mathbf{z}}\downarrow|\hat{\sigma}|\hat{\mathbf{z}}\uparrow\rangle \sin\frac{\Omega t}{2} = ie^{-i\omega t/2}\,\frac{\omega_1}{\Omega}\sin\frac{\Omega t}{2},\end{aligned} \tag{14-75}$$

and the probability that the spin has flipped by time t, shown in Fig. 14-5, is

$$P_\downarrow(t) = |\langle \hat{\mathbf{z}}\downarrow|\Psi(t)\rangle|^2 = \frac{\omega_1^2}{2\Omega^2}(1-\cos\Omega t). \tag{14-76}$$

The maximum value of $P_\downarrow(t)$ is

$$\frac{\omega_1^2}{\Omega^2} = \frac{\omega_1^2}{(\omega-\omega_0)^2+\omega_1^2} \tag{14-77}$$

and occurs after a 180° pulse, i.e., $\Omega t = \pi$. For $\omega - \omega_0 \gg \omega_1$ this maximum probability is very small. However, on resonance, $\omega = \omega_0$, this maximum probability is one; the spin has flipped with certainty. In the flipping process the spin absorbs energy $\hbar\omega_0$ from the r.f. field.

If the spin began pointed down, then the probability that it flipped up is also given by (14–76). In an actual spin resonance experiment[3] one observes only the variation of the net magnetic moment – that of the up spins minus that of the down spins. If one applies only a 90° pulse, $\Omega t = \pi/2$, on resonance, then a spin initially pointing along the z axis is turned with certainty into the x, y plane.

One should notice that we don't ever have a time proportional transition probability as one has in situations where the golden rule is applicable. This is because for a spin initially up there is only one state, spin down, to which it can make transitions; there is neither the continuum of initial or final states that one needs to apply the golden rule.

An interesting application of this spin resonance technique is in an atomic clock. *Very* schematically such a device consists of two resonant cavities through which a beam of spins is projected, as in Fig. 14–6. Initially the spins are up, i.e., along $\hat{\mathbf{z}}$. In each cavity there is an identical r.f. field $\mathcal{H}_1 \cos \omega t$ in the x direction. Let us also suppose that there is a uniform static field $\mathcal{H}_0\hat{\mathbf{z}}$ in both cavities and in the space in between.

To see how this device works in principle, let's calculate the probability that a spin that enters the first cavity up emerges from the second cavity down. We may calculate this by solving the Schrödinger equation, treating the r.f. fields in the cavities as perturbations. The Schrödinger equation for an electron spin is of the form

$$i\hbar \frac{d|\Psi(t)\rangle}{dt} = -\frac{e\hbar}{2mc}[\mathcal{H}_0\sigma_z + \mathcal{H}'(t)\sigma_x]|\Psi(t)\rangle, \tag{14–78}$$

where $\mathcal{H}'(t)$ is the field in the x direction felt by the spin as it moves through the cavities. If we write

$$|\Psi(t)\rangle = e^{i\omega_0 t\sigma_z/2}|\Psi'(t)\rangle, \tag{14–79}$$

where $\omega_0 = e\mathcal{H}_0/mc$, then $|\Psi'(t)\rangle$ obeys

$$\begin{aligned} i\frac{d|\Psi'(t)\rangle}{dt} &= -\frac{e\mathcal{H}'(t)}{2mc} e^{-i\omega_0 t\sigma_z/2}\sigma_x e^{i\omega_0 t\sigma_z/2}|\Psi'(t)\rangle \\ &= -\frac{e\mathcal{H}'(t)}{2mc}\sigma_x e^{i\omega_0 t\sigma_z}|\Psi'(t)\rangle. \end{aligned} \tag{14–80}$$

To first order in $\mathcal{H}'$ we have

[3] See C.P. Slichter, *Principles of Magnetic Resonance* (Harper and Row, New York, 1963).

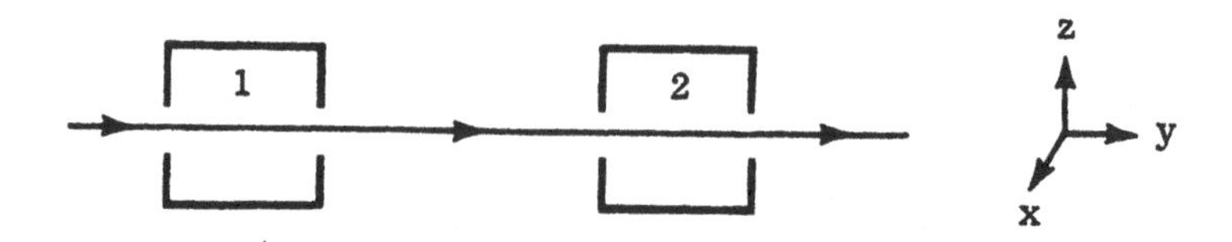

Fig. 14-6
Cavities in an atomic clock.

$$|\Psi'(t)\rangle = |\Psi'(0)\rangle + \frac{ie}{2mc}\int_0^t dt'\, \mathcal{H}'(t')\sigma_x e^{i\omega_0 t'\sigma_z}|\Psi'(0)\rangle \tag{14-81}$$

where t = 0 is the time the spin enters the first cavity. Initially $|\Psi(0)\rangle = |\Psi'(0)\rangle = |\hat{\mathbf{z}}\uparrow\rangle$. If t_2 is the time the spin emerges from the second cavity, then the amplitude for the spin to emerge flipped from up to down is

$$\langle\hat{\mathbf{z}}\downarrow|\Psi(t_2)\rangle = \frac{ie}{2mc}\int_0^{t_2} dt\, \mathcal{H}'(t)e^{i\omega_0 t}. \tag{14-82}$$

Let τ be the time the spin leaves the first cavity, T be the time the spin enters the second cavity and τ the time the spin spends in the second cavity ($t_2 = T + \tau$). Thus

$$\int_0^{t_2} dt\, \mathcal{H}'(t)e^{i\omega_0 t} = \int_0^{\tau} dt\, \mathcal{H}_1 \cos\omega t\, e^{i\omega_0 t} + \int_T^{T+\tau} dt\, \mathcal{H}_1 \cos\omega t\, e^{i\omega_0 t}. \tag{14-83}$$

In evaluating these integrals we may, for $\omega \approx \omega_0$, neglect the counter-rotating component of the r.f. fields, i.e., the part proportional to $e^{i\omega t}$. Then the transition amplitude becomes

$$\langle\hat{\mathbf{z}}\downarrow|\Psi(T+\tau)\rangle = \frac{e\mathcal{H}_1}{4mc}\left(\frac{e^{i(\omega_0-\omega)\tau}-1}{\omega_0-\omega}\right)\left(1+e^{i(\omega_0-\omega)T}\right). \tag{14-84}$$

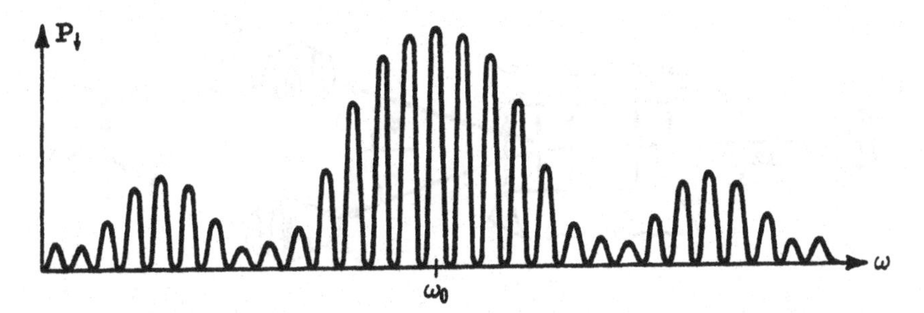

Fig. 14-7

Probability of a spin flip as a function of the frequency of the r.f. field in the two cavities in an atomic clock. The curve is a two-slit diffraction pattern.

The part of this amplitude proportional to $e^{i(\omega_0 - \omega)T}$ is the amplitude for the spin to have flipped in the second cavity; the other part is the amplitude for the spin to have flipped in the first cavity. There is an important interference between these amplitudes. Squaring (14-84) we find that the probability of a spin transition is

$$P_\downarrow = \left(\frac{e\mathcal{H}_1}{2mc}\right)^2 \left(\frac{\sin(\omega_0 - \omega)\tau/2}{(\omega_0 - \omega)/2}\right)^2 \frac{1 + \cos(\omega_0 - \omega)T}{2}; \tag{14-85}$$

the $\cos\,(\omega_0 - \omega)T$ is the interference term. Typically τ is $\ll T$ so that as a function of ω, $P_\downarrow$ looks like Fig. 14-7. This is a standard two-slit diffraction pattern. The enveloping curve $(2\sin(\omega_0 - \omega)\tau/2)^2/(\omega_0 - \omega)^2$ is the standard expression for a one-slit diffraction pattern. The probability of a spin transition varies rapidly with ω and is maximum at $\omega = \omega_0$. Thus by looking at the rate of transitions in the beam as ω is varied, one can accurately tune ω to ω_0. The rate of variation of the interference pattern is inversely proportional to T, the length of time between the spin entering the first cavity and the second. It is the difference in the phase of the r.f. field that the spin sees in the two cavities that gives rise to the rapidly varying interference pattern.

Now in actual practice, the spin used is that of an electron in an atom, such as cesium, and the energy difference between the spin up state and the spin down state is due to the internal fields of the atom, rather than to the $\mathcal{H}_0$ field we put in. The frequency ω_0, which is $\hbar^{-1}$ times the energy difference between the spin up and spin down states, is thus a natural constant, and makes such a clock an extremely

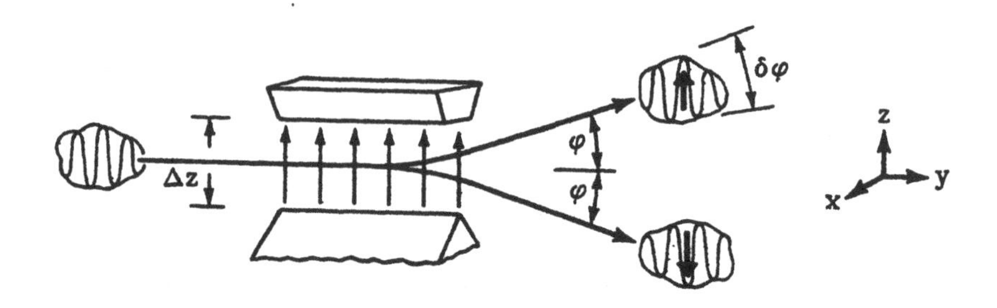

Fig. 14-8
Splitting of a wave packet of a spin ½ atom by an inhomogeneous field in the Stern-Gerlach apparatus. φ is the angle of deflection and $\delta\varphi$ is the angular width of the emerging wave packets.

accurate frequency monitor. Atomic clocks are a very remarkable application of the phenomenon of quantum mechanical interference.

MOTION IN INHOMOGENEOUS MAGNETIC FIELDS

Now let us study the effect of a magnetic field on the spatial motion of a spinning particle. A spin moving through a magnetic field $\mathcal{H}(\mathbf{r}, t)$ has an interaction with the field

$$H_{spin} = -\mathbf{M}_{spin} \cdot \mathcal{H}(\mathbf{r}, t), \tag{14-86}$$

where $\mathbf{M}_{spin}$ is the magnetic moment (operator) belonging to the spin. If the magnetic field is uniform in space, such an interaction will have no direct effect on the spatial motion of the particle. However, if $\mathcal{H}$ depends on $\mathbf{r}$ then this interaction will tend to cause the particle to move to regions of space that minimize the energy (14-86). For example, if the spin magnetic moment is parallel to the field, the particle will be forced by (14-86) toward regions of larger $\mathcal{H}(\mathbf{r}t)$ and if the moment is antiparallel to the field, the particle will be forced toward regions of smaller $\mathcal{H}(\mathbf{r}t)$. The particle experiences a force

$$\mathbf{F} = -\nabla H_{spin} = -\nabla[\mathbf{M}_{spin} \cdot \mathcal{H}(\mathbf{r}, t)]. \tag{14-87}$$

To understand the effect of this force, let us consider the famous Stern-Gerlach experiment in which a beam of silver atoms was passed through an inhomogeneous magnetic field in the z direction, with $d\mathcal{H}_z/dz < 0$ [Fig. 14-8]; what they observed was that the magnetic field separated the beam into two components. Now a silver atom has 47 electrons; 46 of them form a state with both zero spin and zero orbital angular momentum, while the last is in an S state. The total magnetic moment of a silver atom is just that of the spin of this last electron. If this spin is in an eigenstate $|\hat{\mathbf{z}}\uparrow\rangle$ then the magnetic moment is antiparallel to the field and the silver atom will be deflected upward toward smaller field. On the other hand if the spin is in the state $|\hat{\mathbf{z}}\downarrow\rangle$ then the particle will be deflected downward toward stronger field.

What happens to an atom in which the spin is initially in some general state $|\Psi\rangle$? Since we can write $|\Psi\rangle$ as a linear combination of spin up and spin down states, $|\Psi\rangle = |\hat{\mathbf{z}}\uparrow\rangle\langle\hat{\mathbf{z}}\uparrow|\Psi\rangle + |\hat{\mathbf{z}}\downarrow\rangle\langle\hat{\mathbf{z}}\downarrow|\Psi\rangle$, we can look upon the wave packet of the atom before it enters the field as a linear superposition of a wave packet with spin up and a wave packet with spin down. The field then deflects the up spin component of the wave packet upward, and the down spin component downward. Thus the field causes an actual separation in space of the spin up and spin down components of the initial wave packet. Regardless of the initial spin state, if the particle is observed to have been deflected upward after leaving the magnetic field, then its spin will definitely be up; if it is observed to have been deflected downward then its spin will definitely be down. This apparatus is thus useful for preparing a beam of spins all pointing in a given direction.

A spin 1/2 can emerge from the magnetic field in only two possible directions. This is much different from the behavior of a classical magnetic moment which can have a whole continuum of possible final directions depending on the value of $\mathbf{M}\cdot\boldsymbol{\mathcal{H}} = M_z\mathcal{H}_z$. A magnetic moment arising from an orbital angular momentum with total quantum number l can have $2l + 1$ possible deflections. The Stern-Gerlach experiment is thus very good evidence for the quantization of the possible values of a quantum magnetic moment in a given direction. In fact, it is a very practical method for measuring magnetic moments of atoms (as well as nuclei). Also, because a magnetic moment due to an orbital angular momentum must produce an *odd* number of components in the emergent beam, the experiment with silver is good evidence for the existence of the spin degrees of freedom of an electron.

Let us look at the effect of the inhomogeneous field in this experiment in more detail. Suppose, to be specific, that the particle enters the apparatus with its spin along $\hat{\mathbf{x}}$; if its spatial state is a (normalized) wave packet $\psi_0(\mathbf{r}, t)$ then in the z basis the wave function of the particle is originally

$$\Psi(\mathbf{r}, t) = \frac{1}{\sqrt{2}}\begin{pmatrix} \psi_0(\mathbf{r}, t) \\ \psi_0(\mathbf{r}, t) \end{pmatrix} \tag{14-88}$$

When the particle emerges from the apparatus at time T later its wave function is of the form

$$\Psi(\mathbf{r}, t+T) = \frac{1}{\sqrt{2}}\begin{pmatrix} \psi_\uparrow(\mathbf{r}, t+T) \\ \psi_\downarrow(\mathbf{r}, t+T) \end{pmatrix} \tag{14-89}$$

where $\psi_\uparrow$ is the localized wave packet indicated with an up arrow in Fig. 14-8, and $\psi_\downarrow$ is the localized wave packet indicated there by a down arrow. There is no overlap between $\psi_\uparrow$ and $\psi_\downarrow$, i.e., $\psi_\uparrow(\mathbf{r}, t+T) \times \psi_\downarrow(\mathbf{r}, t+T) \approx 0$. Furthermore, from conservation of spin, the total probability for finding the particle with spin up at t + T must equal the probability for finding it up at t. Thus $\int d^3r\, |\psi_\uparrow(\mathbf{r}, t+T)|^2 = \int d^3r\, |\psi_0(\mathbf{r}, t)|^2$ and similarly $\int d^3r\, |\psi_\downarrow(\mathbf{r}, t+T)|^2 = \int d^3r\, |\psi_0(\mathbf{r}, t)|^2$, so that

$$\int d^3r\, |\psi_\uparrow(\mathbf{r}, t+T)|^2 = \int d^3r\, |\psi_\downarrow(\mathbf{r}, t+T)|^2. \tag{14-90}$$

Initially the expectation value of $\mathbf{S}$ at $\mathbf{r}$ and t, is, from (14-88),

$$\langle \mathbf{S} \rangle_{\mathbf{r},t} = \Psi^\dagger(\mathbf{r}, t)\frac{\hbar\boldsymbol{\sigma}}{2}\Psi(\mathbf{r}, t) = \frac{\hbar}{2}\hat{\mathbf{x}}\,|\psi_0(\mathbf{r}, t)|^2;$$

thus the net spin expectation value is

$$\langle \mathbf{S} \rangle_t = \int d^3r\, \langle \mathbf{S} \rangle_{\mathbf{r},t} = \frac{\hbar}{2}\hat{\mathbf{x}}. \tag{14-91}$$

Similarly the spin density in the final state is

$$\frac{\hbar}{2}\hat{\mathbf{z}}\,[\,|\psi_\uparrow(\mathbf{r}, t+T)|^2 - |\psi_\downarrow(\mathbf{r}, t+T)|^2] \tag{14-92}$$

(there is no interference between the two wave packets since they don't overlap), so that the net expectation value of the spin in the final state, the integral of (14-92) over all space, vanishes by (14-90):

$$\langle \mathbf{S} \rangle_{t+T} = 0. \tag{14-93}$$

Now pretend for a moment that the field was uniform. Then we know from (14-62) that at time t + T we would have instead:

$$\langle \mathbf{S} \rangle_{t+T} = \frac{\hbar}{2}\hat{\mathbf{x}}\cos\omega_0 T - \frac{\hbar}{2}\hat{\mathbf{y}}\sin\omega_0 T; \tag{14-94}$$

the field would simply rotate the polarization vector $[\langle\sigma\rangle = 2\langle \mathbf{S}\rangle/\hbar]$ in a negative sense about $\hat{\mathbf{z}}$ with frequency ω_0. Compare this with (14-93). The question is, how is it that when we make the field inhomogeneous the polarization vector of the spins disappears as the particle passes through the apparatus? The answer lies in the fact that because the field is inhomogeneous the rate ω_0 at which the spin rotates is uncertain. Spins that pass through near the top of the apparatus will precess more slowly than those that pass through near the bottom. There are different amplitudes for different amounts of rotation. The resulting $\langle \mathbf{S}\rangle_{t+T}$ is thus essentially an average of (14-94) over the various possible ω_0. The inhomogeneous field *dephases* the spin. But, in order for the average to vanish we must have

$$(\Delta\omega_0)T \gtrsim 2\pi \tag{14-95}$$

where $\Delta\omega_0$ is the uncertainty in ω_0; were $\Delta\omega_0$ much smaller, $\sin \omega_0 T$ and $\cos \omega_0 T$ would not average to zero. This is a very useful result, for it places a lower limit on the magnitude of the inhomogeneity that will split a beam in the Stern-Gerlach apparatus. Indeed, from (14-60) we find that

$$\Delta\omega_0 = \frac{ge}{2mc}\,\Delta\mathcal{H} \approx \frac{ge}{2mc}\,\frac{\partial\mathcal{H}}{\partial z}\,\Delta z$$

where Δz is the height of the gap between the pole pieces of the magnet [Fig. 14-8]; hence the condition for the apparatus to work is

$$\left|\frac{ge}{2mc}\,\frac{\partial\mathcal{H}}{\partial z}\right| T \gtrsim \frac{2\pi}{\Delta z} \tag{14-96}$$

This, however, is precisely the condition that the angular separation of the beams, 2φ, be greater than the angular width of the beams $\delta\varphi$, and therefore that the two outcoming wave packets don't overlap each other. To see this we note that the angle of deflection φ is essentially the component of momentum p_z that the field gives to the particle, divided by p_y, the incoming momentum of the particle. Now $p_z \approx FT$ where

$$F = \left|\frac{ge}{2mc}\,\frac{\partial\mathcal{H}}{\partial z}\right| \frac{\hbar}{2}$$

is the magnitude of the force in the z direction exerted by the inhomogeneous field. Thus

$$\varphi \approx \left| \frac{ge}{2mc} \frac{\hbar}{2} \frac{\partial \mathcal{H}}{\partial z} \right| \frac{T}{p_y}. \tag{14-97}$$

The angular spread of the wave packet as it leaves the apparatus is given by the diffraction theory result

$$\delta\varphi \approx \frac{\lambda}{\Delta z} = \frac{2\pi\hbar}{p_y \Delta z} \tag{14-98}$$

where λ is the deBroglie wavelength of the incident particle. Thus (14-96) is the statement

$$\delta\varphi \lessdot 2\varphi; \tag{14-99}$$

the condition that the field completely dephase the spin is equivalent to the condition that it split the beam completely in two.

More generally, suppose that the magnetic moment can take on several values, M_z', M_z'', ... We may write the force on the particle due to the inhomogeneous field as

$$F(M_z') = -\frac{\partial E(M_z')}{\partial z} \tag{14-100}$$

where $E(M_z')$ is the energy of the magnetic moment in the field. In order then for the diffraction not to wash out the separation of the component with M_z' from the component with M_z'' we need

$$(\Delta z) \frac{\partial}{\partial z} |E(M_z') - E(M_z'')| T \gtrdot 2\pi\hbar. \tag{14-101}$$

The quantity $\Delta z\, \partial |E(M_z') - E(M_z'')|/\partial z$ is the uncertainty in the energy difference $|E(M_z') - E(M_z'')|$ of the states M_z' and M_z'' arising from the variation of the magnetic field across the wave packet of the particle. The inequality (14-10) thus says that if the apparatus is capable of separating the two components, i.e., if it is capable of distinguishing two different energy states, then the uncertainty in the energy resolution times the time T spent in the field must exceed $2\pi\hbar$.

The Stern-Gerlach apparatus can separate states because the energy difference $E(M_z') - E(M_z'')$ of the states depends on z due to the inhomogeneity of the magnetic field. Now the energies of the various excited states of an atom in an electric field depend on the strength of the electric field. Thus one can similarly use a Stern-Gerlach-like apparatus with an inhomogeneous *electric* field to sort out different internal states of an atomic beam. For example, if one

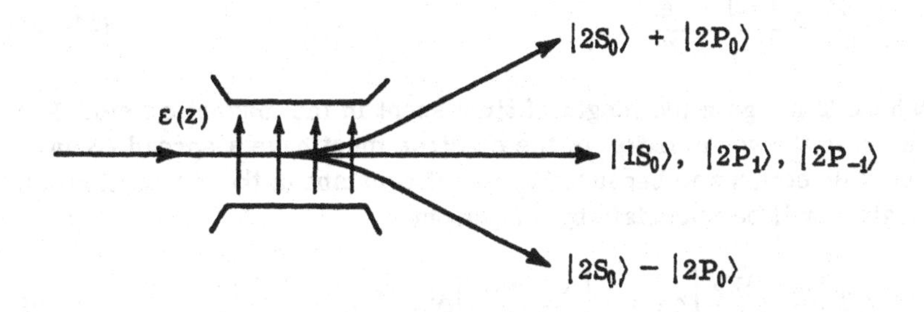

Fig. 14-9
Splitting of a beam of H atoms in n = 1 and 2 states by an inhomogeneous electric field.

passes a beam of hydrogen atoms in a mixture of n = 1 and n = 2 states through an inhomogeneous electric field, then according to the theory of the first-order Stark effect, one will find three beams emerging from the apparatus [Fig. 14-9]. First there will be an undeflected component which is a mixture of $1S_0$, $2P_1$, and $2P_{-1}$ states; one of the deflected components will be composed of atoms all in the state $(|2S_0\rangle + |2P_0\rangle)/\sqrt{2}$ and the other deflected component will consist of atoms in the state $(|2S_0\rangle - |2P_0\rangle)/\sqrt{2}$. If the electric field is sufficiently strong then one expects a fine splitting of the undeflected beam due to the second-order Stark effect.

Can the Stern-Gerlach experiment be done with electrons directly? Unfortunately it cannot, for the following reason. The divergence of a magnetic field must, according to Maxwell, be zero. Thus if $\partial\mathcal{H}_z/\partial z$ is nonzero, then $\partial\mathcal{H}_x/\partial x + \partial\mathcal{H}_y/\partial y$ is also nonzero. Just to simplify the argument, let's suppose that $\partial\mathcal{H}_y/\partial y = 0$. Then there must also be a component of the field in the x direction that satisfies

$$\frac{\partial\mathcal{H}_x}{\partial x} = -\frac{\partial\mathcal{H}_z}{\partial z}. \tag{14-102}$$

An electron moving through the magnet in the y direction with velocity **v** thus feels a Lorentz force in the z direction of magnitude

$$F_L = \frac{ev_y\mathcal{H}_x}{c}. \tag{14-103}$$

This force will deflect the beam in the y, z plane away from the y direction by an angle whose magnitude is

$$\theta \approx \frac{\Delta p_z}{p_y} = \frac{F_L T}{m v_y} = \frac{e}{mc} \mathcal{H}_x \tag{14-104}$$

where T is again the length of time spent in the magnetic field. Now because the wave packet of the electron must have a spread Δx in the x direction and because $\mathcal{H}_x$ is not constant in the x direction, the angle θ will be uncertain by an amount

$$\delta\theta \approx T \left| \frac{e}{mc} \frac{\partial \mathcal{H}_x}{\partial x} \right| \Delta x = T \left| \frac{e}{mc} \frac{\partial \mathcal{H}_z}{\partial z} \right| \Delta x. \tag{14-105}$$

In order for the Stern-Gerlach experiment to separate the electron beam into two components, $\delta\theta$ must be smaller than the angle of separation, 2φ, of the two components. But from (14-97) we have $(g = 2)$,

$$\delta\theta \gtrsim T \left| \frac{e}{mc} \frac{\partial \mathcal{H}_z}{\partial z} \right| \frac{\hbar}{\Delta p_x} \approx 2\varphi \frac{p_y}{\Delta p_x}.$$

Clearly the uncertainty Δp_x must be small compared with p_y for a beam headed in the y direction. Thus $\delta\theta > 2\varphi$ and we see that the angular spread of the beam due to the Lorentz force must always be greater than the separation angle of the two components. The Lorentz force thus washes out the separation effect.

The Lorentz force is never a problem, of course, if one does the experiment with electrically neutral particles. It isn't even a problem when one uses ions, since the mass in (14-104) is the ionic mass, whereas the mass in (14-97) is the electron mass if the magnetic moment is electronic in origin; then $\delta\theta \ll 2\varphi$.

PROBLEMS

1. (a) Find the eigenstates of σ_x, σ_y, and σ_z in the basis in which σ_z is diagonal. Do the same in the basis in which σ_x is diagonal.
   (b) Using the spin representation of rotations find an explicit expression for the vector $\hat{\mathbf{n}}$ that is gotten by rotating the vector $\hat{\mathbf{m}}$ by $|\alpha|$ in a right-handed sense about the direction $\hat{\boldsymbol{\alpha}}$.
2. Consider the spin resonance experiment.
   (a) Calculate $\mathbf{S}(t)$ explicitly in the Heisenberg representation.
   (b) What is the probability as a function of time that a spin initially along $\hat{\mathbf{z}}$ points along $\hat{\mathbf{x}}$? along $\hat{\mathbf{y}}$?
   (c) Calculate, in both the Schrödinger and Heisenberg representations the effects of a 90° r.f. pulse.

3. The spin algebra is very useful for problems involving transitions between only two states of a system such as an atom or molecule. Suppose then that one has an atom with only two states $|+\rangle$, $|-\rangle$ with energies ε_+ and ε_-. What is the Hamiltonian H_0 of the two-level atom? Suppose also that the expectation value of the electric dipole moment, ez, in both $|+\rangle$ and $|-\rangle$ is zero, but that the matrix element $\langle +|ez|-\rangle$ is equal to a real constant μ. Write down the Hamiltonian of the system in the presence of an electric field $\mathbf{E}\cos\omega t$ in the z direction. Solve for the electric polarization of the atom initially in the state $|-\rangle$ as a (nonlinear) function of E, ω, and t, by neglecting the "counterrotating component" of the electric field. What is this polarization to first order in E?

Chapter 15
ADDITION OF ANGULAR MOMENTA

Classically, if we have a system with angular momentum $\mathbf{J}_1$ and a second system with angular momentum $\mathbf{J}_2$ then the total angular momentum of the combined systems is simply $\mathbf{J}_1 + \mathbf{J}_2$. The corresponding situation in quantum mechanics is where we have one system in a state with a total angular momentum quantum number j_1 and component m_1 along some axis, and a second system in a state with total angular momentum quantum number j_2 and component m_2 along some axis; the problem is, what can we say about the total angular momentum of the two systems in this situation. The answer, unfortunately, is not as simple as it is in classical mechanics.

Before launching into the general problem, let's look at the simple case of two systems each with spin $^1/_2$; for example, the two spins could be that of the neutron and the proton in a deuteron. Say that the operator for first spin is $\mathbf{S}_1$ and the operator for the second spin is $\mathbf{S}_2$. Because $\mathbf{S}_1$ and $\mathbf{S}_2$ refer to different particles, they commute:

$$[\mathbf{S}_1, \mathbf{S}_2] = 0. \tag{15-1}$$

Since there are two linearly independent orientations for each spin, there are four linearly independent states of the two spins. Choosing the quantization axis along $\hat{\mathbf{z}}$ we can write these states as

$$|\uparrow\uparrow\rangle, \quad |\uparrow\downarrow\rangle, \quad |\downarrow\uparrow\rangle, \quad |\downarrow\downarrow\rangle \tag{15-2}$$

where the first symbol refers to the eigenvalue of S_{1z}, and the second to the eigenvalue of S_{2z}.

Now consider the total spin angular momentum of the two systems

$$\mathbf{S} = \mathbf{S}_1 + \mathbf{S}_2. \tag{15-3}$$

Because $\mathbf{S}_1$ and $\mathbf{S}_2$ individually obey the angular momentum commutation relations and because $\mathbf{S}_1$ commutes with $\mathbf{S}_2$, $\mathbf{S}$ itself obeys the angular momentum commutation relations

$$[S_i, S_j] = i\varepsilon_{ijk}S_k. \tag{15-4}$$

[We shall generally set $\hbar = 1$ from here on; learning to put the $\hbar$'s back in at the end of a calculation is a vital skill.] Because $\mathbf{S}$ is an angular momentum we can find simultaneous eigenstates of S^2 and S_z. Because there are four linearly independent states of the two spins, we expect to have four linearly independent eigenstates of S^2 and S_z. The spin states $|s, m\rangle$ labeled by the eigenvalues of S^2 and S_z, where $S^2|s, m\rangle = s(s+1)|s, m\rangle$, also form a basis for all the possible states of the two spins. The problem is to construct the states $|s, m\rangle$ in terms of the states (15-2).

Each of the states (15-2) is an eigenfunction of S_z, since $S_z = S_{1z} + S_{2z}$. Thus

$$S_z|\uparrow\uparrow\rangle = \left(\frac{1}{2}+\frac{1}{2}\right)|\uparrow\uparrow\rangle = |\uparrow\uparrow\rangle$$

$$S_z|\uparrow\downarrow\rangle = \left(\frac{1}{2}-\frac{1}{2}\right)|\uparrow\downarrow\rangle = 0$$

$$S_z|\downarrow\uparrow\rangle = \left(-\frac{1}{2}+\frac{1}{2}\right)|\downarrow\uparrow\rangle = 0$$

$$S_z|\downarrow\downarrow\rangle = \left(-\frac{1}{2}-\frac{1}{2}\right)|\downarrow\downarrow\rangle = -|\downarrow\downarrow\rangle. \tag{15-5}$$

The eigenvalues of S_z are clearly 0 and ± 1. We expect then to find a triplet of states with $S_z = +1, 0, -1$ corresponding to total spin one and a singlet state with $S_z = 0$ corresponding to total spin 0.

The state $|\uparrow\uparrow\rangle$ cannot be a zero spin state, and therefore it must have spin one, i.e.,

$$S^2|\uparrow\uparrow\rangle = s(s+1)|\uparrow\uparrow\rangle = 2|\uparrow\uparrow\rangle \tag{15-6}$$

where $s = 1$. To prove (15-6) we write

$$S^2 = (\mathbf{S}_1+\mathbf{S}_2)^2 = S_1^2 + S_2^2 + 2\mathbf{S}_1\cdot\mathbf{S}_2 = \frac{3}{2} + 2S_{1z}S_{2z} + S_{1+}S_{2-} + S_{1-}S_{2+}, \tag{15-7}$$

where

$$S_{1\pm} = S_{1x} \pm iS_{1y}, \qquad S_{2\pm} = S_{2x} \pm iS_{2y} \tag{15-8}$$

and $S_1^2 = S_2^2 = \frac{1}{2}(\frac{1}{2}+1) = \frac{3}{4}$. Thus $S^2|\uparrow\uparrow\rangle = (\frac{3}{2} + 2\cdot\frac{1}{2}\cdot\frac{1}{2})|\uparrow\uparrow\rangle = 2|\uparrow\uparrow\rangle$, since S_{1+} acting on a state with the first spin up gives zero, and S_{2+} acting on a state with the second spin up gives zero.

By the same argument $|\downarrow\downarrow\rangle$ is also a state with $s = 1$. To construct the state with $s = 1$, and $S_z = 0$ we use the fact that $\mathbf{S}$ obeys the angular momentum commutation relations; thus from (6-48):

$$S_-|s,m\rangle = \sqrt{s(s+1)-m(m-1)}\;|s,m-1\rangle \tag{15-9}$$

so that

$$|s=1,m=0\rangle = \frac{1}{\sqrt{2}}S_-|\uparrow\uparrow\rangle = \frac{1}{\sqrt{2}}(S_{1-}+S_{2-})|\uparrow\uparrow\rangle = \frac{1}{\sqrt{2}}(|\downarrow\uparrow\rangle+|\uparrow\downarrow\rangle), \tag{15-10}$$

since $S_{1-}|\uparrow\uparrow\rangle = \sqrt{\frac{1}{2}(\frac{1}{2}+1) - \frac{1}{2}(\frac{1}{2}-1)}\,|\downarrow\uparrow\rangle = |\downarrow\uparrow\rangle$, etc. The state with $s = 0$, $m = 0$, must be orthogonal to (15-10) and can therefore only be the state

$$|s=0,m=0\rangle = \frac{1}{\sqrt{2}}(|\uparrow\downarrow\rangle - |\downarrow\uparrow\rangle). \tag{15-11}$$

The three states with $s = 1$ are thus

$$|\uparrow\uparrow\rangle,\qquad \frac{1}{\sqrt{2}}(|\uparrow\downarrow\rangle+|\downarrow\uparrow\rangle),\qquad |\downarrow\downarrow\rangle;$$

these are called the *triplet* states. The spins in a deuteron are in a triplet state. The state (15-11) is called the *singlet*. [See p. 208.]

We can also write the original states (15-2) as linear combinations of the $|s, m\rangle$ states as follows

$$|\uparrow\uparrow\rangle = |1,1\rangle,\qquad |\uparrow\downarrow\rangle\quad \frac{1}{\sqrt{2}}(|1,0\rangle + |0,0\rangle)$$

$$|\downarrow\uparrow\rangle = \frac{1}{\sqrt{2}}(|1,0\rangle - |0,0\rangle),\qquad |\downarrow\downarrow\rangle = |1,-1\rangle. \tag{15-12}$$

We have thus solved the problem of how to write the eigenstates of S^2 and S_z in terms of those of S_{1z} and S_{2z}, and vice versa.

Two spins $\frac{1}{2}$'s added together can give a total angular momentum one or zero. One can think of the spins as being parallel in the s = 1 states and antiparallel in the s = 0 states. The $|s, m\rangle$ states are eigenfunctions of the operator $\mathbf{S}_1 \cdot \mathbf{S}_2$. To see this we write

$$\mathbf{S}_1 \cdot \mathbf{S}_2 = \frac{1}{2}\left[(\mathbf{S}_1 + \mathbf{S}_2)^2 - S_1^2 - S_2^2\right] = \frac{S^2}{2} - \frac{3}{4}. \tag{15-13}$$

Thus acting on a triplet state

$$\mathbf{S}_1 \cdot \mathbf{S}_2 |s = 1, m\rangle = \frac{1}{4}|s = 1, m\rangle, \tag{15-14}$$

and on a singlet state

$$\mathbf{S}_1 \cdot \mathbf{S}_2 |s = 0, m = 0\rangle = -\frac{3}{4}|s = 0, m = 0\rangle. \tag{15-15}$$

If one has an interaction between two spins of the form

$$V = v\mathbf{S}_1 \cdot \mathbf{S}_2 \tag{15-16}$$

then V is diagonalized in the basis of eigenstates of S^2. If such an interaction is attractive in a triplet state, it is -3 times as strong, and repulsive in a singlet state.

The operator

$$P_1 = \frac{3}{4} + \mathbf{S}_1 \cdot \mathbf{S}_2 \tag{15-17}$$

annihilates a singlet state and multiplies a triplet state by one; acting on any state it therefore projects out the triplet part of the state. Similarly the operator

$$P_0 = 1 - P_1 = \frac{1}{4} - \mathbf{S}_1 \cdot \mathbf{S}_2 \tag{15-18}$$

is a projection operator for the singlet state.

ADDITION OF TWO ANGULAR MOMENTA

Let us now turn to the general problem of adding two angular momenta, $\mathbf{J}_1$ and $\mathbf{J}_2$. We assume that $\mathbf{J}_1$ and $\mathbf{J}_2$ commute with each other. Then we can construct eigenstates of J_1^2, J_2^2, J_{1z}, J_{2z} which we write as $|j_1 j_2 m_1 m_2\rangle$. Thus

$$J_1^2 |j_1 j_2 m_1 m_2\rangle = j_1(j_1 + 1)|j_1 j_2 m_1 m_2\rangle$$

$$J_{1z}|j_1j_2m_1m_2\rangle = m_1|j_1j_2m_1m_2\rangle, \tag{15-19}$$

etc. Now the total angular momentum

$$\mathbf{J} = \mathbf{J}_1 + \mathbf{J}_2 \tag{15-20}$$

obeys the angular momentum commutation relations

$$[J_i, J_j] = i\epsilon_{ijk}J_k; \tag{15-21}$$

we can therefore look for eigenstates of J^2 and J_z. Furthermore, since ${J_1}^2$ and ${J_2}^2$ are scalars they commute with $\mathbf{J}$. Therefore we can look for states that are also eigenstates of ${J_1}^2$ and ${J_2}^2$ in addition to J^2 and J_z; we call these states $|j_1j_2jm\rangle$. Thus

$$J^2|j_1j_2jm\rangle = j(j+1)|j_1j_2jm\rangle$$

$$J_z|j_1j_2jm\rangle = m|j_1j_2jm\rangle$$

$$J_1^2|j_1j_2jm\rangle = j_1(j_1+1)|j_1j_2jm\rangle, \tag{15-22}$$

etc. Notice however that $[J_{1z}, J^2] \neq 0$ and $[J_{2z}, J^2] \neq 0$ so that we cannot in general specify J_{1z} and J_{2z} individually if we specify J^2; only $J_z = J_{1z} + J_{2z}$ can be specified.

The collection of states $|j_1j_2m_1m_2\rangle$ forms a basis, as does the collection of states $|j_1j_2jm\rangle$. The problem of adding two angular momenta is that of finding out how to express the $|j_1j_2jm\rangle$ states as linear combinations of the $|j_1j_2m_1m_2\rangle$ states and vice versa. Because the states $|j_1j_2jm\rangle$ form a basis we can write the state $|j_1j_2jm\rangle$ as

$$|j_1j_2jm\rangle = \sum_{\substack{j_1'j_2' \\ m_1m_2}} |j_1'j_2'm_1m_2\rangle\langle j_1'j_2'm_1m_2|j_1j_2jm\rangle. \tag{15-23}$$

The coefficients $\langle j_1'j_2'm_1m_2|j_1j_2jm\rangle$ are called *Clebsch-Gordan* coefficients. Since

$$\begin{aligned}\langle j_1'j_2'm_1m_2|J_1^2|j_1j_2jm\rangle &= j_1'(j_1'+1)\langle j_1'j_2'm_1m_2|j_1j_2jm\rangle \\ &= \langle j_1'j_2'm_1m_2|j_1j_2jm\rangle j_1(j_1+1)\end{aligned} \tag{15-24}$$

the Clebsch-Gordan coefficient vanishes unless $j_1 = j_1'$ and also $j_2 = j_2'$. Also since $J_z = J_{1z} + J_{2z}$ we have

$$\begin{aligned}\langle j_1j_2m_1m_2|J_z|j_1j_2jm\rangle &= (m_1+m_2)\langle j_1j_2m_1m_2|j_1j_2jm\rangle \\ &= \langle j_1j_2m_1m_2|j_1j_2jm\rangle m\ .\end{aligned} \tag{15-25}$$

so that the Clebsch-Gordan coefficient vanishes unless $m = m_1 + m_2$. Thus the only nonvanishing coefficients are of the form $\langle j_1j_2m_1m_2|j_1j_2j, m = m_1 + m_2\rangle$, and we can write the expansion (15-23) as

$$|j_1j_2jm\rangle = \sum_{\substack{m_1 m_2 \\ (m_1+m_2=m)}} |j_1j_2m_1m_2\rangle\langle j_1j_2m_1m_2|j_1j_2jm\rangle. \qquad (15\text{-}26)$$

For fixed j_1 and j_2 there are $2j_1 + 1$ possible m_1 values and $2j_2 + 1$ possible m_2 values; thus there are $(2j_1 + 1)(2j_2 + 1)$ linearly independent states of the form $|j_1j_2jm\rangle$. Let us notice that there is only one state with $m_1 + m_2 = j_1 + j_2$; there are two states with $m_1 + m_2 = j_1 + j_2 - 1$; these are $m_1 = j_1$, $m_2 = j_2 - 1$ and $m_1 = j_1 - 1$, $m_2 = j_2$. In general, there are as many states with a fixed m as we can choose different m_1 and m_2 to add up to m. For $m \geq |j_1 - j_2|$ there are $j_1 + j_2 - m + 1$ different ways, while for $-|j_1 - j_2| < m < |j_1 - j_2|$, there are $j_1 + j_2 - |j_1 - j_2| + 1$ different ways, and for $m < -|j_1 - j_2|$ there are $j_1 + j_2 - |m| + 1$ different ways. For example, for $j_1 = 2$, $j_2 = 1$ we have 1 state with $m = \pm 3$, 2 states with $m = \pm 2$, 3 states with $m = \pm 1$, and 3 states with $m = 0$, giving a total of $15 = (2 \cdot 2 + 1)(2 \cdot 1 + 1)$ states.

What are the possible j values that we can form from states with fixed j_1 and j_2? This is easy to answer once we realize that if we can form a state $|j_1j_2jm\rangle$, then we can form all the other states $|j_1j_2jm'\rangle$ with $|m'| \leq j$, by acting with the operators

$$J_\pm = J_{1\pm} + J_{2\pm}, \qquad (15\text{-}27)$$

as we did in considering two spin $^1/_2$'s. All we must do to determine the possible j values is to break the $(2j_1 + 1)(2j_2 + 1)$ states up into multiplets. Since $m = j_1 + j_2$ is the maximum m value, $j_1 + j_2$ must also be the maximum j value possible, for if we could form a higher j value we would also have to find a state with $m = j$ present. Thus among the $(2j_1 + 1)(2j_2 + 1)$ states is a multiplet of $2(j_1 + j_2) + 1$ states with $j = j_1 + j_2$.

What is left over? The highest m value left over is $j_1 + j_2 - 1$ and thus there must be a $j = j_1 + j_2 - 1$ multiplet among the states with fixed j_1 and j_2. Keeping up this argument we find that j can take on all values $j_1 + j_2$, $j_1 + j_2 - 1$, $j_1 + j_2 - 2, \ldots,$ down to $|j_1 - j_2|$. Thus the possible j must obey the triangle rule

$$|j_1 - j_2| \leq j \leq j_1 + j_2, \qquad (15\text{-}28)$$

which is equivalent to the condition on the possible lengths of the third side of a triangle the first two sides of which are of length j_1

and j_2. [(15-28) is also the condition classically on the magnitude of total angular momentum formed from two angular momenta of magnitudes j_1 and j_2.] The possible j values must differ by *integers.* Thus, for example, if one adds an integral angular momentum to a half-integral angular momentum, all the possible resultant angular momenta are half-integral. As a check on our argument for the possible j values we note that each j multiplet has 2j + 1 states and

$$\sum_{j=|j_1-j_2|}^{j_1+j_2} (2j+1) = (2j_1+1)(2j_2+1). \tag{15-29}$$

CLEBSCH-GORDAN TECHNOLOGY

The Clebsch-Gordan coefficients obey simple orthogonality relations. Because the $|j_1j_2jm\rangle$ form a basis we can write the expansion

$$|j_1j_2m_1m_2\rangle = \sum_{\substack{j \\ (m=m_1+m_2)}} |j_1j_2jm\rangle\langle j_1j_2jm|j_1j_2m_1m_2\rangle, \tag{15-30}$$

and also

$$\langle j_1j_2m_1'm_2'|j_1j_2m_1m_2\rangle = \sum_{j,m} \langle j_1j_2m_1'm_2'|j_1j_2jm\rangle\langle j_1j_2jm|j_1j_2m_1m_2\rangle. \tag{15-31}$$

Since the left side vanishes unless $m_1 = m_1'$, $m_2 = m_2'$ we find the orthogonality relation

$$\sum_{j,m} \langle j_1j_2m_1'm_2'|j_1j_2jm\rangle\langle j_1j_2jm|j_1j_2m_1m_2\rangle = \delta_{m_1m_1'}\delta_{m_2m_2'}. \tag{15-32}$$

We shall see that the Clebsch-Gordan coefficients can all be taken to be real, so that

$$\langle j_1j_2jm|j_1j_2m_1m_2\rangle = \langle j_1j_2m_1m_2|j_1j_2jm\rangle.$$

[The standard order one finds tabulated is $\langle j_1j_2m_1m_2|j_1j_2jm\rangle$.]

From (15-26) we can derive a second orthogonality relation

$$\sum_{m_1m_2} \langle j_1j_2jm|j_1j_2m_1m_2\rangle\langle j_1j_2m_1m_2|j_1j_2j'm'\rangle = \langle j_1j_2jm|j_1j_2j'm'\rangle = \delta_{jj'}\delta_{mm'}. \tag{15-33}$$

For $j = j'$, $m = m'$, this reduces to

$$\sum_{\substack{m_1 m_2 \\ m_1 + m_2 = m}} \langle j_1 j_2 j m | j_1 j_2 m_1 m_2 \rangle^2 = 1. \tag{15-34}$$

Our next problem is to determine the Clebsch-Gordan coefficients explicitly. The basic trick is to use the recursion relations we find from the matrix element $\langle j_1 j_2 m_1 m_2 | J_\pm | j_1 j_2 j m \rangle$. Acting to the right J_+ raises m by one and J_- lowers it by one. Thus

$$\begin{aligned} &\langle j_1 j_2 m_1 m_2 | J_\pm | j_1 j_2 j m \rangle \\ &\qquad = \sqrt{j(j+1) - m(m \pm 1)}\, \langle j_1 j_2 m_1 m_2 | j_1 j_2 j, m \pm 1 \rangle . \end{aligned} \tag{15-35}$$

Now when the J_{1+} in J_+ acts to the left it reduces m_1 by one; it acts exactly as J_{1-} does acting to the right. Thus

$$\begin{aligned} \langle j_1 j_2 m_1 m_2 | J_\pm | j_1 j_2 j m \rangle &= \langle j_1 j_2 m_1 m_2 | (J_{1\pm} + J_{2\pm}) | j_1 j_2 j m \rangle \\ &= \sqrt{j_1(j_1+1) - m_1(m_1 \mp 1)}\, \langle j_1 j_2, m_1 \mp 1, m_2 | j_1 j_2 j m \rangle \\ &\quad + \sqrt{j_2(j_2+1) - m_2(m_2 \mp 1)}\, \langle j_1 j_2 m_1, m_2 \mp 1 | j_1 j_2 j m \rangle . \end{aligned} \tag{15-36}$$

Combining (15-35) and (15-36) we find the two recursion relations

$$\begin{aligned} &\sqrt{j(j+1) - m(m \pm 1)}\, \langle j_1 j_2 m_1 m_2 | j_1 j_2 j, m \pm 1 \rangle \\ &\qquad = \sqrt{j_1(j_1+1) - m_1(m_1 \mp 1)}\, \langle j_1 j_2 m_1 \mp 1, m_2 | j_1 j_2 j m \rangle \\ &\qquad\quad + \sqrt{j_2(j_2+1) - m_2(m_2 \mp 1)}\, \langle j_1 j_2 m_1, m_2 \mp 1 | j_1 j_2 j m \rangle . \end{aligned} \tag{15-37}$$

These relations, together with (15-34), enable us to determine completely the Clebsch-Gordan coefficients.

The procedure is the following. Let us look at Fig. 15-1, a graph of possible m_1 and m_2 values for fixed j_1 and j_2. The lines of fixed $m = m_1 + m_2$ run from lower right to upper left. The upper left of the recursion relations (15-37) says that if we know the coefficients $\langle j_1 j_2 m_1 m_2 | j_1 j_2 m j \rangle$ for fixed j and for m_1 and m_2 at two corners of a triangle of the form

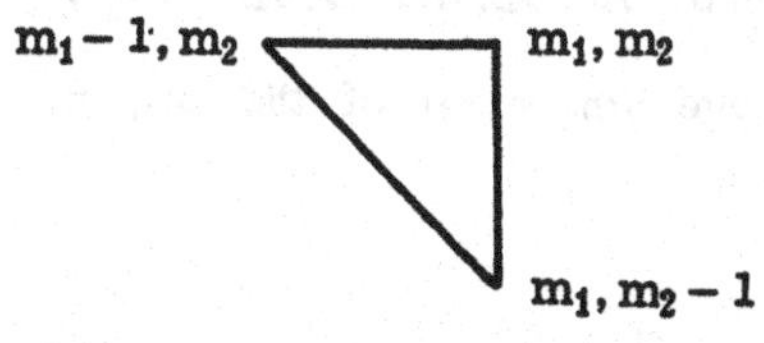

$m_1 - 1, m_2$ ——— m_1, m_2

$m_1, m_2 - 1$ (15-38)

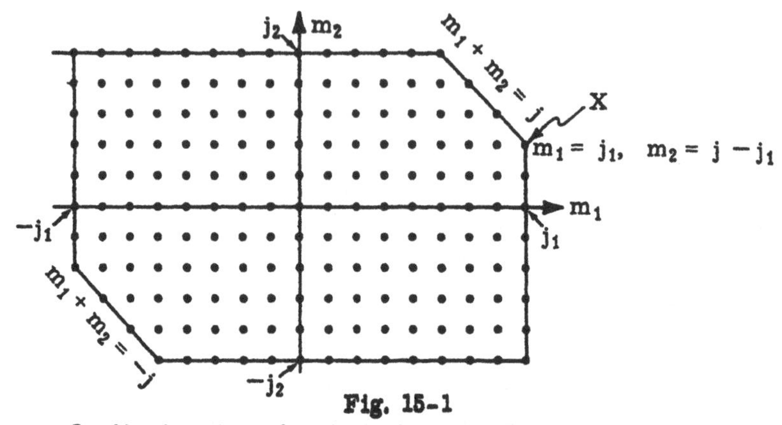

Fig. 15-1
Possible values of m_1 and m_2 for fixed j_1 and j_2. The point X is the starting point for calculating the Clebsch-Gordan coefficients from the recursion relations.

on the m_1, m_2 graph then we know the coefficient at the third corner. Similarly, the lower relation (15-37) relates the coefficients for fixed j at the corners of the triangle

$$\begin{array}{ll} m_1, m_2+1 & \\ m_1, m_2 & m_1+1, m_2 \end{array} \tag{15-39}$$

on the m_1, m_2 graph. Now to begin we assume a *positive* value (a choice of phase of $|j_1j_2jj\rangle$) for $\langle j_1j_2m_1 = j_1, m_2 = j - j_1|j_1j_2jj\rangle$, the point X on the graph. To get the coefficient $\langle j_1j_2, m_1 = j_1, m_2 = j - j_1 - 1|j_1j_2j, j-1\rangle$ we use the lower of the relations (15-37) taking $m_1 = j_1$, $m_2 = j - j_1 - 1$. The coefficient $\langle j_1, j_2, m_1 = j_1 + 1, m_2 = j - j_1 - 1|j_1j_2jm\rangle$ must vanish since $j_1 + 1$ is not a possible m_1 value. Thus for these m_1 and m_2 values

$$\sqrt{2j}\,\langle j_1j_2j_1, j - j_1 - 1|j_1j_2j, j-1\rangle$$

$$= \sqrt{j_2(j_2+1) - (j - j_1 - 1)(j - j_1)}\,\langle j_1j_2j_1, j - j_1|j_1j_2j, j-1\rangle$$

and in general, along the right edge ($m_1 = j_1$) of the m_1, m_2 graph

$$\sqrt{j(j+1)-m(m-1)}\,\langle j_1j_2j_1m_2|j_1j_2j,m-1\rangle$$
$$= \sqrt{j_2(j_2+1)-m_2(m_2+1)}\,\langle j_1j_2j_1,m_2+1|j_1j_2jm\rangle . \tag{15-40}$$

Thus we can work our way down to the lower right corner of the m_1, m_2 graph. Using the triangle relations (15-38) we can then find the bottom edge, and so on. Finally, to fix the original coefficient $\langle j_1j_2m_1 = j,\ m_2 = j - j_1|j_1j_2jj\rangle$, assumed real and positive, we use the sum rule (15-34). This completes the method of calculating the coefficients. Beyond the first few j_1 and j_2 the calculation becomes quite cumbersome and it is easier to look the Clebsch-Gordan coefficients up in books of tables. With some trivial modification they also go under the aliases of vector addition coefficients and Wigner 3j symbols.

Let us consider as an example the important case of a spin $^1/_2$ added to another angular momentum. For example, this might be the spin of an electron added to its orbital angular momentum. For $j_1 = 0$, $j_2 = {}^1/_2$ the only possible j value is $^1/_2$, and trivially

$$\langle j_1 = 0,\ j_2 = \frac{1}{2},\ m_1,\ m_2|j_1,\ j_2 = \frac{1}{2},\ j = \frac{1}{2},\ m\rangle = \delta_{m,\,m_1+m_2}. \tag{15-41}$$

For $j_1 \neq 0$, $j_2 = {}^1/_2$ the possible j values are $j_1 \pm {}^1/_2$. Let us write $j_1 = l$ and calculate the Clebsch-Gordan coefficients for $j = l + {}^1/_2$. Since there is only one state with $m = l + {}^1/_2$,

$$|l,\frac{1}{2},m_1 = l,\ m_2 = \frac{1}{2}\rangle = |l,\frac{1}{2},l+\frac{1}{2},\ l+\frac{1}{2}\rangle ;$$

and thus

$$\langle l,\frac{1}{2},l,m_2 = \frac{1}{2}|l,\frac{1}{2},l+\frac{1}{2},\ l+\frac{1}{2}\rangle = 1. \tag{15-42}$$

From the relation (15-40) we find

$$\langle l,\frac{1}{2},m_1 = m-\frac{1}{2},\ m_2 = \frac{1}{2}|l,\frac{1}{2},l+\frac{1}{2},\ m\rangle$$
$$= \sqrt{\frac{l+m+1/2}{l+m+3/2}}\,\langle l,\frac{1}{2},m_1 = m+\frac{1}{2},\ m_2 = \frac{1}{2}|l,\frac{1}{2},l+\frac{1}{2},\ m+1\rangle .$$

Iterating this relation and using (15-42) we then find

$$\langle l,\frac{1}{2},m_1 = m-\frac{1}{2},\ m_2 = \frac{1}{2}|l,\frac{1}{2},l+\frac{1}{2},\ m\rangle = \sqrt{\frac{l+m+1/2}{2l+1}}. \tag{15-43}$$

In general

$$\langle l, \frac{1}{2}, m_1 = m \pm \frac{1}{2}, m_2 = \pm\frac{1}{2} | l, \frac{1}{2}, l + \frac{1}{2}, m \rangle = \sqrt{\frac{l \pm m + 1/2}{2l+1}} \tag{15-44}$$

and

$$\langle l, \frac{1}{2}, m_1 = m \pm \frac{1}{2}, m_2 = \pm\frac{1}{2} | l, \frac{1}{2}, l - \frac{1}{2}, m \rangle = \mp\sqrt{\frac{l \mp m + 1/2}{2l+1}}. \tag{15-45}$$

These Clebsch-Gordan coefficients enable us to find the eigenstates $|lsjm\rangle$ of the total angular momentum, orbital plus spin, of a particle of spin $1/2$. One useful feature of these states is that they are eigenfunctions of the quantity $\mathbf{L} \cdot \mathbf{S}$ which appears in the spin-orbit interaction. This is because $J^2 = (L+S)^2 = L^2 + S^2 + 2\mathbf{L} \cdot \mathbf{S}$ so that $\mathbf{L} \cdot \mathbf{S} = (J^2 - L^2 - S^2)/2$. Thus

$$\mathbf{L} \cdot \mathbf{S} |lsjm\rangle = \frac{1}{2}\left(j(j+1) - l(l+1) - \frac{3}{4}\right)|lsjm\rangle. \tag{15-46}$$

For $j = l + 1/2$ we have

$$\mathbf{L} \cdot \mathbf{S} |lsjm\rangle = \frac{l}{2} |lsjm\rangle \tag{15-47}$$

and for $j = l - 1/2$ we have

$$\mathbf{L} \cdot \mathbf{S} |lsjm\rangle = \frac{-l-1}{2} |lsjm\rangle. \tag{15-48}$$

The eigenvalue is positive for $\mathbf{L}$ parallel to $\mathbf{S}$ and negative for $\mathbf{L}$ antiparallel to $\mathbf{S}$.

SCATTERING OF SPIN $1/2$ PARTICLES WITH SPINLESS PARTICLES

An important application of the addition of a spin $1/2$ and an orbital angular momentum is in the theory of the scattering of a spinless particle by a spin $1/2$ particle, for example, the scattering of a π meson, which has spin zero, by a proton. The orbital angular momentum is the relative angular momentum of the proton and pion, and the spin $1/2$ is that of the proton.

It is most convenient from the point of view of making measurements to describe the proton-pion system by eigenstates $|l, s, m_l, m_s\rangle$

of L^2, $S^2 = {}^3/_4$, L_z and S_z. For example, one can polarize the proton spin before the scattering. Also, a pion beam aimed at a target of protons has no component of orbital angular momentum about the direction of the beam.

Now the interaction between the pion and nucleon is rotationally invariant; mathematically this means that the operator H_{int} for this interaction commutes with the total angular momentum $\mathbf{J} = \mathbf{L} + \mathbf{S}$, i.e.,

$$[H_{int}, \mathbf{J}] = 0. \tag{15-49}$$

As a consequence the *total* angular momentum of the pion-nucleon system is conserved in the scattering. However, the z components of $\mathbf{L}$ and $\mathbf{S}$ are not individually conserved; for instance, the spin of the proton can flip in the scattering, transferring its angular momentum to the orbital motion. Thus if the system is initially in a state $|l\,sjm\rangle$ the final state after the scattering must also be of the form $|l'sjm\rangle$. [We will see in a moment that for this system l' must equal l.]

To see this, look for example at the scattering amplitude in the Born approximation for scattering from $|l\,sjm\rangle$ to $|l'sj'm'\rangle$; this amplitude is proportional to the matrix element

$$\langle l'sj'm'|H_{int}|l\,sjm\rangle .$$

Now since $[H_{int}, J_z] = 0$ we have

$$\langle l'sj'm'|[H_{int}, J_z]|l\,sjm\rangle = (m - m')\langle l'sj'm'|H_{int}|l\,sjm\rangle = 0.$$

Thus $\langle l'sj'm'|H_{int}|l\,sjm\rangle$ vanishes unless $m' = m$. Similarly, since $[H_{int}, J^2] = 0$ we conclude that the matrix element also vanishes unless $j' = j$. Thus if the initial state is $|l\,sjm\rangle$ the final state must be $|l'sjm\rangle$. This argument is readily generalized to the exact scattering amplitude; the total angular momentum is conserved in the scattering, since the interaction connects only states with the same j and m.

The possible j values for a given l are $j = l \pm {}^1/_2$. Conversely, if we specify j, the possible l values are $j \pm {}^1/_2$. If the initial orbital angular momentum is l and the final is l', then since $l' = j \pm {}^1/_2$, the possible l' values are $l - 1$, l, and $l + 1$. On the other hand the strong interaction forces between the proton and pion are invariant under reflection, that is,

$$[H_{int}, P] = 0 \tag{15-50}$$

where P is the parity operation. Thus the parity of the state $|l\,sjm\rangle$ before scattering must be the same as the parity of the state $|l'sjm\rangle$ after scattering. But the parity of a state with orbital angular momentum[1] l is $(-1)^l$. Thus to conserve parity[2] $(-1)^l = (-1)^{l'}$ or

$$l' - l = 0, \pm 2, \pm 4, \ldots . \tag{15-51}$$

Thus the only possibility consistent with conservation of parity and conservation of total angular momentum is $l' = l$. [This argument doesn't work if, instead of a spin $^1/_2$, we had a spin one. Then it would be possible to change l in the scattering by two.]

Our conclusion then is that the only nonvanishing matrix elements of the interaction are $\langle l\text{sjm}|H_{\text{int}}|l\,\text{sjm}\rangle$. We can go one step further and notice that because of the rotational invariance of H_{int}, this matrix element can't depend on the orientation in space of the total angular momentum, that is, it can't depend on m. The scattering amplitude to all orders in H_{int} only depends on l and j. We write

$$\langle l\text{sjm}|H_{\text{int}}|l\text{sjm}\rangle \equiv B(l, j). \tag{15-52}$$

Suppose then that we want to calculate the amplitude for scattering from a state $|l\,\text{sm}_l\text{m}_s\rangle$ to a state $|l'\text{sm}_l'\text{m}_s'\rangle$. In the Born approximation, this will involve the matrix element $\langle l'\text{sm}_l'\text{m}_s'|H_{\text{int}}|l\,\text{sm}_l\,\text{m}_s\rangle$, which, using the completeness of the states $|l\,\text{sjm}\rangle$ we can write as

$$\langle l'\text{sm}_l'\text{m}_s'|H_{\text{int}}|l\text{sm}_l\text{m}_s\rangle$$

$$= \sum_{\substack{jj' \\ mm'}} \langle l'\text{sm}_l'\text{m}_s'|l'\text{sj}'\text{m}'\rangle\langle l'\text{sj}'\text{m}'|H_{\text{int}}|l\text{sjm}\rangle\langle l\text{sjm}|l\text{sm}_l\text{m}_s\rangle \tag{15-53}$$

$$= \delta_{ll'}\delta_{\text{m}_l+\text{m}_s,\text{m}_l'+\text{m}_s'} \sum_{j=l\pm 1/2} \langle l\text{sm}_{l'}\text{m}_{s'}|l\text{sjm}\rangle\langle l\text{sm}_l\text{m}_s|l\text{sjm}\rangle B(lj).$$

Thus for a given l all the various matrix elements, $\langle l\text{sm}_l'\text{m}_s'|H_{\text{int}}|l\text{sm}_l\text{m}_j\rangle$, depend only on two numbers $B(l,\ l \pm {}^1/_2)$ and, of course, lots of Clebsch-Gordan coefficients (which are all tabulated). One can interpret (15-53) as saying that the scattering can go through the l, $j = l + {}^1/_2$ *channel* or through the l, $j = l - {}^1/_2$ channel. $\langle l\text{sm}_l\text{m}_s|l\,\text{sjm}\rangle$ is the *amplitude for* the state $|l\text{sm}_l\text{m}_s\rangle$ to have total angular momentum

[1] Note that spin, an angular momentum, is untransformed by the parity operation: $P^{-1}SP = S$.

[2] The intrinsic parity of the pion is -1; this means that P applied to a state of the pion with zero momentum changes the sign of the state. The conservation of parity equation should then read $(-1)(-)^l = (-1)(-1)^{l'}$; the result is still (15-51). The intrinsic parity of the proton can similarly be ignored.

quantum numbers j and m; $B(l, j)$ is the amplitude for scattering in the l, j channel, and $\langle l s m_l' m_s' | l s j m\rangle$ is the amplitude for the scattered state $|l s j m\rangle$ to have $L_z = m_l'$ and $S_z = m_s'$. The exact scattering amplitude also has the form (15-53); it can be written in terms of exact amplitudes for the various l, j channels and the same Clebsch-Gordan coefficients. For low-energy pion-proton scattering, only the scattering amplitudes for the first few l are nonnegligible.

THREE ANGULAR MOMENTA

Next let us look at the problem of adding *three* angular momenta $\mathbf{J}_1$, $\mathbf{J}_2$, and $\mathbf{J}_3$. The question is, how do we construct eigenstates of $J^2 = (\mathbf{J}_1 + \mathbf{J}_2 + \mathbf{J}_3)^2$ and $J_z = J_{1z} + J_{2z} + J_{3z}$ in terms of the eigenstates $|j_1 j_2 j_3 m_1 m_2 m_3\rangle$ of J_1^2, J_2^2, J_3^2, J_{1z}, J_{2z}, and J_{3z}? One natural way to go about solving this problem is first to add $\mathbf{J}_1$ to $\mathbf{J}_2$, letting $\mathbf{J}_{12} = \mathbf{J}_1 + \mathbf{J}_2$, and then construct the eigenstates of J_1^2, J_2^2, J_{12}^2, J_{12z}, J_3^2, J_{3z}. We write such a state as $|j_1 j_2 j_{12} j_3 m_{12} m_3\rangle$. Such a state is given in terms of Clebsch-Gordan coefficients and $|j_1 j_2 j_3 m_1 m_2 m_3\rangle$ by

$$|j_1 j_2 j_{12} j_3 m_{12} m_3\rangle = \sum_{m_1, m_2} |j_1 j_2 j_3 m_1 m_2 m_3\rangle \langle j_1 j_2 m_1 m_2 | j_1 j_2 j_{12} m_{12}\rangle, \tag{15-54}$$

since the addition of $\mathbf{J}_1$ and $\mathbf{J}_2$ doesn't depend on j_3 and m_3. Next we add $\mathbf{J}_{12}$ and $\mathbf{J}_3$ together, forming an eigenstate $|(j_1 j_2) j_{12} j_3;\, j, m\rangle$ of J_1^2, J_2^2, J_{12}^2, J_3^2, J^2 and J_z. These are given in terms of the $|j_1 j_2 j_{12} j_3 m_{12} m_3\rangle$ by

$$|(j_1 j_2) j_{12} j_2;\, jm\rangle = \sum_{m_{12}, m_3} |j_1 j_2 j_{12} j_3 m_{12} m_3\rangle \langle j_{12} j_3 m_{12} m_3 | j_{12} j_3 j m\rangle. \tag{15-55}$$

Using (15-55) this becomes

$$|(j_1 j_2) j_{12} j_3;\, jm\rangle \tag{15-56}$$

$$= \sum_{m_1 m_2 m_3} |j_1 j_2 j_3, m_1 m_2 m_3\rangle \sum_{m_{12}} \langle j_1 j_2 m_1 m_2 | j_1 j_2 j_{12} m_{12}\rangle \langle j_{12} j_3 m_{12} m_3 | j_{12} j_3 j m\rangle.$$

Thus the analogous addition coefficients for three angular momenta are products of Clebsch-Gordan coefficients.

Choosing to add $\mathbf{J}_1$ and $\mathbf{J}_2$ first was not a unique choice. We could equally as well have added $\mathbf{J}_2$ to $\mathbf{J}_3$ or $\mathbf{J}_1$ to $\mathbf{J}_3$, but we would have ended up with states different than (15-56). For example if we added

$\mathbf{J}_2$ to $\mathbf{J}_3$ we would have ended up with states

$$|j_1(j_2j_3)j_{23}; jm\rangle$$

$$= \sum_{m_1m_2m_3} |j_1j_2j_3m_1m_2m_3\rangle \langle j_2j_3m_2m_3|j_2j_3j_{23}m_{23}\rangle \langle j_1j_{23}m_1m_{23}|j_1j_{23}jm\rangle . \qquad (15\text{-}57)$$

One can then ask, how does one transform the basis of states of the form (15-56) to those of the form (15-57)? To do the transformation we need the coefficients $\langle(j_1j_2)j_{12}j_3;\ jm|j_1(j_2j_3)j_{23};\ jm\rangle$. As we see by taking the scalar product of (15-36) with (15-57), these coefficients (which turn out to be independent of m) involve sums of products of four Clebsch-Gordan coefficients. They are exceedingly unpleasant to calculate, but they are tabulated; with minor modifications they are called Racah W coefficients or Wigner 6j symbols.

A simple example of the addition of three angular momenta is in a proton-neutron system. The proton and neutron each have a spin $^1/_2$ and the third angular momentum is their relative orbital angular momentum, $\mathbf{L}$. Let $\mathbf{S}_1$ be the proton spin and $\mathbf{S}_2$ the neutron spin. To construct eigenstates for the total angular momentum $\mathbf{J} = \mathbf{S}_1 + \mathbf{S}_2 + \mathbf{L}$, it is most convenient first to add the two spins together. This gives three triplet states with $s = 1$, and a singlet state with $s = 0$. Then one combines the total spin states with the orbital angular momentum states. If $s = 0$, then clearly $j = l$. For $s = 1$, $l = 0$, j is simply 1, while for $s = 1$, $l \geq 1$, j can take on the values l, $l \pm 1$. One denotes such states by the notation $^{2s+1}l_j$. Thus $^3P_2$ is the triplet state with $l = 1$ and $j = 2$.

PROBLEMS

1. Calculate all the Clebsch-Gordan coefficients for $j_1 = 1$, $j_2 = 1$.
2. Show that the wave function of a $j = 1$ state formed from two (spinless) particles in orbital P states is antisymmetric in the coordinates of the two particles.

Chapter 16
ISOTOPIC SPIN

The theory of the addition of quantum mechanical angular momenta has found an interesting application in the study of strongly interacting particles. Basically the proton and neutron are very similar particles. Their masses are nearly equal, they each have spin $^1/_2$, and when interacting with themselves or other particles via strong interactions (or nuclear forces) they behave very similarly. Of course, the one striking difference between the neutron and proton is that the proton has a charge whereas the neutron has none. This means that they have very different electromagnetic interactions; e.g., an electron will be attracted to a proton but will feel rather indifferent about a neutron. However, the strong forces are several orders of magnitude stronger, at small distances, than electromagnetic forces, and in studying such forces, one can to a first approximation neglect the electromagnetic forces. Thus from the point of view of strong interactions, the neutron and the proton are practically identical.

In fact, one can say that the neutron and proton *are* the same particle — the *nucleon*, N — which has an internal degree of freedom which can take on two possible values — protonliness or neutronliness. This is analogous to our thinking of a spin up electron and a spin down electron as being the same kind of particle — the electron — in different states of the internal degree of freedom — spin. By analogy, one calls the internal degree of freedom of the nucleon *isotopic spin*, or isospin for short. The "isospin up" state of the nucleon is the proton and the "isospin down" state is the neutron. Let us call the operator analogous to the z component of the spin $I_3^{(N)}$. Then the proton is the nucleon in an eigenstate of I_3 with eigenvalue $^1/_2$ and the neutron is the nucleon in an eigenstate of I_3 with eigenvalue[1] $-^1/_2$. If we denote these states of the nucleon by $|p\rangle$ and

[1] Warning: Nuclear physicists occasionally use the opposite convention: $I_z = +½$ for the neutron, $-½$ for the proton.

$|n\rangle$ then (suppressing mention of the space and spin degrees of freedom)

$$I_3^{(N)}|p\rangle = \frac{1}{2}|p\rangle \tag{16-1}$$

$$I_3^{(N)}|n\rangle = -\frac{1}{2}|n\rangle . \tag{16-2}$$

Let us also introduce the operators

$$I_\pm^{(N)} = I_1^{(N)} \pm iI_2^{(N)}, \tag{16-3}$$

analogous to $S_x \pm iS_y$, that raise and lower the I_3 eigenvalue:

$$I_-^{(N)}|p\rangle = |n\rangle , \qquad I_-^{(N)}|n\rangle = 0 \tag{16-4}$$

$$I_+^{(N)}|n\rangle = |p\rangle , \qquad I_+^{(N)}|p\rangle = 0. \tag{16-5}$$

If, for example, $|p\rangle$ is a state in which there is a proton with the wave function $\psi_{s_z'}$ [s_z' is the S_z eigenvalue], then the first equation of (16-4) means that $I_-^{(N)}|p\rangle$ is a state in which there is definitely a neutron with the same space and spin wave function $\psi_{s_z'}(\mathbf{r}, t)$. Because $I_1(N)$, $I_2(N)$, and $I_3(N)$ act completely analogously to S_x, S_y, and S_z, they must also obey the angular momentum commutation relations

$$[I_i^{(N)}, I_j^{(N)}] = i\varepsilon_{ijk} I_k^{(N)} \quad (i, j, k = 1, 2, 3). \tag{16-6}$$

The isotopic spin operator $\mathbf{I}^{(N)} = (I_1^{(N)}, I_2^{(N)}, I_3^{(N)})$ has all the formal properties of a spin $^1/_2$ angular momentum operator. The nucleon is also an eigenstate of $[\mathbf{I}^{(N)}]^2 = [I_1^{(N)}]^2 + [I_2^{(N)}]^2 + [I_3^{(N)}]^2$ with total isotopic spin eigenvalue $I = {}^1/_2$:

$$\left[\mathbf{I}^{(N)}\right]^2 |p\rangle = \frac{1}{2}\left(\frac{1}{2}+1\right)|p\rangle .$$

$$\left[\mathbf{I}^{(N)}\right]^2 |n\rangle = \frac{1}{2}\left(\frac{1}{2}+1\right)|n\rangle . \tag{16-7}$$

[One should remember that isotopic spin is only formally like an angular momentum; it doesn't correspond to anything actually going round and round in real space.]

Notice that the operator

$$Q_N = I_3^{(N)} + \frac{1}{2}$$

is the operator for the charge of the nucleon (measured in units of e):

$$Q_N|p\rangle = |p\rangle, \quad Q_N|n\rangle = 0 \tag{16-8}$$

The eigenstates of $I_1^{(N)}$ are the states

$$\frac{1}{\sqrt{2}}(|p\rangle \pm |n\rangle) \tag{16-9}$$

in which there is amplitude $1/\sqrt{2}$ for the particle to be a proton and amplitude $\pm 1/\sqrt{2}$ for it to be a neutron. This is exactly analogous to the situation we studied in Chapter 2 on neutral K mesons. There we had states in which there were various amplitudes for the particle in question to be a K^0 or a $\overline{K}^0$.

One can also play the same game with the three different π mesons, π^+, π^0, π^-. These three particles, aside from their different charges, have very similar properties. One can say that they are all the same particle — the π meson — which has an internal degree of freedom which can take on *three* possible values, corresponding to the three different charges of pions. This internal degree of freedom is analogous to an angular momentum with total angular momentum eigenvalue j = 1. The operator analogous to the z component of angular momentum we call $I_3^{(\pi)}$. Then denoting the three different states of the π by $|\pi^+\rangle$, $|\pi^0\rangle$, and $|\pi^-\rangle$ we have

$$I_3^{(\pi)}|\pi^+\rangle = |\pi^+\rangle$$

$$I_3^{(\pi)}|\pi^0\rangle = 0$$

$$I_3^{(\pi)}|\pi^-\rangle = -|\pi^-\rangle. \tag{16-10}$$

The operator $I_3^{(\pi)}$ is just the operator Q_π for the charge of the pion.

Let us introduce the other two components $I_1^{(\pi)}$, $I_2^{(\pi)}$ of the pion isotopic spin operator by saying that the combinations

$$I_\pm^{(\pi)} = I_1^{(\pi)} \pm iI_2^{(\pi)} \tag{16-11}$$

act as raising and lowering operators for the eigenvalue of $I_3^{(\pi)}$ exactly as $J_\pm$ does for real angular momentum. Thus, for example,

$$I_\pm^{(\pi)}|\pi^0\rangle = \sqrt{2}\,|\pi^\pm\rangle,$$

the factor $\sqrt{2}$ being analogous to $\sqrt{j(j+1) - m(m+1)}$ for j = 1, m = 0. The operators $\mathbf{I}^{(\pi)} = (I_1^{(\pi)}, I_2^{(\pi)}, I_3^{(\pi)})$ act on the states exactly as $\mathbf{J}$ acts on states with j = 1. Thus the pion isotopic spin operators obey the angular momentum commutation relations

$$[I_i^{(\pi)}, I_j^{(\pi)}] = i\varepsilon_{ijk} I_k^{(\pi)}. \tag{16-12}$$

The pion is a particle with total isotopic spin quantum number I = 1:

$$[\mathbf{I}^{(\pi)}]^2 |\pi\rangle = I(I+1)|\pi\rangle = 2|\pi\rangle \tag{16-13}$$

Before showing why all this isotopic spin formalism is useful, let's work out a small mathematical exercise. Suppose that we have a state with both a pion and a nucleon: a π^0 and a proton, just to be definite, and we represent the state of the two particles as $|\pi^0 p\rangle$. This state is an eigenstate of $[\mathbf{I}^{(\pi)}]^2$ with eigenvalue I = 1, an eigenstate of $I_3^{(\pi)}$ with eigenvalue $I_3^{(\pi)} = 0$, an eigenstate of $[\mathbf{I}^{(N)}]^2$ with eigenvalue $I = \frac{1}{2}$, and an eigenstate of $I_3^{(N)}$ with eigenvalue $I_3^{(N)} = \frac{1}{2}$. It is also an eigenstate of the total I_3, where

$$\mathbf{I} = \mathbf{I}^{(\pi)} + \mathbf{I}^{(N)} \tag{16-14}$$

with eigenvalue $\frac{1}{2}$. However, it is not, as one can easily verify, an eigenstate of the total isotopic spin

$$I^2 = \left(\mathbf{I}^{(\pi)} + \mathbf{I}^{(N)}\right)^2. \tag{16-15}$$

The question is, how can we construct states of a pion and a nucleon that are eigenstates of I^2? Now $\mathbf{I}^{(\pi)}$ and $\mathbf{I}^{(N)}$ have all the formal properties of angular momenta, and therefore we go about answering this question in the exact same way we did for real angular momenta. It is just the problem of adding a spin $\frac{1}{2}$ to an angular momentum with j = 1. Let us denote the possible states of a pion and a nucleon by their $I_3^{(\pi)}$ and $I_3^{(N)}$ eigenvalues, viz., $|I_3^{(\pi)}, I_3^{(N)}\rangle$. Thus for example $|\pi^+, n\rangle = |I_3^{(\pi)} = 1, I_3^{(N)} = -\frac{1}{2}\rangle$. The eigenstates of total I^2 and total I_3 we denote by $|I, I_3\rangle$. [The I_3 value is the total charge plus $\frac{1}{2}$.] Then these states are linear combinations of the $|I_3^{(\pi)}, I_3^{(N)}\rangle$ states. Because isotopic spin has all the mathematical properties of angular momentum the linear combination is given by Eq. (15-26):

$$|I, I_3\rangle = \sum_{I_3^{(\pi)}, I_3^{(N)}} |I_3^{(\pi)}, I_3^{(N)}\rangle \langle 1, \frac{1}{2}, I_3^{(\pi)}, I_3^{(N)} | 1, \frac{1}{2}, I, I_3\rangle, \tag{16-16}$$

where $\langle 1, \frac{1}{2}, I_3^{(\pi)}, I_3^{(N)} | 1, \frac{1}{2}, I, I_3\rangle$ is a Clebsch-Gordan coefficient! This coefficient is given by Eqs. (15-44) and (15-45). The possible I values are $\frac{3}{2}$ and $\frac{1}{2}$, and

$$\langle 1, \tfrac{1}{2}, I_3 \mp \tfrac{1}{2}, \pm \tfrac{1}{2} | 1, \tfrac{1}{2}, I = \tfrac{3}{2}, I_3 \rangle = \sqrt{\frac{1}{2} \pm \frac{I_3}{3}}$$

$$\langle 1, \tfrac{1}{2}, I_3 \mp \tfrac{1}{2}, \pm \tfrac{1}{2} | 1, \tfrac{1}{2}, I = \tfrac{1}{2}, I_3 \rangle = \mp\sqrt{\frac{1}{2} \mp \frac{I_3}{3}}. \tag{16-17}$$

Working out (16-16) explicitly we find that the $|I, I_3\rangle$ states of a nucleon and a pion are given for $I = {}^3/_2$ by

$$\left|\tfrac{3}{2}, \tfrac{3}{2}\right\rangle = \left|\pi^+ p\right\rangle$$

$$\left|\tfrac{3}{2}, \tfrac{1}{2}\right\rangle = \sqrt{\tfrac{1}{3}}\left|\pi^+ n\right\rangle + \sqrt{\tfrac{2}{3}}\left|\pi^0 p\right\rangle$$

$$\left|\tfrac{3}{2}, -\tfrac{1}{2}\right\rangle = \sqrt{\tfrac{1}{3}}\left|\pi^- p\right\rangle + \sqrt{\tfrac{2}{3}}\left|\pi^0 n\right\rangle$$

$$\left|\tfrac{3}{2}, -\tfrac{3}{2}\right\rangle = \left|\pi^- n\right\rangle \tag{16-18}$$

and for $I = {}^1/_2$ by

$$\left|\tfrac{1}{2}, \tfrac{1}{2}\right\rangle = \sqrt{\tfrac{2}{3}}\left|\pi^+ n\right\rangle - \sqrt{\tfrac{1}{3}}\left|\pi^0 p\right\rangle$$

$$\left|\tfrac{1}{2}, -\tfrac{1}{2}\right\rangle = \sqrt{\tfrac{1}{3}}\left|\pi^0 n\right\rangle - \sqrt{\tfrac{2}{3}}\left|\pi^- p\right\rangle. \tag{16-19}$$

Physically if the pion and nucleon are in a state $|{}^1/_2, {}^1/_2\rangle$, for example, then we don't know whether we have a π^+ or π^0 or a neutron or proton. The probability *amplitude* for having a π^+ and a neutron is $\sqrt{2/3}$ while the amplitude for having a π^0 and a proton is $-\sqrt{1/3}$; all other amplitudes, such as that for π^+ *and* proton, are zero.

Similarly, we can invert the linear combinations (16-18) and (16-19) to write the initial pion nucleon states as linear combinations of eigenstates of I^2:

$$\left|\pi^+ p\right\rangle = \left|\tfrac{3}{2}, \tfrac{3}{2}\right\rangle$$

$$\left|\pi^0 p\right\rangle = \sqrt{\tfrac{2}{3}}\left|\tfrac{3}{2}, \tfrac{1}{2}\right\rangle - \sqrt{\tfrac{1}{3}}\left|\tfrac{1}{2}, \tfrac{1}{2}\right\rangle$$

$$\left|\pi^- p\right\rangle = \sqrt{\tfrac{1}{3}}\left|\tfrac{3}{2}, -\tfrac{1}{2}\right\rangle - \sqrt{\tfrac{2}{3}}\left|\tfrac{1}{2}, -\tfrac{1}{2}\right\rangle$$

$$\left|\pi^+ n\right\rangle = \sqrt{\frac{1}{3}}\left|\frac{3}{2},\frac{1}{2}\right\rangle + \sqrt{\frac{2}{3}}\left|\frac{1}{2},\frac{1}{2}\right\rangle$$

$$\left|\pi^0 n\right\rangle = \sqrt{\frac{2}{3}}\left|\frac{3}{2},-\frac{1}{2}\right\rangle + \sqrt{\frac{1}{3}}\left|\frac{1}{2},-\frac{1}{2}\right\rangle, \quad \left|\pi^- n\right\rangle = \left|\frac{3}{2},-\frac{3}{2}\right\rangle. \tag{16-20}$$

The reason all this formalism is useful is the experimental fact that the strong interaction forces acting between the pion and nucleon in a state with a definite total I and I_3 quantum numbers cannot change the I_3 or I values and furthermore are independent of the I_3 quantum number, that is, are independent of the total charge of the pion and nucleon. This is a general property of strong interactions called *charge independence:* the strong forces acting between any sets of particles in a state with definite total isotopic spin are independent of the charge of the state and conserve the total I and I_3 quantum numbers. It is equally valid for systems of two nucleons, for nuclei, for strongly interacting strange particles, etc.

To state the charge independence of strong interactions mathematically, let's consider $\langle I', I_3'|H_{int}|I'', I_3''\rangle$, the matrix element, between eigenstates of total I^2 and I_3, of the operator for the interaction between the particles. First of all, because charge is conserved in the interaction, I_3' must equal I_3''. Charge independence makes the stronger statements that the I value is conserved in the interaction, so that the matrix element vanishes unless $I' = I''$; furthermore the matrix element $\langle I', I_3'|H_{int}|I', I_3'\rangle$ is independent of I_3, the charge of the states. Thus the matrix element has the form

$$\langle I', I_3'|H_{int}|I'', I_3''\rangle = \delta_{I', I''}\delta_{I_3', I_3''} b_{I'} \tag{16-21}$$

where $b_{I'}$ depends on the space and spin degrees of freedom.

Equivalently, one can characterize charge independence by the statement that the isospin operator commutes with H_{int}:

$$[H_{int}, \mathbf{I}] = 0. \tag{16-22}$$

For example, from the conservation of charge we have

$$\langle I', I_3'|[H_{int}, I_3]|I'', I_3''\rangle = (I_3'' - I_3')\langle I', I_3'|H_{int}|I'', I_3''\rangle = 0,$$

because the matrix element of H_{int} vanishes unless $I_3' = I_3''$. Thus all matrix elements of $[H_{int}, I_3]$ vanish, so that $[H_{int}, I_3] = 0$. Similarly, conservation of total I means that $[H_{int}, \mathbf{I}^2] = 0$. The independence of the matrix element on the charge then implies $[H_{int}, I_\pm] = 0$, for from (16-21) we have

$$\langle I', I_3' | [H_{int}, I_\pm] | I'', I_3''\rangle = \langle I', I_3' | H_{int} | I'', I_3'' \pm 1\rangle \sqrt{I''(I''+1) - I_3''(I_3'' \pm 1)}$$

$$- \langle I', I_3' \mp 1 | H_{int} | I'', I_3''\rangle \sqrt{I'(I'+1) - I_3'(I_3' \mp 1)}$$

$$= \delta_{I', I''}\delta_{I_3', I_3'' \pm 1} b_{I'} \left(\sqrt{I''(I''+1) - I_3''(I_3'' \pm 1)} - \sqrt{I'(I'+1) - I_3'(I_3' \mp 1)}\right) = 0. \tag{16-23}$$

Thus all matrix elements of $[H_{int}, I_\pm]$ vanish and so

$$[H_{int}, I_\pm] = 0. \tag{16-24}$$

Charge independence thus means that the interaction must be an isotopic spin scalar. It can, for example, contain terms proportional, for the pion-nuclear interaction, to $\mathbf{I}^{(\pi)} \cdot \mathbf{I}^{(N)}$, but no terms like $I_1^{(\pi)} I_3^{(N)}$.

Let us see what predictions charge independence makes about pion-nucleon scattering. Consider the following interactions of protons and charged pions:

$$\pi^+ + p \rightarrow \pi^+ + p \tag{16-25a}$$

$$\pi^- + p \rightarrow \pi^- + p \tag{16-25b}$$

and the charge exchange interaction

$$\pi^- + p \rightarrow \pi^0 + n. \tag{16-25c}$$

Suppose that the proton is at rest initially and the pion energy is the same in these three reactions. Let us ask for the amplitude, A, for observing the final nucleon coming off in a given solid angle in each of the three reactions.

The initial state in (16-25a) is an $I = 3/2$, $I_3 = 3/2$ state. Because I is unchanged by the interaction, the final state must also be an $I = 3/2$, $I_3 = 3/2$ state, and the amplitude for the process is determined just by the forces for $I = 3/2$. Let us call this amplitude $a_{3/2}$. Thus for the reaction (16-25a) we have

$$A_a = a_{3/2}. \tag{16-26}$$

Now the initial state $\pi^- + p$ is a linear combination of $I = 3/2$, $I_3 = -1/2$ and $I = 1/2$, $I_3 = -1/2$ states. The $I = 3/2$ component will undergo scattering with the scattering amplitude $a_{3/2}$ determined by the forces in the $I = 3/2$ state (it is the same $a_{3/2}$ as before because the scattering doesn't depend on I_3); the scattering of the $I = 1/2$ component will be determined by the forces in the $I = 1/2$ state, which will have an amplitude $a_{1/2}$. From (16-20) we see that before the interaction the state is

$$|\pi^- p\rangle = \sqrt{\frac{1}{3}}\left|\frac{3}{2}, -\frac{1}{2}\right\rangle - \sqrt{\frac{2}{3}}\left|\frac{1}{2}, -\frac{1}{2}\right\rangle.$$

The amplitude for the initial state to be $|3/2, -1/2\rangle$ is $\sqrt{1/3}$ and $a_{3/2}$ is amplitude for the $|3/2, -1/2\rangle$ to undergo the given scattering; thus we have an amplitude $\sqrt{1/3}\,a_{3/2}$ for the final state to be an $I = 3/2$ state and an amplitude $-\sqrt{2/3}\,a_{1/2}$ for the final state to be an $I = 1/2$ state. Now what is the amplitude for observing the final nucleon to be a proton, as in (16-25b). The final $I = 3/2$, $I_3 = -1/2$ state has, from (16-18), an amplitude $\sqrt{1/3}$ for being $\pi^- + p$; thus, from scattering of $\pi^- + p$ in the $I = 3/2$ channel we have a net amplitude of $\sqrt{1/3}\,a_{3/2}\,\sqrt{1/3}$ for the final state to be $\pi^- + p$. Similarly, from (16-19) the final $I = 1/2$, $I_3 = -1/2$ state has an amplitude $-\sqrt{2/3}$ for being a $\pi^- + p$. The $I = 1/2$ channel contributes a net amplitude $(-\sqrt{2/3}\,a_{1/2})(-2/3)$ to the process (16-25b). Adding together the amplitudes for both channels we have the net amplitude for (16-25b):

$$A_b = \frac{1}{3}a_{3/2} + \frac{2}{3}a_{1/2}. \tag{16-27}$$

The $I = 3/2$, $I_3 = -1/2$ final state also has an amplitude $\sqrt{2/3}$ for being a $\pi^0 + n$ state, while the $I = 1/2$, $I_3 = -1/2$ final state has an amplitude $\sqrt{1/3}$ for being a $\pi^0 + n$ state. Thus if we start with $\pi^- + p$ the net amplitude for observing a $\pi^0 + n$ after the scattering is

$$A_c = \left(\sqrt{\frac{1}{3}}\,a_{3/2}\right)\sqrt{\frac{2}{3}} + \left(-\sqrt{\frac{2}{3}}\,a_{1/2}\right)\sqrt{\frac{1}{3}} = \frac{\sqrt{2}}{3}(a_{3/2} - a_{1/2}). \tag{16-28}$$

The probabilities of the processes (16-25), are given by the square of their respective amplitudes. Thus the differential cross sections for these processes are in the ratio

$$\sigma_a : \sigma_b : \sigma_c = |a_{3/2}|^2 : \frac{1}{9}|a_{3/2} + 2a_{1/2}|^2 : \frac{2}{9}|a_{3/2} - a_{1/2}|^2. \tag{16-29}$$

It turns out that at 154 MeV center of mass energy the scattering takes place predominantly in the $I = 3/2$, $l = 1$, $j = 3/2$ channel. Then we can neglect $a_{1/2}$ compared with $a_{3/2}$, and we find from (16-29) that

$$\sigma_a : \sigma_b : \sigma_c = 9 : 1 : 2. \tag{16-30}$$

The charge independence of nuclear forces thus leads to simple, but well confirmed, relations for the relative strengths of the three seemingly different reactions (16-25). One can similarly write down the amplitude for any reaction of the form

$$\pi + N \rightarrow \pi + N \qquad (16\text{-}31)$$

in terms of just the two amplitudes $a_{1/2}$ and $a_{3/2}$. For example, at 154 MeV the cross section for $\pi^+ + n \rightarrow \pi^+ + n$ would be $^1/_9$ of the cross section for $\pi^+ + p \rightarrow \pi^+ + p$, while the cross section for the charge exchange reaction $\pi^+ + n \rightarrow \pi^0 + p$ would be $^2/_9$ of that for $\pi^+ + p \rightarrow \pi^+ + p$.

Thus by introducing the internal degree of freedom, isotopic spin, and assuming that the interaction is symmetric under rotations in isotopic spin space, i.e., it commutes with the total isotopic spin, **I**, we are led to many relations among all the possible pion-nucleon scatterings. At present people are studying fancier internal degrees of freedom and corresponding symmetries of the strong interactions among elementary particles, such as the SU(3) symmetries, in order to make some sense out of the vast array of strongly interacting elementary particles.[2]

PROBLEMS

1. Assume to a first approximation in pion-nucleon scattering, at 154 MeV center of mass kinetic energy, that all the scattering is through the $l = 1$, $j = {}^3/_2$, $I = {}^3/_2$ channel. Suppose that one sends a beam of π^- in the positive y direction in the laboratory at a target of protons at rest whose spins are all pointed in the z direction. Assume that the center of mass energy is 154 MeV.
   (a) Calculate the relative total number of scattered protons observed per solid angle in the lab frame. Also calculate the polarization, $\langle \sigma \rangle$, of the scattered protons as a function of direction.
   (b) Repeat part (a) for the scattered neutrons arising from the charge exchange reaction.
2. The deuteron d is an isospin singlet. Calculate the ratio of the scattering cross sections for the reactions $p + p \rightarrow d + \pi^+$ and $p + n \rightarrow d + \pi^0$, assuming charge independence of the interactions.
3. The Λ^0 particle has spin $^1/_2$, isotopic spin 0 and charge 0. It is unstable and can decay into a pion and a nucleon: $\Lambda^0 \rightarrow \pi + N$.

---

[2]See, for example, M. Gell-Mann and Y. Ne'eman, *The Eightfold Way* [W.A. Benjamin, New York, 1964].

Charge and angular momentum are conserved in the decay, but the total isotopic spin quantum number I changes by $1/2$. Neglect all relativistic effects.

(a) What is the ratio of protons to neutrons observed in the decay?

(b) Consider $\Lambda^0 \rightarrow \pi^- + p$. The pion has spin zero. What are the possible values of the relative orbital angular momentum of the π^- and p?

(c) Let $\alpha_l(q)$ be the matrix element of the interaction responsible for the decay, between the initial Λ^0 state and the final $\pi^- p$ orbital state that is an eigenstate of the total angular momentum and in which the magnitude of the relative $\pi^- p$ momentum is q and the relative $\pi^- p$ orbital angular momentum is l. Assume that the decay process is accurately described by lowest-order perturbation theory. Assume also that the spin of the Λ^0 is in the z direction in the frame in which the Λ^0 is at rest. Write down in this frame the normalized angular and spin part of the wave function of the final proton in terms of the α_l. What does the radial wave function of the $\pi^- p$ system look like at large relative separation?

(d) Calculate in terms of the α_l, the angular distribution of the final protons in the center of mass frame.

(e) Calculate as a function of angles the expectation value of the component of the spin of the proton along the direction of motion of the proton.

(f) If one happens to study only the decay of completely unpolarized Λ^0's, what measurements would tell one the parameters needed in (d) and (e)?

(g) Repeat parts (b)-(f) for the decay $\Lambda^0 \rightarrow \pi^0 + n$.

(h) Calculate the total lifetime of the Λ^0 in terms of the α_l and the energy Q liberated in the decay.

4. Pion-nucleon scattering at low energies can be qualitatively described by an effective interaction potential of the form

$$V = \frac{f^2}{4\pi} I^{(\pi)} \cdot I^{(N)} \frac{e^{-\mu r}}{r}$$

[due to the exchange of a ρ meson (mass μ)]; f is a constant, and r is the relative pion-nucleon coordinate.

(a) Calculate the ratio of the $I = 3/2$ and $I = 1/2$ scattering amplitudes.

(b) Calculate in the Born approximation, the low-energy s wave cross sections for the three reactions (16-25).

5. The K^- meson and the $\overline{K^0}$ are an isospin $1/2$ doublet, the Λ^0 an isospin singlet, and the three Σ particles, Σ^+, Σ^0, Σ^- form an

I = 1 triplet. Calculate the ratios of the cross sections for the four strong reactions

$$K^- + p \rightarrow \pi^0 + \Lambda^0,\ \pi^0 + \Sigma^0,\ \pi^{\pm} + \Sigma^{\mp},$$

assuming that the first reaction is described by the same amplitudes in states of total I as the latter three.

Chapter 17
ROTATIONS AND TENSOR OPERATORS

REPRESENTATIONS OF ROTATIONS

If **J** is the total angular momentum of a system, the operator

$$R_\alpha = e^{-i\mathbf{J}\cdot\boldsymbol{\alpha}}, \tag{17-1}$$

acting to the right on a state of the system, rotates it in a positive sense about the axis $\hat{\boldsymbol{\alpha}}$ by an angle $|\boldsymbol{\alpha}|$. Suppose that we act with R_α on an eigenstate $|jm\rangle$ of J^2 and J_z. Under the rotation, the state remains an eigenstate of J^2 with the same eigenvalue; a rotation can't change the total angular momentum of a system. Mathematically, we can write

$$[R_\alpha, J^2] = 0 \tag{17-2}$$

so that

$$J^2 R_\alpha |jm\rangle = R_\alpha J^2 |jm\rangle = j(j+1) R_\alpha |jm\rangle. \tag{17-3}$$

However under the rotation the state is generally no longer an eigenstate of J_z with eigenvalue m. Only rotations about the z axis will leave the z component of the angular momentum unchanged. Because the set of states $|j'm'\rangle$ form a complete set we can express the rotated state as a linear combination of the $|j'm'\rangle$; only $j' = j$ enters the sum because of (17-3). Thus

$$R_\alpha |jm\rangle = \sum_{m''=-j}^{j} |jm''\rangle d^{(j)}_{m''m}(\alpha). \tag{17-4}$$

Multiplying (17-4) on the left by a state $\langle jm'|$, and using the orthonormality of the angular momentum eigenstates we find

$$d^{(j)}_{m'm}(\alpha) = \langle jm'|e^{-iJ\cdot\alpha}|jm\rangle. \tag{17-5}$$

Thus we can associate with each rotation a 2j + 1 by 2j + 1-dimensional matrix $\mathbf{d}^{(j)}(\alpha)$ whose matrix elements are $d^{(j)}_{m'm}(\alpha)$. These matrix elements don't depend at all on the dynamics of the system; they are determined entirely by the properties of the angular momentum operator **J**. This shall become obvious as we go along.

The matrices $\mathbf{d}^{(j)}$ have a very important property. Two rotations performed in a row, say α followed by β, are equivalent to a single rotation, γ. Thus

$$R_\gamma = R_\beta R_\alpha. \tag{17-6}$$

Taking matrix elements of both sides of (17-6) and putting in a complete set of $|j'm'\rangle$ states between R_β and R_α we find

$$\langle jm|R_\gamma|jm'\rangle = \sum_{m''} \langle jm|R_\beta|jm''\rangle\langle jm''|R_\alpha|jm'\rangle, \tag{17-7}$$

or

$$d^{(j)}_{mm'}(\gamma) = \sum_{m''} d^{(j)}_{mm''}(\beta)d^{(j)}_{m''m'}(\alpha), \tag{17-8}$$

or equivalently

$$\mathbf{d}^{(j)}(\gamma) = \mathbf{d}^{(j)}(\beta)\mathbf{d}^{(j)}(\alpha). \tag{17-9}$$

This equation says that the matrix $\mathbf{d}^{(j)}(\gamma)$ associated with the product of the two rotations β and α, is just the product of the matrices associated with the individual rotations. A set of matrices associated with rotations having the property (17-9) is called a *representation* of the rotation group.[1]

The matrix that represents no rotation at all, $\alpha = 0$, has matrix elements $\langle jm'|jm\rangle = \delta_{m'm}$; it is just the 2j + 1-dimensional unit matrix. The matrices $\mathbf{d}^{(j)}$ are unitary, since

$$[\mathbf{d}^{(j)}(\alpha)^\dagger]_{m'm} = d^{(j)}_{mm'}(\alpha)^* = \langle jm'|e^{iJ\cdot\alpha}|jm\rangle = d^{(j)}_{m'm}(-\alpha). \tag{17-10}$$

Thus

[1] The sets with half-integer j are a double valued representation of the group of rotations. They are single valued representations of the larger group SU(2).

$$d^j(\alpha)^\dagger = d^j(-\alpha) \tag{17-11}$$

and from (17-9)

$$d^j(\alpha)^\dagger d^j(\alpha) = 1, \tag{17-12}$$

the unit matrix.

The rotation operators R_α act on the set of states $|jm\rangle$ for fixed j, in an *irreducible* fashion. To see what this means, let's consider the effect of rotations on the set of eight states for $j = 1$ and $j = 2$. Under any rotation a $j = 1$ state becomes a linear combination of $j = 1$ states, with no $j = 2$ components; conversely a $j = 2$ state becomes a linear combination of $j = 2$ states with no $j = 1$ components. Thus this set of eight states breaks up into two groups, the $j = 1$ states and the $j = 2$ states which transform each among themselves under rotation with no mixing. One says that the rotations act on these eight states *reducibly*. On the other hand, for a set of states all with the same j, there is no smaller subset of states that transforms privately among itself under all rotations; the rotations are said to act irreducibly. Put another way, if we start with any state, $|jm\rangle$, then we can rotate it $2j + 1$ different ways into $2j + 1$ linearly independent states, and therefore there can't be any subspace of j states that transforms among itself under rotations. One can prove this in detail starting from the fact that one can generate all the $|jm\rangle$ states starting from $|jj\rangle$ by applying J_- enough times, but we won't work it out here.

Because the rotations act on the $|jm\rangle$ states, for fixed j, irreducibly, the matrices $d^{(j)}$ have the property that there is *no* basis in which they *all* have the form

$$d^{(j)} = \begin{matrix} k \\ 2j+1-k \end{matrix}\left\{\left(\begin{array}{c|c} \overbrace{\blacksquare}^{k} & \overbrace{0}^{2j+1-k} \\ \hline 0 & \blacksquare \end{array}\right)\right. \tag{17-13}$$

for if there was such a basis, then clearly the k states belonging to the upper left-hand part of (17-13) would transform only among themselves under rotation and so would the $2j + 1 - k$ states connected with the lower right-hand corner of (17-13). Because there is no basis in which the matrices $d^{(j)}$ all have the form (17-13), one says that the representation, that is, the set of matrices $d^{(j)}(\alpha)$, is *irreducible.*

It is very convenient to specify rotations in terms of *Euler angles.* Euler noted that one could perform any arbitrary rotation in three simple steps. First one rotates about the z axis, Fig. 17-1, by angle

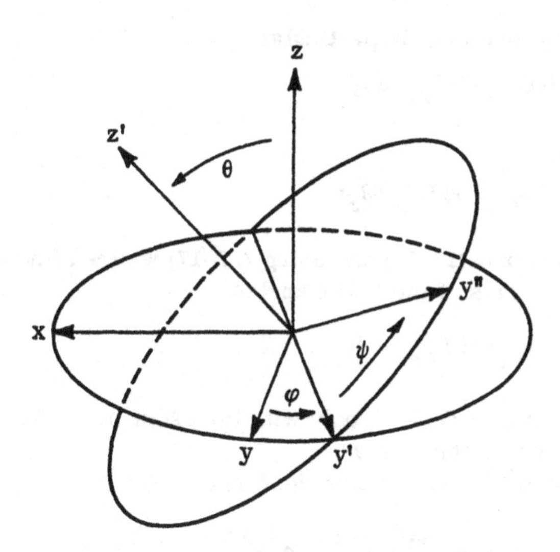

Fig. 17-1
The three Euler angles θ, φ, and ψ.

φ; then about the axis $\hat{y}'$ (which is the y axis rotated about **z** by φ) by angle θ, and then about the axis $\hat{z}'$ (which is the z axis rotated about $\hat{y}'$ by θ) by angle ψ. The operator that represents these three rotations is just

$$R_{\varphi,\theta,\psi} = e^{-i\psi J_{z'}} e^{-i\theta J_{y'}} e^{-i\varphi J_z} \tag{17-14}$$

where $J_{y'}$ and $J_{z'}$ are the components of the angular momentum along $\hat{y}'$ and $\hat{z}'$.

Now as we have seen [e.g., Eq. (14-18)], the component of **J** along $\hat{y}'$ is given in terms of the component of **J** along $\hat{y}$ by

$$J_{y'} = e^{-i\varphi J_z} J_y e^{i\varphi J_z}. \tag{17-15}$$

Thus

$$J_{y'}^2 = e^{-i\varphi J_z} J_y e^{i\varphi J_z} e^{-i\varphi J_z} J_y e^{i\varphi J_z} = e^{-i\varphi J_z} J_y^2 e^{i\varphi J_z},$$

and in general

$$f(J_{y'}) = e^{-i\varphi J_z} f(J_y) e^{i\varphi J_z}, \tag{17-16}$$

where f is any function. In particular

$$e^{-i\theta J_{y'}} = e^{-i\varphi J_z} e^{-i\theta J_y} e^{i\varphi J_z}. \tag{17-17}$$

Similarly

$$e^{-i\psi J_{z'}} = e^{-i\theta J_{y'}} e^{-i\psi J_z} e^{i\theta J_{y'}}. \tag{17-18}$$

Putting (17-18) into (17-14) and using (17-17) we then find the simple form for R in terms of the Euler angles

$$R_{\varphi,\theta,\psi} = e^{-i\varphi J_z} e^{-i\theta J_y} e^{-i\psi J_z}. \tag{17-19}$$

[This looks just like (17-14), only with the order of the operators reversed, and without the primes.]

The matrix $d^{(j)}(\varphi, \theta, \psi)$ takes the form

$$\begin{aligned} d^{(j)}_{mm'}(\varphi, \theta, \psi) &= \langle jm|e^{-i\varphi J_z} e^{-i\theta J_y} e^{-i\psi J_z}|jm'\rangle \\ &= e^{-im\varphi} e^{-im'\psi} \langle jm|e^{-i\theta J_y}|jm'\rangle = e^{-im\varphi} e^{-im'\psi} d^{(j)}_{mm'}(\theta). \end{aligned} \tag{17-20}$$

Thus to construct d we need only know the matrix for a rotation about the y axis by θ. The φ and ψ dependence is trivial. The matrix elements $d^{(j)}_{m,m'}(\theta)$ turn out to be real [see (17-138)], and to have the simple property

$$d^{(j)}_{-m,-m'}(\theta) = (-1)^{2j-m-m'} d^{(j)}_{m,m'}(\theta). \tag{17-21}$$

For the particular case of j being an integer and $m' = 0$ we have

$$d^{(j)}_{m,0}(\varphi, \theta, \psi) = \sqrt{\frac{4\pi}{2j+1}}\, Y_{jm}^*(\theta, \varphi). \tag{17-22}$$

We can see this from the following argument. First we note that the rotation $R_{\varphi,\theta,0}$ takes the original z axis in Fig. 17-1 into the axis whose spherical coordinates (in terms of the original x, y, z direction) are θ, φ. If we let the states refer to the orbital angular momentum of a particle and let the state $|\theta, \varphi\rangle$ be, as in Chapter 6, the basis state in which the coordinates of the particle are definitely θ and φ, we then have the simple result

$$R_{\varphi,\theta,0}|\theta = 0, \varphi = 0\rangle = |\theta, \varphi\rangle. \tag{17-23}$$

Thus

$$\sum_{m'} \langle jm|R_{\varphi,\theta,0}|jm'\rangle \langle jm'|\theta = 0, \varphi = 0\rangle = \langle jm|\theta, \varphi\rangle$$

or

$$\sum_{m'} d^{(j)}_{mm'}(\varphi, \theta, 0) Y_{jm'}{}^*(0, 0) = Y_{jm}{}^*(\theta, \varphi), \tag{17-24}$$

using (6-58). However the spherical harmonics evaluated at the north pole vanish unless $m' = 0$ in which case they equal $\sqrt{(2j+1)/4\pi}$:

$$Y_{jm'}(0, 0) = \delta_{m',0}\sqrt{\frac{2j+1}{4\pi}}. \tag{17-25}$$

This is trivial to show from (6-61). Thus (17-24) becomes

$$d^{(j)}_{m0}(\varphi, \theta, 0) = \sqrt{\frac{4\pi}{2j+1}}\, Y_{jm}{}^*(\theta, \varphi),$$

and (17-22) follows since from (17-20), $d^{(j)}_{m0}$ is independent of ψ.

The addition theorem for spherical harmonics is just a special case of the law for the transformation of spherical harmonics under rotation of their arguments. For integer j we find on multiplying Eq. (17-4) on the left by $\langle\theta, \varphi|$ that

$$Y_{jm}(\theta', \varphi') = \sum_{m'=-j}^{j} Y_{jm'}(\theta, \varphi) d^{(j)}_{m'm}(\alpha) \tag{17-26}$$

where θ', φ' is the direction θ, φ rotated by angle $-|\alpha|$ about $\hat{\alpha}$. For $m = 0$ we use (17-22); then the transformation law (17-26) becomes:

$$Y_{j0}(\theta', 0) = \sqrt{\frac{4\pi}{2j+1}} \sum_{m=-j}^{j} Y_{jm}(\theta, \varphi) Y_{jm}{}^*(\theta_\alpha, \varphi_\alpha) \tag{17-27}$$

where φ_α and θ_α are the first two Euler angles corresponding to the rotation α. [These angles are not the spherical coordinates of the vector $\hat{\alpha}$.] All that remains to establish (17-27) as the addition theorem is to show that θ' is the angle between the directions whose spherical coordinates are θ, φ and θ_α, φ_α; this is left as an exercise.

Consider now the eigenstates $|j_1j_2m_1m_2\rangle$, for fixed j_1 and j_2, of two angular momenta $\mathbf{J}_1$ and $\mathbf{J}_2$. To perform a rotation of these states we act with the operator $e^{-i\alpha\cdot\mathbf{J}}$ where $\mathbf{J} = \mathbf{J}_1 + \mathbf{J}_2$. The rotation operators do not act irreducibly on the states $|j_1j_2m_1m_2\rangle$, since we know that we can write the $|j_1j_2m_1m_2\rangle$ states as linear combinations of the eigenstates $|j_1j_2jm\rangle$ of J^2 and J_z, and that under a rotation, states with different j values don't mix. Thus the set of $(2j_1 + 1)(2j_2 + 1)$ states $|j_1j_2m_1m_2\rangle$ break up into groups of states for the different possible j values, and the states with given j transform only among themselves under rotation. How does $|j_1j_2m_1m_2\rangle$ transform under rotation? Since

$$e^{-i\mathbf{J}\cdot\boldsymbol{\alpha}} = e^{-i\mathbf{J}_1\cdot\boldsymbol{\alpha}}\, e^{-i\mathbf{J}_2\cdot\boldsymbol{\alpha}}$$

the $\mathbf{J}_1$ and $\mathbf{J}_2$ parts transform independently and from (17-4) we have

$$e^{-i\mathbf{J}\cdot\boldsymbol{\alpha}}|j_1j_2m_1m_2\rangle = \sum_{m_1'm_2'} |j_1j_2m_1'm_2'\rangle\, d^{(j_1)}_{m_1'm_1}(\boldsymbol{\alpha})\, d^{(j_2)}_{m_2'm_2}(\boldsymbol{\alpha}) \qquad (17\text{-}28)$$

or equivalently

$$\langle j_1j_2m_1'm_2'|e^{-i\mathbf{J}\cdot\boldsymbol{\alpha}}|j_1j_2m_1m_2\rangle = d^{(j_1)}_{m_1'm_1}(\boldsymbol{\alpha})\, d^{(j_2)}_{m_2'm_2}(\boldsymbol{\alpha}). \qquad (17\text{-}29)$$

We can regard the left side as being the $m_1'm_2'$, m_1m_2 matrix element of a $(2j_1 + 1)(2j_2 + 1)$ by $(2j_1 + 1)(2j_2 + 1)$-dimensional matrix, which one usually writes as $\mathbf{d}^{(j_1)}(\boldsymbol{\alpha}) \otimes \mathbf{d}^{(j_2)}(\boldsymbol{\alpha})$, and calls the *direct product* of the two matrices $\mathbf{d}^{(j_1)}(\boldsymbol{\alpha})$ and $\mathbf{d}^{(j_2)}(\boldsymbol{\alpha})$. Thus

$$[\mathbf{d}^{(j_1)}(\boldsymbol{\alpha}) \otimes \mathbf{d}^{(j_2)}(\boldsymbol{\alpha})]_{m_1'm_2',m_1m_2} \equiv d^{(j_1)}_{m_1'm_1}(\boldsymbol{\alpha})\, d^{(j_2)}_{m_2'm_2}(\boldsymbol{\alpha}). \qquad (17\text{-}30)$$

The set of direct product matrices is reducible, since in the basis of states $|j_1j_2jm\rangle$ they all take the form

$$\mathbf{d}^{(j_1)} \otimes \mathbf{d}^{(j_2)} = \begin{pmatrix} \blacksquare & & & & 0 \\ & \blacksquare & & & \\ & & \blacksquare & & \\ & & & \blacksquare & \\ 0 & & & & \blacksquare \end{pmatrix} \qquad (17\text{-}31)$$

where each submatrix on the diagonal refers to a different j value. The transformation from the $|j_1j_2m_1m_2\rangle$ basis to the $|j_1j_2jm\rangle$ basis is given by the Clebsch-Gordan coefficients. Thus

$$\langle j_1j_2m_1'm_2'|e^{-i\mathbf{J}\cdot\boldsymbol{\alpha}}|j_1j_2m_1m_2\rangle = \qquad (17\text{-}32)$$

$$\sum_{jm'm} \langle j_1j_2m_1'm_2'|j_1j_2jm'\rangle\langle j_1j_2jm'|e^{-i\mathbf{J}\cdot\boldsymbol{\alpha}}|j_1j_2jm\rangle\langle j_1j_2jm|j_1j_2m_1m_2\rangle.$$

But

$$\langle j_1 j_2 j m' | e^{-iJ\cdot\alpha} | j_1 j_2 j m \rangle = d^{(j)}_{m'm}(\alpha) \tag{17-33}$$

so that combining (17-32), (17-33), and (17-29) we have

$$d^{(j_1)}_{m_1'm_1}(\alpha) d^{(j_2)}_{m_2'm_2}(\alpha) = \sum_j \sum_{m'm} \langle j_1 j_2 m_1' m_2' | j_1 j_2 j m' \rangle \langle j_1 j_2 m_1 m_2 | j_1 j_2 j m \rangle d^{(j)}_{m'm}(\alpha), \tag{17-34}$$

or using the orthogonality relations for Clebsch-Gordan coefficients

$$\sum_{m_1'm_1} \sum_{m_2'm_2} \langle j_1 j_2 m_1' m_2' | j_1 j_2 j' m' \rangle \langle j_1 j_2 m_1 m_2 | j_1 j_2 j m \rangle d^{(j_1)}_{m_1'm_1}(\alpha) d^{(j_2)}_{m_2'm_2}(\alpha) = d^{(j)}_{m'm}(\alpha)\delta_{jj'}. \tag{17-35}$$

This latter equation is just (17-31) spelled out in matrix elements. The left side is the jm', j'm matrix element of the direct product matrix $d^{(j_1)} \otimes d^{(j_2)}$ in the "$|j_1 j_2 j m\rangle$" basis. Equation (17-34) is the same relation in the "$|j_1 j_2 m_1 m_2\rangle$" basis.

A special case of (17-34) is for j_1 and j_2 both integer and m_1 and m_2 both zero; then m must also be zero. Using (17-22), and complex conjugating, we find the useful formula

$$Y_{j_1 m_1}(\theta, \varphi) Y_{j_2 m_2}(\theta, \varphi) = \sum_{jm} \sqrt{\frac{(2j_1+1)(2j_2+1)}{4\pi(2j+1)}} \langle j_1 j_2 m_1 m_2 | j_1 j_2 j m \rangle \langle j_1 j_2 0 0 | j_1 j_2 j 0 \rangle Y_{jm}(\theta, \varphi). \tag{17-36}$$

This equation enables us to write the product of two spherical harmonics as a linear combination of single spherical harmonics.

Intuitively if one averages over all rotations one gets something spherically symmetric. For example, if we take a time exposure of a weirdly shaped body, rotating the body in all ways (about a fixed point) during the exposure, then the picture looks like that of a spherically symmetric object. We can make this into a precise mathematical statement as follows. Let us define the average over all rotations of a function $f(\omega)$ of the Euler angles $\omega = (\varphi, \theta, \psi)$ by

$$\int d\omega\, f(\omega) = \int_{-1}^{1} \frac{d\cos\theta}{2} \int_0^{4\pi} \frac{d\varphi}{4\pi} \int_0^{4\pi} \frac{d\psi}{4\pi} f(\varphi, \theta, \psi). \tag{17-37}$$

Clearly,

$$\int d\omega = 1, \tag{17-38}$$

so that (17-37) is an average over all rotations. The reason for the 4π's is that for half-integer j, such as spin, the rotation matrices are double valued; rotating by an extra angle 2π produces an extra minus sign, i.e., $d(2\pi) = -1$. Thus a rotation about an axis by φ isn't quite the same as a rotation by $\varphi + 2\pi$; one must go twice around, $\varphi + 4\pi$, to come back to the original rotation. Now, using (17-20), we find that the average of a rotation matrix over all rotations is

$$\int d\omega\, d^{(j)}_{mm'}(\varphi, \theta, \psi) = \int_{-1}^{1} \frac{d\cos\theta}{2} \int_0^{4\pi} \frac{d\varphi}{4\pi}\, d^{(j)}_{mm'}(\varphi, \theta, 0) \int_0^{4\pi} \frac{d\psi}{4\pi}\, e^{-im'\psi}.$$

But the ψ integral vanishes unless $m' = 0$, which can only be if j is an integer. Thus from (17-22)

$$\int d\omega\, d^{(j)}_{mm'}(\varphi, \theta, \psi) = \delta_{j,\,\text{integer}}\delta_{m',0} \int_{-1}^{1} \frac{d\cos\theta}{2} \int_0^{4\pi} \sqrt{\frac{4\pi}{2j+1}}\, Y_{jm}^*(\theta, \varphi)$$

However the average of a spherical harmonic over all angles vanishes unless $j = 0$. Thus

$$\int d\omega\, d^{(j)}_{mm'}(\varphi, \theta, \psi) = \delta_{j,0}\delta_{m',0}\delta_{m,0}; \tag{17-39}$$

the average over all rotations vanishes for all but $j = 0$. Equation (17-39) is just the matrix element of the operator equation

$$\int d\omega\, R_{\varphi,\theta,\psi} = P_0, \tag{17-40}$$

where P_0 is the projection operator onto the $j = 0$, that is, the spherically symmetric component of a state.

Let us apply (17-39) by averaging both sides of (17-34) over all rotations. This gives

$$\int d\omega\, d^{(j_1)}_{m_1'm_1}(\omega) d^{(j_2)}_{m_2'm_2}(\omega) = \langle j_1 j_2 m_1 m_2 | j_1 j_2 00\rangle \langle j_1 j_2 m_1' m_2' | j_1 j_2 00\rangle. \tag{17-41}$$

Now

$$\langle j_1 j_2 m_1 m_2 | j_1 j_2 00\rangle = \delta_{j_1 j_2}\delta_{m_1,-m_2} \frac{(-1)^{j_1 - m_1}}{2j_1 + 1}, \tag{17-42}$$

so that (17-41) becomes, on letting $m_1 \to -m_1$, $m_1' \to -m_1'$,

$$\int d\omega \; d^{(j_1)}_{-m_1',-m_1}(\omega)(-1)^{2j_1-m_1-m_1'} \; d^{(j_2)}_{m_2'm_2}(\omega) = \frac{\delta_{j_1,j_2}\delta_{m_1,m_2}\delta_{m_1'm_2'}}{2j_1+1}. \tag{17-43}$$

But from Eqs. (17-20) and (17-21), we have

$$d^{(j_1)}_{-m_1',-m_1}(\omega)(-1)^{2j_1-m_1-m_1'} = d^{(j_1)}_{m_1'm_1}(\omega)^*, \tag{17-44}$$

so that (17-43) reduces to the *orthogonality condition* for the rotation matrices

$$\int d\omega \, d^{(j_1)}_{m_1',m_1}(\omega)^* \, d^{(j_2)}_{m_2',m_2}(\omega) = \frac{\delta_{j_1,j_2}\delta_{m_1,m_2}\delta_{m_1',m_2'}}{2j_1+1}. \tag{17-45}$$

As a special example, let

$$\chi^{(j)}(\omega) = \mathrm{tr}\, d^{(j)}(\omega); \tag{17-46}$$

the function $\chi^{(j)}(\omega)$ is called the *character* of the representation; then letting $m_1 = m_1'$, $m_2 = m_2'$ and summing over m_1 and m_2 in (17-45) we find the orthogonality condition for the characters of the representations,

$$\int d\omega \, \chi^{(j_1)}(\omega)^* \chi^{(j_2)}(\omega) = \delta_{j_1,j_2}. \tag{17-47}$$

TENSOR OPERATORS

The set of $2j + 1$ states $|jm\rangle$, for fixed j, transform into themselves under rotation, according to the transformation law (17-4). We can similarly study the transformation properties of operators under rotations. By the transformation of an operator A under a rotation ω we mean the unitary transformation:

$$A \to R_\omega A R_\omega^{-1} \tag{17-48}$$

(Cf. Eq. (6-15), for a similar transformation by the inverse rotation.) The expectation value of an operator A in a state $|\Phi\rangle$ is the same as the expectation value of the "rotated" operator $R_\omega A R_\omega^{-1}$ in the rotated state $R_\omega|\Phi\rangle$.

Certain operators are *scalars* under rotation; this means that they commute with the total angular momentum $\mathbf{J}$ and are unchanged by the transformation (17-48). Then there are operators, such as the three components of the position operator x, y, z, that transform among themselves under rotation like the components of a *vector*. Other operators, such as $p_x^5 z + y$, for example, have no nice transformation properties under rotation. The types of operators having simple transformation properties under rotation are known generally as *tensor operators*. By an *irreducible* tensor operator $\mathbf{T}^{(k)}$ of order k we shall mean a set of 2k + 1 operators $T_q^{(k)}$, $q = -k, -k+1, \ldots, k-1, k$, that transform among themselves under rotation according to the transformation law

$$R_\omega T_q^{(k)} R_\omega^{-1} = \sum_{q'=-k}^{k} T_{q'}^{(k)} d_{q'q}^{(k)}(\omega); \tag{17-49}$$

This transformation law is the equivalent for operators of Eq. (17-4). An irreducible tensor $\mathbf{T}^{(0)}$ is clearly a scalar, while one of order 1 is a vector; the components $T_q^{(1)}$ are linear combinations (which we shall construct shortly) of the x, y, and z components of the vector. The $d_{q'q}^{(k)}(\omega)$ coefficients are the matrix elements of the irreducible representation of the rotation group of dimension 2k + 1. It follows from the irreducibility of the $\mathbf{d}^{(k)}$ matrices that there is no subset of components, $T_q^{(k)}$, or linear combination of components, that transforms among itself privately under rotations; this is why a tensor operator transforming according to (17-49) is called *irreducible*. [Later we shall meet reducible tensor operators.]

If we consider an infinitesimal rotation $\boldsymbol{\varepsilon}$, then

$$R_{\boldsymbol{\varepsilon}} = e^{-i\mathbf{J}\cdot\boldsymbol{\varepsilon}} \approx 1 - i\mathbf{J}\cdot\boldsymbol{\varepsilon}, \tag{17-50}$$

and the transformation law (17-49) becomes, to first order in $\boldsymbol{\varepsilon}$:

$$T_q^{(k)} - i\left[\mathbf{J}\cdot\boldsymbol{\varepsilon}, T_q^{(k)}\right] = \sum_{q'} T_{q'}^{(k)} \langle kq'|(1 - i\mathbf{J}\cdot\boldsymbol{\varepsilon})|kq\rangle$$

$$= T_q^{(k)} - i\boldsymbol{\varepsilon}\cdot\sum_{q} T_{q'}^{(k)} \langle kq'|\mathbf{J}|kq\rangle .$$

Comparing coefficients of $\boldsymbol{\varepsilon}$ we see that tensor operators must obey the commutation relation with the angular momentum:

$$[\mathbf{J}, T_q^{(k)}] = \sum_{q'} T_{q'}^{(k)} \langle kq'|\mathbf{J}|kq\rangle . \tag{17-51}$$

The z component of this relation is

$$[J_z, T_q^{(k)}] = qT_q^{(k)} \tag{17-52}$$

while

$$[J_\pm, T_q^{(k)}] = \sum_{q'} T_{q'}^{(k)} \langle kq'|J_\pm|kq\rangle = T_{q\pm 1}^{(k)} \sqrt{k(k+1) - q(q\pm 1)}\,. \tag{17-53}$$

Equations (17-52) and (17-53) describe the transformation properties of irreducible tensor operators under infinitesimal rotations, and are fully equivalent to the transformation law (17-49). These latter equations are much easier to use in practice than (17-49) to determine if one has a tensor operator or not.

For example, a vector **V** obeys the commutation relation with the angular momentum

$$[J_i, V_j] = i\varepsilon_{ijk} V_k \tag{17-54}$$

where i, j, k = x, y, z. This commutation relation must be equivalent to (17-52) and (17-53) and it therefore tells us how to define the q components of a vector operator (k = 1) in terms of the x, y, and z components. The two sets of components are related by

$$V_{q=1} = -\frac{V_x + iV_y}{\sqrt{2}}$$

$$V_{q=0} = V_z$$

$$V_{q=-1} = \frac{V_x - iV_y}{\sqrt{2}}, \tag{17-55}$$

as one can verify by using (17-54) to evaluate the commutators $[J, V_q]$.

This relation between the "q components" or *spherical components* of a vector and its x, y, z components may look a little strange, but actually it is a familiar one. For example, if we let **V** = **r**, then

$$r_1 = -\frac{x+iy}{\sqrt{2}}, \qquad r_0 = z, \qquad r_{-1} = \frac{x-iy}{\sqrt{2}}. \tag{17-56}$$

Writing x, y, and z in polar coordinates, $z = r\cos\theta$, $x = r\sin\theta\cos\varphi$, $y = r\sin\theta\sin\varphi$, we find

$$r_1 = -\frac{r}{2} \sin\theta\, e^{i\varphi}, \qquad r_0 = r\cos\theta, \qquad r_{-1} = \frac{r}{2}\sin\theta\, e^{-i\varphi}. \tag{17-57}$$

Now if we recall the form of the spherical harmonics for $l = 1$ we see that (17-57) says simply

$$r_q = \sqrt{\frac{4\pi}{3}}\, rY_{1q}(\theta, \varphi) \tag{17-58}$$

The quantity

$$P_{1,q}(x, y, z) = rY_{1q}(\theta, \varphi) \tag{17-59}$$

is just a first-order homogeneous polynomial in x, y, z, viz.,

$$P_{1,1} = -\sqrt{\frac{3}{8\pi}}\,(x + iy)$$

$$P_{1,0} = \sqrt{\frac{3}{4\pi}}\, z$$

$$P_{1,-1} = \sqrt{\frac{3}{8\pi}}\,(x - iy). \tag{17-60}$$

For an arbitrary vector operator **V** we have simply,

$$V_q = \sqrt{\frac{4\pi}{3}}\, P_{1,q}(V_x, V_y, V_z). \tag{17-61}$$

Now we may generalize this relation to generate tensors of arbitrary integer order from a vector operator. First, let us show that

$$P_{l,m}(x, y, z) = r^l Y_{l,m}(\theta, \varphi) \tag{17-62}$$

is a homogeneous polynomial of order l. To see this we note from Eq. (6-56) that

$$r^l Y_{ll}(\theta, \varphi) = c(x + iy)^l, \tag{17-63}$$

which is certainly such a polynomial. The other $r^l Y_{lm}$ can be generated from (17-63) by acting $l - m$ times with

$$\begin{aligned} L_- = L_x - iL_y &= (x - iy)\,ip_z - iz(p_x - ip_y) \\ &= (x - iy)\frac{\partial}{\partial z} - z\left(\frac{\partial}{\partial x} - i\frac{\partial}{\partial y}\right) \end{aligned} \tag{17-64}$$

Clearly L_- acting on a homogeneous polynomial of order l produces a homogeneous polynomial of order l.

The point is that if **V** is a vector operator, then the set of $2l + 1$ polynomials $P_{lm}(V_x, V_y, V_z)$ formed of the components of **V** is an irreducible tensor operator of order l. To prove this we use that fact that because P_{lm} is a homogeneous polynomial we have

$$R_\alpha P_{lm}(V_x, V_y, V_z) R_\alpha^{-1} = P_{lm}(V_{x''}, V_{y''}, V_{z''}), \tag{17-65}$$

where

$$V_{x''} = R_\alpha V_x R_\alpha^{-1} \tag{17-66}$$

is the component of **V** along the axis $\hat{x}''$, which is the x axis rotated about $\boldsymbol{\alpha}$ by angle $-|\alpha|$, etc. From the transformation law (17-26) of the spherical harmonics, we can then write

$$P_{lm}(V_{x''}, V_{y''}, V_{z''}) = \sum_{m'} P_{lm'}(V_x, V_y, V_z) d^{(l)}_{m'm}(\alpha). \tag{17-67}$$

Thus

$$R_\alpha P_{lm} R_\alpha^{-1} = \sum_{m'} P_{m'} d^{(l)}_{m'm}(\alpha), \tag{17-68}$$

which is the correct transformation law for an irreducible tensor operator.

For example, the five components of this tensor for $l = 2$ are

$$P_{2,\pm 2} = \sqrt{\frac{15}{8\pi}}\, V_{\pm 1}^2$$

$$P_{2,\pm 1} = \sqrt{\frac{15}{16\pi}}\, (V_0 V_{\pm 1} + V_{\pm 1} V_0)$$

$$P_{2,0} = \sqrt{\frac{5}{16\pi}}\, (2V_0^2 + V_1 V_{-1} + V_{-1} V_1). \tag{17-69}$$

Tensor operators have many simple properties. For example, if we act with $T_q^{(k)}$ on a state $|\alpha, j_1, m_1\rangle$ of a system (here the α refers to other quantum numbers, such as radial quantum numbers, etc.) we get a state whose z component of angular momentum is $q + m_1$. To prove this, let us consider the transformation properties of the state $T_q^{(k)}|\alpha j_1 m_1\rangle$ under rotation about the z axis by φ:

$$\begin{aligned} R_\varphi T_q^{(k)} |\alpha j_1 m_1\rangle &= R_\varphi T_q^{(k)} R_\varphi^{-1} R_\varphi |\alpha j_1 m_1\rangle \\ &= \sum_{q'} \left[T_{q'}^{(k)} d^{(k)}_{q'q}(\varphi) \right] \sum_{m'} |\alpha j_1 m_1'\rangle d^{(j_1)}_{m_1' m_1}(\varphi). \end{aligned} \tag{17-70}$$

But from (17-20)

$$d^{(j)}_{m'm}(\varphi) = \delta_{m'm} e^{-im\varphi}, \tag{17-71}$$

so that

$$R_\varphi T^{(k)}_q |\alpha j_1 m_1\rangle = e^{-i(q+m_1)\varphi} T^{(k)}_q |\alpha j_1 m_1\rangle. \tag{17-72}$$

Thus $T^{(k)}_q|\alpha j_1 m_1\rangle$ transforms by the phase factor $e^{-i(q+m_1)\varphi}$ under rotation; this is precisely the transformation law for an eigenstate of J_z with eigenvalue $q + m_1$. Thus $T^{(k)}_q$ is an operator that *increases* the eigenvalue of J_z by q. One can also prove this directly from the commutation relation (17-52).

The states $T^{(k)}_q|\alpha j_1 m_1\rangle$ are not, however, eigenstates of J^2. We can construct such eigenstates out of these states in the same way as we found eigenstates of the sum of two angular momenta; we must take linear combinations with Clebsch-Gordan coefficients. Acting with a tensor operator on an angular momentum eigenstate is analogous to adding together two angular momenta with eigenvalues k, q and j_1, m_1. In fact, the state

$$|\tilde{\alpha} jm\rangle = \sum_{q,m_1} T^{(k)}_q |\alpha j_1 m_1\rangle \langle k j_1 q m_1 | k j_1 j m\rangle \tag{17-73}$$

is an eigenstate of J^2 with eigenvalue $j(j+1)$ and of J_z with eigenvalue m. It clearly is an eigenstate of J_z with eigenvalue m, since from (17-72) it is a linear combination of eigenstates of J_z with eigenvalue $q + m_1$, which must equal m, else the Clebsch-Gordan coefficient vanishes.

To show explicitly that (17-73) transforms as a state with quantum numbers j and m, let us act on (17-73) with a rotation R. Then, using (17-49) and (17-4)

$$\begin{aligned} R|\tilde{\alpha} jm\rangle &= \sum_{q'm_1} (R T^{(k)}_q R^{-1}) R |\alpha j_1 m_1\rangle \langle k j_1 q m_1 | k j_1 j m\rangle \\ &= \sum_{q'm_1'} T^{(k)}_{q'} |\alpha j_1 m_1'\rangle \sum_{qm_1} d^{(k)}_{q'q} d^{(j_1)}_{m_1'm_1} \langle k j_1 q m_1 | k j_1 j m\rangle. \end{aligned} \tag{17-74}$$

If we use (17-34) and the orthogonality relation for the Clebsch-Gordan coefficients, (17-74) becomes

$$\begin{aligned} R|\tilde{\alpha} jm\rangle &= \sum_{q'm_1'm'} T^{(k)}_{q'} |\alpha j_1 m_1'\rangle \langle k j_1 q' m_1' | k j_1 j m'\rangle d^{(j)}_{m'm} \\ &= \sum_{m'} |\tilde{\alpha} jm'\rangle d^{(j)}_{m'm}. \end{aligned} \tag{17-75}$$

Thus we see that the state $|\tilde{\alpha}jm\rangle$ transforms like an angular momentum eigenstate with quantum numbers j and m.

Let us take the scalar product of $|\tilde{\alpha}jm\rangle$ with another angular momentum eigenstate $\langle\alpha'j'm'|$. The scalar product $\langle\alpha'j'm'|\tilde{\alpha}jm\rangle$ vanishes unless $j' = j$ and $m' = m$, since eigenstates belonging to different eigenvalues are orthogonal. The scalar product is also independent of m; it doesn't depend on the orientation of the angular momentum vector. To see this let us take the matrix element of

$$\int d\omega\, R_{\omega}^{-1}R_{\omega} = 1. \tag{17-76}$$

Thus

$$\int d\omega \langle\alpha'j'm'|R_{\omega}^{-1}R_{\omega}|\tilde{\alpha}jm\rangle = \langle\alpha'j'm'|\tilde{\alpha}jm\rangle. \tag{17-77}$$

Using (17-4) and the unitarity of the rotation matrices, (17-77) becomes

$$\sum_{\bar{m}\bar{m}'} \int d\omega\, d^{(j')}_{\bar{m}'m'}(\omega)^{*}\, d^{(j)}_{\bar{m}m}(\omega)\langle\alpha'j'\bar{m}'|\tilde{\alpha}j\bar{m}\rangle = \langle\alpha'j'm'|\tilde{\alpha}jm\rangle; \tag{17-78}$$

Then Eq. (17-45) for the integral implies

$$\langle\alpha'j'm'|\tilde{\alpha}jm\rangle = \delta_{jj'}\delta_{mm'} \sum_{\bar{m}} \frac{\langle\alpha'j'\bar{m}|\tilde{\alpha}j'\bar{m}\rangle}{2j'+1}; \tag{17-79}$$

in this form the diagonal matrix elements are independent of the m value.

Thus from Eq. (17-73) we find

$$\sum_{qm_1} \langle\alpha'j'm'|T_q^{(k)}|\alpha j_1m_1\rangle\langle kj_1qm_1|kj_1jm\rangle = \frac{\delta_{jj'}\delta_{mm'}}{2j'+1} \sum_{\bar{m}} \langle\alpha'j'\bar{m}|\tilde{\alpha}j'\bar{m}\rangle, \tag{17-80}$$

and the orthogonality relation for Clebsch-Gordan coefficients yields

$$\langle\alpha'j'm'|T_q^{(k)}|\alpha jm\rangle = \sum_{\bar{m}} \frac{\langle\alpha'j'\bar{m}|\tilde{\alpha}j'\bar{m}\rangle}{2j'+1}\langle kjqm|kjj'm'\rangle. \tag{17-81}$$

This is a remarkable result; it shows that matrix element of a tensor operator between two angular momentum eigenstates factors into a Clebsch-Gordan coefficient containing all the angular details but independent of the dynamics times a coefficient *independent* of the different m, m', and q values, containing all the dynamics. This coefficient depends only on αj, $\alpha' j'$ and on $T^{(k)}$. It is customary to write Eq. (17-81) in the form:

$$\langle \alpha' j' m' | T_q^{(k)} | \alpha j m \rangle = \frac{\langle \alpha' j' \| T^{(k)} \| \alpha j \rangle}{\sqrt{2j'+1}} \langle kjqm | kjj'm' \rangle . \tag{17-82}$$

[The $\sqrt{2j'+1}$ is arbitrary, but conventional.] The quantity $\langle \alpha' j' \| T^{(k)} \| \alpha j \rangle$ is known as the *reduced matrix element*. The result (17-82) is the *Wigner-Eckart theorem*. We shall apply this theorem often in studying atoms.

Actually we've already seen a simple example of the Wigner-Eckart theorem in action when we evaluated dipole matrix elements for one-electron atoms [Eq. (13-117)]; there the matrix element factored into a radial integral, containing all the dynamics, times an angular integral independent of the dynamics. We shall return to this example a little later.

The simplest application of the Wigner-Eckart theorem is to a scalar operator, $S^{(0)}$; then $k = 0$, $q = 0$ and

$$\langle \alpha' j' m' | S^{(0)} | \alpha j m \rangle = \delta_{jj'} \delta_{mm'} \frac{\langle \alpha' j \| S^{(0)} \| \alpha j \rangle}{\sqrt{2j+1}} \tag{17-83}$$

since $\langle 0j0m | 0jj'm \rangle = \delta_{jj'} \delta_{mm'}$.

As a second example, let us evaluate the matrix elements of J_q, the components of the angular momentum operator. From (17-82)

$$\langle \alpha' j' m' | J_q | \alpha j m \rangle = \frac{\langle \alpha' j' \| J \| \alpha j \rangle}{\sqrt{2j'+1}} \langle 1jqm | 1jj'm' \rangle . \tag{17-84}$$

Now to evaluate the reduced matrix element we can choose any convenient value of q. Let us take $q = 0$; from (17-55), $J_0 = J_z$, and

$$\begin{aligned} \langle \alpha' j' m' | J_0 | \alpha j m \rangle &= \delta_{jj'} \delta_{\alpha\alpha'} \delta_{mm'} m \\ &= \frac{\langle \alpha' j' \| J \| \alpha j \rangle}{\sqrt{2j'+1}} \langle 1j0m | 1jj'm' \rangle . \end{aligned} \tag{17-85}$$

Thus, the reduced matrix element vanishes unless $j' = j$. Using the fact that

$$\langle 1j0m | 1jjm \rangle = \frac{m}{\sqrt{j(j+1)}} \tag{17-86}$$

we find then

$$\frac{\langle \alpha' j' \| J \| \alpha j \rangle}{\sqrt{2j'+1}} = \delta_{jj'} \delta_{\alpha\alpha'} \sqrt{j(j+1)} ; \tag{17-87}$$

the reduced matrix element is like $\langle\sqrt{J^2}\rangle$. Putting this back in Eq. (17-84) we find

$$\langle\alpha' j' m'|J_q|\alpha j m\rangle = \delta_{jj'}\delta_{\alpha\alpha'}\sqrt{j(j+1)}\,\langle 1jqm|1jjm'\rangle, \tag{17-88}$$

which is a rather fancy way of writing the matrix elements of **J**.

Note that the product of two tensor operators, $T_q^{(k)}W_{q'}^{(k')}$ is not an irreducible tensor operator. However, the linear combination

$$Z_m^{(j)} = \sum_{q,q'} T_q^{(k)} W_{q'}^{(k')} \langle kk'qq'|kk'jm\rangle \tag{17-89}$$

is an irreducible tensor operator. [Compare (17-89) with Eq. (15-26).] To prove this we write

$$RZ_m^{(j)}R^{-1} = \sum_{qq'} (RT_q^{(k)}R^{-1})(RW_{q'}^{(k')}R^{-1})\langle kk'qq'|kk'jm\rangle$$

$$= \sum_{\bar{q}\bar{q}'}\sum_{qq'} T_{\bar{q}}^{(k)} W_{\bar{q}'}^{(k')} d_{\bar{q}q}^{(k)} d_{\bar{q}'q'}^{(k')} \langle kk'qq'|kk'jm\rangle .$$

Using (17-34) this simplifies to

$$RZ_m^{(j)}R^{-1} = \sum_{m'} Z_{m'}^{(j)} d_{m'm}^{(j)}, \tag{17-90}$$

which is the correct transformation law.

We can also invert (17-89) to write

$$T_q^{(k)}W_{q'}^{(k')} = \sum_{jm} Z_m^{(j)}\langle kk'qq'|kk'jm\rangle, \tag{17-91}$$

which expresses the product of two tensor operators as a linear combination of irreducible tensor operators.

Let's consider as an example, the product of two vector operators **V** and **U**. First in terms of x, y, z components we can write the product in the form

$$V_iU_j = \frac{\mathbf{V}\cdot\mathbf{U}}{3}\delta_{ij} + \frac{V_iU_j - V_jU_i}{2} + \left(\frac{V_iU_j + V_jU_i}{2} - \frac{\mathbf{V}\cdot\mathbf{U}}{3}\delta_{ij}\right). \tag{17-92}$$

The first term on the right has the transformation properties of a scalar. The second term can be written as

$$\frac{V_iU_j - V_jU_i}{2} = \varepsilon_{ijk}(\mathbf{V}\times\mathbf{U})_k; \tag{17-93}$$

this clearly has the transformation properties of a vector — the cross product. The last term, which has five independent components, has the transformation properties of a tensor of rank 2. Notice that the last term is symmetric in i and j and has vanishing trace, that is, if we set j = i and sum over i we get zero.

Evaluating (17-89) for the product of two vectors we find after some calculation

$$Z_0^{(0)} = -\frac{\mathbf{V}\cdot\mathbf{U}}{3} \tag{17-94}$$

as the scalar component;

$$Z_m^{(1)} = \frac{(\mathbf{V}\times\mathbf{U})_m}{i\sqrt{2}} \tag{17-95}$$

as the vector component; and

$$Z_{\pm 2}^{(2)} = V_{\pm 1}U_{\pm 1}$$

$$Z_{\pm 1}^{(2)} = \frac{V_{\pm 1}U_0 + V_0U_{\pm 1}}{\sqrt{2}}$$

$$Z_0^{(2)} = \frac{V_1U_{-1} + 2V_0U_0 + V_{-1}U_1}{\sqrt{6}}, \tag{17-96}$$

as the second-order tensor component. Note that the tensor (17-69) is just a special case of (17-96) for **U** = **V**.

MULTIPOLE RADIATION

Let us now apply all this lore to the problem of the multipole emission of light by an atom. The energy eigenstates of an atom, taking into account the spin-orbit interaction, are not usually eigenstates of the z component of the total orbital angular momentum, but are eigenstates of the total angular momentum, spin plus orbital. The probability of emission of radiation of wave vector **k** and polarization $\boldsymbol{\lambda}$, while the atom makes a transition from an initial state $|\alpha jm\rangle$ to a final state $|\bar{\alpha}\bar{j}\bar{m}\rangle$ is determined by the matrix element

$$\langle\alpha jm|\mathbf{J}(-\mathbf{k})\cdot\boldsymbol{\lambda}|\bar{\alpha}\bar{j}\bar{m}\rangle, \tag{17-97}$$

where

$$\mathbf{J}(-\mathbf{k}) = \int d^3r\, e^{i\mathbf{k}\cdot\mathbf{r}}\mathbf{J}(\mathbf{r}) \tag{17-98}$$

is the Fourier component of the paramagnetic current operator for the atom. Let us rewrite the operator $\mathbf{J}(-\mathbf{k})\cdot\boldsymbol{\lambda}$ as follows. First we expand $e^{i\mathbf{k}\cdot\mathbf{r}}$ in spherical harmonics. Then

$$\mathbf{J}(-\mathbf{k})\cdot\boldsymbol{\lambda} = 4\pi \sum_{lm} i^l Y_{lm}{}^*(\Omega_k) \int d^3r\, Y_{lm}(\Omega_r) j_l(kr)\mathbf{J}(\mathbf{r})\cdot\boldsymbol{\lambda}. \tag{17-99}$$

Now in terms of spherical (or q) components

$$\boldsymbol{\lambda}\cdot\mathbf{J}(\mathbf{r}) = \sum_q (-1)^q \lambda_{-q} J_q(\mathbf{r}) = \sum_q (\boldsymbol{\lambda}^*)_q{}^* J_q(\mathbf{r}). \tag{17-100}$$

Thus

$$\mathbf{J}(-\mathbf{k})\cdot\boldsymbol{\lambda} = 4\pi \sum_{qlm} i^l [Y_{lm}(\Omega_k)(\boldsymbol{\lambda}^*)_q]^* \int d^3r\, Y_{lm}(\Omega_r) j_l(kr) J_q(\mathbf{r}). \tag{17-101}$$

Using the orthogonality relation (15-32) for the Clebsch-Gordan coefficients we can rewrite (17-101) as

$$\mathbf{J}(-\mathbf{k})\cdot\boldsymbol{\lambda} = \sum_{ll'm'} 4\pi i^l\, \Phi_{m'}^{(l')}(l)^* \mathcal{J}_{m'}^{(l')}(l), \tag{17-102}$$

where

$$\mathcal{J}_{m'}^{(l')}(l) = \sum_{q''m''} \int d^3r\, Y_{lm''}(\Omega_r) j_l(kr) J_{q''}(\mathbf{r}) \langle 1lq''m''|1ll'm'\rangle \tag{17-103}$$

and

$$\Phi_{m'}^{(l')}(l,\Omega_k,\boldsymbol{\lambda}) = \sum_{qm} Y_{lm}(\Omega_k)(\boldsymbol{\lambda}^*)_q \langle 1lqm|1ll'm'\rangle. \tag{17-104}$$

The quantities $\mathcal{J}$ and Φ have a very simple interpretation. We leave it as an exercise to show that $\mathcal{J}_{m'}^{(l')}(l)$ transforms as the m' component of an irreducible tensor of order l'. According to the Clebsch-Gordan coefficient $\langle 1lq''m''|1ll'm'\rangle$, the possible l values for fixed l' are $l = l'$ and $l = l' \pm 1$. Let us look at the long wavelength limit, $ka_0 \ll 1$. Then we can approximate the $j_l(kr)$ in (17-103) as $(kr)^l/(2l+1)!!$

For $l = l'$, the operator $\mathcal{J}_{m'}^{(l')}(l')$ is essentially $k^{l'}$ times the magnetic "$2^{l'}$-pole" operator, e.g., $k^{-2}\mathcal{J}_{m'}^{(2)}(2)$ is the magnetic quadrupole operator, etc.

For $l = l' - 1$, the operator $\mathcal{J}_{m'}^{(l')}(l'-1)$ is $k^{l'-1}$ times what is essentially the electric "$2^{l'}$-pole" operator. The operator $\mathcal{J}_{m'}^{(l')}(l'+1)$

also has the character of an electric $2^{l'}$-pole operator; however it has two more powers of k than $\mathcal{J}_{m'}^{(l')}(l'-1)$ and thus its matrix elements are generally negligible compared with those of $\mathcal{J}_{m'}^{(l')}(l'-1)$.

For illustration we calculate $\mathcal{J}_{m'}^{(1)}(1)$, the magnetic dipole operator:

$$\mathcal{J}_{m'}^{(1)}(1) = \sum_{qm} \int d^3r \frac{kr}{3} Y_{1m}(\Omega_r) J_q(r) \langle 11qm|111m' \rangle .$$

But $rY_{1m}(\Omega_r) = r_m\sqrt{3/4\pi}$, from (17-58); thus from (17-89) and (17-95),

$$\sum_{qm} r_m J_q(\mathbf{r})\langle 11qm|111m'\rangle = \frac{(\mathbf{r}\times\mathbf{J}(\mathbf{r}))_{m'}}{i\sqrt{2}} . \tag{17-105}$$

Hence

$$\mathcal{J}_{m'}^{(1)}(1) = \frac{k}{i\sqrt{24\pi}} \int d^3r\ (\mathbf{r}\times\mathbf{J}(\mathbf{r}))_{m'} , \tag{17-106}$$

and is clearly proportional to the magnetic dipole operator.

Thus we can interpret the sum over l' in (17-99) as a sum over the different $2^{l'}$-pole radiation processes; the $l = l'$ terms correspond to magnetic radiation, while the $l = l' \pm 1$ terms are electric radiation terms.

The expression $\Phi_{m'}^{(l')}(l, \Omega_k, \lambda)$ is the amplitude for finding the photon, in an l, l' emission process, in the solid angle Ω_k and with polarization λ; roughly speaking it is the angular part of the "wave function" of the emitted photon. Under simultaneous rotation of the direction Ω_k and λ, Φ transforms like an eigenfunction of angular momentum with total quantum number l' and z component m'. Thus the photon emitted in the l, l' radiation channel has total angular momentum l' and z component m'. The total angular momentum of a photon is a sum of its orbital angular momentum about the origin and its intrinsic angular momentum — the spin of the photon associated with its polarization vector — which is one. We see, from the Y_{lm} in (17-104), that the orbital angular momentum of the photon is l, and that the orbital and polarization parts are added together to give a total angular momentum $l'm'$.

Notice that $\Phi_0^{(0)}$ must be zero, since from (17-89) and (17-94), it is proportional to $\mathbf{k}\cdot\boldsymbol{\lambda}^*$ which is zero for a photon. Thus one can't have a photon in a total angular momentum zero state. There are no $l' = 0$, or monopole radiation, processes.

Taking the matrix element (17-97) of (17-102) we have

$$\langle \alpha jm|\mathbf{J}(-\mathbf{k})\cdot\boldsymbol{\lambda}|\bar{\alpha}\bar{j}\bar{m}\rangle = \sum_{ll'm'} 4\pi i^l \Phi_{m'}^{(l')}(l)^* \langle \alpha jm|\mathcal{J}_{m'}^{(l')}(l)|\bar{\alpha}\bar{j}\bar{m}\rangle . \tag{17-107}$$

We can now apply the Wigner-Eckart theorem to write (17-107) as

$$\langle \alpha j m | \mathbf{J}(-\mathbf{k}) \cdot \boldsymbol{\lambda} | \bar{\alpha} \bar{j} \bar{m} \rangle$$
$$= \sum_{l l' m'} 4\pi i^l \Phi_{m'}^{(l')}(l)^* \frac{\langle \alpha j \| \mathscr{J}^{(l')}(l) \| \bar{\alpha} \bar{j} \rangle}{\sqrt{2j+1}} \langle l' \bar{j} m' \bar{m} | l' \bar{j} j m \rangle. \tag{17-108}$$

All the dynamics are contained in the reduced matrix element, and for given initial and final states one need calculate just one number $\langle \alpha j \| \mathscr{J}^{l'}(l) \| \bar{\alpha} \bar{j} \rangle$ for each different electric or magnetic multipole transition.

The form (17-108) shows us very clearly the selection rules for $2^{l'}$-pole electric and magnetic transitions. On the one hand we have the selection rules dictated by the Clebsch-Gordan coefficient. First,

$$m - \bar{m} = m' \tag{17-109}$$

this expresses the fact that the change in the z component of the angular momentum of the atom in the transition is just the z component of the angular momentum carried away by the photon.

Then second, we have the restriction on the allowed changes in j:

$$|l' - j| \leq \bar{j} \leq l' + j$$

or equivalently

$$|l' - \bar{j}| \leq j \leq l' + \bar{j}. \tag{17-110}$$

The physical origin of the selection rules (17-109) and (17-110) is simple. The atom emits a photon. Consequently the total angular momentum of the photon and the atom after the emission must equal that of the atom before emission. The Clebsch-Gordan coefficient in (17-108) is just the one for adding the angular momentum of the photon to that of the atom after emission to form a state with the same angular momentum as the atom before emission.

Conservation of parity gives us a further selection rule. Let P be the parity operator; then under a spatial inversion the coordinate $\mathbf{r}_i$ and momentum $\mathbf{p}_i$ of each particle transform into minus themselves:

$$P\mathbf{r}_i P^{-1} = -\mathbf{r}_i, \qquad P\mathbf{p}_i P^{-1} = -\mathbf{p}_i. \tag{17-111}$$

Thus

$$PJ(\mathbf{r})P^{-1} = P \sum_i \left[\frac{\mathbf{p}_i}{2m} \delta(\mathbf{r} - \mathbf{r}_i) + \delta(\mathbf{r} - \mathbf{r}_i) \frac{\mathbf{p}_i}{2m} \right] P^{-1}$$
$$= - \sum_i \left[\frac{\mathbf{p}_i}{2m} \delta(\mathbf{r} + \mathbf{r}_i) + \delta(\mathbf{r} + \mathbf{r}_i) \frac{\mathbf{p}_i}{2m} \right] = -\mathbf{J}(-\mathbf{r}). \tag{17-112}$$

Using (17-112) it is easy to show, from Eq. (6-64), that the operator $\mathscr{J}_{m'}^{(l')}(l)$ acting on a state with a definite parity changes the parity of the state by $(-1)^{l+1}$. Thus in an l', l transition between states of definite parity the reduced matrix element $\langle \alpha j \| \mathscr{J}^{(l')}(l) \| \bar{\alpha}\bar{j} \rangle$ vanishes unless the final parity is $(-1)^{l+1}$ times the initial parity.

To illustrate these selection rules, let us look again at electric dipole radiation, $l = 0$, $l' = 1$. The selection rule (17-110) then tells us that

$$\bar{j} = j \pm 1 \quad \text{or} \quad \bar{j} = j, \tag{17-113}$$

while conservation of parity implies that the parity of the final state must be minus that of the initial state. In a one-electron atom neglecting spin the parity of the initial state is $(-1)^j$ and that of the final state is $(-1)^{\bar{j}}$. For this case we must have $\bar{j} = j \pm 1$, as we found when we studied electric dipole radiation.

ANGULAR MOMENTUM AND THE HARMONIC OSCILLATOR

Schwinger[2] has looked at the theory of angular momentum from a point of view that enables one to carry out very elegant calculations of the various quantities occurring in the theory of angular momentum. Recall that we can describe a harmonic oscillator in terms of two operators a and $a^\dagger$, which obey the commutation relation $[a, a^\dagger] = 1$. The eigenvalues of the operator $N = a^\dagger a$ are integers $0, 1, 2, \ldots$, which are the number of excited "quanta" of the oscillators. The normalized eigenstate of $a^\dagger a$ with eigenvalue n has the form $|n\rangle = (a^\dagger)^n/\sqrt{n!}\,|0\rangle$, where $|0\rangle$ is the ground state of the oscillator, the state with zero quanta.

Now suppose that we have two independent harmonic oscillators — one described by operators a_+ and $a_+^\dagger$, and the other described by operators a_- and $a_-^\dagger$. Because the oscillators are independent the "+" operators commute with the "−" operators.

$$[a_+, a_-^\dagger] = 0 = [a_+, a_-], \qquad \text{etc.}$$

The eigenstates of the two oscillators are specified by the eigenvalue n_+, of $N_+ = a_+^\dagger a_+$, and the eigenvalue n_-, of $N_- = a_-^\dagger a_-$. Such an eigenstate has the form

---

[2] J. Schwinger, *On Angular Momentum*, AEC Report NYO-3071 (1952). This method derives from Wigner; see E.P. Wigner, *Group Theory and Its Application to the Quantum Mechanics of Atomic Spectra* [Academic Press, New York, 1959], Chapter 14.

$$|n_+, n_-\rangle = \frac{(a_+{}^\dagger)^{n_+}}{\sqrt{n_+!}} \frac{(a_-{}^\dagger)^{n_-}}{\sqrt{n_-!}} |0\rangle, \tag{17-114}$$

where $|0\rangle$ is the state with $n_+ = n_- = 0$.

Two harmonic oscillators have a simple connection with angular momentum. Let us define three operators formed from $a_\pm$ and $a_\pm{}^\dagger$:

$$J_x = \frac{1}{2}(a_+{}^\dagger a_- + a_-{}^\dagger a_+)$$

$$J_y = \frac{1}{2i}(a_+{}^\dagger a_- - a_-{}^\dagger a_+)$$

$$J_z = \frac{1}{2}(a_+{}^\dagger a_+ - a_-{}^\dagger a_-). \tag{17-115}$$

It is a straightforward exercise to verify that these operators obey the angular momentum commutation relations:

$$[J_i, J_j] = i\varepsilon_{ijk} J_k, \quad i, j, k = x, y, z.$$

Since these operators obey the angular momentum commutation relations, whatever properties they have will generally be true of all angular momenta. The system of two harmonic oscillators can thus be used as a model for angular momentum; we can use our knowledge of the harmonic oscillator to learn about angular momentum.

By a simple calculation we find that

$$J^2 = J_x{}^2 + J_y{}^2 + J_z{}^2 = \frac{N}{2}\left(\frac{N}{2} + 1\right), \tag{17-116}$$

where

$$N = a_+{}^\dagger a_+ + a_-{}^\dagger a_- \tag{17-117}$$

is the operator for the total number of quanta. The eigenvalues of J^2 are clearly of the form $j(j + 1)$ where $j = n/2$ can take on the values $0, 1/2, 1, 3/2, \ldots$. From (17-115) and (17-116) we see that an eigenstate $|n_+, n_-\rangle$ of the two oscillators is also an eigenstate of J^2 and J_z. In fact,

$$J_z|n_+, n_-\rangle = \frac{1}{2}(n_+ - n_-)|n_+, n_-\rangle$$

$$J^2|n_+, n_-\rangle = \frac{n}{2}\left(\frac{n}{2} + 1\right)|n_+, n_-\rangle \tag{17-118}$$

where $n = n_+ + n_-$. Thus the state $|n_+, n_-\rangle$ is an eigenstate of J^2 and

J_z with the eigenvalues j, m given by

$$j = \frac{n_+ + n_-}{2}$$

$$m = \frac{n_+ - n_-}{2}. \tag{17-119}$$

If we fix $n_+ + n_-$ then there are clearly $n_+ + n_- + 1 = 2j + 1$ different possible values of $m = n_+ - n_-$. They are $j, j-1, \ldots, -j+1, -j$.

This structure of the eigenvalues suggests a simple interpretation. We can think of each (+) quantum as being associated with plus one-half unit of angular momentum in the z direction and each (−) quantum as being associated with minus one-half unit angular momentum in the z direction. The total number of units $(1/2)n_+ + (1/2)n_-$, is just the total j value. The net angular momentum along z associated with these units is $(1/2)n_+ - (1/2)n_-$, which is just the m value. This picture is further supported if we notice that the operators $J_\pm = J_x \pm iJ_y$ have the form

$$J_+ = a_+{}^\dagger a_-, \qquad J_- = a_-{}^\dagger a_+; \tag{17-120}$$

The operator J_+ increases m by one leaving j fixed. It is clear from (17-120) that it does this by reducing the number of − quanta by one and increasing the number of + quanta by one. Similarly J_- reduces n_+ by one and increases n_- by one.

It is more convenient to label the states $|n_+, n_-\rangle$ by their j, m values, given by (17-115). Since $n_+ = j + m$ and $n_- = j - m$, the eigenstate $|j, m\rangle$ is given by (17-114) as

$$|j, m\rangle = \frac{(a_+{}^\dagger)^{j+m}}{\sqrt{(j+m)!}} \; \frac{(a_-{}^\dagger)^{j-m}}{\sqrt{(j-m)!}} |0\rangle. \tag{17-121}$$

These are simple explicit forms for the eigenstates for the angular momentum operators. The states (17-121) are normalized to 1.

To illustrate how this approach works let us operate with J_+ on $|jm\rangle$:

$$J_+|jm\rangle = a_+{}^\dagger a_- |jm\rangle = \frac{(a_+{}^\dagger)^{j+m+1}}{\sqrt{(j+m)!}} \; \frac{(a_-)(a_-{}^\dagger)^{j-m}}{\sqrt{(j-m)!}} |0\rangle. \tag{17-122}$$

Now from the commutation properties of the a_- and $a_-{}^\dagger$ we have

$$a_-(a_-{}^\dagger)^{j-m} = (a_-{}^\dagger)^{j-m} a_- + (j-m)(a_-{}^\dagger)^{j-m-1}.$$

The a_- acting on $|0\rangle$ gives zero, and therefore (17-122) becomes

$$J_+|jm\rangle = \sqrt{(j+m+1)(j-m)}\,\frac{(a_+{}^\dagger)^{j+m+1}}{\sqrt{(j+m+1)!}}\;\frac{(a_-{}^\dagger)^{j-m-1}}{\sqrt{(j-m-1)!}}\,|0\rangle$$

$$= \sqrt{j(j+1)-m(m+1)}\,|j, m+1\rangle. \qquad (17\text{-}123)$$

This equation then tells us that J_+ increases m by one, and it also tells us the correct numerical factor multiplying $|j, m+1\rangle$.

It is interesting to note that within this model we can introduce operators

$$K_+ = a_+{}^\dagger a_-{}^\dagger \qquad (17\text{-}124)$$

and

$$K_- = a_+ a_- \qquad (17\text{-}125)$$

that serve to increase, or decrease, j by one, leaving m fixed. For example,

$$K_+|j, m\rangle = \sqrt{(j+m+1)(j-m+1)}\,|j+1, m\rangle. \qquad (17\text{-}126)$$

Now let us carry out the evaluation of the rotation matrix elements $d_{m,m'}{}^{(j)}(\varphi, \theta, \psi)$. As we have seen, the φ and ψ dependence is in simple exponential factors, and the hard part is to find

$$d^{(j)}_{m,m'}(\theta) = \langle jm|e^{-i\theta J_y}|jm'\rangle. \qquad (17\text{-}127)$$

To evaluate (17-127) we take the explicit form (17-121) for the states and for J_y in this model. Then (17-127) becomes an expectation value of lots of $a_\pm$ and $a_\pm{}^\dagger$'s in the state $|0\rangle$. To evaluate this expectation value we use a generating function trick. Let x_+ and x_- be parameters and let us form the sum

$$G(x_+, x_-) = \sum_{m'} \frac{x_+^{j+m'}\,x_-^{j-m'}}{\sqrt{(j+m')!\,(j-m')!}}\,d^{(j)}_{mm'}(\theta)$$

$$= \langle jm|e^{-iJ_y\theta} \sum_{m'} \frac{(x_+a_+{}^\dagger)^{j+m'}}{(j+m')!}\;\frac{(x_-a_-{}^\dagger)^{j-m'}}{(j-m')!}\,|0\rangle. \qquad (17\text{-}128)$$

To write the right side we used the explicit form (17-121) for $|jm'\rangle$. The sum over m' is, according to the binomial theorem just

$$\frac{(x_+a_+{}^\dagger + x_-a_-{}^\dagger)^{2j}}{(2j)!}.$$

Thus

$$G(x_+, x_-) = \frac{1}{(2j)!} \langle jm | e^{-iJ_y\theta} (x_+ a_+^\dagger + x_- a_-^\dagger)^{2j} | 0 \rangle . \qquad (17\text{-}129)$$

Now

$$e^{-iJ_y\theta} (x_+ a_+^\dagger + x_- a_-^\dagger)^{2j} = \left[e^{-iJ_y\theta} (x_+ a_+^\dagger + x_- a_-^\dagger) e^{iJ_y\theta} \right]^{2j} e^{-iJ_y\theta}, \qquad (17\text{-}130)$$

and $e^{-iJ_y\theta} |0\rangle = |0\rangle$ since the state with $j = 0$ is an eigenfunction of J_x, J_y, and J_z with eigenvalue zero. Thus to evaluate (17-129) we need to know how $a_\pm^\dagger$ transform under the rotation $e^{-iJ_y\theta}$.

To discover $e^{-iJ_y\theta} a^\dagger e^{iJ_y\theta}$ we note that

$$\frac{\partial}{\partial\theta} \left(e^{-iJ_y\theta} a^\dagger e^{iJ_y\theta} \right) = \frac{1}{i} e^{-iJ_y\theta} [J_y, a_\pm^\dagger] e^{iJ_y\theta} \qquad (17\text{-}131)$$

and according to (17-115)

$$[J_y, a_\pm^\dagger] = \frac{1}{2i} [a_+^\dagger a_- - a_-^\dagger a_+, a_\pm^\dagger] = \mp \frac{1}{2i} a_\mp^\dagger, \qquad (17\text{-}132)$$

so that (17-131) becomes

$$\frac{\partial}{\partial\theta} \left(e^{-iJ_y\theta} a_\pm^\dagger e^{iJ_y\theta} \right) = \pm \frac{1}{2} \left(e^{-iJ_y\theta} a_\mp^\dagger e^{iJ_y\theta} \right). \qquad (17\text{-}133)$$

These two equations have the solution

$$e^{-iJ_y\theta} a_+^\dagger e^{iJ_y\theta} = a_+^\dagger \cos\frac{\theta}{2} + a_-^\dagger \sin\frac{\theta}{2}$$

$$e^{-iJ_y\theta} a_-^\dagger e^{iJ_y\theta} = a_-^\dagger \cos\frac{\theta}{2} - a_+^\dagger \sin\frac{\theta}{2}. \qquad (17\text{-}134)$$

Then combining (17-134), (17-130), and (17-129), we find

$$\begin{aligned} G(x_+, x_-) &= \frac{1}{(2j)!} \langle jm | \left[x_+ \left(a_+^\dagger \cos\frac{\theta}{2} + a_-^\dagger \sin\frac{\theta}{2} \right) \right. \\ &\qquad \left. + x_- \left(a_-^\dagger \cos\frac{\theta}{2} - a_+^\dagger \sin\frac{\theta}{2} \right) \right]^{2j} |0\rangle \\ &= \frac{1}{(2j)!} \langle jm | \left[a_+^\dagger \left(x_+ \cos\frac{\theta}{2} - x_- \sin\frac{\theta}{2} \right) \right. \\ &\qquad \left. + a_-^\dagger \left(x_- \cos\frac{\theta}{2} + x_+ \sin\frac{\theta}{2} \right) \right]^{2j} |0\rangle . \end{aligned} \qquad (17\text{-}135)$$

Now let us use the binomial theorem in reverse to write

$$G(x_+, x_-) = \sum_{m'=-j}^{j} \langle jm| \frac{\left[a_+^\dagger\left(x_+ \cos\frac{\theta}{2} - x_- \sin\frac{\theta}{2}\right)\right]^{j+m'}}{(j+m')!}$$

$$\times \frac{\left[a_-^\dagger\left(x_- \cos\frac{\theta}{2} + x_+ \sin\frac{\theta}{2}\right)\right]^{j-m'}}{(j-m')!} |0\rangle$$

$$= \sum_{m'=-j}^{j} \frac{\left(x_+ \cos\frac{\theta}{2} - x_- \sin\frac{\theta}{2}\right)^{j+m'}}{\sqrt{(j+m')!}} \frac{\left(x_- \cos\frac{\theta}{2} + x_+ \sin\frac{\theta}{2}\right)^{j-m'}}{\sqrt{(j-m')!}} \langle jm|jm'\rangle$$

$$= \frac{\left(x_+ \cos\frac{\theta}{2} - x_- \sin\frac{\theta}{2}\right)^{j+m}}{\sqrt{(j+m)!}} \frac{\left(x_- \cos\frac{\theta}{2} + x_+ \sin\frac{\theta}{2}\right)^{j-m}}{\sqrt{(j-m)!}}. \tag{17-136}$$

Let us choose the special values

$$x_+ = -\sin\frac{\theta}{2}\cos\frac{\theta}{2}, \qquad x_- = t - \cos^2\frac{\theta}{2}.$$

Then Eq. (17-136) becomes, on using (17-128):

$$\sum_{m''} \frac{(-1)^{j+m''}\left(\sin\frac{\theta}{2}\cos\frac{\theta}{2}\right)^{j+m''}\left(t - \cos^2\frac{\theta}{2}\right)^{j-m''}}{\sqrt{(j+m'')!\,(j-m'')!}} d^{(j)}_{mm''}(\theta)$$

$$= \frac{\left(\sin\frac{\theta}{2}\right)^{j+m}\left(\cos\frac{\theta}{2}\right)^{j-m}}{\sqrt{(j+m)!\,(j-m)!}} t^{j+m}(1-t)^{j-m}. \tag{17-137}$$

To extract $d_{mm'}^{(j)}(\theta)$ we differentiate both sides with respect to t, $j - m'$ times and then set $t = \cos^2(\theta/2)$. This gives

$$d^{(j)}_{mm'}(\theta) = (-1)^{j+m'} \sqrt{\frac{(j+m')!}{(j-m')!\,(j+m)!\,(j-m)!}} \left(\sin\frac{\theta}{2}\right)^{m'-m} \left(\cos\frac{\theta}{2}\right)^{m+m'}$$

$$\times \left\{ (1-t)^{m-m'} t^{-m-m'} \left(\frac{d}{dt}\right)^{j-m'} \left[t^{j+m} (1-t)^{j-m} \right] \right\}_{t=\cos^2(\theta/2)} . \tag{17-138}$$

The polynomial in $\cos^2(\theta/2)$ in braces is known, to within a constant, as a Jacobi polynomial. One should use (17-138) to verify Eq. (17-22) directly. It is obvious from (17-138) that $d_{mm'}{}^{(j)}(\theta)$ is real.

Isotopic spin, which has all the formal properties of an angular momentum, can be described in terms of this two-harmonic oscillator model. Thus an isotopic spin state $|I, I_3\rangle$, which would correspond to a pion or nucleon, is related in this model, to a state of the two oscillators with the numbers of quanta, $n_+ = I + I_3$, $n_- = I - I_3$. Perhaps, one might ask are these quanta physically real things — kinds of elementary particles — and are nucleons and pions just bound states of these new particles? In fact, a somewhat more complicated version of Schwinger's scheme involving three harmonic oscillators can be used to generate the SU(3) symmetries of the strongly interacting particles — the generalizations of isotopic spin that we mentioned before. The quanta of the three different oscillators are simply the *quarks* that compose the strongly interacting particles.

PROBLEMS

1. What are the Euler angles corresponding to a rotation about an axis $\hat{\alpha}$ by an angle $|\alpha|$?
2. Complete the demonstration that (17-28) is the addition theorem.
3. Verify Eq. (17-22) directly from (17-138).

Chapter 18
IDENTICAL PARTICLES

In following the motion of a collection of identical classical particles – say billiard balls – it is always possible for us to keep track of individual particles. For example, to find out where a given particle, "number 4," is at any later time we merely have to paint a "4" on that particle (carefully enough so that the paint won't affect its dynamics) and then notice at the later time which particle is labeled "4." It is not, however, possible to do this when studying a system of identical quantum mechanical particles, such as electrons. The problem is that quantum mechanical particles are too small for one to attach physical labels; they don't have enough degrees of freedom to enable one to mark each particle differently.

Now one might say that it really isn't necessary actually to label the particles in order to keep track of individuals; all that's necessary is to take a movie of the particles and then trace out the trajectories of the individual particles. This works perfectly well with classical particles, but every time one observes a quantum mechanical system one disturbs the system in some uncontrollable fashion; if the wave functions of the individual particles overlap at all then it is not possible, in principle, to keep track of the individual trajectories of the particles without seriously affecting their motion.

Identical quantum mechanical particles are completely indistinguishable from one another. If we have a state with one particle localized about the origin and a second identical particle localized about some other position $\mathbf{r}$, then the state with the second particle localized about the origin and the first about $\mathbf{r}$ is *not* a different state from the first, since there is no way of distinguishing the two situations. We would now like to study the consequences of this fundamental indistinguishability of identical quantum mechanical particles.

Suppose that $|\Psi\rangle$ is a state of n identical particles. Then the wave function of the state is

$$\psi(1,2,\ldots,n) = \langle 1,2,\ldots,n|\Psi\rangle,$$

where we use the index i to refer to both space and spin coordinates. We shall use the words "first particle" to refer to the particle described by the first *argument* of ψ, "second particle" to refer to the particle described by the second argument of ψ, etc. Thus $\psi(1, 2, \ldots, n)$ is the amplitude for observing the first particle at $\mathbf{r}_1$ with spin value s_1 (e.g., up or down for spin $1/2$), the second particle at $\mathbf{r}_2$ with spin s_2, etc.[1] [Question: what does it mean to refer to the "first" or "second" particle – isn't this labeling the particles, and didn't we agree that this wasn't possible? The point is that in describing the particles we can give them names, this doesn't require physically labeling the particles; but the mathematics, as well as all experimental apparatus, must treat all the particles absolutely equivalently.]

To say that two particles are identical means that there are no interactions that can distinguish them. Thus any operator corresponding to a physical observable of a collection of n identical particles must treat all the particles on the same footing, and therefore it must be a symmetrical function of the coordinates of the particles. For example, the Hamiltonian of n free particles of mass m is

$$H_0(1, 2, \ldots, n) = \frac{p_1^2}{2m} + \frac{p_2^2}{2m} + \cdots + \frac{p_n^2}{2m}, \tag{18-1}$$

and is clearly symmetric under the interchange of any $\mathbf{p}_i$ with $\mathbf{p}_j$. The subscripts refer to the particles; $\mathbf{p}_i$ is the momentum operator for the particle in the ith slot in the wave function. The momentum operator of the particles

$$\mathbf{P} = \mathbf{p}_1 + \mathbf{p}_2 + \cdots + \mathbf{p}_n, \tag{18-2}$$

the density operator

$$\rho(\mathbf{r}) = \sum_i \delta(\mathbf{r} - \mathbf{r}_i), \tag{18-3}$$

and the current operator

$$\mathbf{j}(\mathbf{r}) = \sum_i \left[\frac{\mathbf{p}_i}{2m}\,\delta(\mathbf{r} - \mathbf{r}_i) + \delta(\mathbf{r} - \mathbf{r}_i)\,\frac{\mathbf{p}_i}{2m}\right] \tag{18-4}$$

are all symmetric under the interchange of the coordinates and momenta of the ith and j^{th} particles. In general, any observable

[1] Do not confuse the subscripts used to label the positions and spin with the names of the particles. For example $\psi(r_2 s_2, r_1 s_1)$ is the amplitude for observing the first particle at r_2 with spin s_2, and the second particle at r_1 with spin s_1, etc.

$A(1, 2, \ldots, n)$ must be a symmetric function of the indices $1, 2, \ldots n$. One calls such an operator a *symmetric* operator.

PERMUTATIONS AND SYMMETRY

It is convenient to introduce operators that carry out permutations of the particles. Let us define the permutation operator P_{ij} by saying that when it acts on a state $|\Psi\rangle$ it interchanges the ith and j^{th} particles in the state. In terms of wave functions

$$\langle 1,2, \ldots, i, \ldots, j, \ldots, n|P_{ij}|\Psi\rangle = \langle 1,2, \ldots, j, \ldots, i, \ldots, n|\Psi\rangle \quad (18\text{-}5)$$

which we can write (using the same notation for P_{ij} and its representation in the position basis)

$$P_{ij}\psi(1,2, \ldots, i, \ldots, j, \ldots, n) = \psi(1,2, \ldots, j, \ldots, i, \ldots, n). \quad (18\text{-}6)$$

In other words, P_{ij} produces the amplitude for observing the jth particle at $\mathbf{r}_i$ with spin s_i and the ith particle at $\mathbf{r}_j$ with spin s_j. We can also define fancier permutations $P_{abc\ldots k}$, by saying that when $P_{abc\ldots k}$ acts on a state it replaces particle b with particle a, particle c with particle b, etc. and finally particle a with particle k; a is then where b used to be, b where c used to be, etc., so that in terms of wave functions $P_{abc\ldots k}$ replaces the ath argument of wave function by the bth, the bth, by the cth, etc. For example,

$$P_{123}\psi(1,2,3, \ldots, n) = \psi(2,3,1, \ldots, n).$$

For instance, if $\psi(1, 2, 3) = r_1^2(r_2^2 - r_3^2)$ then $P_{12}\psi(1,2,3) = r_2^2(r_1^2 - r_3^2)$ and $P_{123}\psi(1, 2, 3) = r_2^2(r_3^2 - r_1^2)$.

Any permutation can be written as a product of simple interchanges P_{ij}. For example,

$$P_{12}P_{13}\psi(123) = P_{12}\psi(321) = \psi(231) = P_{123}\psi(123)$$

so that $P_{123} = P_{12}P_{13}$. One should notice that permutations in general do not commute with one another, so the order in which one writes the P's is important. For example, $P_{12}P_{13}\psi(123) = \psi(231)$ while $P_{13}P_{12}\psi(123) = \psi(312)$.

If $A(1, 2, \ldots, n)$ is *any* operator, not necessarily symmetric, then

$$P_{ij}A(1,2, \ldots, i, \ldots, j, \ldots, n)$$
$$= A(1,2, \ldots, j, \ldots, i, \ldots, n)P_{ij}. \qquad (18\text{-}7)$$

Consider first the following example. Let A be the momentum operator p_1, and let $n = 2$. Then for any $|\Psi\rangle$,

$$\langle r_1s_1, r_2s_2|P_{12}p_1|\Psi\rangle = \langle r_2s_2, r_1s_1|p_1|\Psi\rangle$$
$$= \frac{\hbar}{i}\nabla_2\langle r_2s_2, r_1s_1|\Psi\rangle,$$

since p_1 acting to the left differentiates the argument in the first slot. But from (18-5),

$$\frac{\hbar}{i}\nabla_2[\langle r_2s_2, r_1s_1|\Psi\rangle] = \frac{\hbar}{i}\nabla_2[\langle r_1s_1, r_2s_2|P_{12}|\Psi\rangle] = \langle r_2s_2, r_1s_1|p_2P_{12}|\Psi\rangle.$$

Thus we must have $P_{12}p_1 = p_2P_{12}$. More generally,

$$P_{ij}A(\ldots, i, \ldots, j, \ldots)\psi(\ldots, i, \ldots, j, \ldots)$$

$$= A(\ldots, j, \ldots, i, \ldots)\psi(\ldots, j, \ldots, i, \ldots)$$

$$= A(\ldots, j, \ldots, i, \ldots)P_{ij}\psi(\ldots, i, \ldots, j, \ldots)$$

where in this equation A is the representation of the operator A in the position basis. Since this equation is true for arbitrary ψ, Eq. (18-7) follows. In particular if A is a symmetric operator, then from (18-7),

$$PA = AP \quad \text{or} \quad PAP^{-1} = A, \qquad (18\text{-}8)$$

where P is any permutation. The inverse of a permutation $P_{abc\ldots k}$ is just $P_{k\ldots cba}$; e.g., $P_{ij}^{-1} = P_{ji} = P_{ij}$.

It is important to notice that the expectation value of a symmetric operator in *any* state is independent of the order of labeling the particles, that is

$$\sum_{s_1 \ldots s_n} \int d^3r_1 \ldots d^3r_n\, \psi^*(1,2, \ldots, n)A(1,2, \ldots, n)\psi(1,2, \ldots, n)$$

is independent of the order of the points $1, 2, \ldots, n$ in ψ. This means that as long as we use only symmetric operators, no physical results can change if we interchange two particles, no matter *what* the symmetry properties of the state are under interchange of particles.

Suppose that $|\Psi\rangle$ is an eigenstate of a symmetric many particle Hamiltonian $H(1, 2, \ldots, n)$, with energy E. Then because H commutes with any permutation P we have

$$HP|\Psi\rangle = PH|\Psi\rangle = PE|\Psi\rangle = EP|\Psi\rangle . \tag{18-9}$$

Thus $P|\Psi\rangle$ is also an eigenstate of H with the same energy E. If $P|\Psi\rangle$ is not the same state as $|\Psi\rangle$ (to within a multiplicative constant) the energy level, E, of the Hamiltonian is degenerate. This phenomenon is known as *exchange degeneracy*. However, as we shall see, not all the degenerate states are physically allowed.

Let us consider a system consisting of just two identical particles. Then P_{12} commutes with $H(1, 2)$, and we can ask for the eigenfunctions of H that are simultaneously eigenfunctions of P_{12}. Since $P_{12}{}^2 = 1$ its eigenvalues are ± 1. If $\psi(1, 2)$ is any eigenfunction of H with energy E, then

$$\psi_s(1,2) = \psi(1,2) + \psi(2,1) \tag{18-10}$$

is an eigenfunction of H with the same eigenvalue and it is also an eigenfunction of P_{12} with eigenvalue +1:

$$P_{12}\psi_s(1,2) = \psi(2,1) + \psi(1,2) = \psi_s(1,2). \tag{18-11}$$

Similarly, the antisymmetric combination

$$\psi_a(1,2) = \psi(1,2) - \psi(2,1) \tag{18-12}$$

is an eigenfunction of P_{12} with eigenvalue -1, and an eigenfunction of H with eigenvalue E.

Now it is an experimental fact that a pair of identical particles will always be found to have a wave function that is also an eigenstate of P_{12}, and furthermore the eigenvalue, ± 1, depends *only* on the kind of particles involved! For example, the wave function of two electrons must *always* be antisymmetric

$$P_{12}\psi_a(1,2) = \psi_a(2,1) = -\psi_a(1,2), \tag{18-13}$$

while the wave function of two π^0 mesons must always be symmetric:

$$P_{12}\psi_s(1,2) = \psi_s(2,1) = \psi_s(1,2). \tag{18-14}$$

Particles that must have symmetric states are called *bosons;* pions, photons, and He^4 nuclei are such particles. Particles whose states must be antisymmetric under interchange of the two particles are called *fermions;* electrons, protons, and neutrons are examples of fermions.

In fact, there is a general connection between the spin of the particle and the possible symmetry of the states under interchange of the particles. Particles with integer spin ($s = 0, 1, 2, \ldots$) are always bosons while particles with half-integer spin ($s = 1/2, 3/2, 5/2, \ldots$) are always fermions. It is not at all obvious from the point of view of nonrelativistic quantum mechanics why this relation, called the *connection between spin and statistics,* is true, and one must look within the framework of relativistic quantum field theory to understand it fully theoretically. We shall get a first glimpse of this connection when we consider relativistic wave mechanics.

It is also an experimental fact that the wave function $\psi(1, 2, \ldots, n)$ of a system of n identical fermions must be completely antisymmetric under interchange of the coordinates of *any* two of the particles

$$\begin{aligned} P_{ij}\psi(1, \ldots, i, \ldots, j, \ldots, n) &= \psi(1, \ldots, j, \ldots, i, \ldots, n) \\ &= -\psi(1, \ldots, i, \ldots, j, \ldots, n), \end{aligned} \tag{18-15}$$

or $P_{ij}|\Psi\rangle = -|\Psi\rangle$ for all i, j. The amplitude for finding particle j at $\mathbf{r}_i$ with spin s_i and particle i at $\mathbf{r}_j$ with spin s_j is minus the amplitude for finding particle i at $\mathbf{r}_i$ with s_i and particle j at $\mathbf{r}_j$ with s_j. Thus the wave function of the 47 electrons in a silver atom is a completely antisymmetric function of the (space and spin) coordinates of the electrons. The wave function of an assembly of n identical bosons must be a completely symmetric function; it remains unchanged under the interchange of any two particles:[2]

$$P_{ij}\psi(1, \ldots, i, \ldots, j, \ldots, n) = \psi(1, \ldots, i, \ldots, j, \ldots, n) \tag{18-16}$$

or $P_{ij}|\Psi\rangle = |\Psi\rangle$ for all i, j.

It is important to notice that a system of n identical particles starting out in a completely symmetric, or antisymmetric, state must always remain in such a state. This is because any perturbation $V(1, 2, \ldots, n)$ that can act to change the state must be a

[2]Note that (18-15) and (18-16) are not the only possibilities for a collection of n particles with $n \geq 3$. Since the P_{ij} don't commute with each other in general, one can't form a complete basis out of states, such as (18-15) and (18-16), that are eigenfunctions of all the P_{ij} simultaneously. Other possible symmetry schemes for identical particles can't be ruled out on logical grounds alone, since the expectation value of symmetric observables are independent of the interchange of any two particles even if the state has no symmetry properties whatsoever. One different possibility, called *parastatistics,* has been discussed by O.W. Greenberg and A.M.L. Messiah, *Phys. Rev.* **138**, B 1115 (1965).

completely symmetric function of the coordinates of the n particles, and therefore commutes with any P. Thus if $P|\Psi\rangle = \pm|\Psi\rangle$ then

$$PV|\Psi\rangle = VP|\Psi\rangle = \pm V|\Psi\rangle, \tag{18-17}$$

so that $V|\Psi\rangle$ has the same symmetry as $|\Psi\rangle$.

Now we have a problem. Mustn't we, in studying the properties of a single electron in a hydrogen atom located in Urbana, take into account the fact that there are lots of other electrons in the world, and that the total wave function of all these electrons, including our one, must be a completely antisymmetric function? How is it that the simple one-electron wave functions that we wrote down for an electron in a hydrogen atom correctly described the properties of the atom? The answer is that if there is never any overlap between the wave function of our one electron and the wave functions of any other electrons, then we needn't worry about having to construct an antisymmetric wave function, including the other electrons, in order to describe correctly our electron. To see this let's suppose that our electron is described by a normalized wave function $\varphi_U(\mathbf{r})$ (neglecting the spin for simplicity) and that there is another electron in London described by a normalized wave function $\varphi_L(\mathbf{r})$, and let us suppose that φ_U and φ_L don't overlap, that is,

$$\varphi_U(\mathbf{r})\,\varphi_L(\mathbf{r}) = 0 \text{ for all } \mathbf{r}. \tag{18-18}$$

The correctly antisymmetrized and normalized wave function that describes the two electrons is

$$\psi(\mathbf{r}_1, \mathbf{r}_2) = \frac{1}{\sqrt{2}}\,[\varphi_U(\mathbf{r}_1)\varphi_L(\mathbf{r}_2) - \varphi_U(\mathbf{r}_2)\varphi_L(\mathbf{r}_1)]. \tag{18-19}$$

Let us ask, for example, for the probability of observing an electron at point $\mathbf{r}$. This is the probability $\int d^3r_2\,|\psi(\mathbf{r}, \mathbf{r}_2)|^2$, that electron 1 is at $\mathbf{r}$, regardless of where electron 2 is, plus $\int d^3r_1\,|\psi(\mathbf{r}_1, \mathbf{r})|^2$, the probability that electron 2 is at $\mathbf{r}$ regardless of where electron 1 is. Using the wave function (18-19) we find that this total probability is then

$$\begin{aligned} P(\mathbf{r}) &= \int d^3r_2\,|\psi(\mathbf{r}, \mathbf{r}_2)|^2 + \int d^3r_1\,|\psi(\mathbf{r}_1, \mathbf{r})|^2 \\ &= |\varphi_U(\mathbf{r})|^2 \int d^3r_1\,|\varphi_L(\mathbf{r}_1)|^2 + |\varphi_L(\mathbf{r})|^2 \int d^3r_2\,|\varphi_L(\mathbf{r}_2)|^2 \\ &\qquad - 2\mathrm{Re}[\varphi_U(\mathbf{r})\,\varphi_L^*(\mathbf{r}) \int d^3r_1\,\varphi_U^*(\mathbf{r}_1)\,\varphi_L(\mathbf{r}_1)]. \end{aligned} \tag{18-20}$$

Notice that the last term, the interference between the two terms of (18-19), always vanishes when φ_U and φ_L don't overlap. Since φ_U and φ_L are normalized, (18-20) becomes

$$P(\mathbf{r}) = |\varphi_U(\mathbf{r})|^2 + |\varphi_L(\mathbf{r})|^2. \tag{18-21}$$

If the point $\mathbf{r}$ is in Urbana, then simply

$$P(\mathbf{r}) = |\varphi_U(\mathbf{r})|^2 \tag{18-22}$$

since an electron in London has zero probability of being found in Urbana. But (18-22) is exactly the result we would have found had we not worried about the other electron in the first place. It is generally true that as long as the wave functions φ_U and φ_L don't overlap neglecting to construct an antisymmetric wave function will have no effect. The rule is that one must antisymmetrize the wave function, or symmetrize for bosons, for only the relevant particles. If there is no chance of the wave functions overlapping then symmetrization or antisymmetrization is unimportant. For example, in the classical limit, the wave functions of particles are sharply localized in space, and so one expects no effects of symmetrization for classical systems.

What kind of wave functions must systems of identical *composite particles* such as atoms, have? Consider for example the wave function, $\psi(\mathbf{r}_{e_1}, \mathbf{r}_{p_1}; \mathbf{r}_{e_2}, \mathbf{r}_{p_2})$ that describes two hydrogen atoms. $\mathbf{r}_{e_1}$ is the position of the electron in the first atom, $\mathbf{r}_{p_1}$ the position of the proton in the first atom, etc., and we suppress the spin coordinates. Now ψ must change sign under the interchange of the two electron coordinates, since electrons are fermions. It must also change sign under the interchange of the two proton coordinates, since they are also fermions. It need not change sign however under the interchange of a proton coordinate, $\mathbf{r}_{p_1}$ say, with an electron coordinate, $\mathbf{r}_{e_2}$ say, since electrons and protons are not identical particles. Now if we interchange the two hydrogen atoms the wave function becomes

$$\begin{aligned} \psi(\mathbf{r}_{e_2}, \mathbf{r}_{p_2}; \mathbf{r}_{e_1}, \mathbf{r}_{p_1}) &= -\psi(\mathbf{r}_{e_1}, \mathbf{r}_{p_2}; \mathbf{r}_{e_2}, \mathbf{r}_{p_1}) \\ &= \psi(\mathbf{r}_{e_1}, \mathbf{r}_{p_1}; \mathbf{r}_{e_2}, \mathbf{r}_{p_2}); \end{aligned} \tag{18-23}$$

that is, the wave function remains unchanged. Thus hydrogen atoms behave as bosons. It is easy to see that the general rule is that composite particles containing an even number of fermions and any number of bosons (and therefore having integer spin), must be bosons, while composite particles containing an odd number of fermions and any number of bosons (and therefore having half-integer spin) must be fermions. For example, He^4 atoms, containing two protons, two neutrons, and two electrons are bosons, while He^3 atoms, with only one neutron, are fermions.

The antisymmetry of the wave function for fermions implies immediately that identical fermions must obey the *exclusion principle.* Since $\psi(2, 1, 3, \ldots) = -\psi(1, 2, 3, \ldots)$ we see that if $1 = 2$, then ψ vanishes. This means that there is zero amplitude for finding two fermions having the same values of their coordinates. Two identical fermions cannot occupy the same space point *if* they have the same spin orientation, nor can they have the same value of momentum if they have the same spin orientation; in general, two identical fermions cannot occupy the same state.

Consider two spin ½ fermions. They can have four different total spin states, one singlet and three triplet states. The singlet spin state is

$$\chi_{0,0}(s_1, s_2) = \frac{1}{\sqrt{2}}[\chi_\uparrow(s_1)\chi_\downarrow(s_2) - \chi_\downarrow(s_1)\chi_\uparrow(s_2)] \tag{18-14}$$

where $\chi_\uparrow(s)$ is the state with spin definitely up and $\chi_\downarrow(s)$ is the state with spin definitely down [$\chi_\uparrow(s = \uparrow) = 1$, $\chi_\uparrow(s = \downarrow) = 0$, etc.]; it is clearly antisymmetric under the interchange of s_1 and s_2. In order that the total wave function $\psi(\mathbf{r}_1, \mathbf{r}_2)\chi_{0,0}(s_1, s_2)$ of two particles in a singlet state be antisymmetric under $\mathbf{r}_1 \leftrightarrow \mathbf{r}_2$, $s_1 \leftrightarrow s_2$, the spatial part of the wave function $\psi(\mathbf{r}_1, \mathbf{r}_2)$ must be symmetric under interchange of $\mathbf{r}_1$ and $\mathbf{r}_2$. Similarly if the spins are in one of the three triplet states

$$\chi_{1,1}(s_1, s_2) = \chi_\uparrow(s_1)\chi_\uparrow(s_2)$$

$$\chi_{1,0}(s_1, s_2) = \frac{1}{\sqrt{2}}[\chi_\uparrow(s_1)\chi_\downarrow(s_2) + \chi_\downarrow(s_1)\chi_\uparrow(s_2)]$$

$$\chi_{1,-1}(s_1, s_2) = \chi_\downarrow(s_1)\chi_\downarrow(s_2), \tag{18-25}$$

which are symmetric under the interchange of the spin coordinates, then the spatial wave function must be antisymmetric under the interchange of $\mathbf{r}_1$ and $\mathbf{r}_2$. Conversely, if the spatial wave function is symmetric then the spins are in a singlet state, and if it is antisymmetric the spins are in a triplet state. [Notice that spin states of definite total S^2 have definite permutation symmetry.]

STATES OF NONINTERACTING IDENTICAL PARTICLES

It is instructive to write down explicitly wave functions for N identical noninteracting particles in a potential well $V(\mathbf{r})$, as in Fig. (18-1). The Hamiltonian of a single particle in the well is

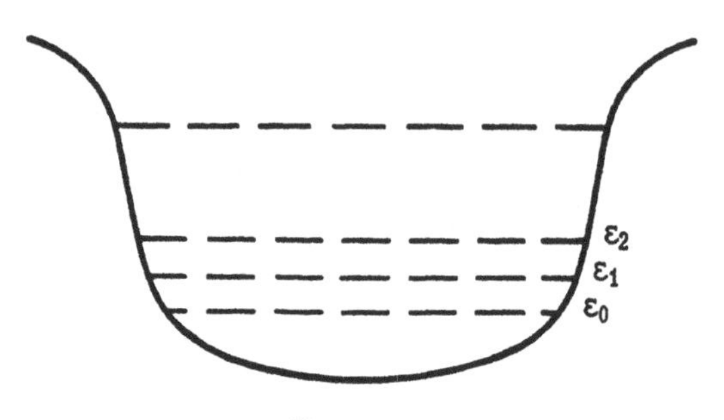

Fig. 18-1

$$H_0(1) = \frac{p_1^2}{2m} + V(\mathbf{r}_1). \tag{18-26}$$

Let us write the orthonormal energy eigenstates of H_0 as $\varphi_0(\mathbf{r}_1)$, $\varphi_1(\mathbf{r}_1)$, $\varphi_2(\mathbf{r}_1)$, ...,with corresponding energies ε_0, ε_1, ε_2, ... The Hamiltonian for N particles in the well is just the sum of their individual energy operators:

$$H = H_0(1) + H_0(2) + \cdots + H_0(N). \tag{18-27}$$

It is easy to write down solutions of the Schrödinger equation

$$H\Psi(1, \ldots, N) = E\Psi(1, \ldots, N). \tag{18-28}$$

For example,

$$\Psi(1, 2, \ldots, N) = \varphi_a(1)\varphi_b(2) \cdots \varphi_n(N) \tag{18-29}$$

is a solution which corresponds to the first particle in state a with energy ε_a, the second in state b with energy ε_b, etc. The total energy of the state is simply

$$E = \varepsilon_a + \varepsilon_b + \cdots + \varepsilon_n. \tag{18-30}$$

Because fermions always have nonzero spin, we must include in the φ's in (18-27), the state of the spin as well as the spatial state. For spin ½ fermions we would have, corresponding to a single level $\varphi_a(\mathbf{r}_1)$ of the well, two possible single particle states, $\varphi_a(\mathbf{r}_1)\chi_\uparrow(s_1)$ and $\varphi_a(\mathbf{r}_1)\chi_\downarrow(s_1)$. We shall assume in this case when we write $\varphi_a(1)$ that the subscript on φ refers to the spin state as well as the energy level of the well, e.g., $\varphi_a(1)$ could refer to $\varphi_a(\mathbf{r}_1)\chi_\uparrow(s_1)$ or $\varphi_a(\mathbf{r}_1)\chi_\downarrow(s_1)$. In general, (18-29) is not an admissible solution for N identical particles since it lacks symmetry under interchange of any two particles. There are many other solutions to (18-28) with the energy (18-30). For instance, the state

$$\Psi'(1,2,\ldots,N) = \varphi_a(2)\varphi_b(1)\cdots\varphi_n(N), \tag{18-31}$$

in which the first particle is in state b and the second is in state a, is a different solution from (18-29) if a and b are different single particle states, but it also has the same energy (18-30). This is just an example of exchange degeneracy. Any permutation of the indices 1, 2, ..., N on the right side of (18-29) yields, in general, another solution to (18-28) with the same energy.

To construct admissible eigenstates for N identical particles we must take the linear combination of states of the form (18-28) that is completely symmetric for bosons, or, for fermions, completely antisymmetric. Thus in the case of bosons, this state is

$$\Psi_s(1,2,\ldots,N) = \sum_P P\varphi_a(1)\varphi_b(2)\cdots\varphi_n(N) = \varphi_a(1)\varphi_b(2)\cdots\varphi_n(N)$$
$$+\varphi_a(2)\varphi_b(1)\cdots\varphi_n(N)+\cdots+\varphi_a(3)\varphi_b(5)\cdots\varphi_n(8)+\cdots \tag{18-32}$$

where the sum is over all N! possible permutations of the indices 1, ..., N. The energy of (18-32) is just $\varepsilon_a+\varepsilon_b+\cdots+\varepsilon_n$. The state (18-32) is completely symmetric, since interchanging any two indices merely rearranges the terms in the sum. For instance, if we interchange 1 and 2, then as we've written (18-32), the first term becomes the second, the second the first, and so on. In general, the rule for constructing a completely symmetric state from *any* arbitrary state $\Psi(1,2,\ldots,N)$ is to sum Ψ over all N! possible permutations of 1, 2, ..., N:

$$\Psi_s(1,2,\ldots,N) = \sum_P P\Psi(1,2,\ldots,N). \tag{18-33}$$

[Note that if Ψ is antisymmetric in any particular pair of indices, then the symmetrization process (18-33) gives zero.]

Now if the particles are fermions we must construct a completely antisymmetric state. Recall that any permutation can be represented as a product of interchanges. Though there are many different ways of doing this, e.g.,

$$P_{123} = P_{12}P_{13} = P_{24}P_{14}P_{24}P_{13}, \text{ etc.,}$$

what all the different ways have in common is that either they *all* involve an even number of interchanges, or else they all involve an odd number of interchanges. If the permutation P is representable by an odd number of interchanges, it is called an *odd* permutation, and for an antisymmetric state Ψ_a,

$$P\Psi_A = -\Psi_A;$$

if the representation requires an even number of interchanges, P is called an *even* permutation, and

$$P\Psi_A = +\Psi_A.$$

To form an antisymmetric state from (18-28) we must, in summing over all permutations include a minus sign in front of all terms gotten from (18-29) by an odd permutation. If we call the *sign* of an even permutation +1 and the sign of an odd permutation −1, and denote the sign by the symbol $(-1)^P$, then we can write the antisymmetric wave function formed from (18-29) as

$$\begin{aligned}\Psi_A(1,2,\ldots,N) &= \sum_P (-1)^P P\varphi_a(1)\varphi_b(2)\cdots\varphi_n(N)\\ &= \varphi_a(1)\varphi_b(2)\cdots\varphi_n(N) - \varphi_a(2)\varphi_b(1)\cdots\varphi_n(N) \pm \cdots\end{aligned} \tag{18-34}$$

The general rule for constructing an antisymmetric state from an arbitrary state $\Psi(1,\ldots,N)$ is to sum over all N! permutations of $1,\ldots,N$, multiplying by the sign of the permutation:

$$\Psi_A(1,\ldots,N) = \sum_P (-1)^P P\Psi(1,\ldots,N). \tag{18-35}$$

Notice now that if any two of the states φ in (18-29) are the same (for both the space and spin degrees of freedom) then the antisymmetrized state (18-34) vanishes. This is just a statement again of the exclusion principle – two fermions cannot be put in the same state. They can, of course, have the same spatial wave function, as long as they have opposite spin. Thus one can put two fermions, of opposite spin, in each state of the well. N spin $1/2$ fermions must therefore occupy at least N/2 different states of the well.

Another way of writing the sum in (18-34) is as a determinant:

$$\Psi_A(1,2,\ldots,N) = \begin{vmatrix} \varphi_a(1) & \varphi_a(2) & \ldots & \varphi_a(N) \\ \varphi_b(1) & \varphi_b(2) & \ldots & \varphi_b(N) \\ \vdots & \vdots & & \vdots \\ \varphi_n(1) & \varphi_n(2) & \ldots & \varphi_n(N) \end{vmatrix}. \tag{18-36}$$

This determinant is often called a *Slater determinant.*

Let us calculate the normalization of the state (18-34). To do this we must integrate the absolute square of (18-34) over all N position vectors, $\mathbf{r}_i$, and sum over all N spin variables, s_i ($s_i = \uparrow$ and $s_i = \downarrow$). This gives

$$\langle \Psi_A | \Psi_A \rangle = \sum_{s_1 \ldots s_N} \int d^3r_1 \cdots d^3r_N \sum_{PP'} (-1)^P (-1)^{P'} \times [P\varphi_a^*(1) \cdots \varphi_n^*(N)][P'\varphi_a(1) \cdots \varphi_n(N)]. \tag{18-37}$$

Terms corresponding to $P \neq P'$ in (18-37) will give zero, since the different φ's are orthogonal. For example a term proportional to $\varphi_a^*(1)\varphi_c(1)$ clearly gives zero when integrated over $\mathbf{r}_1$ and summed over s_1. Thus (18-37) becomes

$$\langle \Psi_A | \Psi_A \rangle = \sum_{s_1 \ldots s_N} \int d^3r_1 \cdots d^3r_N \sum_P P|\varphi_a(1)|^2|\varphi_b(2)|^2 \cdots |\varphi_n(N)|^2.$$

Since the φ's are normalized to one, each separate permuted term is just one, and we are left with the sum, over all N! permutations, of one. Thus

$$\langle \Psi_A | \Psi_A \rangle = N!; \tag{18-38}$$

to normalize (18-34) to unity we must divide it by $\sqrt{N!}$ We leave it as an exercise to verify that the corresponding result for the Bose wave function (18-32) is

$$\langle \Psi_s | \Psi_s \rangle = \frac{N!}{N_a! N_b! \cdots N_n!} \tag{18-39}$$

where N_i is the number of times the single particle state φ_i occurs in (18-31).

As an example, let us consider just two states of the well $\varphi_0(\mathbf{r})$ and $\varphi_1(\mathbf{r})$ and see what the possible states are for two particles. First of all if the particles are distinguishable, we can construct four distinct states

$$\varphi_0(\mathbf{r}_1)\varphi_0(\mathbf{r}_2), \qquad \varphi_1(\mathbf{r}_1)\varphi_1(\mathbf{r}_2), \tag{18-40}$$

$$\varphi_0(\mathbf{r}_1)\varphi_1(\mathbf{r}_2), \qquad \varphi_0(\mathbf{r}_2)\varphi_1(\mathbf{r}_1). \tag{18-41}$$

The possible energies are $2\varepsilon_0$, $2\varepsilon_1$, and $\varepsilon_0 + \varepsilon_1$; the latter energy level is twofold degenerate. If the particles are identical bosons, both states (18-40) are allowable, but the two states (18-41) are physically indistinguishable; in fact, the only state of two identical bosons with one particle in φ_0 and the other in φ_1 is

$$\varphi_0(\mathbf{r}_1)\varphi_1(\mathbf{r}_2)+\varphi_0(\mathbf{r}_2)\varphi_1(\mathbf{r}_1). \tag{18-42}$$

Thus the degeneracy of the level $\varepsilon_0+\varepsilon_1$ isn't present for identical bosons. Finally if the two particles are fermions, then if they have the same spin orientation, their only possible state is

$$\varphi_0(\mathbf{r}_1)\varphi_1(\mathbf{r}_2)-\varphi_0(\mathbf{r}_2)\varphi_1(\mathbf{r}_1). \tag{18-43}$$

The particles can be put in the states (18-40) only if they have different spin orientations.

Next let us look at the differences between the ground state of N bosons in the well and that of N fermions. The ground state of N bosons is extremely simple; they all occupy the lowest level φ_0, of the well. The normalized ground state wave function is

$$\Psi_0(1,2,\ldots,N)=\varphi_0(1)\varphi_0(2)\cdots\varphi_0(N) \tag{18-44}$$

and the ground state energy is $N\varepsilon_0$. Even if we have 10^{23} bosons in a box, in their ground state they all occupy the lowest level of the box; this macroscopic occupation of one single particle state is eventually responsible for the superfluid behavior of liquid He^4.

Such a state would, of course, violate the exclusion principle for fermions. In fact, the ground state of N spin ½ fermions contains only two particles (of opposite spin) in φ_0, two in φ_1, etc., and finally two in $\varphi_{N/2}$ if N is even, or if N is odd, there are two particles in $\varphi_{(N-1)/2}$ but only one in $\varphi_{(N+1)/2}$. This difference in the ground states of even numbers and odd numbers of fermions is unimportant in systems of enormous numbers of fermions, such as electrons in a metal; however, it leads to some quite different behavior of even and odd nuclei.

The ground state energy for N fermions is $2(\varepsilon_0+\varepsilon_1+\cdots+\varepsilon_{N/2})$ for N even, and $2(\varepsilon_0+\cdots+\varepsilon_{(N-1)/2})+\varepsilon_{(N+1)/2}$ for N odd. This energy is always greater than the ground state energy of a corresponding number of bosons in the well; this additional zero-point energy is an important consequence of the exclusion principle.

SCATTERING OF IDENTICAL PARTICLES

The scattering of two identical particles exhibits striking effects of the symmetry of their wave function. Consider first the scattering of two spinless bosons, say an α particle by another α particle. The wave function of the two α's can be written as

$$\Psi(\mathbf{r}_1,\mathbf{r}_2)=e^{i\mathbf{p}\cdot(\mathbf{r}_1+\mathbf{r}_2)}\psi(\mathbf{r}), \tag{18-45}$$

where $2\mathbf{p}$ is the momentum of the center of mass of the two particles, and $\mathbf{r} = \mathbf{r}_1 - \mathbf{r}_2$ is their relative coordinate. The wave function, $e^{i\mathbf{p}\cdot(\mathbf{r}_1+\mathbf{r}_2)}$, of the center of mass is obviously symmetric under interchange of $\mathbf{r}_1$ and $\mathbf{r}_2$. Therefore, in order that $\Psi(\mathbf{r}_1, \mathbf{r}_2) = \Psi(\mathbf{r}_2, \mathbf{r}_1)$, we must have $\psi(\mathbf{r}_1 - \mathbf{r}_2) = \psi(\mathbf{r}_2 - \mathbf{r}_1)$, or

$$\psi(\mathbf{r}) = \psi(-\mathbf{r}); \tag{18-46}$$

that is, $\psi(\mathbf{r})$ has even parity. This implies that the only possible eigenstates of angular momentum of the two particles are those with *even* l, since $Y_{lm}(-\hat{\mathbf{r}}) = (-1)^l Y_{lm}(\hat{\mathbf{r}})$.

Neglecting the symmetry we would write the relative wave function asymptotically as

$$e^{i\mathbf{k}\cdot\mathbf{r}} + f(\theta)\frac{e^{ikr}}{r}. \tag{18-47}$$

The correct asymptotic wave function is the symmetrized version of (18-47),

$$\psi(\mathbf{r}) = e^{i\mathbf{k}\cdot\mathbf{r}} + e^{-i\mathbf{k}\cdot\mathbf{r}} + [f(\theta) + f(\pi - \theta)]\frac{e^{ikr}}{r}. \tag{18-48}$$

This involves $f(\pi - \theta)$ since under interchange of the two particles, $\mathbf{r} \to -\mathbf{r}$, we have $\theta \to \pi \;\; \theta$, and $|\mathbf{r}| \to |\mathbf{r}|$. The first two terms in (18-48) represent the incident waves in the center of mass frame; we cannot distinguish the target particle from the incident particle and hence either particle has equal amplitude for being either. The coefficient of e^{ikr}/r is the scattering amplitude:

$$f_s(\theta) = f(\theta) + f(\pi - \theta). \tag{18-49}$$

We may give a simple interpretation to the presence of the two terms in the scattering amplitude: the process in Fig. 18-2(a) in which the incident particle is scattered by angle θ, is indistinguishable from the process in Fig. 18-2(b) in which the incident particle is scattered by $\pi - \theta$. In both processes one particle emerges at angle θ and an identical one emerges at angle $\pi - \theta$. The total amplitude for particles to emerge at θ and $\pi - \theta$ is just the sum of amplitudes for these two processes, $f(\theta) + f(\pi - \theta)$.

The differential cross section is

$$\frac{d\sigma}{d\Omega} = |f(\theta) + f(\pi - \theta)|^2 = |f(\theta)|^2 + |f(\pi - \theta)|^2 + 2\,\mathrm{Re}\, f^*(\theta) f(\pi - \theta). \tag{18-50}$$

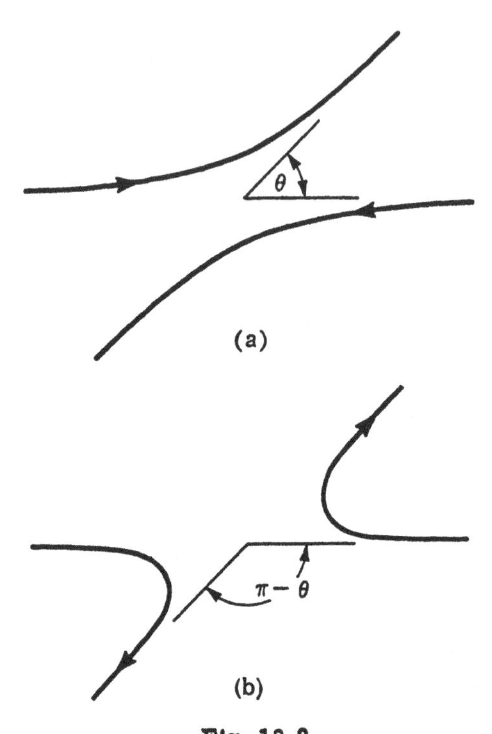

Fig. 18-2
Two indistinguishable processes in the scattering of identical particles.

The first two terms are just the cross sections for the individual scattering processes pictured above. They would be the total cross section for observing one of the two particles (either one) at θ, if the particles were distinguishable:

$$\left(\frac{d\sigma}{d\Omega}\right)_{\text{distinguishable}} = |f(\theta)|^2 + |f(\pi - \theta)|^2. \tag{18-51}$$

The last term in (18-50) is the effect of the exchange symmetry; one may think of it as being due to interference between the two processes. This exchange term generally makes the differential cross section more wiggly. If, for example, $f(\theta)$ were something like $e^{i\lambda\theta}$ then the cross section for scattering of two distinguishable particles would be a constant, whereas the exchange term in (18-50) would behave as $\cos \lambda(\pi - 2\theta)$.

Consider scattering through 90°. Then $f(\theta) = f(\pi - \theta) = f(\pi/2)$. If the particles are distinguishable, the cross section for observing a scattered particle (either one) at 90° is $2|f(\pi/2)|^2$. On the other hand, if the particles are indistinguishable, (18-50) gives

$$\left(\frac{d\sigma}{d\Omega}\right)_{90°} = 4\left|f\left(\frac{\pi}{2}\right)\right|^2, \tag{18-52}$$

which is *twice* the result for distinguishable particles! This simple effect of identity is well confirmed experimentally.

If in terms of partial waves

$$f(\theta) = \sum_l i^l P_l(\cos\theta)(2l+1)f_l \tag{18-53}$$

then

$$f(\theta)+f(\pi-\theta) = 2\sum_{l\text{ even}} i^l (2l+1)P_l(\cos\theta)f_l, \tag{18-54}$$

since

$$P_l(\cos(\pi-\theta)) = P_l(-\cos\theta) = (-1)^l P_l(\cos\theta).$$

Next let's work out the effects of exchange symmetry on the scattering of two spin ½ fermions – the scattering of electrons by electrons, for example. For simplicity, let us suppose that the interaction between the particles doesn't depend on their spin. Then their wave function takes the form

$$\Psi(1,2) = e^{i\mathbf{p}\cdot(\mathbf{r}_1+\mathbf{r}_2)}\psi(\mathbf{r}_1-\mathbf{r}_2)\chi(s_1, s_2), \tag{18-55}$$

where $\chi(s_1, s_2)$ is the spin part of the wave function. If the spins are in a singlet state then the relative wave function $\psi(\mathbf{r})$ must be symmetric. The scattering in singlet spin states is thus identical to that of two spinless bosons, and

$$\left(\frac{d\sigma}{d\Omega}\right)_{\text{singlet}} = |f(\theta)+f(\pi-\theta)|^2 \tag{18-56}$$

where $f(\theta)$ is the scattering amplitude for distinguishable particles. If the spins of the two particles are in a triplet state, the total wave function Ψ will be antisymmetric only if the relative spatial wave function is antisymmetric:

$$\psi(\mathbf{r}) = -\psi(-\mathbf{r}) \tag{18-57}$$

This implies that only *odd l* relative orbital angular momentum states are possible for triplet spin states.

The asymptotic form of the relative wave function is found by antisymmetrizing the wave function for distinguishable particles. Thus asymptotically

$$\psi(\mathbf{r}) = e^{i\mathbf{k}\cdot\mathbf{r}} - e^{-i\mathbf{k}\cdot\mathbf{r}} + [f(\theta) - f(\pi - \theta)]\,\frac{e^{ikr}}{r}; \tag{18-58}$$

the scattering amplitude is

$$f_a = f(\theta) - f(\pi - \theta) \tag{18-59}$$

and the cross section is

$$\begin{aligned}\left(\frac{d\sigma}{d\Omega}\right)_{\text{triplet}} &= |f(\theta) - f(\pi - \theta)|^2 \\ &= |f(\theta)|^2 + |f(\pi - \theta)|^2 - 2\,\mathrm{Re}\, f^*(\theta) f(\pi - \theta).\end{aligned} \tag{18-60}$$

Again we can look at the two terms in the scattering amplitude as representing the two processes in Fig. 18-2; however, Eq. (18-59) now says that instead of adding the amplitudes for these indistinguishable processes, as for bosons and singlet spins, we must *subtract* the individual amplitudes to form the total scattering amplitude $f_a(\theta)$ in (18-59). An interesting consequence of having to subtract the amplitudes before squaring is that fermions in triplet states can never scatter through 90°, since

$$f_a\left(\frac{\pi}{2}\right) = f\left(\frac{\pi}{2}\right) - f\left(\pi - \frac{\pi}{2}\right) = 0.$$

Note that the partial wave expansion of the antisymmetrized scattering amplitude is

$$f(\theta) - f(\pi - \theta) = 2 \sum_{l\ \text{odd}} i^l (2l+1) P_l(\cos\theta) f_l. \tag{18-61}$$

If the spins of either the target or the beam particles are unpolarized, then the spins of the scattering particles are equally likely to be found in each of the four spin states, (18-54) and (18-56). The observed cross section is then an average of three parts triplet, because there are three triplet states, and one part singlet:

$$\begin{aligned}\left(\frac{d\sigma}{d\Omega}\right)_{\text{unpol.}} &= \frac{3}{4}\left(\frac{d\sigma}{d\Omega}\right)_{\text{triplet}} + \frac{1}{4}\left(\frac{d\sigma}{d\Omega}\right)_{\text{singlet}} \\ &= |f(\theta)|^2 + |f(\pi - \theta)|^2 - \mathrm{Re}\, f^*(\theta) f(\pi - \theta).\end{aligned} \tag{18-62}$$

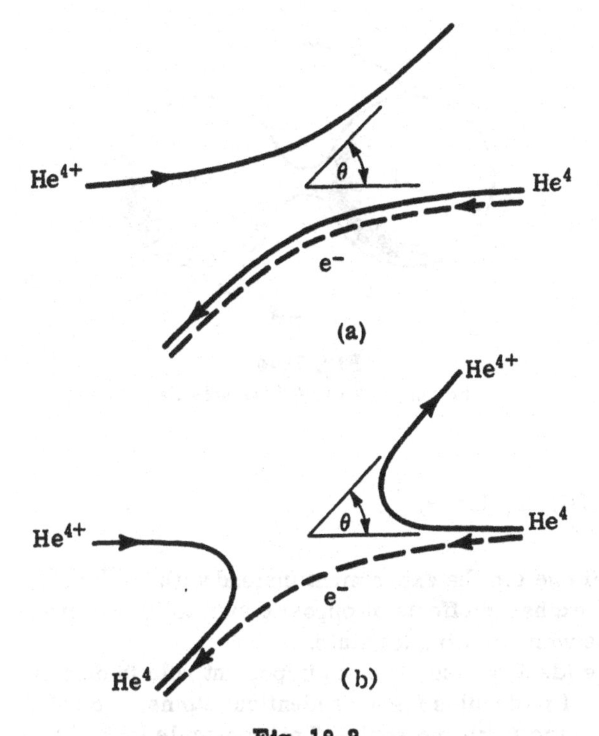

Fig. 18-3

Direct scattering (a) and charge exchange scattering (b) of a He^{4+} ion by a He^4 atom. These two processes are indistinguishable.

Effects of identity can also show up in the scattering of seemingly nonidentical particles. Consider for example, the elastic scattering of a He^{4+} ion from a He^4 atom, both in their ground states. These are certainly distinguishable particles, since they have different charges. However the He^4 atom can be regarded as a composite of a He^{4+} ion and an electron, and this He^{4+} is indistinguishable from the bombarding ion. This means that the direct scattering process, Fig. 18-3(a), in which the incident ion is simply scattered by angle θ (and the electron — the dotted line — and the He^{4+} ion — the solid line — that make up the He^4 atom stay together) is indistinguishable from the charge exchange process, Fig. 18-3(b), in which the incident ion is scattered by $\pi - \theta$ and an electron hops from the He^4 atom to the incident ion. If we let $f(\theta)$ be the amplitude for the direct process and $f_{ex}(\theta)$ be the amplitude for the incident ion to be scattered by θ in the charge exchange process, then the total elastic scattering amplitude is

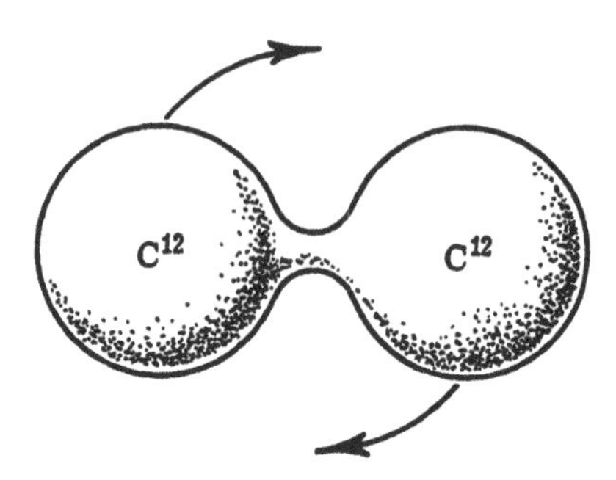

Fig. 18-4
Rotating state of a $(C^{12})_2$ molecule.

$$f_{total}(\theta) = f(\theta) + f_{ex}(\pi - \theta). \tag{18-63}$$

Note that if one did the experiment instead with He^{3+} on He^3 then one would find exchange effects of opposite sign to those here if the nuclear spins were in a triplet state.[3]

Particle identity also plays an important role in determining the properties of molecules made of identical atoms. One of the simplest examples is the $(C^{12})_2$ molecule. This molecule looks like a dumbbell, Fig. 18-4, and its low-lying excited states are the rotations of the dumbbell about its midpoint. The energy of these states is essentially

$$E_{rot} = \hbar^2 \frac{l(l+1)}{2\mathcal{I}} \tag{18-64}$$

where l is the angular momentum of the state and $\mathcal{I}$ is the moment of inertia of the molecule. Now the C^{12} nucleus has zero spin and is a boson. Thus these low-lying states can have only *even* l in order that they be symmetric under the interchange of the two C^{12} nuclei. Odd l states don't appear in the rotational spectrum. [One can essentially ignore the electrons here since they remain in their ground state in these slowly rotating states.]

Another interesting example is the H_2 molecule. The spins of the two protons can be in either a singlet or a triplet state. When they are in a singlet state the molecule is known as *parahydrogen*, and in the triplet state the molecule is called *orthohydrogen*. In the triplet state, the relative orbital wave function for the protons must have

---

[3]See W. Aberth, D.C. Lorents, R.P. Marchi, and F.T. Smith, *Phys. Rev. Letters* **14**, 776 (1965).

odd parity and hence only odd l. Thus the rotational levels of orthohydrogen must have odd l, and similarly those of parahydrogen have even l. Now the point is that it is extremely difficult to convert an ortho molecule to a para, or vice versa. A gas of pure parahydrogen will remain such for many, many months. [There are, however, catalysts such as paramagnetic gases that speed up the conversion.] Thus the properties of a gas of H_2 depends upon the ratio of ortho to para molecules. For example because the rotational spectra, (18-64), differ for the two forms of H_2

$$E_{rot, para} = 0, \frac{3}{\mathscr{I}}, \frac{10}{\mathscr{I}}, \frac{21}{\mathscr{I}}, \ldots$$

$$E_{rot, ortho} = \frac{1}{\mathscr{I}}, \frac{6}{\mathscr{I}}, \frac{15}{\mathscr{I}}, \ldots \tag{18-65}$$

(in units of $\hbar^2$), the specific heat of an H_2 gas depends on the para-ortho ratio. The absolute ground state of the H_2 molecule, in which the protons have no relative rotation, is a para state.

In the simple two-fermion examples we have considered, the total wave function

$$\Psi(1,2) = \psi(\mathbf{r}_1, \mathbf{r}_2)\chi(s_1, s_2) \tag{18-66}$$

is always made anti-symmetric by making ψ symmetric and χ anti-symmetric, or vice versa. When one considers three or more particles there are in fact other possible ways of arranging for overall antisymmetry of Ψ. Consider just three spin $1/2$ fermions. One way to make $\Psi(1, 2, 3)$ antisymmetric is to take a symmetric $\chi(s_1, s_2, s_3)$ [thus $S = 3/2$] times an antisymmetric $\psi(\mathbf{r}_1, \mathbf{r}_2, \mathbf{r}_3)$. The other way around won't work, since it isn't possible to construct a completely antisymmetric spin wave function $\chi(s_1, s_2, s_3)$ from just the two choices, up or down, for each spin. (Try.) There is another possibility though. Suppose that we take a $\chi(s_1, s_2, s_3)$ that is antisymmetric in s_2 and s_3, for example, i.e., s_2 and s_3 in a spin singlet, [$S = 1/2$],

$$\chi(s_1, s_2, s_3) = \chi_\uparrow(s_1)[\chi_\uparrow(s_2)\chi_\downarrow(s_3) - \chi_\downarrow(s_2)\chi_\uparrow(s_3)], \tag{18-67}$$

and multiply it by a spatial wave function $\psi(\mathbf{r}_1, \mathbf{r}_2, \mathbf{r}_3)$ that is symmetric in $\mathbf{r}_2$ and $\mathbf{r}_3$. This yields a total wave function antisymmetric in 2 and 3, but with no symmetry properties under interchange of 1 and 2 or 1 and 3. However the wave function

$$\Psi(1,2,3) = \chi(s_1, s_2, s_3)\psi(\mathbf{r}_1, \mathbf{r}_2, \mathbf{r}_3) + \chi(s_2, s_3, s_1)\psi(\mathbf{r}_2, \mathbf{r}_3, \mathbf{r}_1) + \chi(s_3, s_1, s_2)\psi(\mathbf{r}_3, \mathbf{r}_1, \mathbf{r}_2) \tag{18-68}$$

is totally antisymmetric under interchange of any two sets of space and spin coordinates. (Verifying this is left as an exercise.)

IDENTITY AND ISOTOPIC SPIN

Let us briefly see how the symmetry requirements for identical particles can be worked into the isotropic spin formalism. Neutrons and protons, for example, are distinguishable particles, even though they behave so similarly in strong interactions. Therefore if we have a state of many nucleons, we only require that it be antisymmetric under interchange of two neutrons, or two protons, but it need not be antisymmetric under the interchange of a neutron and a proton. The isotopic spin formalism, on the other hand, treats neutrons and protons as just being different states of a single particle — the nucleon.

Now the question is: how can we simply construct states in the isospin formalism that have the proper antisymmetry under exchanges of protons, and separately under exchanges of neutrons, without at the same time forcing the states also to be antisymmetric under the exchange of a neutron and a proton? The answer is that one requires the entire wave function of several nucleons (or other identical fermions) to be antisymmetric under interchanges of the space, spin *and* isospin coordinates of any two particles.[4] Let's see how this works for two nucleons. The two nucleons can be either in an isospin singlet state

$$\phi_s(1,2) = \frac{1}{\sqrt{2}}\,[\phi_p(1)\,\phi_n(2) - \phi_n(1)\,\phi_p(2)] \tag{18-69}$$

(where $\phi_n(1)$ is the state where particle 1 is a neutron, etc.) or they can be in an isospin triplet state

$$\phi_T(1,2) = \begin{cases} \phi_p(1)\,\phi_p(2) \\[1ex] \dfrac{1}{\sqrt{2}}\,[\phi_p(1)\,\phi_n(2) + \phi_n(1)\,\phi_p(2)] \\[1ex] \phi_n(1)\,\phi_n(2). \end{cases} \tag{18-70}$$

As with ordinary spin the isospin singlet state is antisymmetric and the isospin triplet state is symmetric under interchanges of the coordinates of the two particles. Therefore if we require that the entire wave function

[4] See J.M. Blatt and V.F. Weisskopf, *Theoretical Nuclear Physics* [John Wiley and Sons, New York, 1952], Section III, 5.

$$\Psi(1,2) = \psi_{\substack{\text{space}\\ \text{and spin}}}(1,2)\,\phi_{\text{isospin}}(1,2) \tag{18-71}$$

be antisymmetric, then for an isospin triplet state, the space and spin wave function must be antisymmetric. Two protons, or two neutrons must always be in an isospin triplet state, and so the antisymmetry of their space-spin wave function is guaranteed. However a neutron and a proton can be in either an isospin triplet or isospin singlet, and therefore their space-spin wave function can be either antisymmetric or symmetric. Thus demanding antisymmetry of the entire wave function $\Psi(1, 2)$, while giving the correct antisymmetry for identical particles, places no unnecessary restrictions on the space and spin part of the wave function for distinguishable particles. [This is analogous to the way two fermions with the same spin orientation must have an antisymmetric space function, but can have either spatial symmetry if their spin orientations are different.] Notice that charge independence of nuclear forces between two nucleons is simply the statement that the force between two nucleons in a given space-spin state is independent of whether they are two protons, two neutrons, or a neutron-proton pair.

The deuteron, a bound state of p and n, is a spin triplet, and a mixture of S and D orbital angular momenta. Thus its space and spin wave function is symmetric, and we conclude that it is in an isotopic singlet. It could not be an isospin triplet, for then there would also have to be p-p and n-n deuterons; they could not have the same *symmetric* spin triplet, orbital S and D wave functions of the usual n-p deuteron. [None, incidentally, have ever been observed.]

The symmetry requirement on bosons, such as pions, in the isotopic spin formalism is that the entire wave function (18-71) for two (or more) identical bosons be totally symmetric under interchange of space, spin (integer), and isospin coordinates of any two bosons. Thus two pions in a relative angular momentum eigenstate with even l must have $I = 0$ or $I = 2$, and similarly if in a state with odd l they must have $I = 1$. Given these restrictions on the possible total isotopic spin states for identical bosons, or fermions, the charge independence of strong interactions is still the statement that total I is conserved in the interactions.

PROBLEMS

1. Calculate the differential cross section for the Coulomb scattering of α particles. Compare the result with that for the scattering of two electrons in singlet and triplet states.

2. Verify explicitly that the wave function (18-68) is totally antisymmetric.
3. Using Eqs. (9-71) and (9-72), calculate explicitly the low-energy cross section for elastic scattering of neutrons by (a) unpolarized protons; (b) unpolarized neutrons.
4. Construct the isotopic spin states of two pions. What is the symmetry of the spatial wave function in each case?

Chapter 19
SECOND QUANTIZATION

When dealing with systems of just a few identical particles, it is easy to construct explicitly symmetric or antisymmetric wave functions. This can prove however to be a rather cumbersome task when studying systems with enormous numbers of identical particles, such as the electrons in a metal, or liquid He^4. I would like therefore to describe a very elegant way of accounting for the symmetry of the states and the operators of systems of many identical particles, and illustrate its use in a few simple calculations.

CREATION AND ANNIHILATION OPERATORS

In studying the harmonic oscillator we introduced operators a and $a^\dagger$ that annihilated and created one quantum of excitation of the oscillator. We can introduce similar operators in identical particle systems that remove *particles* from and add particles to the system. The photon creation and annihilation operators that we studied in Chapter 13 are examples of such operators. Suppose that we have a potential well $V(\mathbf{r})$ with single particle energy eigenstates, $\varphi_0(\mathbf{r})$, $\varphi_1(\mathbf{r})$, etc. Consider, for the moment the state of an n boson system in which all n particles sit in the lowest level, $\varphi_0(\mathbf{r})$, of the well. Let us denote this state by $|n\rangle$. Since the particles are bosons n can be any nonnegative integer. For completeness, let $|0\rangle$ denote the state with no particles present.

We can now introduce operators a_0 and $a_0^\dagger$ defined formally by

$$a_0|n\rangle = \sqrt{n}\,|n-1\rangle$$

$$a_0^\dagger|n\rangle = \sqrt{n+1}\,|n+1\rangle. \tag{19-1}$$

These operators relate states of an n particle system with all particles in φ_0 with those of an n ± 1 particle system with all particles in φ_0. a_0 may be thought of as a *particle annihilation operator,* since acting on a state with n particles in the single particle state, φ_0, it produces the state with only n − 1 particles in φ_0. Similarly $a_0^\dagger$ is a *particle creation operator;* it adds a particle to the state φ_0.

The operators a_0 and $a_0^\dagger$ have properties identical to the harmonic oscillator operators. For example, a_0 and $a_0^\dagger$ obey the commutation relation

$$[a_0, a_0^\dagger] = 1, \tag{19-2}$$

since acting on any state $|n\rangle$

$$(a_0 a_0^\dagger - a_0^\dagger a_0)|n\rangle = [(n+1) - n]|n\rangle .$$

It follows immediately from (19-1) that we can write

$$|n\rangle = \frac{(a_0^\dagger)^n}{\sqrt{n!}}|0\rangle . \tag{19-3}$$

The state with n particles in the lowest level can be produced by adding n particles, in φ_0, to the "vacuum," $|0\rangle$.

Also $a_0^\dagger$ is the Hermitian conjugate of a_0. Note carefully that when $a_0^\dagger$ acts to the left it *removes* a particle, since from (19-1),

$$\langle n|a_0^\dagger = \sqrt{n}\,\langle n-1| ; \tag{19-4}$$

similarly a_0 acts as a *creation* operator to the left,

$$\langle n|a_0 = \sqrt{n+1}\,\langle n+1| . \tag{19-5}$$

The operator $N_0 = a_0^\dagger a_0$ measures the number of particles in a state; $a_0^\dagger a_0|n\rangle = a_0^\dagger\sqrt{n}\,|n-1\rangle = n|n\rangle$.

Suppose now that the particles are fermions, and that we only consider states $|n\rangle$ in which all the particles are in the lowest level of the well with spin up. The only such states are $|1\rangle$, the state with one particle, and $|0\rangle$, the state with no particles, since we can't put two fermions in the same state. Again we can introduce creation and annihilation operators $a_0^\dagger$ and a_0 by the definitions

$$a_0|0\rangle = 0, \qquad a_0|1\rangle = |0\rangle$$

$$a_0^\dagger|0\rangle = |1\rangle, \qquad a_0^\dagger|1\rangle = 0. \tag{19-6}$$

Explicitly, in the $|0\rangle$, $|1\rangle$ basis

$$a_0 = \begin{pmatrix} 0 & 1 \\ 0 & 0 \end{pmatrix}, \qquad a_0^\dagger = \begin{pmatrix} 0 & 0 \\ 1 & 0 \end{pmatrix}. \tag{19-7}$$

The condition $a_0^\dagger|1\rangle = 0$ guarantees that we can't put two fermions in the same state. The fermion operators a_0 and $a_0^\dagger$ obey an anti-commutation relation:

$$\{a_0, a_0^\dagger\} \equiv a_0 a_0^\dagger + a_0^\dagger a_0 = 1, \tag{19-8}$$

since

$$(a_0 a_0^\dagger + a_0^\dagger a_0)|1\rangle = (0+1)|1\rangle = |1\rangle$$

$$(a_0 a_0^\dagger + a_0^\dagger a_0)|0\rangle = (1+0)|0\rangle = |0\rangle.$$

Also

$$a_0^2 = 0, \qquad (a_0^\dagger)^2 = 0 \tag{19-9}$$

since

$$a_0 a_0|1\rangle = a_0|0\rangle = 0$$

$$a_0^\dagger a_0^\dagger|0\rangle = a_0^\dagger|1\rangle = 0.$$

The equation $a_0^2 = 0$ says that it is impossible to remove two fermions from the same state. As before, the operator $N_0 = a_0^\dagger a_0$ measures the number of particles in the state φ_0 since $a_0^\dagger a_0|0\rangle = 0$ and $a_0^\dagger a_0|1\rangle = a_0^\dagger|0\rangle = |1\rangle$.

Summing up then, for one single particle level of the well, the boson creation and annihilation operators obey the commutation relations

$$[a_0, a_0^\dagger] = 1, \qquad [a_0, a_0] = [a_0^\dagger, a_0^\dagger] = 0 \tag{19-10}$$

while the fermion operators obey the anticommutation relations

$$\{a_0, a_0^\dagger\} = 1, \qquad \{a_0, a_0\} = \{a_0^\dagger, a_0^\dagger\} = 0. \tag{19-11}$$

Let's consider the situation where we now allow the particles to occupy two levels of the well, say φ_0 and φ_1. A many boson state will have n_0 particles in the state φ_0 and n_1 particles in the state φ_1. Let us denote this state by $|n_0, n_1\rangle$. Again we can introduce creation and annihilation operators defined by

$$a_0|n_0, n_1\rangle = \sqrt{n_0}\,|n_0 - 1, n_1\rangle$$

$$a_0^\dagger|n_0, n_1\rangle = \sqrt{n_0+1}\,|n_0+1, n_1\rangle$$

$$a_1|n_0, n_1\rangle = \sqrt{n_1}\,|n_0, n_1-1\rangle$$

$$a_1^\dagger|n_0, n_1\rangle = \sqrt{n_1+1}\,|n_0, n_1+1\rangle. \tag{19-12}$$

a_0 destroys a particle in the state φ_0, $a_1^\dagger$ creates a particle in the state φ_1, etc. It is trivial to show that from (19-12) that

$$[a_0, a_0^\dagger] = 1, \qquad [a_1, a_1^\dagger] = 1$$

and furthermore that the "0" operators commute with the "1" operators

$$[a_0, a_1] = 0, \qquad [a_0^\dagger, a_1^\dagger] = 0$$

$$[a_0, a_1^\dagger] = 0, \qquad [a_0^\dagger, a_1] = 0,$$

since for bosons it makes no difference in what order one performs an operation such as adding a particle to one level and removing one from the other level.

Again, all the states $|n_0, n_1\rangle$ can be constructed from the "vacuum" $|0, 0\rangle$ by acting with $a_0^\dagger$ and $a_1^\dagger$ repeatedly:

$$|n_0, n_1\rangle = \frac{(a_1^\dagger)^{n_1}}{\sqrt{n_1!}}\,\frac{(a_0^\dagger)^{n_0}}{\sqrt{n_0!}}\,|0, 0\rangle. \tag{19-13}$$

The operator $a_0^\dagger a_0$ is the operator for the number of particles in the state φ_0 and $a_1^\dagger a_1$ measures the number of particles in the state φ_1. Then

$$N = a_0^\dagger a_0 + a_1^\dagger a_1 \tag{19-14}$$

is the *total number operator:*

$$N|n_1, n_2\rangle = (n_1+n_2)\,|n_1, n_2\rangle. \tag{19-15}$$

For fermions occupying the two levels φ_0 and φ_1 (again with their spins up, say) there are four possible states $|n_0, n_1\rangle$:

$$|0, 0\rangle, \qquad |0, 1\rangle, \qquad |1, 0\rangle, \qquad |1, 1\rangle.$$

We first introduce the creation and annihilation operators $a_1^\dagger$ and a_1 defined by the operations

$$a_1^\dagger|0, 0\rangle = |0, 1\rangle, \qquad a_1^\dagger|1, 0\rangle = |1, 1\rangle$$

$$a_1^\dagger|0, 1\rangle = a_1^\dagger|1, 1\rangle = 0 \tag{19-16}$$

$$a_1|0,0\rangle = a_1|1,0\rangle = 0$$

$$a_1|0,1\rangle = |0,0\rangle, \quad a_1|1,1\rangle = |1,0\rangle. \tag{19-17}$$

These operators create or destroy particles with the single particle wave function φ_1. We also define the action of the creation and annihilation operators a_0 and $a_0^\dagger$ on the states with no particles in the state φ_1 by

$$a_0^\dagger|0,0\rangle = |1,0\rangle, \quad a_0^\dagger|1,0\rangle = 0$$

$$a_0|1,0\rangle = |0,0\rangle, \quad a_0|0,0\rangle = 0. \tag{19-18}$$

Now we must take some care in defining how a_0 and $a_0^\dagger$ act on the states $|0, 1\rangle$ and $|1, 1\rangle$ which already have a particle in φ_1. The reason is that we want to build into the operator language the concept that if we interchange two fermions in a state, the state changes sign. How do we use the a's and a†'s to interchange the two particles in the state $|1, 1\rangle$? First we remove one from the state φ_1, using a_1:

$$|1,1\rangle \rightarrow |1,0\rangle = a_1|1,1\rangle,$$

then transfer the remaining particle from φ_0 to φ_1 by applying a_0 followed by $a_1^\dagger$:

$$|1,0\rangle \rightarrow |0,1\rangle = a_1^\dagger a_0|1,0\rangle,$$

and then put the leftover particle back into φ_0 by using $a_0^\dagger$. This gives a state

$$a_0^\dagger a_1^\dagger a_0 a_1|1,1\rangle = a_0^\dagger|0,1\rangle$$

which we want to have opposite sign from the original state. Thus we must require

$$a_0^\dagger|0,1\rangle = -|1,1\rangle, \tag{19-19}$$

in order that we get properly antisymmetrized states.

To complete the definition of a_0 and $a_0^\dagger$ we write

$$a_0^\dagger|1,1\rangle = 0 = a_0|0,1\rangle \tag{19-20}$$

and

$$a_0|1,1\rangle = -|0,1\rangle. \tag{19-21}$$

This latter equation, which is necessary for a_0 to be the Hermitian conjugate of $a_0^\dagger$, simply says that a_0 undoes the operation of $a_0^\dagger$.

It is trivial to show from the definitions (19-16) - (19-21) that the creation and annihilation operators obey the anticommutation relations

$$\{a_0, a_0^\dagger\} = 1$$

$$\{a_1, a_1^\dagger\} = 1$$

$$\{a_0, a_0\} = \{a_1, a_1\} = 0$$

$$\{a_0^\dagger, a_0^\dagger\} = \{a_1^\dagger, a_1^\dagger\} = 0, \qquad (19\text{-}22)$$

and furthermore the "0" operators anticommute with the "1" operators:

$$\{a_0, a_1\} = \{a_0^\dagger, a_1^\dagger\} = 0$$

$$\{a_0, a_1^\dagger\} = \{a_0^\dagger, a_1\} = 0. \qquad (19\text{-}23)$$

These anticommutation relations are a consequence of the antisymmetry of fermion states under the interchange of two particles. The states can all be constructed from the ground state by operating with $a_0^\dagger$ and $a_1^\dagger$:

$$|n_0, n_1\rangle = (a_1^\dagger)^{n_1}(a_0^\dagger)^{n_0}|0, 0\rangle. \qquad (19\text{-}24)$$

Note that the $a_0^\dagger$ acts first. There are no factorials in (19-24) since $n! = 1$ for $n = 0$ or 1.

It is completely straightforward to generalize the above to the situation where we allow the particles to occupy the complete set of states of the well, and to have all spin orientations. We specify the possible states by stating how many particles n_i there are in a given level of the well (and with a given spin orientation, if the particles have spin). The states then look like $|n_0, n_1, n_2, \ldots\rangle$. We have a creation and an annihilation operator, $a_i^\dagger$ and a_i, for each different single particle state.

For bosons the a_i and $a_i^\dagger$ obey the commutation relations

$$[a_i, a_j^\dagger] = \delta_{ij}$$

$$[a_i, a_j] = 0 = [a_i^\dagger, a_j^\dagger]. \qquad (19\text{-}25)$$

We write the state $|n_0, n_1, \ldots\rangle$ in terms of the $a^\dagger$'s as

$$|n_0, n_1, n_2, \ldots\rangle = \cdots \frac{(a_2{}^\dagger)^{n_2}}{\sqrt{n_2!}} \frac{(a_1{}^\dagger)^{n_1}}{\sqrt{n_1!}} \frac{(a_0{}^\dagger)^{n_0}}{\sqrt{n_0!}} |0\rangle, \tag{19-26}$$

where $|0\rangle$ is short for $|0, 0, 0, \ldots\rangle$, the vacuum.

The photon annihilation and creation operators that we introduced in studying the interaction of radiation with matter are just like the little a's except for trivial numerical factors.

For fermions

$$|n_0, n_1, n_2, \ldots\rangle = \cdots (a_2{}^\dagger)^{n_2}(a_1{}^\dagger)^{n_1}(a_0{}^\dagger)^{n_0}|0\rangle \tag{19-27}$$

and the operators obey anticommutation relations

$$\{a_i, a_j{}^\dagger\} = \delta_{ij}$$

$$\{a_i, a_j\} = 0 = \{a_i{}^\dagger, a_j{}^\dagger\}. \tag{19-28}$$

In either case, the number of particles in the single particle state i is measured by $a_i{}^\dagger a_i$, and

$$N = \sum_i a_i{}^\dagger a_i \tag{19-29}$$

measures the total number of particles. For both fermions and bosons

$$[a_i{}^\dagger a_i, a_j{}^\dagger a_j] = 0. \tag{19-30}$$

As an example, let the complete set of states be plane waves in a box, using periodic boundary conditions. Then the normalized wave functions are of the form

$$\varphi_{\mathbf{p}}(\mathbf{r}) = \frac{e^{i\mathbf{p}\cdot\mathbf{r}}}{\sqrt{V}}. \tag{19-31}$$

The p's are restricted to values

$$p_x = \frac{2\pi n_x}{L_x}, \quad n_x = 0, \pm 1, \pm 2, \ldots, \tag{19-32}$$

etc. The creation operators $a_{ps}{}^\dagger$ adds a particle with momentum p and spin orientation s to the box, while a_{ps} removes a particle with momentum p and spin orientation s from the box.

The amplitude at the point r' for finding the particle added by a $a_{ps}{}^\dagger$ is just $e^{i\mathbf{p}\cdot\mathbf{r}'}/\sqrt{V}$. Now the operator

$$\psi_s^\dagger(\mathbf{r}) \equiv \sum_{\mathbf{p}} \frac{e^{-i\mathbf{p}\cdot\mathbf{r}}}{\sqrt{V}} a_{\mathbf{p}s}^\dagger \tag{19-33}$$

adds a particle to the system in a superposition of momentum states with amplitude $e^{-i\mathbf{p}\cdot\mathbf{r}}/\sqrt{V}$; therefore the amplitude at the point $\mathbf{r}'$ for finding the particle added by $\psi_s^\dagger(\mathbf{r})$ is a coherent sum of amplitudes $e^{i\mathbf{p}\cdot\mathbf{r}'}/\sqrt{V}$ with coefficients $e^{-i\mathbf{p}\cdot\mathbf{r}}/\sqrt{V}$. This net amplitude is thus

$$\sum_{\mathbf{p}} \frac{e^{-i\mathbf{p}\cdot\mathbf{r}}}{\sqrt{V}} \frac{e^{i\mathbf{p}\cdot\mathbf{r}'}}{\sqrt{V}} = \delta(\mathbf{r}-\mathbf{r}'). \tag{19-34}$$

[This equation is nothing but the usual statement of Fourier series

$$f(\mathbf{r}') = \frac{1}{V}\sum_{\mathbf{p}} e^{i\mathbf{p}\cdot\mathbf{r}'} \int d^3r''\, e^{-i\mathbf{p}\cdot\mathbf{r}''} f(\mathbf{r}'')$$

applied to the function $f(\mathbf{r}') = \delta(\mathbf{r}-\mathbf{r}')$.] In other words, the operator $\psi_s^\dagger(\mathbf{r})$ adds all the amplitude at point $\mathbf{r}$; we can say that $\psi_s^\dagger(\mathbf{r})$ adds a particle at point $\mathbf{r}$ (with spin orientation s).

Similarly, the operator

$$\psi_s(\mathbf{r}) \equiv \sum_{\mathbf{p}} \frac{e^{i\mathbf{p}\cdot\mathbf{r}}}{\sqrt{V}} a_{\mathbf{p}s}, \tag{19-35}$$

which is the Hermitian adjoint of $\psi_s^\dagger(\mathbf{r})$, removes a particle from the point $\mathbf{r}$. The ψ's and $\psi^\dagger$'s are called *field operators*.

The commutation relations of the ψ's and $\psi^\dagger$'s are easy to compute from those of the $a_{\mathbf{p}s}$'s and $a_{\mathbf{p}s}^\dagger$'s. Since $a_{\mathbf{p}s}a_{\mathbf{p}'s'} \mp a_{\mathbf{p}'s'}a_{\mathbf{p}s} = 0$ (the upper sign refers to bosons and the lower to fermions) we find

$$\psi_s(\mathbf{r})\psi_{s'}(\mathbf{r}') \mp \psi_{s'}(\mathbf{r}')\psi_s(\mathbf{r}) = 0, \tag{19-36}$$

and similarly

$$\psi_s^\dagger(\mathbf{r})\psi_{s'}^\dagger(\mathbf{r}') \mp \psi_{s'}^\dagger(\mathbf{r}')\psi_s^\dagger(\mathbf{r}) = 0. \tag{19-37}$$

For bosons, adding a particle at $\mathbf{r}$ is an operation that commutes with adding a particle at $\mathbf{r}'$; for fermions these operations commute except for a change of sign of the state. Finally

$$\begin{aligned}\psi_s(\mathbf{r})\psi_{s'}^\dagger(\mathbf{r}') \mp \psi_{s'}^\dagger(\mathbf{r}')\psi_s(\mathbf{r}) &= \sum_{\mathbf{p}\mathbf{p}'} \frac{e^{i\mathbf{p}\cdot\mathbf{r}}e^{-i\mathbf{p}'\cdot\mathbf{r}'}}{V}(a_{\mathbf{p}s}a_{\mathbf{p}'s'}^\dagger \mp a_{\mathbf{p}'s'}^\dagger a_{\mathbf{p}s}) \\ &= \sum_{\mathbf{p}\mathbf{p}'} \frac{e^{i\mathbf{p}\cdot\mathbf{r}}e^{-i\mathbf{p}'\cdot\mathbf{r}'}}{V}\delta_{\mathbf{p}\mathbf{p}'}\delta_{ss'} = \delta(\mathbf{r}-\mathbf{r}')\delta_{ss'},\end{aligned}$$

so that

$$\psi_s(\mathbf{r})\psi_{s'}^\dagger(\mathbf{r}') \mp \psi_{s'}^\dagger(\mathbf{r}')\psi_s(\mathbf{r}) = \delta(\mathbf{r}-\mathbf{r}')\delta_{ss'}. \tag{19-38}$$

Adding particles commutes (or anticommutes) with removing particles, unless one happens to do the adding and removing at the same point. Then, for example, if there are no particles at $\mathbf{r}$, $\psi_s^\dagger(\mathbf{r})\psi_s(\mathbf{r})$ gives zero – one can't remove a particle if there are none – while $\psi_s(\mathbf{r})\psi_s^\dagger(\mathbf{r})$ won't be zero since the $\psi_s^\dagger(\mathbf{r})$ adds a particle for the $\psi_s(\mathbf{r})$ to remove.

The state

$$|\mathbf{r}_1, \mathbf{r}_2, \ldots, \mathbf{r}_n\rangle = \frac{1}{\sqrt{n!}} \psi^\dagger(\mathbf{r}_n) \cdots \psi^\dagger(\mathbf{r}_2)\psi^\dagger(\mathbf{r}_1)|0\rangle \tag{19-39}$$

(let's suppress the spin indices for simplicity) is the state of n particles with one at $\mathbf{r}_1$, one at $\mathbf{r}_2$, etc. These states form a very convenient basis for systems of many identical particles since as a consequence of the commutation relations of the $\psi^\dagger$'s, (19-39) has the proper symmetry under interchanges of the $\mathbf{r}_i$. For example, for fermions

$$|\mathbf{r}_2, \mathbf{r}_1, \mathbf{r}_3, \ldots, \mathbf{r}_n\rangle = -|\mathbf{r}_1, \mathbf{r}_2, \mathbf{r}_3, \ldots, \mathbf{r}_n\rangle$$

since

$$\psi^\dagger(\mathbf{r}_2)\psi^\dagger(\mathbf{r}_1) = -\psi^\dagger(\mathbf{r}_1)\psi^\dagger(\mathbf{r}_2).$$

Furthermore

$$\psi^\dagger(\mathbf{r})|\mathbf{r}_1, \ldots, \mathbf{r}_n\rangle = \sqrt{n+1}\,|\mathbf{r}_1, \ldots, \mathbf{r}_n, \mathbf{r}\rangle \tag{19-40}$$

so that adding a particle by using a creation operator automatically produces a correctly symmetrized state. This property is really the great advantage of the creation and annihilation operators.

If we act on $|\mathbf{r}_1, \ldots, \mathbf{r}_n\rangle$ with $\psi(\mathbf{r})$ we get

$$\psi(\mathbf{r})|\mathbf{r}_1, \ldots, \mathbf{r}_n\rangle = \frac{1}{\sqrt{n!}} \psi(\mathbf{r})\psi^\dagger(\mathbf{r}_n) \cdots \psi^\dagger(\mathbf{r}_1)|0\rangle$$

$$= \frac{1}{\sqrt{n!}} [\delta(\mathbf{r}-\mathbf{r}_n) \pm \psi^\dagger(\mathbf{r}_n)\psi(\mathbf{r})]\psi^\dagger(\mathbf{r}_{n-1}) \cdots \psi^\dagger(\mathbf{r}_1)|0\rangle.$$

If we continue to commute the $\psi(\mathbf{r})$ with the $\psi^\dagger$'s to its right until it reaches the $|0\rangle$, (and $\psi|0\rangle = 0$) we find

$$\psi(\mathbf{r})|\mathbf{r}_1, \ldots, \mathbf{r}_n\rangle = \frac{1}{\sqrt{n}}[\delta(\mathbf{r}-\mathbf{r}_n)|\mathbf{r}_1, \ldots, \mathbf{r}_{n-1}\rangle$$

$$\pm\delta(\mathbf{r}-\mathbf{r}_{n-1})|\mathbf{r}_1, \ldots, \mathbf{r}_{n-2}, \mathbf{r}_n\rangle \qquad (19\text{-}41)$$

$$+\cdots+(\pm 1)^{n-1}\delta(\mathbf{r}-\mathbf{r}_1)|\mathbf{r}_2, \ldots, \mathbf{r}_n\rangle].$$

Thus removing a particle at $\mathbf{r}$ can work only if $\mathbf{r} = \mathbf{r}_n$, or $\mathbf{r}_n = \mathbf{r}_{n-1}, \ldots,$ or $\mathbf{r} = \mathbf{r}_1$. What remains is the correctly symmetrized combination of $n-1$ particle states.

It is very important to notice that $\psi^\dagger$ adds a particle only when it acts to the right. Acting to the left it *removes* a particle, and ψ acting to the left *adds* a particle. For example, the state $\langle\mathbf{r}_1, \ldots, \mathbf{r}_n|$, which is the row vector conjugate to the state $|\mathbf{r}_1 \ldots \mathbf{r}_n\rangle$, is

$$\langle\mathbf{r}_1 \ldots \mathbf{r}_n| = [\psi^\dagger(\mathbf{r}_n)\cdots\psi^\dagger(\mathbf{r}_1)|0\rangle]^\dagger/\sqrt{n!} = \langle 0|\psi(\mathbf{r}_1)\cdots\psi(\mathbf{r}_n)/\sqrt{n!},$$

since $[\psi^\dagger(\mathbf{r})]^\dagger = \psi(\mathbf{r})$. Thus one builds up the state $\langle\mathbf{r}_1, \ldots, \mathbf{r}_n|$ by acting to the left on $\langle 0|$ with ψ's. Note also that the order of the ψ's in $\langle\mathbf{r}_1 \ldots \mathbf{r}_n|$ is reversed from that of the $\psi^\dagger$'s in $|\mathbf{r}_1, \ldots \mathbf{r}_n\rangle$.

By similar repeated commutations one can calculate the normalization condition on the $|\mathbf{r}_1, \ldots, \mathbf{r}_n\rangle$ basis states:

$$\langle\mathbf{r}_1', \ldots, \mathbf{r}_{n'}'|\mathbf{r}_1, \ldots, \mathbf{r}_n\rangle$$

$$= \frac{\delta_{nn'}}{n!}\sum_P (\pm 1)^P\, P\delta(\mathbf{r}_1-\mathbf{r}_1')\delta(\mathbf{r}_2-\mathbf{r}_2')\cdots\delta(\mathbf{r}_n-\mathbf{r}_n') \qquad (19\text{-}42)$$

where the sum is over all permutations of the coordinates $\mathbf{r}_1', \ldots, \mathbf{r}_n'$, and $(\pm 1)^P = 1$ for bosons and equals the sign of the permutation for fermions. n' must equal n since states with different numbers of particles are orthogonal.

Let us now construct the n particle state $|\Phi\rangle$ in which the particles have a wave function $\varphi(\mathbf{r}_1, \ldots, \mathbf{r}_n)$. This state is simply the coherent sum of localized states $|\mathbf{r}_1, \ldots, \mathbf{r}_n\rangle$ with relative phases $\varphi(\mathbf{r}_1, \ldots, \mathbf{r}_n)$. Thus

$$|\Phi\rangle = \int d^3r_1 \ldots d^3r_n\, \varphi(\mathbf{r}_1, \ldots, \mathbf{r}_n)|\mathbf{r}_1, \ldots, \mathbf{r}_n\rangle. \qquad (19\text{-}43)$$

The state $|\Phi\rangle$ is correctly symmetrized, even if the wave function $\varphi(\mathbf{r}_1, \ldots, \mathbf{r}_n)$ used to construct $|\Phi\rangle$ isn't symmetrized. In fact, we may ask for the amplitude for observing particles at $\mathbf{r}_1', \ldots, \mathbf{r}_n'$ if they are in the state $|\Phi\rangle$. This amplitude is

$$\langle\mathbf{r}_1', \ldots, \mathbf{r}_n'|\Phi\rangle$$

$$= \int d^3r_1\cdots d^3r_n\, \varphi(\mathbf{r}_1, \ldots, \mathbf{r}_n)\langle\mathbf{r}_1', \ldots, \mathbf{r}_n'|\mathbf{r}_1, \ldots, \mathbf{r}_n\rangle,$$

and from (19-42) we then find

$$\langle \mathbf{r}_1', \ldots, \mathbf{r}_n' | \Phi \rangle = \frac{1}{n!} \sum_P (\pm 1)^P \, P\varphi(\mathbf{r}_1', \ldots, \mathbf{r}_n'). \tag{19-44}$$

Thus the "true" wave function of the state, $\langle \mathbf{r}_1', \ldots, \mathbf{r}_n' | \Phi \rangle$, is always properly symmetrized. If φ is already properly symmetrized, then all n! terms on the right side of (19-44) are equal and

$$\langle \mathbf{r}_1', \ldots, \mathbf{r}_n' | \Phi \rangle = \varphi(\mathbf{r}_1', \ldots, \mathbf{r}_n').$$

The state $|\Phi\rangle$ is normalized to one if $\varphi(\mathbf{r}_1, \ldots)$ is symmetrized and is itself normalized to one. To see this we write

$$\begin{aligned}
\langle \Phi | \Phi \rangle &= \int d^3r_1 \cdots d^3r_n \, \varphi^*(\mathbf{r}_1, \ldots, \mathbf{r}_n) \\
&\qquad \times \langle \mathbf{r}_1, \ldots, \mathbf{r}_n | \int d^3r_1' \cdots d^3r_n' | \mathbf{r}_1', \ldots, \mathbf{r}_n' \rangle \varphi(\mathbf{r}_1', \ldots, \mathbf{r}_n') \\
&= \int d^3r_1 \cdots d^3r_n d^3r_1' \cdots d^3r_n' \, \varphi^*(\mathbf{r}_1, \ldots, \mathbf{r}_n) \\
&\qquad \times \varphi(\mathbf{r}_1', \ldots, \mathbf{r}_n') \frac{1}{n!} \sum_P (\pm 1)^P \, P\delta(\mathbf{r}_1 - \mathbf{r}_1') \cdots \delta(\mathbf{r}_n - \mathbf{r}_n') \\
&= \int d^3r_1 \cdots d^3r_n |\varphi(\mathbf{r}_1, \ldots, \mathbf{r}_n)|^2 = 1.
\end{aligned} \tag{19-45}$$

Since $\langle \mathbf{r}_1, \ldots, \mathbf{r}_n | \Phi \rangle$ is always the amplitude for observing particles at $\mathbf{r}_1, \ldots, \mathbf{r}_n$, we can always write $|\Phi\rangle$, (Eq. 19-43), as

$$|\Phi\rangle = \int d^3r_1 \cdots d^3r_n |\mathbf{r}_1, \ldots, \mathbf{r}_n\rangle \langle \mathbf{r}_1, \ldots, \mathbf{r}_n | \Phi \rangle. \tag{19-46}$$

In other words, the operator

$$1_n = \int d^3r_1 \cdots d^3r_n |\mathbf{r}_1, \ldots, \mathbf{r}_n\rangle \langle \mathbf{r}_1, \ldots, \mathbf{r}_n | \tag{19-47}$$

is the unit operator when operating on properly symmetrized n particle states. If $|\Phi\rangle$ is an n particle state then

$$1_{n'} |\Phi\rangle = \delta_{nn'} |\Phi\rangle. \tag{19-48}$$

Thus

$$1 = \sum_{n=0}^{\infty} 1_n = |0\rangle\langle 0| + \sum_{n=1}^{\infty} 1_n \tag{19-49}$$

is the unit operator when acting on properly symmetrized states of any number of particles.

SECOND QUANTIZED OPERATORS

Let us now learn how to write operators for physical observables in this formalism. As a first example, let us show that

$$\rho(\mathbf{r}) = \psi^\dagger(\mathbf{r})\psi(\mathbf{r}) \tag{19-50}$$

is the operator for the density of particles at $\mathbf{r}$. To see this we write the matrix element $\langle\Phi'|\rho(\mathbf{r})|\Phi\rangle$ of $\rho(\mathbf{r})$ between two n particle states in terms of the wave functions of the states:

$$\langle\Phi'|\rho(\mathbf{r})|\Phi\rangle = \langle\Phi'|\psi^\dagger(\mathbf{r})\psi(\mathbf{r})|\Phi\rangle = \langle\Phi'|\psi^\dagger(\mathbf{r})1\psi(\mathbf{r})|\Phi\rangle$$
$$= \langle\Phi'|\psi^\dagger(\mathbf{r})1_{n-1}\psi(\mathbf{r})|\Phi\rangle$$

since ψ acting on an n particle state leaves an $n-1$ particle state. Then using (19-46) and (19-40) we have

$$\langle\Phi'|\rho(\mathbf{r})|\Phi\rangle = \int d^3r_1\cdots d^3r_{n-1}\langle\Phi'|\psi^\dagger(\mathbf{r})|\mathbf{r}_1\ldots\mathbf{r}_{n-1}\rangle\langle\mathbf{r}_1\ldots\mathbf{r}_{n-1}|\psi(\mathbf{r})|\Phi\rangle$$
$$= n\int d^3r_1\cdots d^3r_{n-1}\langle\Phi'|\mathbf{r}_1,\ldots,\mathbf{r}_{n-1},\mathbf{r}\rangle\langle\mathbf{r}_1,\ldots,\mathbf{r}_{n-1},\mathbf{r}|\Phi\rangle.$$

Because the wave functions $\langle\mathbf{r}_1,\ldots,\mathbf{r}_n|\Phi\rangle$ and $\langle\mathbf{r}_1,\ldots,\mathbf{r}_n|\Phi'\rangle$ are symmetrized (or antisymmetrized), this equation is equivalent to

$$\langle\Phi'|\rho(\mathbf{r})|\Phi\rangle = \int d^3r_1\cdots d^3r_n\langle\Phi'|\mathbf{r}_1\cdots\mathbf{r}_n\rangle\sum_i\delta(\mathbf{r}-\mathbf{r}_i)\langle\mathbf{r}_1\cdots\mathbf{r}_n|\Phi\rangle, \tag{19-51}$$

which is nothing but the matrix element of the operator $\sum_i\delta(\mathbf{r}-\mathbf{r}_i)$, our old form for the density operator, between the wave functions $\langle\mathbf{r}_1\ldots\mathbf{r}_n|\Phi'\rangle$ and $\langle\mathbf{r}_1\ldots\mathbf{r}_n|\Phi\rangle$. Thus the operator $\psi^\dagger(\mathbf{r})\psi(\mathbf{r})$ has the same matrix elements as the usual density operator and therefore it is the representation of the density operator in terms of the field operators.

We can think of $\psi^\dagger(\mathbf{r})\psi(\mathbf{r})$ as examining the density of particles at $\mathbf{r}$ by trying to remove a particle from $\mathbf{r}$ and then putting it back. If the particles have spin, then $\psi_s{}^\dagger(\mathbf{r})\psi_s(\mathbf{r})$ is the operator for the density of particles at $\mathbf{r}$ with spin orientation s. The total density is

$$\rho(\mathbf{r}) = \sum_s \psi_s^\dagger(\mathbf{r})\psi_s(\mathbf{r}), \tag{19-52}$$

and the operator for the total number of particles in the system is

$$N = \int d^3r\ \rho(\mathbf{r}). \tag{19-53}$$

Just as a check, let us substitute for the ψ's in terms of the a's. Then (19-53) becomes

$$N = \sum_s \int d^3r \sum_p \frac{e^{-i\mathbf{p}\cdot\mathbf{r}}}{\sqrt{V}}\, a_{ps}^\dagger \sum_{p'} \frac{e^{i\mathbf{p}'\cdot\mathbf{r}}}{\sqrt{V}}\, a_{p's} \tag{19-54}$$

$$= \sum_s \sum_{pp'} a_{ps}^\dagger a_{p's} \int d^3r\ \frac{e^{i(\mathbf{p}'-\mathbf{p})\cdot\mathbf{r}}}{V}.$$

However the $\mathbf{r}$ integral vanishes unless $\mathbf{p} = \mathbf{p}'$, when it is equal to one. Thus (19-54) becomes

$$N = \sum_{ps} a_{ps}^\dagger a_{ps}, \tag{19-55}$$

which is our previous result.

The operator for the kinetic energy of the particles is most easily written down directly in terms of the a_p and $a_p^\dagger$ operators. To measure the kinetic energy of a system we count the number of particles of momentum $\mathbf{p}$, multiply it by $p^2/2m$, the kinetic energy of a particle of momentum $\mathbf{p}$, and then sum over all $\mathbf{p}$. But $a_\mathbf{p}^\dagger a_\mathbf{p}$ is the operator for the number of particles of momentum $\mathbf{p}$, and therefore the kinetic energy operator is

$$T = \sum_{ps} \frac{p^2}{2m}\, a_{ps}^\dagger a_{ps}. \tag{19-56}$$

To express T in terms of the field operators we first invert (19-33) and (19-35), finding

$$a_{ps}^\dagger = \int d^3r\ \frac{e^{i\mathbf{p}\cdot\mathbf{r}}}{\sqrt{V}}\, \psi_s^\dagger(\mathbf{r})$$

$$a_{ps} = \int d^3r\ \frac{e^{-i\mathbf{p}\cdot\mathbf{r}}}{\sqrt{V}}\, \psi_s(\mathbf{r}). \tag{19-57}$$

The first equation says that to add a particle with momentum $\mathbf{p}$ one adds a particle at different points $\mathbf{r}$ with relative amplitude $e^{i\mathbf{p}\cdot\mathbf{r}}/\sqrt{V}$. Substituting (19-57) into (19-53) gives

$$T = \frac{1}{2m}\frac{1}{V}\sum_{\mathbf{p},s}\int d^3r\, d^3r'\, (\nabla e^{i\mathbf{p}\cdot\mathbf{r}})\cdot(\nabla' e^{-i\mathbf{p}\cdot\mathbf{r}'})\psi_s^\dagger(\mathbf{r})\psi_s(\mathbf{r}'),$$

where we have written $\mathbf{p}e^{i\mathbf{p}\cdot\mathbf{r}} = -i\nabla e^{i\mathbf{p}\cdot\mathbf{r}}$. Integrating by parts and doing the sum over $\mathbf{p}$ then yields

$$T = \frac{1}{2m}\int d^3r\, \nabla\psi^\dagger(\mathbf{r})\cdot\nabla\psi(\mathbf{r}). \tag{19-58}$$

Notice how this expression for the kinetic energy *operator* for a *many-particle* system looks, in form, exactly like the expression $(1/2m)\int d\mathbf{r}\, \nabla\varphi^*(\mathbf{r})\cdot\nabla\varphi(\mathbf{r})$ we would write down for the *expectation value* of the kinetic energy for a *single* particle in terms of its wave function, $\varphi(\mathbf{r})$. Similarly the density operator $\psi^\dagger(\mathbf{r})\psi(\mathbf{r})$ looks like the usual wave function expression for the probability density $\varphi^*(\mathbf{r})\varphi(\mathbf{r})$ for finding a single particle with wave function φ at point $\mathbf{r}$. This formal similarity is the reason the creation and annihilation operator formalism is called *second quantization;* one-particle wave functions appear to have become operators which create and annihilate particles, while single particle expectation values appear to have become operators for physical quantities. This is *only* an appearance though; we don't now have a super doubly quantized quantum mechanics — only a new language for the old quantum mechanics.

We can use this similarity to write down other operators. For example, the particle current density operator is

$$\mathbf{j}(\mathbf{r}) = \frac{1}{2im}[\psi^\dagger(\mathbf{r})\nabla\psi(\mathbf{r}) - \nabla\psi^\dagger(\mathbf{r})\cdot\psi(\mathbf{r})]; \tag{19-59}$$

this is the same form as the probability current density for a single particle we studied a long time ago. Also for spin $\frac{1}{2}$ particles, the operator for the density of spin at point $\mathbf{r}$ is

$$\mathbf{S}(\mathbf{r}) = \frac{1}{2}\sum_{ss'}\psi_s^\dagger(\mathbf{r})\boldsymbol{\sigma}_{ss'}\psi_{s'}(\mathbf{r}) \tag{19-60}$$

where $\boldsymbol{\sigma} = (\sigma_x, \sigma_y, \sigma_z)$ are the three Pauli spin matrices.

To develop some feeling for this new formalism, let us examine some properties of a gas of noninteracting spin $\frac{1}{2}$ fermions in their ground state. The ground state $|\Phi_0\rangle$ is characterized by all the momentum states being filled up to some momentum p_f, the Fermi momentum. Then

$$n_{p\uparrow} = \langle \Phi_0 | a^\dagger_{p\uparrow} a_{p\uparrow} | \Phi_0 \rangle = \begin{cases} 1, & |p| \le p_f \\ 0, & |p| \ge p_f \end{cases}, \tag{19-61}$$

and $n_{p\uparrow} = n_{p\downarrow}$. The Fermi momentum is determined by the condition that the total number of particles is given by

$$N = \sum_{s,p} n_{ps} = 2 \sum_{|p| \le p_f} 1.$$

Converting the sum to an integral gives

$$N = 2V \int_0^{p_f} \frac{d^3p}{(2\pi)^3} = \frac{p_f^3}{3\pi^2} V. \tag{19-62}$$

Thus

$$p_f^3 = \frac{3\pi^2 N}{V} = 3\pi^2 n, \tag{19-63}$$

where n is the average particle density.

Next let us consider $\langle \rho(\mathbf{r}) \rangle = \sum_s \langle \Phi_0 | \psi_s^\dagger(\mathbf{r}) \psi_s(\mathbf{r}) | \Phi_0 \rangle$ in the gas. Expressing the ψ's in terms of a's we find

$$\langle \rho(\mathbf{r}) \rangle = \sum_{spp'} \frac{e^{-ip\cdot r} e^{ip'\cdot r}}{V} \langle \Phi_0 | a^\dagger_{ps} a_{p's} | \Phi_0 \rangle .$$

Now the latter expectation value vanishes unless $\mathbf{p} = \mathbf{p}'$, since if we remove a particle of momentum $\mathbf{p}'$ from the ground state, we can only come back to the ground state by adding back a particle of the same momentum $\mathbf{p}'$. Thus

$$\langle \Phi_0 | a^\dagger_{ps} a_{p's} | \Phi_0 \rangle = \delta_{pp'} n_{ps}, \tag{19-64}$$

whereupon

$$\langle \rho(\mathbf{r}) \rangle = \frac{1}{V} \sum_{sp} n_{ps} = n; \tag{19-65}$$

the density in the gas is uniform – a not too surprising result.

A useful quantity to know, as we shall see, is

$$G_s(\mathbf{r} - \mathbf{r}') = \langle \Phi_0 | \psi_s^\dagger(\mathbf{r}) \psi_s(\mathbf{r}') | \Phi_0 \rangle, \tag{19-66}$$

the amplitude for removing a particle at $\mathbf{r}'$ with spin s from the ground

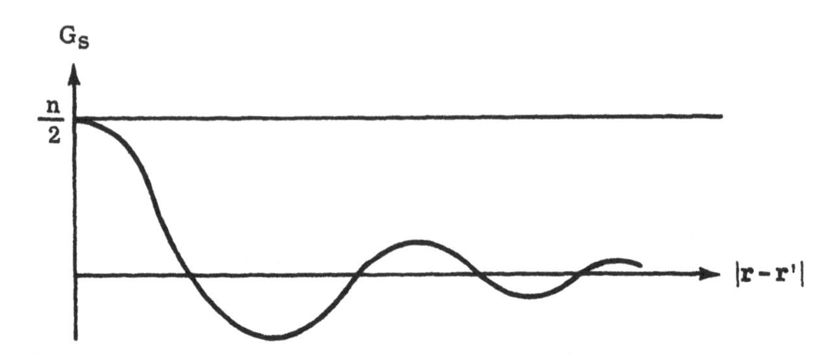

Fig. 19-1
The one-particle density matrix G_S for noninteracting spin ½ fermions.

state and then returning to the ground state by replacing a particle with spin s at point $\mathbf{r}$. Writing the ψ's in terms of the a's, and using (19-64) we find

$$G_S(\mathbf{r}-\mathbf{r}') = \frac{1}{V}\sum_{p} e^{-i\mathbf{p}\cdot(\mathbf{r}-\mathbf{r}')} n_{ps}. \tag{19-67}$$

Converting the sum to an integral we have

$$\begin{aligned} G_S(\mathbf{r}-\mathbf{r}') &= \int_0^{p_f} \frac{d^3p}{(2\pi)^3} e^{-i\mathbf{p}\cdot(\mathbf{r}-\mathbf{r}')} \\ &= \frac{1}{4\pi^2}\int_0^{p_f} p^2\,dp \int_{-1}^{1} d\mu\, e^{-ip|\mathbf{r}-\mathbf{r}'|\mu} \\ &= \frac{3n}{2}\,\frac{\sin x - x\cos x}{x^3}, \end{aligned} \tag{19-68}$$

where $x = p_f|\mathbf{r}-\mathbf{r}'|$ and we have used (19-63). This amplitude is shown as a function of $|\mathbf{r}-\mathbf{r}'|$ in Fig. 19-1. Clearly, for $\mathbf{r} = \mathbf{r}'$, G_S equals the density n/2, of particles with spin orientation s. For small $|\mathbf{r}-\mathbf{r}'|$

$$G_S(\mathbf{r}-\mathbf{r}') = \frac{n}{2}\left[1 - \frac{(p_f|\mathbf{r}-\mathbf{r}'|)^2}{10}\right]. \tag{19-69}$$

G_S is called the *one-particle density matrix.*

PAIR CORRELATION FUNCTIONS

In a gas of fermions there is a certain tendency for particles of the same spin to avoid each other. This is a simple consequence of the exclusion principle: two particles of the same spin can't be at the same point in space, and therefore, the amplitude for their being close together must be relatively small. Let us calculate the relative probability of finding a particle at $\mathbf{r}'$ if we know that there is one at $\mathbf{r}$. One way to formulate this problem is to remove (mathematically) a particle (with spin s) at $\mathbf{r}$ from the system, leaving behind $N-1$ particles in the state $|\Phi'(\mathbf{r}, s)\rangle = \psi_s(\mathbf{r})|\Phi_0\rangle$, and then ask for the density distribution of particles (with spin s') in this new state. This density is

$$\langle \Phi'(\mathbf{r}, s)|\psi_{s'}^\dagger(\mathbf{r}')\psi_{s'}(\mathbf{r}')|\Phi'(\mathbf{r}, s)\rangle = \langle \Phi_0|\psi_s^\dagger(\mathbf{r})\psi_{s'}^\dagger(\mathbf{r}')\psi_{s'}(\mathbf{r}')\psi_s(\mathbf{r})|\Phi_0\rangle$$
$$\equiv \left(\frac{n}{2}\right)^2 g_{ss'}(\mathbf{r}-\mathbf{r}'). \tag{19-70}$$

Another equivalent way of asking the same question is first to remove a particle from $\mathbf{r}$ using $\psi_s(\mathbf{r})$ and then one from $\mathbf{r}'$ using $\psi_{s'}(\mathbf{r}')$; the relative amplitude for ending up in some $N-2$ particle state $|\Phi_i''\rangle$ is $\langle\Phi_i''|\psi_{s'}(\mathbf{r}')\psi_s(\mathbf{r})|\Phi_0\rangle$. If we sum over a *complete* set of $N-2$ particle states, we find that the total probability for removing the two particles is

$$\sum_i |\langle\Phi_i''|\psi_{s'}(\mathbf{r}')\psi_s(\mathbf{r})|\Phi_0\rangle|^2 = \langle\Phi_0|\psi_s^\dagger(\mathbf{r})\psi_{s'}^\dagger(\mathbf{r}')\sum_i|\Phi_i''\rangle\langle\Phi_i''|\psi_{s'}(\mathbf{r}')\psi_s(\mathbf{r})|\Phi_0\rangle$$
$$= \langle\Phi_0|\psi_s^\dagger(\mathbf{r})\psi_{s'}^\dagger(\mathbf{r}')\psi_{s'}(\mathbf{r}')\psi_s(\mathbf{r})|\Phi_0\rangle.$$

This is just the same result as (19-70).

To evaluate $g_{ss'}(\mathbf{r}-\mathbf{r}')$, we expand the ψ's in (19-67) in terms of the a's; this gives

$$\left(\frac{n}{2}\right)^2 g_{ss'}(\mathbf{r}-\mathbf{r}') = \frac{1}{V^2}\sum_{pp'qq'} e^{-i(\mathbf{p}-\mathbf{p}')\cdot\mathbf{r}}\, e^{-i(\mathbf{q}-\mathbf{q}')\cdot\mathbf{r}'}$$
$$\times\langle\Phi_0|a_{ps}^\dagger a_{qs'}^\dagger a_{q's'} a_{p's}|\Phi_0\rangle. \tag{19-71}$$

Now the expectation value vanishes unless the particles we put back have the same momentum and spin as the particles we remove. Thus if $s \neq s'$, $\mathbf{p}'$ must equal $\mathbf{p}$, $\mathbf{q}'$ must equal $\mathbf{q}$, and

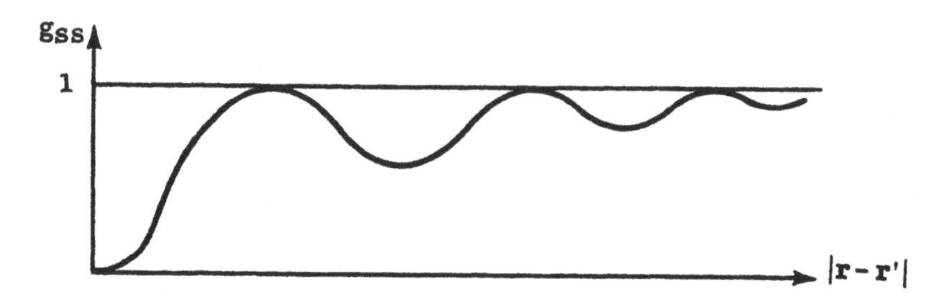

Fig. 19-2
The pair correlation function, for parallel spin, for noninteracting spin ½ fermions.

$$\langle\Phi_0|a^\dagger_{ps}\, a^\dagger_{qs'}\, a_{qs'}\, a_{ps}|\Phi_0\rangle = \langle\Phi_0|a^\dagger_{ps}\, a_{ps}\, a^\dagger_{qs'}\, a_{qs'}|\Phi_0\rangle = n_{ps}n_{qs'}. \tag{19-72}$$

Then (19-71) becomes

$$\left(\frac{n}{2}\right)^2 g_{ss'}(\mathbf{r}-\mathbf{r}') = \frac{1}{V^2}\sum_{pq} n_{ps}n_{qs'} = n_s n_{s'}$$

or

$$g_{ss'}(\mathbf{r}-\mathbf{r}') = 1, \qquad \text{for } s \neq s'. \tag{19-73}$$

This says that the relative probability for finding particles at $\mathbf{r}$ and $\mathbf{r}'$ with different spin is independent of the distance $|\mathbf{r}-\mathbf{r}'|$; this is the same as the result for a classical noninteracting gas. The exclusion principle doesn't effect particles of opposite spin.

If the spins are the same, $s = s'$, then there are two possibilities: $p = p'$, $q = q'$ or $p = q'$, $q = p'$. [If $p' = q'$ then the expectation value vanishes since $a_{p's}{}^2 = 0$.] Thus

$$\begin{aligned}\langle\Phi_0|a^\dagger_{ps}\, a^\dagger_{qs}\, a_{q's}\, a_{p's}|\Phi_0\rangle &= \delta_{pp'}\delta_{qq'}\langle\Phi_0|a^\dagger_{ps}\, a^\dagger_{qs}\, a_{qs}\, a_{ps}|\Phi_0\rangle \\ &\quad + \delta_{pq'}\delta_{qp'}\langle\Phi_0|a^\dagger_{ps}\, a^\dagger_{qs}\, a_{ps}\, a_{qs}|\Phi_0\rangle \\ &= (\delta_{pp'}\delta_{qq'} - \delta_{pq'}\delta_{qp'})\langle\Phi_0|a^\dagger_{ps}\, a_{ps}\, a^\dagger_{qs}\, a_{qs}|\Phi_0\rangle \\ &= (\delta_{pp'}\delta_{qq'} - \delta_{pq'}\delta_{qp'})n_{ps}n_{qs},\end{aligned} \tag{19-74}$$

since for $q \neq p$ the a_{ps} anticommutes with $a_{qs}^\dagger$ and a_{qs}, while for $q = p$, the expectation value vanishes. Plugging (19-74) into (19-71) we find

$$\left(\frac{n}{2}\right)^2 g_{ss}(r - r') = \frac{1}{V^2} \sum_{pq} [1 - e^{-i(p-q)\cdot(r-r')}] n_{ps} n_{qs}$$
$$= \left(\frac{n}{2}\right)^2 - [G_s(r - r')]^2, \tag{19-75}$$

where G_s is given by (19-67) and (19-68). Thus

$$g_{ss}(r - r') = 1 - \frac{9}{x^6} (\sin x - x \cos x)^2, \tag{19-76}$$

where $x = p_f |r - r'|$. The function $g_{ss}(r - r')$ is graphed in Fig. 19-2. We see that there is a substantial reduction in the probability for finding two fermions of the same spin at distances $\lesssim p_f^{-1}$. The exclusion principle causes large *correlations* in the motion of particles of the same spin. It is almost as if fermions of the same spin repelled each other at short distances. This effective "repulsion" arises just from the exchange symmetry of the wave function — not from any real forces between the particles. At large separation, g_{ss} approaches one, the same value as for opposite spins.

The function $g_{ss'}(r - r')$ is called the *pair correlation function.* If we use (19-75), (19-73), and (19-65), we can write our result for this function as

$$\langle \Phi_0 | \psi_s^\dagger(r) \psi_{s'}^\dagger(r') \psi_{s'}(r') \psi_s(r) | \Phi_0 \rangle$$
$$= \langle \Phi_0 | \psi_s^\dagger(r) \psi_s(r) | \Phi_0 \rangle \langle \Phi_0 | \psi_{s'}^\dagger(r') \psi_{s'}(r) | \Phi_0 \rangle \tag{19-77}$$
$$- \langle \Phi_0 | \psi_s^\dagger(r) \psi_{s'}(r') | \Phi_0 \rangle \langle \Phi_0 | \psi_{s'}^\dagger(r') \psi_s(r) | \Phi_0 \rangle .$$

[This factorization of the pair correlation function depends *only* on the wave function of the N-particle system being a Slater determinant of single particle orbitals; for example, it is therefore valid for a system of noninteracting fermions in any potential well.]

Let us now evaluate the pair correlation function for a system of noninteracting spinless bosons in the state

$$|\Phi\rangle = |n_{p_0}, n_{p_1}, \ldots\rangle . \tag{19-78}$$

The density in this state is

$$\langle \Phi | \psi^\dagger(\mathbf{r})\psi(\mathbf{r}) | \Phi \rangle = \frac{1}{V} \sum_{\mathbf{p}} n_{\mathbf{p}} \equiv n. \tag{19-79}$$

The calculation of the pair correlation function begins with Eq. (19-71), whose form is equally valid for bosons. The expectation value $\langle \Phi | a_{\mathbf{p}}^\dagger a_{\mathbf{q}}^\dagger a_{\mathbf{q}'} a_{\mathbf{p}'} | \Phi \rangle$ is nonvanishing only if $\mathbf{p} = \mathbf{p}'$, $\mathbf{q} = \mathbf{q}'$ or $\mathbf{p} = \mathbf{q}'$, $\mathbf{q} = \mathbf{p}'$. These are not distinct cases though if $\mathbf{p} = \mathbf{q}$. Thus we have

$$\begin{aligned}
\langle \Phi | a_{\mathbf{p}}^\dagger a_{\mathbf{q}}^\dagger a_{\mathbf{q}'} a_{\mathbf{p}'} | \Phi \rangle &= (1 - \delta_{\mathbf{pq}})(\delta_{\mathbf{pp}'}\delta_{\mathbf{qq}'} \langle \Phi | a_{\mathbf{p}}^\dagger a_{\mathbf{q}}^\dagger a_{\mathbf{q}} a_{\mathbf{p}} | \Phi \rangle \\
&\quad + \delta_{\mathbf{pq}'}\delta_{\mathbf{qp}'} \langle \Phi | a_{\mathbf{p}}^\dagger a_{\mathbf{q}}^\dagger a_{\mathbf{p}} a_{\mathbf{q}} | \Phi \rangle) + \delta_{\mathbf{pq}}\delta_{\mathbf{pp}'}\delta_{\mathbf{qq}'} \langle \Phi | a_{\mathbf{p}}^\dagger a_{\mathbf{p}}^\dagger a_{\mathbf{p}} a_{\mathbf{p}} | \Phi \rangle \\
&= (1 - \delta_{\mathbf{pq}})(\delta_{\mathbf{pp}'}\delta_{\mathbf{qq}'} + \delta_{\mathbf{pq}'}\delta_{\mathbf{qp}'}) n_{\mathbf{p}} n_{\mathbf{q}} + \delta_{\mathbf{pq}}\delta_{\mathbf{pp}'}\delta_{\mathbf{qq}'} n_{\mathbf{p}}(n_{\mathbf{p}} - 1).
\end{aligned} \tag{19-80}$$

Putting this into (19-71) we find

$$\begin{aligned}
&\langle \Phi | \psi^\dagger(\mathbf{r})\psi^\dagger(\mathbf{r}')\psi(\mathbf{r}')\psi(\mathbf{r}) | \Phi \rangle \\
&= n^2 + \left| \frac{1}{V} \sum_{\mathbf{p}} n_{\mathbf{p}} e^{-i\mathbf{p}\cdot(\mathbf{r}-\mathbf{r}')} \right|^2 - \frac{1}{V^2} \sum_{\mathbf{p}} n_{\mathbf{p}}(n_{\mathbf{p}} + 1).
\end{aligned} \tag{19-81}$$

This result differs from the fermion result in two respects: the sign of the second term is positive (a consequence of the exchange symmetry of boson wave functions), and the presence of the last term, which arises because one can have many bosons in the same state.

For example, if *all* the particles are in only one state $\mathbf{p}_0$, then (19-81) becomes

$$n^2 + n^2 - \left[\frac{1}{V^2} N(N+1) \right] = \frac{N(N-1)}{V^2}. \tag{19-82}$$

This says simply that the relative amplitude for removing the first particle is N/V, while the amplitude for removing the second is $(N-1)/V$, since there are only $N-1$ particles left after removing the first.

Consider next the case that $n_{\mathbf{p}}$ is a smoothly varying distribution. To be definite let us take

$$n_{\mathbf{p}} = c e^{-\alpha(\mathbf{p}-\mathbf{p}_0)^2/2}, \tag{19-83}$$

which essentially represents a beam of particles of momentum centered, with a Gaussian spread, about $\mathbf{p}_0$. If we take the limit of

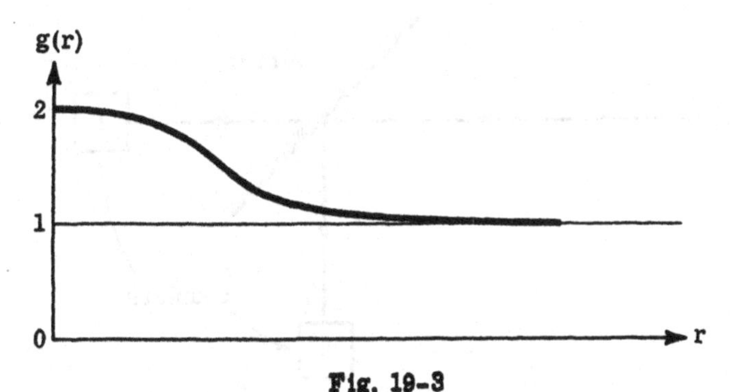

Fig. 19-3
The pair correlation function for noninteracting spin zero bosons.

large volume, keeping n fixed, then the last term in (19-81) is of the order 1/V smaller than the first two terms, and we can drop it. Converting the sums to integrals, (19-81) becomes

$$\langle \Phi | \psi^\dagger(\mathbf{r})\psi^\dagger(\mathbf{r}')\psi(\mathbf{r}')\psi(\mathbf{r}) | \Phi \rangle \equiv n^2 g(\mathbf{r}-\mathbf{r}')$$
$$= n^2 + \left| \int \frac{d^3p}{(2\pi)^3} n_{\mathbf{p}} e^{-i\mathbf{p}\cdot(\mathbf{r}-\mathbf{r}')} \right|^2 = n^2\left(1 + e^{-(\mathbf{r}-\mathbf{r}')^2/\alpha}\right). \tag{19-84}$$

The $e^{-(\mathbf{r}-\mathbf{r}')^2/\alpha}$ term is the effect of exchange. We see that it *increases* the probability for two bosons to be found at small separations. In fact, the probability for finding two bosons right on top of each other, $\mathbf{r} = \mathbf{r}'$, is *twice* the value for finding two at a large $|\mathbf{r}-\mathbf{r}'|$, as in Fig. 19-3.

THE HANBURY-BROWN AND TWISS EXPERIMENT

The *Hanbury-Brown and Twiss experiment* [1] provides a simple way of observing this tendency of bosons to clump together. Basically, the experiment measures the probability of observing two photons simultaneously at different points in a beam of incoherent light (which as we've seen, can be described in terms of the occupation numbers of the photon states). The actual measuring apparatus uses a half silvered mirror, Fig. 19-4, to split the beam into two identical beams; this avoids the problem of one detector

[1] *Nature* 177, 27 (1956); 178, 1447 (1956).

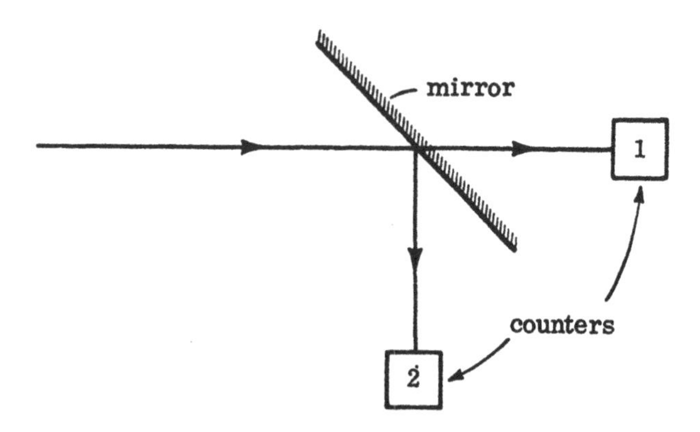

Fig. 19-4
The half-silvered mirror and counters in the Hanbury-Brown and Twiss experiment.

casting a shadow on the other. The amplitude for a photon to be transmitted, or reflected by the mirror, is $1/\sqrt{2}$. Hanbury-Brown and Twiss measured the light intensities $I_1(t)$ observed in detector 1 at time t, and $I_2(t+\tau)$ observed in detector 2 at a later time $t+\tau$, and averaged the product of the intensities over t, keeping τ fixed. This is equivalent to determining the relative probability of observing two photons at two points separated by a distance $c\tau$ in the beam, where c is the speed of light. The observed average correlated intensities $\overline{I_1(t)I_2(t+\tau)}$, as a function of τ, turned out to have just the form we derived for $g(r)$, in Fig. 19-3, with $r = c\tau$.

This experiment looks like a fine verification of the laws of quantum mechanics for identical bosons. On the contrary, it can be understood completely in terms of classical electromagnetism. What the experiment teaches us is that the boson nature of the photon is already contained in the superposition principle obeyed by classical electromagnetic fields. To see this, let us suppose, as in Fig. 19-5, that in the source of the beam there are just two emitters, A and B. Assume that A emits coherent light with amplitude α and wave num-

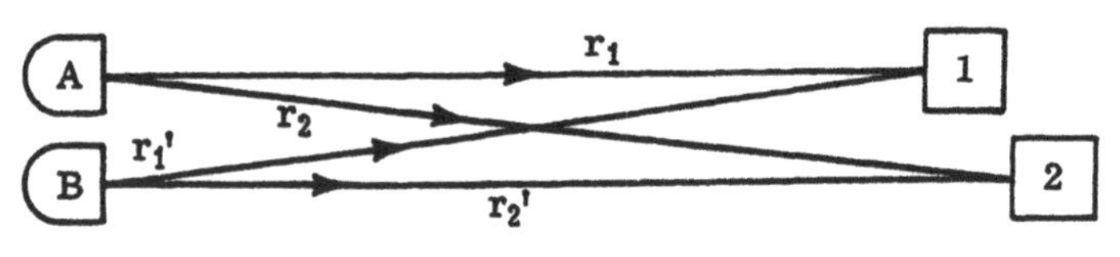

Fig. 19-5

ber k, B emits coherent light with amplitude β and wave number k', that the relative phase of these two sources is random, and that the light from each has the same polarization. The light from A falling on detector 1 has amplitude αe^{ikr_1} where r_1 is the distance to detector 1 from A; the light from B on 1 has amplitude $\beta e^{ik'r_1'}$ where r_1' is the distance from B to 1. Thus the total amplitude falling on 1, according to the superposition principle, is

$$a_1 = \alpha e^{ikr_1} + \beta e^{ik'r_1'} \tag{19-85}$$

(times some polarization vector) while the intensity is

$$I_1 = |\alpha|^2 + |\beta|^2 + 2 \text{ Re } \alpha^* \beta^{i(k'r_1' - kr_1)}. \tag{19-86}$$

If we average over the relative phase of α and β (equivalent to averaging over t in the Hanbury-Brown and Twiss experiment) we find

$$\bar{I}_1 = |\alpha|^2 + |\beta|^2. \tag{19-87}$$

Similarly the amplitude falling on the second counter is

$$a_2 = \alpha e^{ikr_2} + \beta e^{ik'r_2'} \tag{19-88}$$

(times a polarization vector) where r_2 is the distance from A to 2 and r_2' is the distance from B to 2. Thus

$$I_2 = |\alpha|^2 + |\beta|^2 + 2 \text{ Re } \alpha^* \beta \; e^{i(k'r_2' - kr_2)} \tag{19-89}$$

and averaged,

$$\bar{I}_2 = |\alpha|^2 + |\beta|^2. \tag{19-90}$$

The product of the averaged intensities, $\bar{I}_1 \bar{I}_2$ is independent of the distance between detectors 1 and 2. However, the product of the intensities is

$$I_1 I_2 = |a_1 a_2|^2 = |\alpha^2 e^{ik(r_1 + r_2)} + \beta^2 e^{ik'(r_1' + r_2')} + \alpha\beta(e^{ikr_1} e^{ik'r_2'} + e^{ik'r_1'} e^{ikr_2})|^2; \tag{19-91}$$

multiplying this out and averaging over the relative phase of α and β (which eliminates the terms proportional to $\alpha\beta|\alpha|^2$, $\alpha\beta|\beta|^2$, etc.) we find

$$\overline{I_1 I_2} = |\alpha|^4 + |\beta|^4 + |\alpha|^2|\beta|^2 |e^{ikr_1} e^{ik'r_2'} + e^{ik'r_1'} e^{ikr_2}|^2$$
$$= \bar{I}_1\bar{I}_2 + 2|\alpha|^2|\beta|^2 \cos[k'(r_1' - r_2') - k(r_1 - r_2)]. \tag{19-92}$$

For a well collimated beam, $r_1 - r_2 \approx r_1' - r_2'$ so that (19-93) becomes

$$\overline{I_1 I_2} = \bar{I}_1\bar{I}_2 + 2|\alpha|^2|\beta|^2 \cos[(k' - k)(r_1 - r_2)]. \tag{19-93}$$

Thus we find a term in the correlated intensities that depends on the relative separation of the two detectors; this term is maximum when the two detectors are at the same point. Now finally we should average the result (19-93) over all the different k and k' present in the beam. Then we find, for a Gaussian distribution, exactly the form (19-84). The photon bunching effect is thus a consequence of the superposition principle for light applied to noisy sources.[2]

From a quantum mechanical point of view we can interpret the three terms on the right side (19-91) as follows. The α^2 term is the amplitude for the two observed photons both to have come from A; this leads to the $|\alpha|^4$ term in (19-92). The β^2 term is the amplitude for them both to have come from B; this produces the $|\beta|^4$ term in (19-92). The $\alpha\beta$ term is the amplitude for one of the photons to have come from A and the other from B. There are two ways for this to occur — the photon from A can strike 1 while the photon from B strikes 2, or vice versa. These two ways are indistinguishable, and it is just the interference between them that leads to the cos term in (19-92).

Try to imagine the results of the Hanbury-Brown and Twiss experiment if it were performed with a beam of electrons.

THE HAMILTONIAN

There is still one very important operator we have not yet learned how to write in second quantized language — the Hamiltonian. If the particles interact by means of a two-body potential $v(\mathbf{r} - \mathbf{r}')$ then the interaction energy operator is

$$\mathcal{V} = \frac{1}{2}\sum_{ss'}\int d^3r\, d^3r'\, v(\mathbf{r} - \mathbf{r}')\psi_s^\dagger(\mathbf{r})\psi_{s'}^\dagger(\mathbf{r}')\psi_{s'}(\mathbf{r}')\psi_s(\mathbf{r}). \tag{19-94}$$

[2] This is discussed further by E. Purcell in *Nature* 178, 1449 (1956).

Note carefully the order of the operators. The order is the same as that used in (19-70) to determine the pair distribution function. It is left as an exercise to verify (19-94) by writing out a matrix element $\langle \Phi' | \mathcal{V} | \Phi \rangle$ of (19-94) in terms of the wave functions of the states. We can interpret the potential energy operator (19-94) as first trying to remove particles from $\mathbf{r}$ and $\mathbf{r}'$; if it is successful it counts a $v(\mathbf{r}-\mathbf{r}')$ and then replaces the particles, replacing the last removed particle first. It then sums over all *pairs* of points $\mathbf{r}$ and $\mathbf{r}'$, whence the factor $1/2$.

The second quantized Hamiltonian for particles of mass m acting pairwise is, using (19-58), thus

$$
\begin{aligned}
H = \sum_s \int d^3r \frac{1}{2m} \nabla\psi_s^\dagger(\mathbf{r}) \cdot \nabla\psi_s(\mathbf{r}) \\
+ \frac{1}{2} \sum_{ss'} \int d^3r\, d^3r'\, v(\mathbf{r}-\mathbf{r}')\psi_s^\dagger(\mathbf{r})\psi_{s'}^\dagger(\mathbf{r}')\psi_{s'}(\mathbf{r}')\psi_s(\mathbf{r}).
\end{aligned}
\tag{19-95}
$$

Let us evaluate the ground state energy of a gas of spin $1/2$ fermions, treating the interaction v as a perturbation. To lowest order the energy is simply the kinetic energy,

$$
E^{(0)} = \sum_{ps} \frac{p^2}{2m} = 2\int_0^{p_f} V \frac{d^3p}{(2\pi)^3} \frac{p^2}{2m} = \frac{3}{5}\frac{p_f^2}{2m} N. \tag{19-96}
$$

The average kinetic energy per particle is $3/5$ of the Fermi energy. The first-order change $E^{(1)}$ in the energy is simply the expectation value of $\mathcal{V}$ in the unperturbed ground state. Thus

$$
\begin{aligned}
E^{(1)} &= \frac{1}{2}\int d^3r\, d^3r'\, v(\mathbf{r}-\mathbf{r}') \sum_{ss'} \langle \Phi_0 | \psi_s^\dagger(\mathbf{r})\psi_{s'}^\dagger(\mathbf{r}')\psi_{s'}(\mathbf{r}')\psi_s(\mathbf{r}) | \Phi_0 \rangle \\
&= \frac{1}{2}\int d^3r\, d^3r'\, v(\mathbf{r}-\mathbf{r}') \sum_{ss'} \left(\frac{n}{2}\right)^2 g_{ss'}(\mathbf{r}-\mathbf{r}'),
\end{aligned}
\tag{19-97}
$$

where $g_{ss'}(\mathbf{r}-\mathbf{r}')$ is the pair correlation function. Using (19-73), and (19-75), we find

$$
E^{(1)} = \frac{1}{2}\int d^3r\, d^3r'\, v(\mathbf{r}-\mathbf{r}')[n^2 - \sum_s G_s(\mathbf{r}-\mathbf{r}')^2]. \tag{19-98}
$$

The n^2 term gives $Nnv_0/2$, where $v_0 = \int d^3r\, v(\mathbf{r})$; it represents the average interaction of a uniform density of particles with itself, leaving out all correlation effects. This energy is called the direct, or Hartree, energy. The second term, called the *exchange energy*,

$$E_{ex} = -\frac{1}{2}\int d^3r\, d^3r'\, v(\mathbf{r}-\mathbf{r}') \sum_s G_s(\mathbf{r}-\mathbf{r}')^2, \tag{19-99}$$

is the correction to the direct energy due to exchange. It accounts for the fact that particles of the same spin tend to stay apart; for this reason the effects of the short-ranged part of $v(\mathbf{r}-\mathbf{r}')$ are overcounted in the direct energy and the exchange energy subtracts out this overcounting, as well as the self-interactions included in the direct term. From (19-68) we find that the exchange energy is given by

$$\frac{E_{ex}}{N} = -\frac{9n}{4}\int d^3r\, \frac{(\sin p_f r - p_f r \cos p_f r)^2}{(p_f r)^6}\, v(r). \tag{19-100}$$

Thus to first order, the ground state energy per particle is

$$E_0 = \frac{3}{5}\frac{p_f^2}{2m} + \frac{nv_0}{2} + \frac{E_{ex}}{N}. \tag{19-101}$$

As an example, we consider a gas of electrons of average density n interacting through a Coulomb interaction

$$v(\mathbf{r}-\mathbf{r}') = \frac{e^2}{|\mathbf{r}-\mathbf{r}'|}. \tag{19-102}$$

The conduction electrons in a metal form such a gas. In any physical situation, one never has an isolated gas, but rather, there are always enough positive charges present to make the overall system electrically neutral. To a first approximation, in a metal or plasma, one can replace the positive ions by a uniform background of positive charge of density $+ne$. The electrostatic self-energy of this background, $(1/2)\int d^3r\, d^3r'\, (e^2n^2/|\mathbf{r}-\mathbf{r}'|)$, plus the average electrostatic interaction between the positive background and the electrons, $-\int d^3r\, d^3r'\, (e^2n^2/|\mathbf{r}-\mathbf{r}'|)$, exactly cancels the Hartree energy of the electrons. [This cancellation is not accidental, the electrostatic energy of an overall neutral system can only be proportional to the volume, in the limit of a large system — not to a higher power of the volume.] Thus the net interaction energy of the electron gas, to first order, is just the exchange energy:

$$\frac{E_{ex}}{N} = -\frac{9\pi ne^2}{p_f^2}\int_0^\infty \frac{dx}{x^5}(\sin x - x\cos x)^2 = -\frac{3}{4\pi}p_f e^2. \tag{19-103}$$

How valid is the perturbation expansion we have begun? What is small? The only dimensionless parameter (called r_s) that one can construct for the electron gas in its ground state is the ratio of the average interparticle spacing, d, to the Bohr radius, $a_0 = \hbar^2/me^2$. Defining d by $(4\pi d^3/3)n = 1$, so that

$$d = \frac{(9\pi/4)^{1/3}}{p_f} \tag{19-104}$$

we have

$$r_s = \frac{d}{a_0} = \left(\frac{9\pi}{4}\right)^{1/3}\left(\frac{me^2}{p_f}\right). \tag{19-105}$$

Expressed in terms of r_s, the energy per particle, (19-96) plus (19-103), is

$$E = \left(\frac{2.21}{r_s^2} - \frac{0.916}{r_s}\right)\left(\frac{e^2}{2a_0}\right). \tag{19-106}$$

The first term is the kinetic energy and the second the exchange energy; $e^2/2a_0$ is the Rydberg. (19-106) must be an upper bound to the energy by the Rayleigh-Ritz variation principle, since it is the expectation value of H in the unperturbed ground state. This estimate is valid for small r_s, or dense gases.

In actual metals, $1.8 \leq r_s \leq 5.5$. For $r_s \leq 2.3$, (19-106) is negative indicating, since it is an upper bound, that the system binds together. The exclusion principle plays an important role in this binding, keeping apart parallel spin particles, and thereby lowering their electrostatic energy. Actually the energy should be still lower than (19-106) due to the fact that even electrons of opposite spin tend to stay apart, because of the repulsive Coulomb interactions. The exact expansion of the energy in r_s begins as[3]

$$E = \left(\frac{2.21}{r_s^2} - \frac{0.916}{r_s} + 0.0622 \ln r_s - 0.094 + \cdots\right)\frac{e^2}{2a_0}. \tag{19-107}$$

The difference between (19-107) and (19-106) is called the *correlation energy*. The ln r_s means that the energy is not a simple analytic function of r_s. One can see, from the relative size of the terms in (19-107), that this expansion of E in terms of r_s is not valid for metallic densities.

It is often useful to write the interaction energy operator in terms of the a_p operators. Writing the ψ's in terms of the a_p we find

[3] M. Gell-Mann and K. Brueckner, *Phys. Rev.* 106, 364 (1957).

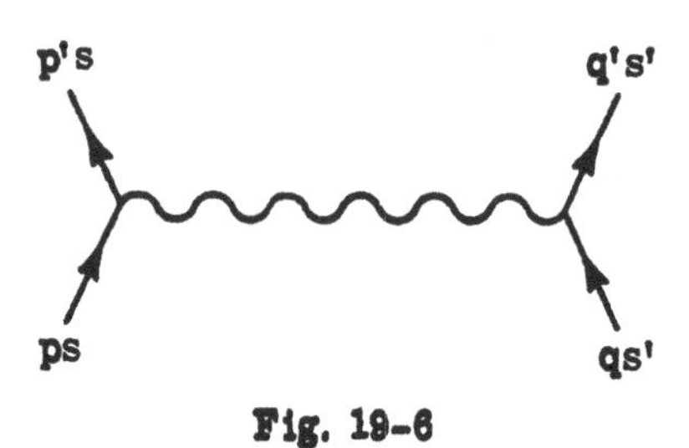

Fig. 19-6

$$\mathcal{V} = \frac{1}{2V} \sum_{pp'qq'} \sum_{ss'} v_{p'-p} \delta_{p+q,p'+q'} a^{\dagger}_{p's} a^{\dagger}_{q's'} a_{qs'} a_{ps} \tag{19-108}$$

where $v_k = \int d^3r\, e^{-ik\cdot r}\, v(r)$. One can think of the interaction operator written this way as a sum over scattering processes of the form shown in Fig. 19-6. The momentum $p+q$ of the scattering particles is conserved and the amplitude for the scattering is $v_{p'-p}$.

Finally, let us consider briefly the second quantized operators in the Heisenberg representation. Recall that in this representation the equation of motion of any operator X(t) not dependent explicitly on time is

$$i \frac{\partial X(t)}{\partial t} = [X(t), H(t)]. \tag{19-109}$$

When H doesn't depend explicitly on time, (19-109) is equivalent to

$$X(t) = e^{iHt} X e^{-iHt}. \tag{19-110}$$

The commutation relations (19-36), (19-37), and (19-38) remain valid as long as all the operators are at the *same* time. Then by simple evaluation of commutators, one can verify that $\psi_s(rt)$ obeys the equation of motion, for H given by (19-95):

$$i \frac{\partial \psi_s(rt)}{\partial t} = -\frac{\nabla^2}{2m} \psi_s(rt) + \left[\sum_{s'} \int d^3r'\, v(r-r')\psi_{s'}{}^{\dagger}(r')\psi_{s'}(r')\right] \psi_s(rt). \tag{19-111}$$

This equation has the same structure as the Schrödinger equation, only the ψ's are operators. The term in square brackets is roughly the operator for the potential energy felt by a particle at $\mathbf{r}$ due to the other particles. This term is an operator, and not a simple numerical function because a particle in a many particle system constantly affects the potential it feels from the other particles. As a consequence, Eq. (19-111) is far more difficult to solve than a single particle Schrödinger equation, and one can usually only solve it approximately.

PROBLEMS

1. Construct explicit 4×4 matrices to represent the fermion creation and annihilation operators a_0, $a_0^\dagger$, a_1, and $a_1^\dagger$ for two levels. Check the anticommutation relations.
2. (a) Calculate, to first order in the interparticle interaction, the energy of an $N + 1$ particle system of spin $1/2$ fermions with one particle of momentum $\mathbf{p}$ outside an N-particle Fermi sea (quasi-particle state). Repeat for the state of $N - 1$ particles with a particle of momentum $\mathbf{p}$ removed from an N-particle Fermi sea (hole state). Measure the energies from the N-particle ground state energy.
   (b) Evaluate the quasi-particle and hole energies for a Coulomb interaction (remember the uniform positive background).
3. Suppose that the wave function of an N-fermion system is a Slater determinant of orthonormal functions φ_i. Using the second quantized formalism show that the pair correlation function of the state factors as for plane waves, Eq. (19-77). [The operators $a_i = \int d^3r\ \varphi_i^*(\mathbf{r})\psi(\mathbf{r})$ play a useful role.]
4. Two electrons are in plane wave states in a box. Calculate to first order in the Coulomb interaction the energy difference of parallel and antiparallel spin alignments [exchange interaction].

Chapter 20
ATOMS

Accurate calculations of the properties of atoms are enormously difficult. There are two sources of this difficulty: the Coulomb interaction between the electrons, and the spin-orbit interaction between the electron spins and the electric fields in the atom. I shall therefore only describe the general features of atomic structure and merely outline briefly how one does calculations.[1,2]

TWO-ELECTRON ATOMS

The simplest atom after the hydrogen atom is one containing just two electrons, such as He, H^-, Li^+, and Be^{++}. (Ions have the same general structure as neutral atoms, and we can discuss them simultaneously.) The spin-orbit interaction can be regarded as a weak perturbation in the lighter atoms, and to a first approximation it can be neglected entirely. The Hamiltonian of a two-electron atom is then

$$H = \frac{p_1^2}{2m} + \frac{p_2^2}{2m} - \frac{Ze^2}{r_1} - \frac{Ze^2}{r_2} + \frac{e^2}{|\mathbf{r}_1 - \mathbf{r}_2|}. \qquad (20\text{-}1)$$

where Z is the charge of the nucleus, and $\mathbf{r}_1$ and $\mathbf{r}_2$ are the positions of the two electrons measured from the nucleus. We have assumed the nucleus to be fixed in space, but to be really accurate one must take into account the nuclear motion as well.

[1] Fuller details can be found in E.U. Condon and G.H. Shortley, *The Theory of Atomic Spectra* [Cambridge Univ. Press, Cambridge, 1959]; H.Bethe and R. Jackiw, *Intermediate Quantum Mechanics, Second Edition* [W.A. Benjamin, New York, 1968]; J. Slater, *Quantum Theory of Atomic Structure* [McGraw-Hill, New York, 1960].

[2] Generally $\hbar = 1$ in this Chapter.

Because of the electron–electron interaction term in (20–1) it is impossible to find the energies and wave functions exactly. [In fact, it is only comparatively recently that mathematicians have shown rigorously that the Schrödinger equation for this Hamiltonian has well behaved solutions.] Let us consider then some approximate methods for finding the ground state energy and wave function of H.

The simplest approximation is to treat the electron-electron interaction as a perturbation. To zero order, then, each electron finds itself in a Coulomb potential, $-Ze^2/r$, and the state of lowest energy is the one in which both electrons are in the hydrogenic 1S states of this potential:

$$\psi_0(1,2) = \psi_{1S}(\mathbf{r}_1)\psi_{1S}(\mathbf{r}_2)\chi_{0,0}(s_1, s_2), \tag{20–2}$$

where

$$\psi_{1S}(\mathbf{r}_1) = \frac{1}{\pi^{1/2}}\left(\frac{Z}{a_0}\right)^{3/2} e^{-Zr/a_0} \tag{20–3}$$

and $\chi_{0,0}$ is the singlet spin state. The spins must be in a singlet state because the spatial part of (20–2) is symmetric in $\mathbf{r}_1$ and $\mathbf{r}_2$. The energy of each electron in this Coulomb potential is $-(Ze)^2/2a_0 = -mZ^2e^4/2$, so that the ground state energy to zero order is

$$E_0^{(0)} = -\frac{Z^2e^2}{a_0}. \tag{20–4}$$

The first-order change in the energy is then the expectation value of the perturbation in the state (20–2):

$$\Delta E = \int d^3r_1\, d^3r_2 \frac{e^2}{|\mathbf{r}_1 - \mathbf{r}_2|}\left(\frac{Z}{a_0}\right)^6 \frac{1}{\pi^2}\, e^{-2Z(r_1+r_2)/a_0}. \tag{20–5}$$

To do the integral we can use the expression

$$\frac{1}{|\mathbf{r}_1 - \mathbf{r}_2|} = \int \frac{d^3k}{(2\pi)^3}\, e^{i\mathbf{k}\cdot(\mathbf{r}_1-\mathbf{r}_2)}\,\frac{4\pi}{k^2}; \tag{20–6}$$

(20–5) then becomes

$$\Delta E = \int \frac{d^3k}{2\pi^4}\left(\frac{Z}{a_0}\right)^6 \frac{e^2}{k^2}\left|\int d^3r\, e^{i\mathbf{k}\cdot\mathbf{r}-2Zr/a_0}\right|^2. \tag{20–7}$$

Now

$$\int d^3r\, e^{i\mathbf{k}\cdot\mathbf{r}-2Zr/a_0} = \frac{16\pi Z/a_0}{[k^2+(2Z/a_0)^2]^2} \tag{20–8}$$

so that (20-7) reduces to

$$\Delta E = \frac{4Ze^2}{\pi a_0} \int_0^\infty \frac{dx}{(x^2+1)^4} = \frac{5}{8}\,\frac{Ze^2}{a_0}\,. \tag{20-9}$$

The ground state energy to first order is then

$$E_0 = -\left(Z - \frac{5}{8}\right)\frac{Ze^2}{a_0}\,. \tag{20-10}$$

The result must be an upper bound to the actual ground state energy since doing first-order perturbation theory for the energy is completely equivalent to using the Rayleigh-Ritz variational principle, taking the unperturbed ground state wave function (20-2) as a trial function.

One can experimentally determine the ground state energy for a two-electron atom by measuring the ionization energy, E_i, i.e., the minimum energy required to remove one of the electrons to infinity. Then

$$E_i = E_0' - E_0 \tag{20-11}$$

where

$$E_0' = -\frac{Z^2e^2}{2a_0} \tag{20-12}$$

is the ground state energy of the remaining one-electron atom; Eq. (20-12) is exact nonrelativistically.

For He, $E_i = 1.807$ Rydbergs, experimentally, so that

$$E_0 = -5.807 \text{ Ry}, \tag{20-13}$$

while (20-10) gives −5.5 Ry. This is not very accurate. One reason is that the wave function (20-2) ignores the fact that one electron tends to screen the charge on the nucleus as seen by the other electron. In other words, when an electron looks at the nucleus it sees not only the positive charge of the nucleus but also some negative charge density around the nucleus from the other electron. This suggests that one can improve the accuracy by choosing a trial wave function of the form (20-2) with an effective Z value in ψ_{1s} somewhat less than the actual Z of the nucleus. Let us take in (20-2)

$$\psi_{1s}(r) = \frac{1}{\pi^{1/2}}\left(\frac{Z'}{a_0}\right)^{3/2} e^{-Z'r/a_0} \tag{20-14}$$

and treat Z' as a variational parameter.

To evaluate the expectation value of H in this state we note first of all that the expectation value of the electron-electron interaction is $5Z'e^2/8a_0$, from (20-9). The value of the kinetic energy of *each* electron is just that for an electron in the ground state of a Coulomb potential $Z'e^2/r$, namely $Z'^2e^2/2a_0$. The expectation value of the nuclear Coulomb potential energy for each electron is

$$\left\langle\frac{-Ze^2}{r}\right\rangle = \frac{Z}{Z'}\left\langle\frac{-Z'e^2}{r}\right\rangle.$$

For the wave function (20-14), $\langle -Z'e^2/r\rangle$ is the potential energy of an electron in the ground state of the Coulomb potential $-Z'e^2/r$; this energy is $-Z'^2e^2/a_0$. Thus the expectation value of the total energy of the two-electron atom with this trial wave function is

$$E_0(Z') = \frac{5}{8}\frac{Z'e^2}{a_0} + 2\left(\frac{Z'^2e^2}{2a_0} - \frac{ZZ'e^2}{a_0}\right) = \left(Z'^2 - 2ZZ' + \frac{5Z'}{8}\right)\frac{e^2}{a_0}. \qquad (20\text{-}15)$$

$E_0(Z')$ has a minimum for

$$Z' = Z - \frac{5}{16}; \qquad (20\text{-}16)$$

as we expected the effective charge is somewhat less than Z. The minimum value of $E_0(Z')$ is

$$E_0 = -\left(Z - \frac{5}{16}\right)^2 \frac{e^2}{a_0}. \qquad (20\text{-}17)$$

This is 0.2 Ry lower than (20-10) – a large improvement for He. However it is still not a good enough approximation for Z = 1 since it predicts that the H^- ion has a higher energy than a separated electron and hydrogen atom, whereas experimentally H^- is stable.

One can get improved accuracy by taking many-parameter trial wave functions. Eventually one must be careful to include the effects of the nuclear motion. Bethe and Jackiw describe such calculations in some detail.

HARTREE APPROXIMATION

In larger atoms we can expect that to a first approximation an electron sees some average field due to all the other electrons and the nucleus, and that the electron moves in some hydrogen-like orbit in this average field. The problem then is to determine both the average field and the single particle wave functions for an electron in this field. These are not independent projects, though, since the average field felt by one electron depends on the states of all the other electrons, which in turn depend on the average fields felt by those electrons.

To begin with let us assume that the wave function of the electrons is a simple product of one electron wave functions, or *orbitals:*

$$\Psi(1, \ldots, N) = \varphi_1(1)\varphi_2(2)\cdots\varphi_N(N). \tag{20-18}$$

Such a wave function isn't properly antisymmetrized, but we can partially include the exclusion principle by requiring that all the φ_i's in (20-18) be different and orthogonal, either in space or spin. Then no two electrons will be in the same state.

Let us now find the *average* potential that an electron moves in. The average charge density at a point $\mathbf{r}'$ due to the jth electron is $-e|\varphi_j(\mathbf{r}')|^2$; thus from Coulomb's law, another electron at $\mathbf{r}$ will have a potential energy

$$\int d^3r' \frac{e^2}{|\mathbf{r}-\mathbf{r}'|} |\varphi_j(\mathbf{r}')|^2$$

due to its interaction with the jth electron. The ith electron will therefore move in an average potential

$$V_i(\mathbf{r}) = \int d\mathbf{r}' \frac{e^2}{|\mathbf{r}-\mathbf{r}'|} \sum_{j(\neq i)} |\varphi_j(\mathbf{r}')|^2 - \frac{Ze^2}{r} \tag{20-19}$$

arising from all the other electrons and the nucleus. The orbital $\varphi_i(\mathbf{r})$ of the ith electron is then determined by the single particle Schrödinger equation

$$\left[-\frac{\nabla^2}{2m} + V_i(\mathbf{r})\right] \varphi_i(\mathbf{r}) = \varepsilon_i \varphi_i(\mathbf{r}). \tag{20-20}$$

The set of coupled equations (20-19) and (20-20) can only be solved numerically, usually by iteration. When $V_i(\mathbf{r})$ depends on i the eigenfunctions $\varphi_i(\mathbf{r})$ are generally not orthogonal. It is a very reasonable approximation to replace $V_i(\mathbf{r})$ by its average over all angles. Then the solutions $\varphi_i(\mathbf{r})$ will be angular momentum eigenstates.

This method of treating atoms was originally developed by Hartree. The Hartree equations (20-19) and (20-20) can be derived from the variational principle, using (20-18) as a trial function and varying the φ_i. We leave it as an exercise to verify that the expectation value of the energy in the state (20-18) is

$$E = \sum_i \varepsilon_i - \frac{1}{2} \sum_{\substack{i,j \\ (i \neq j)}} \int d^3r\, d^3r' \frac{e^2}{|\mathbf{r} - \mathbf{r}'|} |\varphi_i(\mathbf{r})|^2 |\varphi_j(\mathbf{r}')|^2. \tag{20-21}$$

FERMI-THOMAS APPROXIMATION

The general features of the self-consistent potential $V_i(\mathbf{r})$ can be determined by a semiclassical approach due to Thomas and Fermi. First of all, in a large atom, the average potential (20-19) felt by the ith electron is approximately equal to the potential

$$V(\mathbf{r}) = \int d^3r' \frac{e^2}{|\mathbf{r} - \mathbf{r}'|} \sum_j |\varphi_j(\mathbf{r}')|^2 \quad - \quad \frac{Ze^2}{r} \tag{20-22}$$

felt by an external test charge. This is because the one term $j = i$ in (20-22) is relatively unimportant if there are very many electrons in the atom. The quantity

$$\sum_j |\varphi_j(\mathbf{r}')|^2 = n(\mathbf{r}') \tag{20-23}$$

is just the density of electrons at $\mathbf{r}'$. We would like then to find out how this density depends on the potential $V(\mathbf{r})$.

The Fermi-Thomas method is based on the observation that in a very large atom most electrons have a large kinetic energy and therefore $V(\mathbf{r})$ varies slowly in space compared with the rate of spatial variation of the electronic wave functions. Thus to a first approximation the wave functions $\varphi_i(\mathbf{r})$ will be locally like plane waves:

$$\varphi_i(\mathbf{r}) \sim e^{i\mathbf{p}_i(\mathbf{r}) \cdot \mathbf{r}} \tag{20-24}$$

where the magnitude of the local momentum $p_i(\mathbf{r})$ is related to the energy ε_i of the electron by

$$\varepsilon_i = \frac{p_i^2(\mathbf{r})}{2m} + V(\mathbf{r}). \tag{20-25}$$

The semiclassical form (20-24) is valid in regions where

$$\varepsilon_i \gg V(\mathbf{r});$$

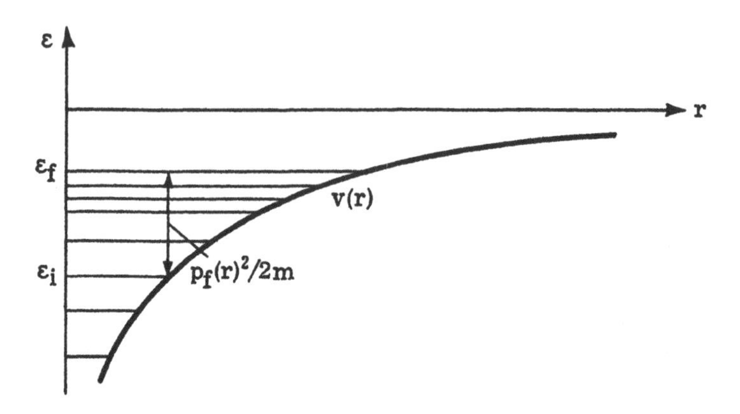

Fig. 20-1
Electron distribution in the ground state in the Fermi-Thomas model.

an electron in the state φ_i will be spread out uniformly in the classically allowable regions but will have only small probability of being in the classically forbidden regions.

Thus we can picture the electron distribution in the atom in its ground state as in Fig. 20-1. The electrons occupy states up to some maximum energy, ε_f. At each point $\mathbf{r}$ one has electrons with local momentum ranging from 0 up to a maximum

$$p_f(\mathbf{r}) = (2m[\varepsilon_f - V(r)])^{1/2}. \tag{20-26}$$

Only electrons with $\varepsilon_i > V(\mathbf{r})$ have appreciable probability of being at $\mathbf{r}$. One can regard the electrons locally as forming a noninteracting Fermi gas. The density of electrons at $\mathbf{r}$ is determined by the density of momentum states at $\mathbf{r}$. Since all the states up to $p_f(\mathbf{r})$ are filled the density $n(\mathbf{r})$ is given by the usual relation

$$n(r) = \frac{p_f^3(r)}{3\pi^2}. \tag{20-27}$$

Thus from (20-26) we find that the local density is related to V(r) by

$$n(r) = \frac{[2m(\varepsilon_f - V(r))]^{3/2}}{3\pi^2}. \tag{20-28}$$

Equations (20-28), (20-24), and (20-22) then determine V(r).

The most convenient way to solve for V(r) is to convert (20-22) into Poisson's equation by acting with ∇^2. Since $\nabla^2|\mathbf{r}-\mathbf{r}'|^{-1} = -4\pi\delta(\mathbf{r}-\mathbf{r}')$, (20-22) becomes for $r > 0$,

$$\nabla^2 V(\mathbf{r}) = -4\pi e^2 n(\mathbf{r}). \tag{20-29}$$

Assuming V to be spherically symmetric we have then the differential equation for V(r):

$$\frac{1}{r^2}\frac{\partial}{\partial r}r^2\frac{\partial}{\partial r}[-V(r)] = 4e^2\,\frac{[2m(\varepsilon_f - V(r))]^{3/2}}{3\pi}. \tag{20-30}$$

The boundary conditions on this differential equation are first, that as r approaches zero, V(r) approaches $-Ze^2/r$, the nuclear potential. Also, at the outer radius R of the atom, where $\varepsilon_f - V(R) = 0$, the electron density falls to zero. For a neutral atom $V(R) = 0$ if R is beyond the electronic density, and therefore $\varepsilon_f = 0$, while for an ion of charge $(Z-N)e$, $V(r) = -(Z-N)e^2/r$, for $r > R$.

Let us consider only the neutral case. Then making the substitutions

$$V(r) = -\frac{Ze^2}{r}\,\Phi(x), \qquad r = \frac{bx}{Z^{1/3}}$$

$$b = \frac{1}{2}\left(\frac{3\pi}{4}\right)^{2/3} a_0 = 0.8853\; a_0 \tag{20-31}$$

we find that (20-29) reduces to

$$x^{1/2}\,\frac{d^2\Phi(x)}{dx^2} = \Phi^{3/2}(x), \tag{20-32}$$

and the boundary condition on Φ is $\Phi(0) = 1$. This equation has been solved numerically and the solution has the form

$$\Phi(x) = \begin{cases} 1 - 1.59x, & x \to 0 \\[2ex] \dfrac{144}{x^3}, & x \to \infty. \end{cases} \tag{20-33}$$

The only place where $\Phi = 0$ and therefore $V = 0$ is at ∞. This means that the electron distribution in the Fermi-Thomas atom extends to infinity, a rather defective property of the model. For small r

$$V(r) = -\frac{Ze^2}{r} + 1.80\; Z^{4/3}\,\frac{e^2}{a_0}. \tag{20-34}$$

The second term is the potential at $r = 0$ due to the electrons.

This model makes several simple predictions about the structure of large atoms. First, since $\Phi(x)$ does not depend on any atomic parameters the *shape* of the effective potential is the same for all atoms. We see from (20-31) that the scale of lengths in the atom is proportional to $Z^{-1/3}$; as Z grows the atom decreases in size. The size of the potential at a given value of x is proportional to $Z^{4/3}$. From (20-30) we may conclude that the density at a given value of x is proportional to Z^2. Thus the average mean electronic momentum is proportional to $Z^{2/3}$.

There is another interesting prediction that one can make. Imagine that we want to find angular momentum eigenstates in the potential V(r). In the Schrödinger equation we find a centrifugal barrier term $l(l+1)/mr^2$ added onto V(r). A bound state of a given l will not exist unless the combination

$$V_{eff}(r) = V(r) + \frac{l(l+1)}{mr^2} \tag{20-35}$$

has a minimum below zero energy. Now from the solution to (20-32) one discovers that for $Z \leq 4$, V_{eff} has a minimum only for $l = 0$; for $4 < Z \leq 19$ it has a minimum for $l = 0$ and $l = 1$, while for $19 < Z \leq 53$ it has a minimum only for $l = 0$, 1, and 2. This means that for $Z \leq 4$ one can have only s electrons in the ground state, for $Z \leq 19$ only s and p electrons and for $Z \leq 53$ only s, p, and d electrons. Experimentally p electrons first appear at Z = 5, d electrons at Z = 20, and f electrons at Z = 58.

We should notice finally that the Fermi-Thomas approach breaks down at small distances $\sim a_0/Z$, since V(r) varies too rapidly there, and it also breaks down at large distances $\sim a_0$ where the electronic kinetic energies are no longer large. Most of the electrons in this model are found within these limits though. The Fermi-Thomas approximation neglects the effects of exchange in the effective potential V(r). One can include these effects with little difficulty; the calculations are described in Bethe and Jackiw.

HARTREE-FOCK

Let us go back though and see how we can include exchange effects in the Hartree approximation. To do this we take as a trial function a fully antisymmetrized determinantal wave function

$$\Psi(1,2,\ldots,N) = \frac{1}{\sqrt{N!}} \begin{vmatrix} \varphi_1(1) & \ldots & \varphi_1(N) \\ \cdot & & \cdot \\ \cdot & & \cdot \\ \cdot & & \cdot \\ \varphi_N(1) & \ldots & \varphi_N(N) \end{vmatrix} \tag{20-36}$$

where the φ_i are orthonormal. This is the *Hartree-Fock approximation.* The expectation value of the Hamiltonian in this state is evaluated as follows. The electronic kinetic energy is

$$-\sum_{i=1}^{N} \int d^3r \, \frac{|\nabla\varphi_i(r)|^2}{2m} \tag{20-37}$$

since there is one electron in each state φ_i. Similarly the potential energy due to interaction of the electrons with the nucleus is

$$-\sum_i \int d^3r \, |\varphi_i(r)|^2 \frac{Ze^2}{r} \tag{20-38}$$

The expectation value of the electron-electron interaction is, in second quantized language,

$$\frac{1}{2} \sum_{s,s'} \int d^3r \, d^3r' \, \frac{e^2}{|r-r'|} \langle \psi_s^\dagger(r)\psi_{s'}^\dagger(r')\psi_{s'}(r')\psi_s(r)\rangle . \tag{20-39}$$

Now for a determinantal wave function the pair correlation function becomes [(19-77), also problem 3, Chapter 19]:

$$\begin{aligned}\langle \psi_s^\dagger(r)\psi_{s'}^\dagger(r')\psi_{s'}(r')\psi_s(r)\rangle &= \langle \psi_s^\dagger(r)\psi_s(r)\rangle\langle \psi_{s'}^\dagger(r')\psi_{s'}(r')\rangle \\ &\quad -\delta_{ss'}\langle \psi_s^\dagger(r)\psi_s(r')\rangle\langle \psi_s^\dagger(r')\psi_s(r)\rangle .\end{aligned} \tag{20-40}$$

The density of particles at r is

$$\sum_s \langle \psi_s^\dagger(r)\psi_s(r)\rangle = \sum_i |\varphi_i(r)|^2 \tag{20-41}$$

while

$$\langle \psi_s^\dagger(r)\psi_s(r')\rangle = \sum_{\substack{i \text{ with} \\ \text{spin } s}} \varphi_i^*(r)\varphi_i(r'). \tag{20-42}$$

Thus putting (20-37)–(20-42) together we find that the expectation value of the Hamiltonian in the state (20-36) is

$$\langle H\rangle = -\sum_i \int d^3r \left[\frac{|\nabla\varphi_i(\mathbf{r})|^2}{2m} + \frac{Ze^2}{r}|\varphi_i(\mathbf{r})|^2\right]$$

$$+\frac{1}{2}\sum_{ij}\int d^3r\, d^3r'\, \frac{e^2}{|\mathbf{r}-\mathbf{r}'|}|\varphi_i(\mathbf{r})|^2|\varphi_j(\mathbf{r}')|^2 \tag{20-43}$$

$$-\frac{1}{2}\sum_{ij}\delta_{s_i,s_j}\int d^3r\, d^3r'\, \frac{e^2}{|\mathbf{r}-\mathbf{r}'|}\varphi_i^*(\mathbf{r})\varphi_i(\mathbf{r}')\varphi_j^*(\mathbf{r}')\varphi_j(\mathbf{r}).$$

The δ_{s_i,s_j} means that the sum is only over j with the same spin quantum number as i. The last term in (20-43) is the exchange energy. The term $i = j$ doesn't in fact appear in (20-43).

Now to find equations for the φ_i we must minimize (20-43) subject to the condition that the φ_i remain orthonormal. After a few technical maneuvers, which are described in the standard books, one finds that the φ_i are determined by the nonlinear equations:

$$\left(-\frac{\nabla^2}{2m}-\frac{Ze^2}{r}\right)\varphi_i(\mathbf{r}) + \int d^3r'\, \frac{e^2}{|\mathbf{r}-\mathbf{r}'|}\sum_j \varphi_j^*(\mathbf{r}')[\varphi_j(\mathbf{r}')\varphi_i(\mathbf{r}) \tag{20-44}$$

$$-\varphi_j(\mathbf{r})\varphi_i(\mathbf{r}')\delta_{s_i,s_j}] = \varepsilon_i\varphi_i(\mathbf{r}).$$

These are the Hartree-Fock equations. They differ from the Hartree equations (20-20) in having the $\varphi_j(\mathbf{r})\varphi_i(\mathbf{r}')$ term on the left side. This exchange term, involving an integral over $\varphi_i(\mathbf{r}')$, acts as a nonlocal potential in the Schrödinger equation (20-44). The combination $\varphi_j(\mathbf{r}')\varphi_i(\mathbf{r}) - \varphi_j(\mathbf{r})\varphi_i(\mathbf{r}')$ is the amplitude that electrons in the states i and j (with the same spin) will be found at $\mathbf{r}$ and $\mathbf{r}'$. Because the potential term in (20-44) contains this correctly antisymmetrized amplitude, the correlations between particles of the same spin arising from the exclusion principle are included in the Hartree-Fock equations.

The energy eigenvalue ε_i is approximately (minus) the energy required to remove a particle in the ith state from the atom. To see this we multiply (20-44) on the left by $\varphi_i^*(\mathbf{r})$ and integrate over all $\mathbf{r}$. This gives

$$\varepsilon_i = -\int d^3r\left(\frac{|\nabla\varphi_i(\mathbf{r})|^2}{2m} + \frac{Ze^2}{r}|\varphi_i(\mathbf{r})|^2\right) \tag{20-45}$$

$$+\sum_j\int d^3r\, d^3r'\, \frac{e^2}{|\mathbf{r}-\mathbf{r}'|}\varphi_i^*(\mathbf{r})\varphi_j^*(\mathbf{r}')\,[\varphi_j(\mathbf{r}')\varphi_i(\mathbf{r}) - \varphi_j(\mathbf{r})\varphi_i(\mathbf{r}')\delta_{s_j,s_i}].$$

Comparing this equation with Eq. (20-43), we see that ε_i is just the amount by which $\langle H \rangle$ is decreased if we cross out all terms in (20-43) referring to a particular i. Thus $-\varepsilon_i$ is the energy required to remove a particle in the ith orbital, *provided* that the wave functions of the remaining particles are unchanged. They will, of course, be somewhat modified since the remaining electrons no longer have the ith electron to interact with, so the $-\varepsilon_i$ are only approximately the ionization energies.

The ground state energies calculated from the Hartree-Fock equations are lower than the Hartree energies, (20-21). They are still too high by about 1 eV per pair of electrons. This remaining energy difference is the correlation energy. The Hartree-Fock equations fail to include the correlations between electrons of either spin that result from their repulsive Coulomb interaction. [The calculation we did in the last section of the energy of an electron gas [Eq. (19-102)] is a special case of the Hartree-Fock method in which the wave functions φ_i are plane waves.]

Let us now consider the general nature of the solutions to the Hartree or Hartree-Fock equations for an atom. It is a good first approximation to assume that the potential each electron feels is spherically symmetric. Then the solutions φ_i can be chosen as eigenstates of orbital angular momentum, specified by quantum numbers l and m_l and spin, with quantum number m_s. Different φ_i with the same l, m_l, and m_s can be labeled by a principal quantum number $n(\geq l+1)$ as in the hydrogen atom. Thus

$$\varphi_i = R_{nl}(r)\, Y_{l,m_l}(\theta, \varphi)\, \chi_{m_s}(s). \tag{20-46}$$

$n-l-1$ is the number of nodes of the radial wave function. The energy ε_i increases as n increases. However, unlike in the hydrogen atom, this energy also depends on l. This is because the average potential in an atom with more than one electron is no longer a pure $1/r$ potential.

One can have $(2s+1)(2l+1) = 4l+2$ different possible orbitals with the same n and l. This set of $4l+2$ orbitals is called a *shell*. We can then describe the Hartree or Hartree-Fock solution by giving the *electron configuration*, i.e., saying how many electrons occupy each shell. Because of the exclusion principle, there can only be two electrons in a given s shell, six in a p shell, ten in a d shell, fourteen in an f shell, etc. For example, the ground state of the nitrogen atom, N, has two electrons in the 1s shell, two in the 2s shell and three in the 2p shell; one writes this electron configuration as

$$(1s)^2(2s)^2(2p)^3.$$

The superscript denotes the number of electrons in the shell.

Table 20-1

Periodic Table of the Elements

Key:

Element Z
Ground State Configuration
Ground State $^{2S+1}L_J$

	Alkalis										Noble Metals						Halogens	Rare Gases
I	H^1 1s $^2S_{1/2}$																	He^2 $1s^2$ $^1S_0$
II	Li^3 $1s^22s$ $^2S_{1/2}$	Be^4 $1s^22s^2$ $^1S_0$											B^5 $2s^22p$ $^2P_{1/2}$	C^6 $2s^22p^2$ $^3P_0$	N^7 $2p^3$ $^4S_{3/2}$	O^8 $2p^4$ $^3P_2$	F^9 $2p^5$ $^2P_{3/2}$	Ne^{10} $2p^6$ $^1S_0$
III	Na^{11} 3s $^2S_{1/2}$	Mg^{12} $3s^2$ $^1S_0$											Al^{13} $3s^23p$ $^2P_{1/2}$	Si^{14} $3s^23p^2$ $^3P_0$	P^{15} $3p^3$ $^4S_{3/2}$	S^{16} $3p^4$ $^3P_2$	Cl^{17} $3p^5$ $^2P_{3/2}$	A^{18} $3s^23p^6$ $^1S_0$
IV	K^{19} 4s $^2S_{1/2}$	Ca^{20} $4s^2$ $^1S_0$	Sc^{21} $4s^23d$ $^2D_{3/2}$	Ti^{22} $4s^23d^2$ $^3F_2$	V^{23} $4s^23d^3$ $^4F_{3/2}$	Cr^{24} $4s3d^5$ $^7S_3$	Mn^{25} $4s^23d^5$ $^6S_{5/2}$	Fe^{26} $4s^23d^6$ $^5D_4$	Co^{27} $4s^23d^7$ $^4F_{9/2}$	Ni^{28} $4s^23d^8$ $^3F_4$	Cu^{29} $4s3d^{10}$ $^2S_{1/2}$	Zn^{30} $4s^23d^{10}$ $^1S_0$	Ga^{31} $4s^23d^{10}4p$ $^2P_{1/2}$	Ge^{32} $3d^{10}4p^2$ $^3P_0$	As^{33} $3d^{10}4p^3$ $^4S_{3/2}$	Se^{34} $3d^{10}4p^4$ $^3P_2$	Br^{35} $3d^{10}4p^5$ $^2P_{3/2}$	Kr^{36} $4s^24p^6$ $^1S_0$
V	Rb^{37} 5s $^2S_{1/2}$	Sr^{38} $5s^2$ $^1S_0$	Y^{39} $5s^24d$ $^2D_{3/2}$	Zr^{40} $5s^24d^2$ $^3F_2$	Nb^{41} $5s4d^4$ $^6D_{1/2}$	Mo^{42} $5s4d^5$ $^7S_3$	Tc^{43} $5s^24d^5$ $^6S_{5/2}$	Ru^{44} $5s4d^7$ $^5F_5$	Rh^{45} $5s4d^8$ $^4F_{9/2}$	Pd^{46} $4d^{10}$ $^1S_0$	Ag^{47} $5s4d^{10}$ $^2S_{1/2}$	Cd^{48} $5s^24d^{10}$ $^1S_0$	In^{49} $5s^24d^{10}5p$ $^2P_{1/2}$	Sn^{50} $4d^{10}5p^2$ $^3P_0$	Sb^{51} $4d^{10}5p^3$ $^4S_{3/2}$	Te^{52} $4d^{10}5p^4$ $^3P_2$	I^{53} $4d^{10}5p^5$ $^2P_{3/2}$	Xe^{54} $5s^25p^6$ $^1S_0$
VI	Cs^{55} 6s $^2S_{1/2}$	Ba^{56} $6s^2$ $^1S_0$	La^{57} $6s^25d$ $^2D_{3/2}$	Hf^{72} $6s^25d^2$ $^3F_2$	Ta^{73} $6s^25d^3$ $^4F_{3/2}$	W^{74} $6s^25d^4$ $^5D_0$	Re^{75} $6s^25d^5$ $^6S_{5/2}$	Os^{76} $6s^25d^6$ $^5D_4$	Ir^{77} $6s^25d^7$ $^4F_{9/2}$	Pt^{78} $6s5d^9$ $^3D_3$	Au^{79} $6s5d^{10}$ $^2S_{1/2}$	Hg^{80} $6s^25d^{10}$ $^1S_0$	Tl^{81} $6s^25d^{10}6p$ $^2P_{1/2}$	Pb^{82} $6p^2$ $^3P_0$	Bi^{83} $6p^3$ $^4S_{3/2}$	Po^{84} $6p^4$ $^3P_2$	At^{85} $6p^5$ $^2P_{3/2}$	Rn^{86} $6p^6$ $^1S_0$
VII	Fr^{87} 7s $^2S_{1/2}$	Ra^{88} $7s^2$ $^1S_0$	Ac^{89} $7s^26d$ $^2D_{3/2}$															
Lanthanides (Rare Earths)			La^{57} $6s^25d$ $^2D_{3/2}$	Ce^{58} $6s^25d4f$ $^3H_5$	Pr^{59} $6s^24f^3$ $^4I_{9/2}$	Nd^{60} $6s^24f^4$ $^5I_4$	Pm^{61} $6s^24f^5$ $^6H_{5/2}$	Sm^{62} $6s^24f^6$ $^7F_0$	Eu^{63} $6s^24f^7$ $^8S_{7/2}$	Gd^{64} $6s^25d4f^7$ $^9D_2$	Tb^{65} $6s^25d4f^8$	Dy^{66} $6s^24f^{10}$	Ho^{67} $6s^24f^{11}$	Er^{68} $6s^24f^{12}$	Tm^{69} $6s^24f^{13}$ $^2F_{7/2}$	Yb^{70} $6s^24f^{14}$ $^1S_0$	Lu^{71} $6s^25d4f^{14}$ $^2D_{3/2}$	
Actinides			Ac^{89} $7s^26d$ $^2D_{3/2}$	Th^{90} $7s^26d^2$	Pa^{91} $6d^3$	U^{92} $6d5f^3$ $^5L_6$	Np^{93} $5f^5$	Pu^{94} $5f^6$	Am^{95} $5f^7$ $^8S_{7/2}$	Cm^{96} $6d5f^7$	Bk^{97} $5f^9$	Cf^{98} $5f^{10}$	E^{99} $5f^{11}$	Fm^{100} $5f^{12}$	Mv^{101} $5f^{13}$			

THE PERIODIC TABLE

We can now begin to understand the structure of the periodic table of the elements. Under the symbol for each element in the accompanying table is written the ground state configuration. In hydrogen there is a single 1s electron while in He the 1s shell is filled. As we go on to higher Z we fill next the 2s shell, in Li and Be, and *then* the 2p shell, as in B through Ne. States of different l are not degenerate; the smaller l states, for a given n, have lower energy. The reason is that electrons with smaller l can spend more time close to the nucleus; this is energetically a very favorable place to be, for there the electron is *inside* the screening cloud of the other electrons and it feels more strongly the attractive field of the nucleus. The third group, Na through A has filled 1s, 2s, and 2p shells; first the 3s shell is filled and then the 3p.

It is in the fourth row that things become complicated. We might expect next to start filling the 3d shell. However the 4s electrons have roughly the same energy as the 3d, so there develops a competition between them. As we can see from the table, most of the time the 4s electrons win. However Cr has a single 4s electron as does Cu. In Zn the 4s and 3d shells are filled. The row is completed by filling out the 4p shell.

The 3d electrons don't extend as far out from the nucleus as the 4s and 4p electrons. As a consequence the 3d electrons play a minor role in determining the chemical properties of these elements. These properties are determined by the outermost electrons. However it is the incomplete d shells in some of these elements, e.g., Fe, Ni, that is the source of their interesting magnetic properties.

The fifth row basically repeats the fourth; the 5s electrons compete with the 4d electrons now. At Nb the number of 5s electrons drops to one while the 4d shell continues to fill. Finally Pd has a filled 4d shell and no 5s electrons. The 4d shell remains filled for the rest of the row while the 5s shell refills, and then the 5p shell fills.

Row six begins by filling the 6s shell in Cs and Ba. Then La gets a 5d electron. At this point the 4f electrons enter and displace the 5d electron; as one runs through the rare earths, La-Lu, the 4f shell fills fairly regularly, with the exception of a 5d electron again at Gd. The 4f electrons live rather deep inside the atom and so all the rare earths have similar chemistry. After the 4f shell is filled, the 5d shell begins to fill, occasionally displacing one or both 6s electrons. Finally, after the 5d shell fills, the 6s shell refills and then the 6p shell fills.

Lastly, the seventh row is similar to the beginning of the sixth. Now the 7s, 6d, and 5f electrons compete for the ground state. The periodic table ends before the 7p electrons appear.

As we have mentioned, the chemical properties of an atom are determined by its outer electrons. Consequently atoms with similar outer shell configurations, those in a given *column* of the periodic table, have similar chemistry.

In the rightmost column are the rare, or inert gases. Aside from He, these are characterized by filled p shells. To excite a rare gas atom we must increase the principal quantum number of some electron, and this requires a large amount of energy. For example, the $(1s)^2(2s)^2(2p)^5(3s)$ configuration of Ne lies about 16 eV above the $(1s)^2(2s)^2(2p)^6$ ground state. This large excitation energy, together with the fact that all their shells are filled, is why the rare gases are with few exceptions chemically inert. As we shall see later, the closed shells play an important role in making the rare gases inert; consider that the first excited state of hydrogen is 10 eV above the ground state, yet hydrogen is very active chemically.

In the first column on the left are the alkalis with one relatively weakly bound outer s electron. For example, the $(1s)^2(2p)$ configuration of Li has about 2 eV greater energy than the $(1s)^2(2s)$ ground state. This outer electron readily forms chemical bonds. In the second column from the right are the halogens, which need only one electron to fill their outer p shells. We can regard them as having a weakly bound "hole" which also easily forms chemical bonds. Later, when we study molecules, we shall discuss the chemical properties of the elements in greater detail.

SPLITTING OF CONFIGURATIONS

The electronic configuration alone does not uniquely specify the state of an atom. For example, the $(1s)^2(2s)(2p)$ configuration of beryllium, which lies about 2.5 eV above the $(1s)^2(2s)^2$ ground state, could be any of twelve different states. This is because the 2s electron can be in either of two different spin states, while the 2p electron can be in any of six different m_l, m_s states. Different states corresponding to the same configuration have small energy differences. There are two reasons for this: first, the electron-electron interactions differ somewhat in the different states, and second, there is some splitting due to the spin-orbit interaction. In the lighter atoms, the spin-orbit splitting is the smaller of the two effects and can be treated as a weak perturbation after we sort out the first effect.

To see how the states belonging to a given configuration are split, we must look at the constants of the motion of an atom, for we can classify the energy eigenstates by the eigenvalues of other constants of the motion. In the absence of external fields, the total angular momentum **J** is a constant of the motion. When we neglect the spin-orbit interaction, the total orbital angular momentum **L** and the total spin angular momentum **S** are separately constants of the motion. Therefore we can classify the different states of the lighter atoms by their total **L** and **S** quantum numbers. We use the capital letters L, S, M_L, and M_S to denote these quantum numbers.

Even though in the absence of spin-orbit interactions the individual spins of the electrons commute with the Hamiltonian, the atomic states are not eigenstates of the individual spins. This is a result of the antisymmetry of the wave function. For example the antisymmetric spin singlet state

$$\chi_\uparrow(1)\chi_\downarrow(2) - \chi_\downarrow(1)\chi_\uparrow(2)$$

isn't an eigenstate of either s_{1z} or s_{2z}, but it is an eigenstate of $s_z = s_{1z} + s_{2z}$.

States belonging to the same configuration but with different L or S differ in energy by fractions of an electron volt. Without spin-orbit interactions, the multiplet of $(2L+1)(2S+1)$ states with the same L and S are degenerate; the energy of the atom doesn't depend on the orientation of either the total orbital angular momentum vector or the total spin angular momentum vector. Now the spin-orbit interaction breaks the degeneracy of these $(2L+1)(2S+1)$ states and produces the *fine structure* of atomic spectra. Because the total angular momentum is still a constant of the motion, even in the presence of spin-orbit interactions, the L, S multiplets are split into levels with definite J (as well as L and S since the spin-orbit interaction in the lighter atoms is too weak to mix different L, S multiplets). These energy differences are hundredths of an electron volt.

Thus in the lighter atoms we have the following energy level scheme:

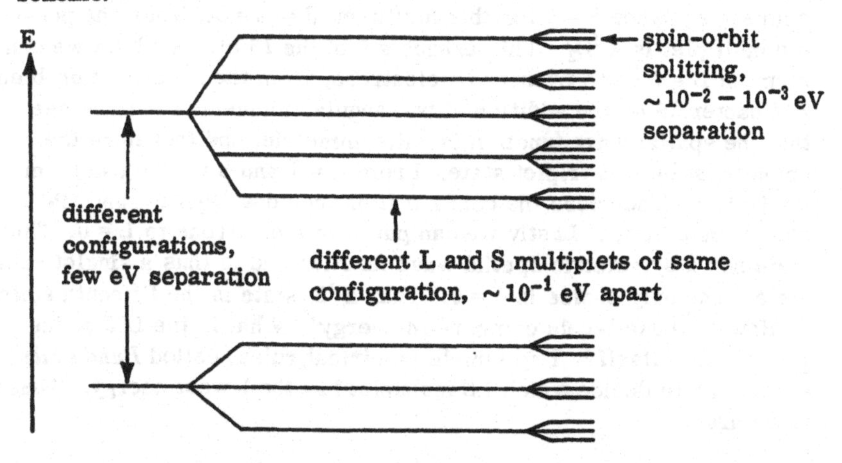

Levels of definite L, S, and J are labeled by spectroscopists by the notation ${}^{2S+1}L_J$. For example, ${}^3P_2$ denotes a level with L = 1, S = 1, J = 2, and is read "triplet P two." An atomic energy level is completely specified by giving *both* the electronic configuration and the LSJ values.

The configurations corresponding to the ground states in the beginning of the periodic table have a unique L, S, and J. For example, H with its 1s electron is a ${}^2S_{1/2}$ level; the J = $1/2$ comes from the spin. The two spins in He are in a singlet state and the two electrons are in s states; therefore the ground state configuration of He can only be a ${}^1S_0$ state. Closed shells always have S = 0, L = 0, and J = 0. Then Li has a 2s electron above a closed 1s shell and therefore is a ${}^2S_{1/2}$ state. Similarly Be has a closed 1s shell and a closed 2s shell and is a ${}^1S_0$ state.

So far all is simple. Now the $(1s)^2(2s)^2(2p)$ configuration of B has L = 1 and S = $1/2$, certainly, but J can be either $3/2$ or $1/2$ (recall that $|L-S| \leq J \leq L+S$). Thus the ground state configuration contains ${}^2P_{1/2}$ and ${}^2P_{3/2}$ levels, or *terms*. The J = $3/2$ level is 2J+1 = 4-fold degenerate while the J = $1/2$ level is doubly degenerate; this accounts for the six different possible ways of putting an electron in the 2p shell. We can also count these six ways by saying that m_l can be 1, 0, −1 while m_s can be $\pm 1/2$. States with definite J are linear combinations of these states with definite m_l and m_s.

Matters are worse in carbon. Here the ground state configuration is $(1s)^2(2s)^2(2p)^2$. There are six states in the 2p shell for the first 2p electron, and, because of the exclusion principle, five for the second. There are therefore $6 \cdot 5/2 = 15$ different states corresponding to the same configuration. How do these break up into LSJ multiplets? The largest L value we can get from two p electrons is clearly two. In this case the electrons have a spatial wave function that is symmetric, e.g., $R_{21}(r)R_{21}(r')Y_{11}(r)Y_{11}(r')$, and therefore they must be in a singlet spin state. Since S = 0 for this multiplet, J = L = 2. Thus one possible multiplet is a ${}^1D_2$. This exhausts 5 of the 15 states. Next we can form an L = 1 state from two p electrons; from the results of problem 2, Chapter 15 on the addition of two angular momenta $l = 1$, we see that the spatial wave function is antisymmetric, and therefore the spins must be in a triplet state. From L = 1 and S = 1 we can form J = 0, 1, 2. Therefore the configuration includes ${}^3P_0$, ${}^2P_1$, and ${}^3P_2$ states, nine in all. Lastly we can pair two p electrons to L = 0. This state has a symmetric spatial wave function and is thus a singlet spin state. The only choice is J = 0. Thus a ${}^1S_0$ state is the fifteenth state.

How do these levels compare in energy? What is the LSJ of the ground state itself? Two simple empirical rules, called *Hund's rules*, enable one to decide which LS multiplet has the lowest energy. These rules are

(1) The LS multiplet with the largest S has the lowest energy.

(2) In case the largest S is associated with several L's, then the largest L has the lowest energy.

Furthermore the spin-orbit splitting of the *ground state* LS multiplet in atoms with a single unfilled shell is determined by the following rule:

(3) If the incomplete shell is not more than half filled then the lowest level has $J = |L - S|$, the smallest possible J. If the shell is more than half filled then the lowest level has $J = L + S$, the maximum value.

We shall discuss the origin of these rules shortly. First, let's apply them.

The ground state configuration of boron contained $^2P_{1/2}$ and $^2P_{3/2}$ terms. The L and S values are determined, and only the third rule is relevant; since the P shell is only one-sixth filled, the actual ground state has the least J, the $^2P_{1/2}$ level. Now the ground state configuration of carbon contains $^1D_2$, $^3P_0$, $^3P_1$, $^3P_2$, and $^1S_0$ terms. The first Hund's rule tells us that the triplet levels have the least energy. $L = 1$ for these so the second rule is inapplicable. The 2p shell is one-third full, so rule 3 implies that the $^3P_0$ level is the actual ground state.

Let us next look at N. Here the three 2p electrons can be in any of $(6 \cdot 5 \cdot 4)/(1 \cdot 2 \cdot 3) = 20$ states. An $L = 3$ state is not possible since the 2p electrons would have to have a symmetric spatial wave function and therefore a completely antisymmetric spin wave function. But it is impossible, for spin $1/2$, to form a completely antisymmetric spin state for three particles. (Two is the limit.) Thus only $L = 2$, 1, and 0 are possible. One can form an antisymmetric wave function for $L = 2$ and 1, only if $S = 1/2$, and for $L = 0$ only if $S = 3/2$. Thus there is an $L = 2$, $S = 1/2$ multiplet which is split by the spin-orbit interaction into $^2D_{3/2}$ and $^2D_{5/2}$ levels. There is an $L = 1$, $S = 1/2$ multiplet split into $^2P_{1/2}$ and $^2P_{3/2}$ levels, and an $L = 0$, $S = 3/2$, or $^4S_{3/2}$ level. One may count that these levels comprise 20 states. The $^4S_{3/2}$ has the lowest energy, since it has the largest S.

Next, in oxygen, there are four 2p electrons. They have available 15 states, which is just the number of ways we can distribute the unoccupied states, or holes. Also, a closed shell is an $L = 0$, $S = 0$ configuration, and therefore the possible angular momentum states of the four particles are the same as those available to the two holes. These in turn are the same as the states available to two p electrons, namely, $^1D_2$, $^3P_0$, $^3P_1$, $^3P_2$, and $^1S_0$. The $S = 1$, $L = 1$ multiplet has lowest energy, and since the p shell is more than half full, the third rule implies that the ground state of oxygen is $^3P_2$.

Another way of looking at the counting argument for oxygen is to notice that any way we arrange the 4 electrons in the 2p shell, there will always be two electrons having opposite spin and opposite m_l. One can think of these two electrons as effectively being paired to $L = 0$ and $S = 0$. The possible LS values of the configuration are then those that can be formed from the remaining two electrons.

Flourine has five 2p electrons, or one 2p hole. Therefore the ground state configuration can be either $^2P_{1/2}$ or $^2P_{3/2}$, as for boron. Because the 2p shell is more than half full, the $^2P_{3/2}$ level is the lowest in energy.

Lastly, in neon, all shells are filled and the ground state is $^1S_0$. We have indicated the ground state angular momenta on the periodic table. With a few exceptions in the unfilled f and d atoms the angular momentum is the same for each atom in a given column of the table.

One cannot give a simple "proof" of Hund's two rules; they are generally true experimentally, and are verified by calculations on individual atoms. But we can understand why they are true by a few simple arguments. The first rule says that the multiplet with the largest S will have the least energy. Now the larger is S, the more symmetric is the spin part of the wave function. Compare, for example, the symmetry of a triplet spin state with the antisymmetry of a singlet spin state. But the more the spin part is symmetric, the more antisymmetric must the space part be, in order that the entire wave function be totally antisymmetric. But when the wave function is more antisymmetric the electrons stay further apart and consequently reduce the contribution of the repulsive electron-electron interaction to the energy. Thus the most antisymmetric spatial wave function will, in general, have the least energy.

The second rule says that given maximum S, the multiplet with the largest L will have the least energy. Consider first a simple mechanical analogy: two balls, at the ends of rigid rods, rotating in the x, y plane with a linear velocity of fixed magnitude v, each a (variable) distance R from the z axis, as shown in Fig. 20-2. The magnitude of the angular momentum l of each ball about the point P depends only on v and not on R. As we increase R the angular momentum M_L of the system about the z axis increases, as does the spacing between the balls. Now let's return to the atom. For a given L all the different M_L states will have the same energy; consider then the state with $M_L = L$. If L is small, the electrons, like the two classical balls, spend much of their time close to the z axis, and the average interelectron distance is small. For larger L the electrons must be further out from the z axis and the average interelectron distance is larger. This has the effect of reducing the average repulsive energy of the electrons. The largest L will have the least energy.

Before going on to consider the spin-orbit interaction in detail we should mention one technical point. To construct the Hartree-Fock orbitals one uses a wave function that is a determinant of the orbitals.

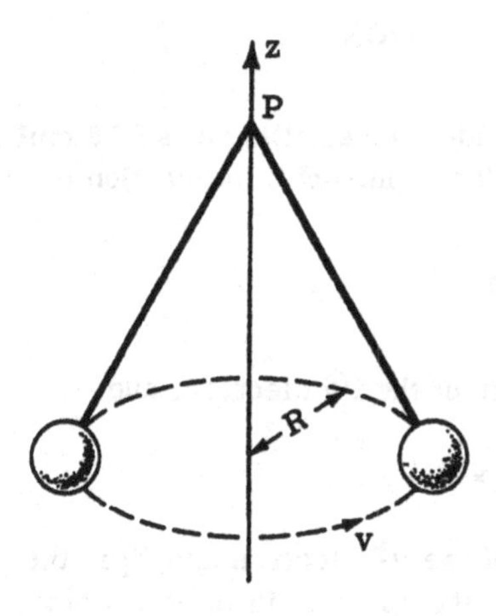

Fig. 20-2

Mechanical analogy for Hund's second rule.

There are, however, many angular momentum eigenstates for a given configuration whose total wave function cannot be written as a single determinant, but is rather a linear combination of determinants. An $L = 0$, $S = 0$ state formed from two p electrons is such a case. The wave function of the p electrons is proportional to an angular and spin part

$$[Y_{11}(1)Y_{1,-1}(2) + Y_{1,-1}(1)Y_{11}(2) - Y_{10}(1)Y_{10}(2)]\,[\chi_\uparrow(1)\chi_\downarrow(2) - \chi_\downarrow(1)\chi_\uparrow(2)],$$

which is the sum of three determinants

$$\det[Y_{11}\chi_\uparrow;\ Y_{1,-1}\chi_\downarrow] - \det[Y_{11}\chi_\downarrow;\ Y_{1,-1}\chi_\uparrow] - \det[Y_{10}\chi_\uparrow;\ Y_{10}\chi_\downarrow].$$

The notation $\det[\varphi, \varphi']$ is shorthand for the determinantal function

$$\begin{vmatrix} \varphi(1) & \varphi'(1) \\ \varphi(2) & \varphi'(2) \end{vmatrix}.$$

Under these circumstances one evaluates the total energy by first constructing the set of Hartree-Fock single particle orbitals of the form (20-46), then forming the correct multiplet wave function as a linear combination of Slater determinants formed by these orbitals, and finally taking the expectation value of the true Hamiltonian in this state.

SPIN-ORBIT INTERACTION

Let us now consider the splitting of an LS multiplet by the spin-orbit interaction. The spin-orbit interaction is (cf. Chapter 14)

$$H_{s.o.} = \sum_i \mathbf{X}(\mathbf{r}_i) \cdot \mathbf{s}_i \tag{20-47}$$

where $\mathbf{s}_i$ is the spin of the ith electron, and

$$\mathbf{X}(\mathbf{r}_i) = -\frac{|e|}{2mc^2}\, \mathbf{v}_i \times \mathbf{E}_i\,; \tag{20-48}$$

$\mathbf{v}_i$ is the velocity of the ith electron and $\mathbf{E}_i$ is the electric field it experiences. To find the splitting in lighter atoms, where the velocities and electric fields are smaller, we can use first-order degenerate perturbation theory. We then need to know the matrix elements

$$\langle NLSM_L M_S | H_{s.o.} | NLSM_L{}' M_S{}' \rangle$$

between two states in the same multiplet; N stands for all the other quantum numbers needed to specify the electronic state. The unperturbed states $|NLSM_L M_S\rangle$ for different M_L and M_S are degenerate.

Since the Hamiltonian in the absence of spin-orbit coupling doesn't depend on the electron spins, the exact wave function of the state $|NLSM_L M_S\rangle$ is some linear combination (for antisymmetrization when necessary) of spatial wave functions $\psi^{(\alpha)}_{NLM_L}(\mathbf{r}_1, \ldots, \mathbf{r}_k)$, which are eigenstates of L^2 and L_z, times spin functions $\chi^{(\alpha)}_{SM_S}(s_1, \ldots, s_k)$, which are eigenstates of S^2 and S_z. Thus the wave function has the form

$$\langle 1, \ldots, k | NLSM_L M_S \rangle = \sum_\alpha c_\alpha \psi^{(\alpha)}_{NLM_L} \chi^{(\alpha)}_{SM_S} \tag{20-49}$$

The matrix element $\langle NLSM_L M_S | \mathbf{X}_i \cdot \mathbf{s}_i | NLSM_L{}' M_S{}' \rangle$ is therefore given by

$$\sum_{\alpha\alpha'} c_\alpha{}^* c_{\alpha'} \langle \alpha NLM_L | \mathbf{X}_i | \alpha' NLM_L{}' \rangle \cdot \langle \alpha SM_S | \mathbf{s}_i | \alpha' SM_S{}' \rangle \tag{20-50}$$

where $\langle \alpha NLM_L | \mathbf{X}_i | \alpha' NLM_L{}' \rangle$ is the matrix element of $\mathbf{X}_i$ between $\psi^{(\alpha)}_{NLM_L}$ and $\psi^{(\alpha')}_{NLM_L{}'}$ and $\langle \alpha SM_S | \mathbf{s}_i | \alpha' SM_S{}' \rangle$ is the matrix element of $\mathbf{s}_i$ between $\chi^{(\alpha)}_{SM_S}$ and $\chi^{(\alpha')}_{SM_S{}'}$.

Now from the Wigner-Eckart theorem, $\langle \alpha S M_S | \mathbf{s}_i | \alpha' S M_S' \rangle$, the matrix element of a vector operator between total spin eigenstates, must be proportional to the matrix element of the total spin $\mathbf{S}$ between the same states:

$$\langle \alpha S M_S | \mathbf{s}_i | \alpha' S M_S' \rangle \sim \langle S M_S | \mathbf{S} | S M_S' \rangle; \tag{20-51}$$

the constant of proportionality doesn't depend on M_S or M_S'. Similarly, the matrix element of $\mathbf{X}_i$, a vector operator, is proportional to the same matrix element of $\mathbf{L}$, the only relevant spatial vector:

$$\langle \alpha N L M_L | \mathbf{X}_i | \alpha' N L M_L' \rangle \sim \langle L M_L | \mathbf{L} | L M_L' \rangle; \tag{20-52}$$

the proportionality constant is independent of M_L and M_L'. Thus from (20-51) and (20-52) we find that

$$\begin{aligned} &\langle NLSM_L M_S | \mathbf{X}_i \cdot \mathbf{s}_i | NLSM_L' M_S' \rangle \\ &\qquad \sim \langle L M_L | \mathbf{L} | L M_L' \rangle \cdot \langle S M_S | \mathbf{S} | S M_S' \rangle; \end{aligned} \tag{20-53}$$

summing over i yields

$$\begin{aligned} &\langle NLSM_L M_S | H_{s.o.} | NLSM_L' M_S' \rangle \\ &\qquad = \zeta(NLS) \langle L M_L | \mathbf{L} | L M_L' \rangle \langle S M_S | \mathbf{S} | S M_S' \rangle \\ &\qquad = \zeta(NLS) \langle NLSM_L M_S | \mathbf{L} \cdot \mathbf{S} | NLSM_L' M_S' \rangle . \end{aligned} \tag{20-54}$$

$\zeta(NLS)$ is some proportionality constant, the same for all states in the LS multiplet. Equation (20-54), which is exact, is a nice example of the usefulness of the Wigner-Eckart theorem.

At this point it is trivial to diagonalize the perturbation by passing to the $|NLSJM\rangle$ basis. Since $\mathbf{J}^2 = (\mathbf{L} + \mathbf{S})^2$, we have

$$\mathbf{L} \cdot \mathbf{S} = \frac{(\mathbf{J}^2 - \mathbf{L}^2 - \mathbf{S}^2)}{2}. \tag{20-55}$$

Thus $\langle NLSJM | H_{s.o.} | NLSJ'M \rangle$ vanishes unless $J = J'$, and $M = M'$, and then

$$\begin{aligned} \langle NLSJM | H_{s.o.} | NLSJM \rangle &= \zeta(NLS) \frac{\langle NLSJM | (\mathbf{J}^2 - \mathbf{L}^2 - \mathbf{S}^2) | NLSJM \rangle}{2} \\ &= \zeta(NLS) \frac{[J(J+1) - L(L+1) - S(S+1)]}{2}. \end{aligned} \tag{20-56}$$

This is the change in energy of the NLSJ level due to the spin-orbit interaction. The difference in energy between adjacent levels J − 1 and J in the same multiplet is thus

$$E_{NLS}(J) - E_{NLS}(J-1) = \zeta(NLS)J. \tag{20-57}$$

This simple formula for the spin-orbit splitting is *Landé's interval rule.* If ζ is positive the multiplet is called regular; if it is negative the multiplet is said to be *inverted.*

The empirical rule for determining the J of the ground state when there is a single incomplete shell is then the statement that if the shell is half or less filled, the multiplet the ground state belongs to is regular; otherwise it is inverted.

SPIN-ORBIT SPLITTING IN HARTREE-FOCK

We can verify the validity of this rule in the Hartree-Fock approximation. Let us assume that the effective potential felt by each electron has spherical symmetry. Then, as in Eq. (14-51), (20-48) becomes

$$\mathbf{X}_i = \frac{1}{2m^2c^2r_i}\frac{dV}{dr_i}\boldsymbol{l}_i, \tag{20-58}$$

where $\boldsymbol{l}_i$ is the angular momentum operator of the ith electron. Then

$$H_{s.o.} = \sum_i \frac{1}{2m^2c^2r_i}\frac{dV}{dr_i}\boldsymbol{l}_i\cdot\mathbf{s}_i. \tag{20-59}$$

To evaluate $\zeta(NLS)$ we note that if in (20-54) we set $M_L = M_L'$ and $M_S = M_S'$ then

$$\langle NLSM_LM_S|H_{s.o.}|NLSM_LM_S\rangle = \zeta(NLS)M_LM_S, \tag{20-60}$$

since the diagonal matrix elements of L_xS_x and L_yS_y vanish. Let us write the wave function of the state $|NLSM_LM_S\rangle$ as a linear combination of Slater determinants, $\sum_\beta a_\beta|N\beta\rangle$, where $\sum_\beta |a_\beta|^2 = 1$. We use the very abbreviated notation $|N\beta\rangle$ to stand for a single Slater determinant of orbitals $\varphi_{nlm_l}\chi_{m_s}$. Recall that to manufacture eigenstates of $\mathbf{L}$ and $\mathbf{S}$ it is often necessary to take linear combinations of Slater determinants. If there is only one unfilled shell, these determinants differ only in the particular set of occupied states in that shell. The radial functions for each shell are the same in each determinant; only the angular functions differ. $|N\beta\rangle$ is an eigenstate of L_z with eigenvalue M_L, and S_z with eigenvalue M_S. We can then write

$$\langle NLSM_LM_S|H_{s.o.}|NLSM_LM_S\rangle = \sum_{\beta\beta'} a_\beta{}^* a_{\beta'}\langle N\beta|\sum_i \frac{1}{2m^2c^2}\frac{dV}{dr_i}\, l_i \cdot s_i|N\beta'\rangle. \tag{20-61}$$

Only the terms with $\beta = \beta'$ can contribute to the sum. This follows because the spin-orbit operator is a sum of single particle operators. Therefore all the particles but i in a given nonvanishing term of (20-61) must be in the same orbital, due to the orthonormality of the set of orbitals. But if all the other particles are in the same orbitals, so must the ith particle in order that $|N\beta\rangle$ and $|N\beta'\rangle$ have the same L_z and S_z eigenvalues.

We evaluate $\langle N\beta|H_{s.o.}|N\beta\rangle$ by noticing that it must be a sum of terms coming from each occupied orbital in the determinant $|N\beta\rangle$:

$$\langle N\beta|H_{s.o.}|N\beta\rangle = \sum_{nlm_lm_s}\int d^3r\, \varphi_{nlm_l}{}^*(\mathbf{r})\, \frac{1}{2m^2c^2r}\, \frac{dV}{dr}\, l\varphi_{nlm_l}(\mathbf{r})\cdot\langle m_s|\mathbf{s}|m_s\rangle. \tag{20-62}$$

The diagonal spin expectation value is zero for s_x and s_y, while $\langle m_s|s_z|m_s\rangle = m_s$. Thus only the l_z term in (20-62) enters, and l_z acting on its own eigenstate gives simply m_l. Hence

$$\langle n\beta|H_{s.o.}|N\beta\rangle = \frac{1}{2m^2c^2}\sum_{nlm_lm_s} \xi_{nl}m_lm_s, \tag{20-63}$$

where

$$\xi_{nl} = \int d^3r|\varphi_{nlm_l}(r)|^2\frac{1}{r}\frac{dV}{dr}; \tag{20-64}$$

ξ_{nl} is independent of m_l because of the spherical symmetry of V. For a filled shell

$$\sum_{m_lm_s} m_lm_s = \sum_{\text{all } m_l} m_l \sum_{\text{all } m_s} m_s = 0,$$

since all m_l and m_s occur. Therefore only terms in (20-63) referring to unfilled shells are nonzero. Let us assume that only one shell nl is partially filled. Then

$$\langle N\beta|H_{s.o.}|N\beta\rangle = \frac{1}{2m^2c^2}\xi_{nl}\sum_{m_lm_s} m_lm_s. \tag{20-65}$$

The sum is over the occupied states in the nl shell, i.e., those entering the determinant $|N\beta\rangle$.

Now the ground state, according to Hund, always has the maximum possible S. Let us choose $M_S = S$. Then in a half- or less-filled shell this state is the one with all spins up, $m_s = ½$. Thus in (20-65)

$$\sum_{m_l m_s} m_l m_s = \sum m_l \sum m_s = M_L S,$$

so that

$$\langle N\beta | H_{s.o.} | N\beta \rangle = \frac{\xi_{nl}}{2m^2c^2} M_L S. \tag{20-66}$$

This is the same for all β since they all have the same M_L and M_S values. Hence from (20-61),

$$\langle NLSM_L S | H_{s.o.} | NLSM_L S \rangle = \frac{\xi_{nl}}{2m^2c^2} M_L S, \tag{20-67}$$

since $\sum_\beta |a_\beta|^2 = 1$. Comparing (20-67) with (20-60) we find the simple result that if $L \neq 0$,

$$\zeta(NLS) = \frac{\xi_{nl}}{2m^2c^2}. \tag{20-68}$$

For $L = 0$, there is no spin-orbit splitting.

We may argue that ξ_{nl} is positive, since V in general increases with r. Thus $\zeta > 0$ for a less than half-filled shell with $L \neq 0$. For a more than half-filled shell, we still choose the state $M_S = S$. Now for each $|N\beta\rangle$

$$\sum_{\substack{\text{occupied}\\ \text{states}}} m_l m_s + \sum_{\substack{\text{unoccupied}\\ \text{states}}} m_l m_s = \sum_{\substack{\text{full}\\ \text{shell}}} m_l m_s = 0; \tag{20-69}$$

but for the state $M_S = S$ all the unoccupied states have spin down and therefore

$$\sum_{\text{unocc.}} m_l m_s = \sum_{\text{unocc}} m_l \sum_{\text{unocc}} m_s = (-M_L)(-M_S), \tag{20-70}$$

since the total M_L or M_S of the unoccupied states is minus that of the occupied states. Thus

$$\sum_{\substack{\text{occ.}\\ \text{states}}} m_l m_s = -M_L M_S, \tag{20-71}$$

and from (20-61) and (20-60) we now find, if $L \neq 0$

$$\zeta(\text{NLS}) = -\frac{\xi_{nl}}{2m^2c^2}. \tag{20-72}$$

If the shell is more than half-filled the ground state belongs to an inverted multiplet.

Finally let's consider the case that the incomplete shell is half-filled. Then in the state $M_S = S$, all the spins are up and to fill the shell half-way all the different m_l must be occupied, or $M_L = 0$; since this is the only possible M_L value, L itself must be zero. Thus there is no spin orbit splitting of a half-filled shell with maximum possible S.

In order of magnitude

$$\xi_{nl} \sim \int |\varphi_{nl}|^2 \frac{e^2}{r^3} \sim \frac{e^2}{(Z^{-1/3}a_0)^3},$$

so that

$$\zeta \sim \frac{Z}{a_0{}^2 m^2 c^2}\left(\frac{e^2}{2a_0}\right) = Z\alpha^2 \text{ Ry}, \tag{20-73}$$

where $\alpha = e^2/\hbar c = 1/137$ is the fine structure constant. ζ measures the size of the fine structure in the spectrum. Thus spin-orbit splittings are of order 10^{-3} of the configuration energy differences, and increase with Z.

For small Z it is a very good approximation to treat the spin-orbit splitting as a small perturbation of the LS multiplets. This description of an atom is called the L-S or *Russell-Saunders coupling scheme.* On the other hand for very large Z the spin-orbit interaction can no longer be treated as a perturbation but must be taken into account at the beginning. One must include the spin-orbit interaction in the original single particle orbitals. This means that they are no longer separate l, m_l, m_s eigenstates but rather l, j, m eigenstates, where j is the total angular momentum of the single electron. One then constructs multiplets by combining these j eigenstates. This procedure is called the *j-j coupling scheme,* and one can read about it in Condon and Shortley.

ZEEMAN EFFECT

We turn now to studying the splitting of an LS multiplet by a uniform magnetic field $\mathcal{H}$. First let us take the case that $\mathcal{H}$ is weak compared to the magnetic fields producing the spin-orbit splitting (fields of about 10^5 gauss). Then we may evaluate the energy shift of the level $|NLSJM\rangle$ by first-order perturbation theory. The magnetic interaction is,[3] to first order in $\mathcal{H}$,

$$H_{mag} = -\frac{e}{2mc}\mathbf{L}\cdot\mathcal{H} - \frac{e}{mc}\mathbf{S}\cdot\mathcal{H}. \tag{20-74}$$

For the field in the z direction the energy shift is given by

$$\Delta E_{LSJ}(M) = \langle NLSJM|H_{mag}|NLSJM\rangle$$

$$= -\frac{e\mathcal{H}}{2mc}\langle NLSJM|(L+2S)_z|NLSJM\rangle, \tag{20-75}$$

$$= -\frac{e\mathcal{H}}{2mc}(M\hbar + \langle NLSJM|S_z|NLSJM\rangle) \tag{20-76}$$

since $(L+S)_z = J_z$ and the state is a J_z eigenstate.

To evaluate the matrix element of the S_z, the z component of a vector operator, we use the Wigner-Eckart theorem to write

$$\frac{\langle NLSJM|S_z|NLSJM\rangle}{\langle NLSJM|J_z|NLSJM\rangle} = \frac{\langle NLSJ\|S\|NLSJ\rangle}{\langle NLS\ \|J\|NLSJ\rangle}; \tag{20-77}$$

the ratio of the reduced matrix elements. Now from the same theorem

$$\langle NLSJM|\mathbf{J}\cdot\mathbf{S}|NLSJM\rangle \sim \langle NLSJM\|J\|NLSJM\rangle\langle NLSJM\|S\|NLSJM\rangle \tag{20-78}$$

where the coefficient is a bunch of Clebsch-Gordan coefficients; similarly

$$\langle NLSJM|\mathbf{J}^2|NLSJM\rangle \sim \langle NLSJM\|J\|NLSJM\rangle^2 \tag{20-79}$$

where the coefficient is the *same* bunch of Clebsch-Gordan coefficients. Thus the ratio of (20-78) to (20-79) equals (20-77):

[3] e is a negative quantity.

$$\frac{\langle NLSJM|S_z|NLSJM\rangle}{\langle NLSJM|J_z|NLSJM\rangle} = \frac{\langle NLSJM|\mathbf{J}\cdot\mathbf{S}|NLSJM\rangle}{\langle NLSJM|J^2|NLSJM\rangle},$$

or

$$\langle NLSJM|S_z|NLSJM\rangle = \frac{M}{J(J+1)}\langle NLSJM|\mathbf{J}\cdot\mathbf{S}|NLSJM\rangle. \tag{20-80}$$

This theorem is valid for any vector operator, not just S. Now

$$\mathbf{J}\cdot\mathbf{S} = \mathbf{S}^2 + \mathbf{L}\cdot\mathbf{S} = \frac{(\mathbf{J}^2 - \mathbf{L}^2 + \mathbf{S}^2)}{2}, \tag{20-81}$$

so that

$$\langle NLSJM|S_z|NLSJM\rangle = \frac{M\hbar}{J(J+1)}\,\frac{[J(J+1)+S(S+1)-L(L+1)]}{2} \tag{20-82}$$

Thus we can write the magnetic field splitting as

$$\Delta E_{NLSJ}(M) = -\frac{e\hbar\mathcal{H}}{2mc}\,gM, \tag{20-83}$$

where the gyromagnetic ratio is

$$g = 1 + \frac{J(J+1)+S(S+1)-L(L+1)}{2J(J+1)}. \tag{20-84}$$

This is the *Landé g-factor* for LS coupling.

The atom behaves as if it has a magnetic moment $ge\hbar/2mc$; each LSJ multiplet splits into $2J+1$ magnetic sublevels. If $S = 0$ the magnetic moment is due to the orbital motion and $g = 1$, while if $L = 0$ the magnetic moment arises purely from the spins and $g = 2$. In general $1 \le g \le 2$. This splitting in a magnetic field is the *Zeeman effect.*

For a field strong compared to the spin-orbit fields, the main splitting of the LS multiplet is due to the external magnetic field. Then a level $|NLSM_LM_S\rangle$ will have a shift in energy

$$\langle NLSM_LM_S|H_{mag}|NLSM_LM_S\rangle = -\frac{e\mathcal{H}}{2mc}(M_L + 2M_S) \tag{20-85}$$

from the magnetic field. Levels with the same $M_L + 2M_S$ are degenerate. The spin-orbit interaction then shifts the energy by

$$\langle NLSM_LM_S|H_{s.o.}|NLSM_LM_S\rangle = \zeta(NLS)M_LM_S;$$

hence

$$\Delta E_{NLS}(M_L, M_S) = -\frac{e\mathcal{H}}{2mc}(M_L + 2M_S) + \zeta(NLS)M_L M_S. \qquad (20\text{-}86)$$

This level splitting in a strong field is called the *Paschen-Back effect.*

If the field is comparable to the spin-orbit fields then one must calculate the splitting of the LS multiplet by diagonalizing the

$$\langle NLSM_L M_S | (H_{mag} + H_{s.o.}) | NLSM_L' M_S' \rangle$$

perturbation matrix. Finally, if the field is very large, one must begin to include the effects of the A^2 term in the Hamiltonian; this term leads to the diamagnetic behavior of atoms.

PROBLEMS

1. (a) Write down explicitly the angular wave functions for the 2p electrons in $^1D_2$, $^3P_0$, $^3P_1$, $^3P_2$, and $^1S_0$ levels of the ground state configuration of carbon.
   (b) Write out the fully antisymmetrized $^1D_2$ wave function as a linear combination of Slater determinants.
2. Find by actual construction the possible L, S states for the four 2p electrons in oxygen.
3. Calculate for the ^2S ground state multiplet and the ^2P multiplet lying above the ground state in an alkali atom, the level splittings due to the spin-orbit interaction together with a uniform magnetic field whose strength is comparable to the spin-orbit fields.
4. Calculate by first-order perturbation theory the energies of the "ortho" and "para" first excited states (1s)(2s), of the Helium atom. Experimentally $E_{para} = -2.146\ e^2/a_0$, $E_{ortho} = -2.175\ e^2/a_0$.

Chapter 21
MOLECULES

The problem of finding the electron orbits in a molecule is much more complicated than in atoms because the effective potential the electrons feel is no longer approximately spherically symmetric.[1] One can picture the nuclei in a molecule as having classical equilibrium positions — points of minimum potential energy — about which they oscillate slowly, while the electrons travel rapidly around in the Coulomb potential of the nuclei. This simple picture of a molecule works because the nuclei are so much heavier than the electrons; the ratio of the mass m of an electron to the mass M of a nucleus is typically

$$\frac{m}{M} \sim 10^{-4} \text{ to } 10^{-5}. \tag{21-1}$$

As a consequence the zero-point motion of the nuclei is far less than that of the electrons. The nuclei not only move much more slowly than the electrons, but they can have fairly well localized positions in the molecule and not violate the uncertainty principle.

From the point of view of the electrons, the nuclei are practically sitting still; the electrons find themselves in an almost static nuclear Coulomb potential and as in an atom they distribute themselves in an eigenstate of this potential. The slow nuclear vibrations have the effect of only adiabatically deforming these electronic eigenstates. In a molecule of size a, we may argue from the uncertainty principle

[1] See as general references L. Pauling, *The Nature of the Chemical Bond,* 3rd ed. [Cornell Univ. Press, Ithaca, New York, 1960]; J.C. Slater, *Quantum Theory of Molecules and Solids* [McGraw-Hill, New York, 1963]; C. J. Ballhausen and H. B. Gray, *Molecular Orbital Theory* [W.A. Benjamin, New York, 1965].

that typical electron momenta are $\sim\hbar/a$ and therefore the various excited states of the electrons will be spaced in energy by about

$$E_{\text{elect}} \sim \frac{\hbar^2}{ma^2}; \tag{21-2}$$

these excitation energies are several electron volts, as in atoms.

Now to the nuclei, the electrons are a blurry cloud. As the nuclei move, they distort the electronic wave functions. This distortion changes the total electronic energy slightly and tends to encourage the nuclei to move toward positions of minimum electronic energy; it is almost as if the nuclei were immersed in an elastic medium, formed of the electrons. The nuclei therefore oscillate about positions of minimum total energy – electronic plus the repulsive Coulomb energy between nuclei. We can estimate the typical frequency ω of these oscillations by saying that a nucleus is in a harmonic oscillator potential $M\omega^2\mathbf{R}^2/2$, where $\mathbf{R}$ is the nuclear position, measured from the equilibrium position. If the nucleus moves a distance $\sim a$ from home then the electronic energy will change by an amount on the order of an electronic excitation energy, $\hbar^2/2ma^2$. Thus we can argue that $M\omega^2 a^2/2 \sim \hbar^2/2ma^2$, or

$$\omega \sim \left(\frac{m}{M}\right)^{1/2} \frac{\hbar}{ma^2}. \tag{21-3}$$

The nuclear vibration energies, $\hbar\omega$, are therefore a factor $(m/M)^{1/2}$ smaller than electronic excitation energies:

$$E_{\text{vib}} \sim \left(\frac{m}{M}\right)^{1/2} E_{\text{elect}}. \tag{21-4}$$

These vibrational energies are tenths or hundredths of an electron volt.

The zero-point momentum of a nucleus in such a harmonic oscillator well is given by $P^2/2M \sim \hbar\omega/2$ so that $P \sim (M/m)^{1/4}(\hbar/a)$; this is about ten times greater than electronic momenta. Thus the typical velocity of a nucleus is $v_N = P/M \sim (m/M)^{3/4}\hbar/ma$; that is, $(m/M)^{3/4}$ smaller than a typical electron velocity. Also, the deviations δ of the nuclei from the equilibrium sites are given by $M\omega^2\delta^2/2 \sim \hbar\omega/2$ or

$$\left(\frac{\delta}{a}\right)^2 \sim \frac{\hbar\omega}{M\omega^2 a^2} \sim \frac{E_{\text{vib}}}{E_{\text{elect}}} \sim \left(\frac{m}{M}\right)^{1/2} \tag{21-5}$$

Thus

$$\left(\frac{\delta}{a}\right) \sim \left(\frac{m}{M}\right)^{1/4} \sim \frac{1}{10}.$$

There is one other important type of excitation of a molecule — a rotation of the molecule as a whole about its center of mass. The energies of these rotations are very small since they involve little internal distortion of the molecule. A rotational excitation with angular momentum $\hbar l$ will have an energy

$$E_{rot} \sim \frac{\hbar^2 l(l+1)}{2\mathscr{I}} \tag{21-6}$$

where $\mathscr{I} \sim Ma^2$ is the moment of inertia of the molecule; thus

$$E_{rot} \sim \left(\frac{m}{M}\right) E_{elect}.$$

The general excited state of a molecule will be a combination of electronic, vibrational, and rotational excitations and will have an energy

$$E = E_{elect} + E_{vib} + E_{rot}. \tag{21-7}$$

These pieces are in relative size

$$E_{elect} : E_{vib} : E_{rot} = 1 : \left(\frac{m}{M}\right)^{1/2} : \frac{m}{M} \tag{21-8}$$

BORN-OPPENHEIMER METHOD

Now let us see how this picture comes out of the Schrödinger equation. The Hamiltonian for the molecule is

$$H = T_e + T_N + V_{ee} + V_{eN} + V_{NN}, \tag{21-9}$$

where

$$T_e = \sum_{\substack{\text{electrons} \\ i}} \frac{p_i^2}{2m} \tag{21-10}$$

is the electronic kinetic energy,

$$T_N = \sum_{\substack{\text{nuclei} \\ \alpha}} \frac{P_\alpha^2}{2M_\alpha} \tag{21-11}$$

is the nuclear kinetic energy, V_{ee} is the repulsive electron-electron Coulomb interaction, V_{eN} is the attractive electron-nucleus Coulomb interaction, and V_{NN} is the repulsive nucleus-nucleus Coulomb interaction. The point is that the nuclear kinetic energy, $\sim 1/M$, is a very small term in this Hamiltonian, and we may calculate the eigenstates and energy eigenvalues of (21-9) by treating T_N as a small perturbation. This procedure is known as the *Born-Oppenheimer* or *adiabatic approximation;* the small parameter characterizing the expansion is $(m/M)^{1/4}$, the ratio of a nuclear vibrational displacement to the spacing between nuclei [Eq. (21-5)].

First let us neglect T_N completely. Then the nuclear positions, which we will denote collectively by R, are no longer dynamical variables in H, but are merely parameters. The Schrödinger equation for the wave function of the molecule is then

$$[T_e + V_{ee}(r) + V_{eN}(r, R)]\varphi_n(r, R) = [\varepsilon_n(R) - V_{NN}(R)]\varphi_n(r, R). \qquad (21\text{-}12)$$

We use r to denote collectively the electron coordinates. Since the nuclear positions are parameters in the potential, the energy eigenvalue ε_n depends on R. The eigenvalue $\varepsilon_n(R)$ is the electronic energy; it includes the constant $V_{NN}(R)$. The solutions of (21-12) form a complete set of states, and we may expand any function of the electronic coordinates in terms of them. In particular, we can expand the r dependence of the actual wave function $\Psi(r, R)$ of the molecule, which includes the nuclear motion, in terms of the $\varphi_n(r, R)$:

$$\Psi(r, R) = \sum_n \Phi_n(R)\varphi_n(r, R). \qquad (21\text{-}13)$$

The problem is to find the expansion coefficients $\Phi_n(R)$; they depend on R since we are expanding only the r dependence of Ψ.

The wave function Ψ solves the Schrödinger equation

$$(T_e + T_N + V_{ee} + V_{eN} + V_{NN})\Psi(r, R) = E\Psi(r, R). \qquad (21\text{-}14)$$

Substituting (21-13) in (21-14), and using (21-12) we find

$$\sum_m (\varepsilon_m(R) + T_N)\Phi_m(R)\varphi_m(r, R) = E\sum_m \Phi_m(R)\varphi_m(r, R). \qquad (21\text{-}15)$$

Then multiplying on the left by $\varphi_n^*(r, R)$, integrating over *all* electron positions, and using the orthonormality of the φ_m, we have

$$\sum_m \int dr\, \varphi_n^*(r, R) T_N \Phi_m(R)\varphi_m(r, R) + \varepsilon_n(R)\Phi_n(R) = E\Phi_n(R). \qquad (21\text{-}16)$$

T_N is a sum of second derivatives, which act on the R dependence of both Φ_m and φ_n. Since $\nabla^2(\Phi\varphi) = (\nabla^2\Phi)\varphi + 2\nabla\Phi\cdot\nabla\varphi + \Phi\nabla^2\varphi$, we can write (21-16) as

$$[T_N + \varepsilon_n(R)]\Phi_n(R) = E\Phi_n(R) - \sum_m A_{nm}\Phi_m(R), \tag{21-17}$$

where

$$A_{nm}\Phi_m(R) = -\hbar^2 \sum_\alpha \frac{1}{2M_\alpha} \int dr\ \varphi_n{}^*(r,R) \times [2\nabla_{R_\alpha}\Phi_m(R) \cdot \nabla_{R_\alpha}\varphi_m(r,R) + \Phi_m(R)\nabla_{R_\alpha}{}^2\varphi_m(r,R)]. \tag{21-18}$$

This is the term that mixes together different n and m and therefore different electronic wave functions in Ψ. We shall see shortly that this term is of relative size $(m/M)^{1/2}$, and may be treated as a perturbation in (21-17).

Neglecting the A_{nm} term we see that

$$[T_N + \varepsilon_n(R)]\Phi_n(R) = E\Phi_n(R); \tag{21-19}$$

this is a simple Schrödinger equation for $\Phi_n(R)$, the amplitude for the nuclei to have positions R when the electrons are in the state φ_n. The total electronic energy $\varepsilon_n(R)$, which includes V_{NN}, plays the role of an effective potential energy for the nuclei. The total effect of the electrons is to couple the nuclei together with "rubber bands" whose force constants depend on the electronic state. In fact, this is essentially the way a real rubber band works. When we describe a rubber band by Hooke's law, we are subsuming all the complicated molecular motion into a simple spring constant. Changing the internal motion, by heating the rubber band for example, changes the spring constant, exactly as changing the electronic state φ_n changes the effective potential felt by the nuclei in a molecule.

To lowest order in A then, there is no mixing of different electronic states and there is only one term in (21-13). We can therefore write the eigenstates of the molecule as

$$\Psi_{n\nu}(r,R) = \Phi_{n\nu}(R)\varphi_n(r,R), \tag{21-20}$$

where ν labels different solutions to (21-19). The energy $E_{n\nu}$ of these solutions is the total energy of the molecule, independent of all coordinates.

Let us consider only the *bound* states of the molecule. Then the potential energy $\varepsilon_n(R)$ will have a set of minima, which are the nuclear equilibrium positions. Actually $\varepsilon_n(R)$ will not have a unique set of minima, since rotating the molecule as a whole about its center of mass, or translating the molecule in space will not change $\varepsilon_n(R)$. To describe the nuclear vibrations one must first separate out the translational and rotational degrees of freedom, exactly as one does classically. Then the ε_n will have minima in the coordinates that describe relative nuclear motion, and to calculate the nuclear vibration energies one must expand ε_n about the equilibrium nuclear separations. The vibrational part of the Schrödinger equation (21-19) then reduces to that of a set of coupled harmonic oscillators, and to solve the problem one must find the classical normal modes. The normal mode oscillators are simple harmonic oscillators; their energy levels are uniformly spaced and they have Gaussian-like wave functions in the normal mode coordinates. The total nuclear wave function $\Phi_{n\nu}$ is a product of translational, rotational, and vibrational parts. We shall not study the nuclear vibration problem any further, but we should note that the nuclear modes and frequencies depend on the particular electronic state φ_n.

We have one bit of unfinished business — showing that the A_{nm} term in (21-17) may be neglected. The $-\hbar^2\nabla_{R_\alpha}\varphi_m/2M_\alpha$ term in (21-18) is of order of magnitude $-(m/M)(\hbar^2\nabla_r^2/2m)\varphi_m$, or m/M times an electronic kinetic energy. This is far smaller than the energy spacing of the different electronic levels n, so it will give negligible mixing of different n. To estimate the other term in A we note that Φ_m has the form of a harmonic oscillator wave function $\sim e^{-(R-R_0)^2M\omega/2\hbar}$ Then

$$\nabla_R\Phi_m \sim |R-R_0|\frac{M\omega}{\hbar}\,\Phi_m \sim \frac{(\delta M\omega)}{\hbar}\,\Phi_m,$$

where δ is a typical nuclear displacement from equilibrium and ω is a typical nuclear vibrational frequency. Now $\delta|\nabla_R\varphi_m| \sim \varphi_m$ so that this entire term in A is on the order of $\hbar\omega$ in magnitude. But the vibrational frequencies are $\sim(m/M)^{1/2}$ smaller than the spacing between electronic levels, and therefore this term also will produce only negligible mixing of different electronic levels in Ψ. Thus to lowest order in $(m/M)^{1/2}$ we can neglect A.

In summary then, one calculates the states of molecules by first solving the electronic problem for fixed nuclear positions, and then uses the total electronic energy as a potential function to describe the nuclear motion. The question then of whether a molecule binds together is that of determining whether this potential function has minima sufficiently deep to hold the nuclei. Let us now study how this all happens in actual molecules.

THE H_2^+ ION

The simplest molecule is the H_2^+ ion. The Hamiltonian, neglecting the nuclear kinetic energy and spin-orbit interaction is

$$H = -\frac{\nabla^2}{2m} - \frac{e^2}{|\mathbf{r}-\mathbf{R}_A|} - \frac{e^2}{|\mathbf{r}-\mathbf{R}_B|} + \frac{e^2}{|\mathbf{R}_A-\mathbf{R}_B|} \tag{21-21}$$

where $\mathbf{r}$ is the electronic position and $\mathbf{R}_A$ and $\mathbf{R}_B$ are the positions of the two protons. It is impossible to solve the electronic problem exactly, and instead one must use approximate methods; the variational method is the most common approach in all molecular electron problems.

A reasonable form to take for a trial wave function in a variational calculation is a linear combination of hydrogen atom 1s states for the separate protons:

$$\psi(\mathbf{r}) = \alpha\psi_A(\mathbf{r}) + \beta\psi_B(\mathbf{r}), \tag{21-22}$$

where $\psi_A(\mathbf{r})$ is the normalized hydrogen 1s state centered on proton A, and $\psi_B(\mathbf{r})$ is the 1s function centered on proton B, e.g.,

$$\psi_A(\mathbf{r}) = (\pi a_0^3)^{-1/2}\, e^{-|\mathbf{r}-\mathbf{R}_A|/a_0}. \tag{21-23}$$

Now because the potential is symmetric about the midpoint of the molecule $(\mathbf{R}_A+\mathbf{R}_B)/2$, we can classify the electron states by their parity under reflection in the plane through the midpoint that is normal to the axis of the molecule. Thus either $\beta = \alpha$ and the state has even parity or $\beta = -\alpha$ and the state has odd parity; we write

$$\psi_\pm(\mathbf{r}) = C_\pm[\psi_A(\mathbf{r}) \pm \psi_B(\mathbf{r})]. \tag{21-24}$$

Let us first evaluate the normalization constant by squaring (21-24) and integrating over all $\mathbf{r}$. Thus we find $1 = C_\pm^2[1+1\pm 2S(R)]$ or

$$C_\pm = 1/\sqrt{2\pm 2S(R)}, \tag{21-25}$$

where $S(R)$ is the *overlap integral:*

$$S(R) = \int d^3r\, \psi_A(\mathbf{r})\psi_B(\mathbf{r}) = \left(1 + \frac{R}{a_0} + \frac{R^2}{3a_0^2}\right) e^{-R/a_0}, \tag{21-26}$$

and $R = |\mathbf{R}_A - \mathbf{R}_B|$.

Integrals like (21-26) are very easy to do in elliptical coordinates:

$$\mu = \frac{(|\mathbf{r}-\mathbf{R}_A| + |\mathbf{r}-\mathbf{R}_B|)}{R}$$

$$\nu = \frac{(|\mathbf{r}-\mathbf{R}_A| - |\mathbf{r}-\mathbf{R}_B|)}{R}; \tag{21-27}$$

then

$$\int d^3r\, f = \int_1^\infty d\mu \int_{-1}^1 d\nu \int_0^{2\pi} d\varphi\, \frac{R^3}{8}(\mu^2-\nu^2) f. \tag{21-28}$$

The expectation value of H in the state $\psi_\pm(\mathbf{r})$ is

$$\begin{aligned}\langle H\rangle_\pm = \varepsilon_\pm(R) &= \frac{\langle A|H|A\rangle + \langle B|H|B\rangle \pm 2\langle A|H|B\rangle}{2 \pm 2S} \\ &= \frac{\langle A|H|A\rangle \pm \langle A|H|B\rangle}{1 \pm S},\end{aligned} \tag{21-29}$$

where

$$\begin{aligned}\langle A|H|A\rangle = \int \psi_A{}^*(\mathbf{r}) H \psi_A(\mathbf{r})\, d^3r &= \epsilon_1 + \frac{e^2}{R} - \int \psi_A{}^2(\mathbf{r}) \frac{e^2}{|\mathbf{r}-\mathbf{R}_B|} d^3r \\ &= \epsilon_1 + \frac{e^2}{R}\left(1 + \frac{R}{a_0}\right) e^{-2R/a_0}.\end{aligned} \tag{21-30}$$

ϵ_1 is minus one Rydberg; this term comes from $-\hbar^2\nabla^2/2m - e^2/|\mathbf{r}-\mathbf{R}_A|$ acting on ψ_A, its eigenstate. Clearly $\langle B|H|B\rangle = \langle A|H|A\rangle$. Also

$$\begin{aligned}\langle A|H|B\rangle &= \int \psi_A{}^*(\mathbf{r}) H \psi_B(\mathbf{r}) \\ &= \left(\epsilon_1 + \frac{e^2}{R}\right) S - \int d^3r\, \psi_A(\mathbf{r})\psi_B(\mathbf{r}) \frac{e^2}{|\mathbf{r}-\mathbf{R}_B|}.\end{aligned} \tag{21-31}$$

The latter integral, the *exchange integral,* is given by

$$\int d^3r\, \psi_A(\mathbf{r})\psi_B(\mathbf{r}) \frac{e^2}{|\mathbf{r}-\mathbf{R}_B|} = \frac{e^2}{a_0}\left(1 + \frac{R}{a_0}\right) e^{-R/a_0}. \tag{21-32}$$

Putting together the pieces and plotting $\varepsilon_\pm(R)$, the total energy, as a function of the proton separation, we find the result in Fig. 21-1. In the odd parity state the effective potential between the nuclei is always repulsive, so the molecule will not hold together. On the other hand, we find a minimum in the even parity state at 1.3 Å, and a minimum energy of -1.76 eV; thus the molecule binds in this state. Experimentally the separation in H_2^+ is 1.06 Å and the binding energy is -2.8 eV.

A single electron wave function, such as (21-22), is called a *molecular orbital.* The particular form of (21-22) is a linear combination of atomic orbitals, called LCAO for short. The state ψ_+ which leads to a binding of the molecule, is called a *bonding orbital,* while ψ_-, which always has a positive energy, is called an *antibonding orbital.*

The reason the odd parity state fails to bind the molecule is that the wave function $\psi_-(\mathbf{r})$ must vanish on the midplane bisecting the molecule; thus the amplitude for the electrons being between the

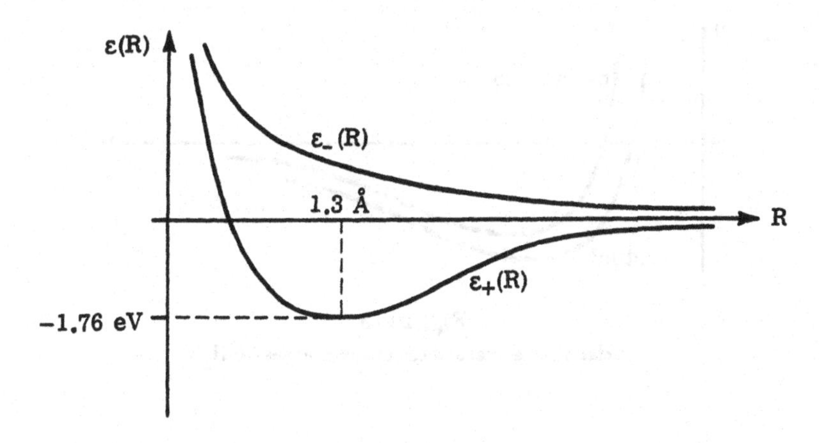

Fig. 21-1

Interaction energy of an H^+ ion and an H atom as a function of proton separation; ϵ_+ is the energy for the bonding orbital and ϵ_- for the antibonding orbital. [The zero of energy is taken as minus one Rydberg here.]

protons, a region of rather negative potential energy, is very small. On the other hand, the wave function of the even parity state is maximum between the two protons, and in this state the electron benefits from the attraction of both protons. If the charge on either nucleus were greater than one, a single electron would be incapable of binding the two nuclei; thus one can't have a $(HHe)^{++}$ ion.

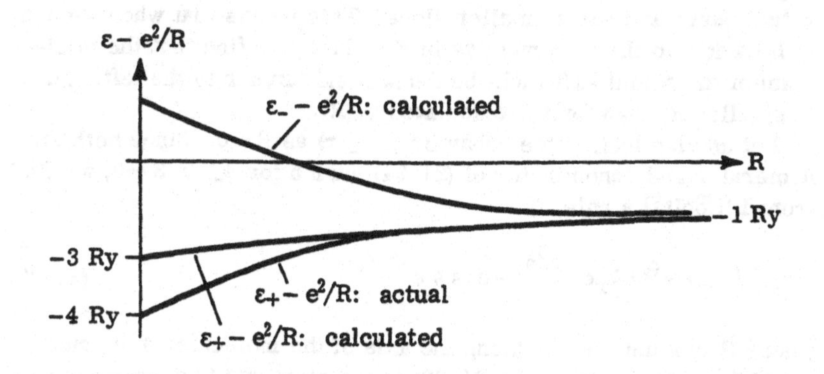

Fig. 21-2

Electronic energy of the $H_2{}^+$ ion as a function of proton separation. As $R \to 0$ the system becomes a He^{2+} ion.

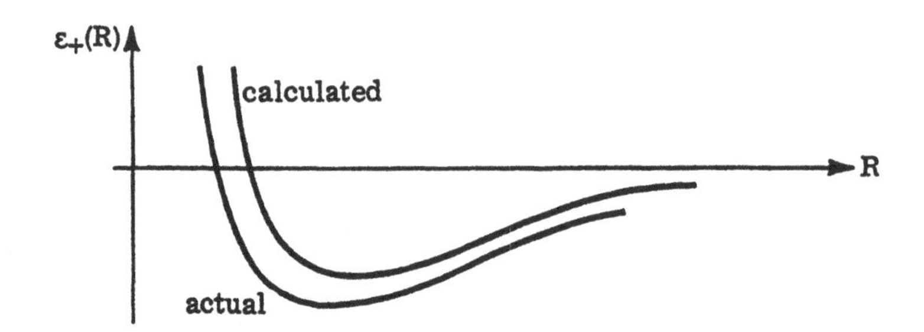

Fig. 21-3
Relation of actual and calculated energies for H_2^+.

The molecular orbital $\psi_+(\mathbf{r})$ should be very close to the actual wave function at very large proton separation. On the other hand, when two protons are very close ($\ll a_0$) then the electron wave function should be practically that of the electron in the 1s state of a He^+ ion. However, as $R \to 0$, $\psi_+(\mathbf{r})$ becomes simply a hydrogen 1s wave function, and therefore ψ_+ is not accurate for small R. This is the reason the binding we found was too weak, and also why the size of the molecule we calculated was too large.

We can see this clearly in Fig. 21-2 where we plot $\varepsilon(R) - e^2/R$; this is the electronic part of the energy. At very large R the molecule becomes a hydrogen atom and a proton, and so both the calculated and actual $\varepsilon_+ - e^2/R$ approach minus one Rydberg, the hydrogen atom ground state energy. But at small distances our calculated $\varepsilon_+ - e^2/R$ approaches -3 Ry, while the He^+ ground state energy is -4 Ry. Thus the calculated curve lies above the actual curve and has a smaller slope. This means that when we add e^2/R back onto these curves, as in Fig. 21-3, we find that the minimum of the actual $\varepsilon_+(R)$ will be deeper and further to the left, i.e., at smaller R, than for the calculated $\varepsilon_+(R)$.

Let us also look at the behavior of $\psi_-(\mathbf{r})$ as $R \to 0$. Since both the numerator and denominator of (21-24) vanish for ψ_- as $R \to 0$, we find, from L'Hôpital's rule,

$$\lim_{R \to 0} \psi_-(\mathbf{r}) \sim \hat{\mathbf{R}} \cdot \nabla_r e^{-r/a_0} \sim \cos\theta\, e^{-r/a_0} \tag{21-33}$$

where $\hat{\mathbf{R}}$ is a unit vector along the axis of the molecule; θ is measured from this axis. Now (21-33) is a p state and therefore the expectation value of the electron energy in (21-33) is raised above the -3 Ry for ψ_+ by the centrifugal barrier energy which as one can calculate is $+8$ Ry. Thus $\varepsilon_-(R) - e^2/R$ is positive for small R, and so when we add e^2/R back on, $\varepsilon_-(R)$ has no minimum.

There is an interesting analogue of the H_2^+ ion that occurs in high-energy physics. A μ^- meson can bind together a proton and a deuteron to form an $(HD)^+$ "mulecule." The size of this molecule is $\sim a_0/207$, since the muon has 207 electron masses. Thus the muon has the effect of bringing the proton and deuteron much closer than, because of their Coulomb repulsion, they would normally come at low energy. At this distance the fusion reaction

$$p + d \to (He^3)^{++} + 5.4 \text{ MeV}$$

is fairly probable, and so the system is unstable. In a sense the muon acts as a catalyst for this fusion reaction. Because $m/M \sim 1/10$ the adiabatic approximation is not valid for this molecule, and the relative p-d motion cannot be neglected.

THE HYDROGEN MOLECULE

The next more complicated molecule is H_2, a two-electron system. The problem is equivalent to that of binding an electron to H_2^+ Neglecting the spin-orbit interaction the energy eigenstates will be either singlet or triplet spin states. There are two simple variational approaches to the calculation of the energy. The first, the molecular orbital method, is to take a trial wave function that is an antisymmetrized product of molecular orbitals of the form (21-24). This method is analogous to the Hartree-Fock method in atoms. To a first approximation we expect that the lowest-energy state will be one in which both electrons are in the bonding orbital ψ_+ and therefore from the exclusion principle have singlet spin. The normalized wave function for such a state is

$$\psi_S(1,2) = \frac{1}{2(1+S(R))} [\psi_A(\mathbf{r}_1) + \psi_B(\mathbf{r}_1)][\psi_A(\mathbf{r}_2) + \psi_B(\mathbf{r}_2)]\chi_{\text{singlet}}. \qquad (21\text{-}34)$$

To form a triplet state we must have the electrons in different space states; the lowest-energy state would have one electron in the bonding orbital ψ_+ and the other in the antibonding orbital ψ_-, and therefore it would have higher energy than (21-34).

The trial wave function (21-34) has two defects. At very small proton separation, the system is like a helium atom, while (21-34) puts the two electrons in s states for Z = 1, instead of 2. Also, if we multiply (21-34) out we see that

$$\psi_S(1,2) \sim [\psi_A(\mathbf{r}_1)\psi_A(\mathbf{r}_2) + \psi_B(\mathbf{r}_1)\psi_B(\mathbf{r}_2)] \\ + [\psi_A(\mathbf{r}_1)\psi_B(\mathbf{r}_2) + \psi_A(\mathbf{r}_2)\psi_B(\mathbf{r}_1)]. \qquad (21\text{-}35)$$

The first two terms represent both electrons revolving around the same proton, and the last two terms represent one electron around each proton. Now for very large separation the amplitude for both electrons to be on the same proton is very small, since a H^- ion plus proton has a higher energy than two hydrogen atoms; the exact wave function at very large separation is

$$\psi(1,2) \sim [\psi_A(\mathbf{r}_1)\psi_B(\mathbf{r}_2) + \psi_B(\mathbf{r}_1)\psi_A(\mathbf{r}_2)].$$

Thus the wave function (21-34) errs in including such a large amplitude $\psi_A(\mathbf{r}_1)\psi_A(\mathbf{r}_2) + \psi_B(\mathbf{r}_1)\psi_B(\mathbf{r}_2)$ at large R.

An alternative method for calculating the energy of the molecule is the *valence bond* or *Heitler-London* method. Here one leaves out the $\psi_A\psi_A + \psi_B\psi_B$ terms and uses the normalized wave function

$$\psi_s(\mathbf{r}_1, \mathbf{r}_2) = \frac{1}{\sqrt{2(1+S^2)}}[\psi_A(\mathbf{r}_1)\psi_B(\mathbf{r}_2) + \psi_B(\mathbf{r}_1)\psi_A(\mathbf{r}_2)] \qquad (21\text{-}36)$$

for the singlet state, and

$$\psi_t(\mathbf{r}_1, \mathbf{r}_2) = \frac{1}{\sqrt{2(1-S^2)}}[\psi_A(\mathbf{r}_1)\psi_B(\mathbf{r}_2) - \psi_B(\mathbf{r}_1)\psi_A(\mathbf{r}_2)] \qquad (21\text{-}37)$$

for the triplet state. These wave functions still have the same inadequacy at small r, and it turns out that (21-34) and (21-36) give very similar results for the binding energy and size of the H_2 molecule.

Let us sketch briefly the Heitler-London calculation of the energy. The expectation values of the Hamiltonian in the states (21-36) and (21-37) is

$$\begin{aligned}\mathcal{E}_\pm(R) &= \langle H\rangle_\pm \\ &= \frac{\langle AB|H|AB\rangle + \langle BA|H|BA\rangle \pm \langle AB|H|BA\rangle \pm \langle BA|H|AB\rangle}{2 \pm 2S^2}\end{aligned} \qquad (21\text{-}38)$$

where the upper sign denotes the singlet state, the lower the triplet, and

$$\langle AB|H|AB\rangle = \int d^3r\, d^3r'\, \psi_A(\mathbf{r})\psi_B(\mathbf{r}')H\psi_A(\mathbf{r})\psi_B(\mathbf{r}')$$

$$\langle BA|H|AB\rangle = \int d^3r\, d^3r'\, \psi_B(\mathbf{r})\psi_A(\mathbf{r}')H\psi_A(\mathbf{r})\psi_B(\mathbf{r}'), \qquad (21\text{-}39)$$

etc. Then because $\langle BA|H|BA\rangle = \langle AB|H|AB\rangle$ and $\langle AB|H|BA\rangle = \langle BA|H|AB\rangle$ we have

$$\mathcal{E}_\pm(R) = \frac{\langle AB|H|AB\rangle \pm \langle BA|H|AB\rangle}{1 \pm S^2}. \qquad (21\text{-}40)$$

Now

$$H = -\frac{\nabla_1^2}{2m} - \frac{\nabla_2^2}{2m} - \frac{e^2}{|\mathbf{r}_1 - \mathbf{R}_A|} - \frac{e^2}{|\mathbf{r}_2 - \mathbf{R}_B|} + \frac{e^2}{|\mathbf{R}_A - \mathbf{R}_B|} + \frac{e^2}{|\mathbf{r}_1 - \mathbf{r}_2|}, \qquad (21\text{-}41)$$

where as before $\mathbf{R}_A$ and $\mathbf{R}_B$ are the positions of the two protons; thus

$$\langle AB|H|AB\rangle = 2\epsilon_1 + \frac{e^2}{R} + V_c(R), \qquad (21\text{-}42)$$

where

$$V_c(R) = \int d^3r_1\, d^3r_2\, \psi_A{}^2(\mathbf{r}_1)\psi_B{}^2(\mathbf{r}_2) \times \left[\frac{e^2}{|\mathbf{r}_1 - \mathbf{r}_2|} - \frac{e^2}{|\mathbf{r}_2 - \mathbf{R}_A|} - \frac{e^2}{|\mathbf{r}_1 - \mathbf{R}_B|}\right]. \qquad (21\text{-}43)$$

Also

$$\langle BA|H|AB\rangle = S^2\left(2\epsilon_1 + \frac{e^2}{R}\right) + V_{ex}(R), \qquad (21\text{-}44)$$

where $V_{ex}(R)$ is the exchange contribution to the electron energy:

$$V_{ex}(R) = \int d^3r_1\, d^3r_2\, \psi_A(\mathbf{r}_1)\psi_B(\mathbf{r}_2)\psi_A(\mathbf{r}_2)\psi_B(\mathbf{r}_1) \times \left[\frac{e^2}{|\mathbf{r}_1 - \mathbf{r}_2|} - \frac{e^2}{|\mathbf{r}_2 - \mathbf{R}_A|} - \frac{e^2}{|\mathbf{r}_1 - \mathbf{R}_B|}\right]. \qquad (21\text{-}45)$$

This integral, whose actual form is complicated, is roughly proportional to S^2, the square of the overlap integral. Substituting (21-42) and (21-43) into (21-40) we find

$$\varepsilon_\pm(R) = 2\epsilon_1 + \frac{e^2}{R} + \frac{V_c(R) \pm V_{ex}(R)}{1 \pm S^2}, \qquad (21\text{-}46)$$

which we can write alternatively as

$$\varepsilon_\pm(R) = 2\epsilon_1 + \frac{(V_c(R) + e^2/R) \pm (V_{ex}(R) + S^2e^2/R)}{1 \pm S^2}. \qquad (21\text{-}47)$$

The term $V_c(R) + e^2/R$ is always positive, while $V_{ex}(R) + S^2e^2/R$ is in general negative. Thus $\varepsilon_+(R)$ lies below $\varepsilon_-(R)$ and one finds that $\varepsilon_-(R) - 2\epsilon_1$ has no minimum, while $\varepsilon_+(R) - 2\epsilon_1$ has one 3 eV deep at $R = 1.5\ a_0$, as in Fig. 21-4. We conclude therefore that the H_2 molecule binds in the singlet state but not in the triplet state, and — a very important point — the strength of the binding is roughly proportional to the amount of overlap of the two electron states $\psi_A(\mathbf{r})$ and $\psi_B(\mathbf{r})$.

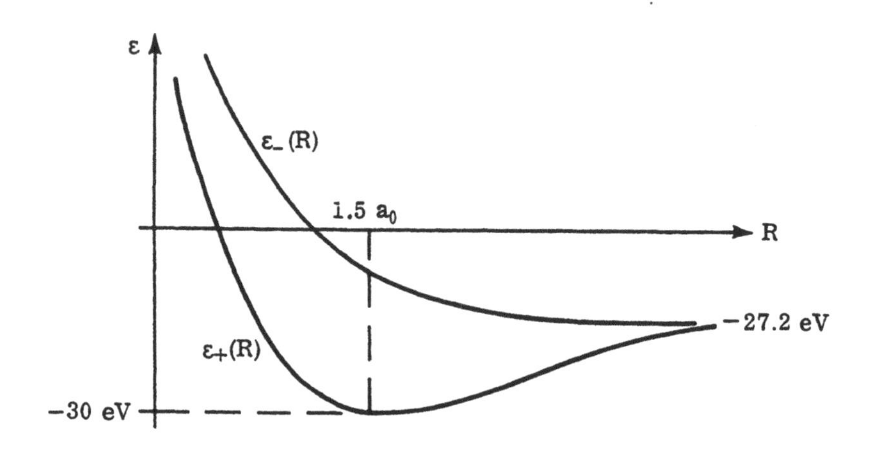

Fig. 21-4
Singlet (+) and triplet (-) energies for the H_2 molecule as calculated by the Heitler-London method.

The reason the triplet state gives a higher energy than the singlet is that the triplet wave function vanishes if both electrons lie on the midplane of the molecule. However, as in H_2^+ this is an energetically favorable place to be. On the other hand, the singlet amplitude is largest when both electrons are on the midplane. In studying the electron gas, and atoms, we argued that the exchange in triplet states kept the electrons further apart and lowered their interaction energy. This is still true in molecules. However in a triplet state the electrons also spend less time in regions where their Coulomb interaction with the nuclei is strongest, and this decrease of attraction to the nuclei more than overcompensates the lower electron-electron energy. Thus the triplet state ends up having higher energy than the singlet state.

At very large R, the system becomes two hydrogen atoms, while at very, very small R, the molecule becomes just like a helium atom. Thus the electronic energies and wave functions for the molecule should lie somewhere between those for two hydrogen atoms and those for a helium atom. The ground state of helium is a singlet, and as we slowly separate the two protons, the wave function turns into that for the ground state of the molecule — still a singlet state. See Fig. 21-5. The helium ground state energy becomes at finite separation $\varepsilon_+(R) - e^2/R$ for the molecule. Since the lowest triplet state of helium lies 1.45 Ry above the singlet ground state, the triplet function $\varepsilon_-(R) - e^2/R$ must lie considerably above the singlet, and

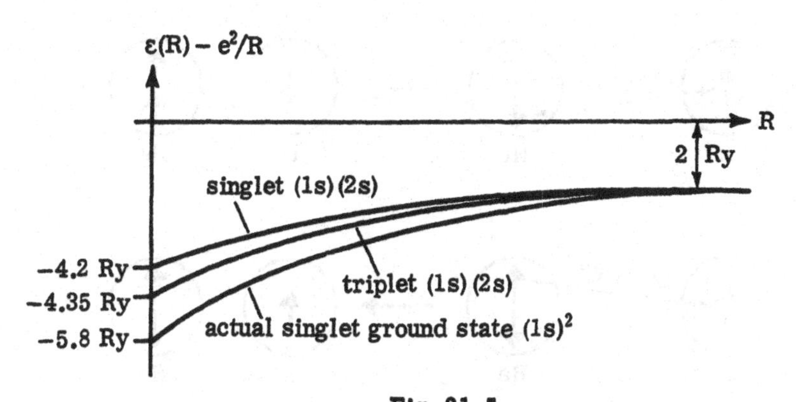

Fig. 21-5
Electron energies of the H_2 molecule labeled in terms of the states of the He^2 atom.

when we add e^2/R back on, the triplet energy has no minimum. Exactly, as in H_2^+, the variational calculation, while exact at very large R, gives too large an energy at very small R, and this has the effect of shifting the calculated minimum toward larger R. [Of course, by the rules of the variational calculation, the calculated $\varepsilon(R)$ always lies above the true $\varepsilon(R)$.]

Actually at very large separation two hydrogen atoms have a $1/R^6$ van der Waals attraction. This attraction first appears in a second-order perturbation theory calculation (p. 235); the wave function (21-36) isn't good enough to include this effect. The van der Waals attraction is about one thousand times weaker than the binding forces we've been calculating here. If we drew a picture of the potential $\varepsilon(R)$ on a scale that would show the van der Waals attraction, then the covalent bonding forces would appear almost as a hard core. The van der Waals attraction is too weak to bind together molecules of only a few atoms. The zero-point energy of the nuclei is greater than the binding energy. But the van der Waals force does play a very important role in the binding of large molecules, as well as crystals of rare gases.

PAIRING OF ELECTRONS

As we saw in the hydrogen molecule, large overlap of the wave functions of two electrons with singlet spin leads to strong attraction, while large overlap in a triplet state leads to repulsion. We can see

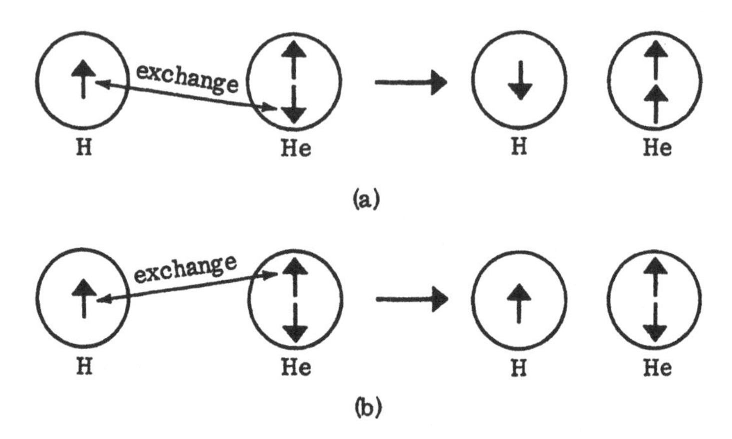

Fig. 21-6

then why a hydrogen atom and a helium atom won't form a HHe molecule. The two electrons in the helium in its ground state are both in 1s states and have opposite spin. Now the electron on the hydrogen can't exchange with the electron of opposite spin on the helium and go into the 1s helium state, as in Fig. 21-6(a), since this would put two electrons of the same spin in that state. The only exchange is between the two electrons of the same spin, as in Fig. 21-6(b) and such exchange in parallel spin states causes a repulsion between the two atoms, as in the triplet state in H_2.

At very, very small separations, the helium atom and hydrogen atom system is just like a lithium atom. Seen from this end the atoms fail to bind because in Li the 1s shell is filled and the third electron must go into the 2s shell where it is rather weakly bound; its ionization energy is 0.4 Ry. Thus the curve for $\varepsilon(R) - 2e^2/R$ must end at $R = 0$ at too high a value for $\varepsilon(R)$ to have the minimum needed for binding.

We may see the mathematical source of the repulsion between H and He if we choose as a trial wave function a Slater determinant

$$\psi(1,2,3) = C\,\det[\psi_A(1)\chi_\uparrow(1),\ \psi_A(2)\chi_\downarrow(2),\ \psi_B(3)\chi_\uparrow(3)] \tag{21-48}$$

where ψ_A is the 1s helium wave function and ψ_B is the 1s hydrogen wave function. In calculating the normalization constant we find after some algebra

$$\begin{aligned} C^{-2} &= 6\int d^3r_1\, d^3r_2\, d^3r_3\, \psi_A(1)\psi_B(3)\psi_A(2)[\psi_A(1)\psi_B(3) - \psi_B(1)\psi_A(3)] \\ &= 6[1 - S^2], \end{aligned} \tag{21-49}$$

where S is the overlap integral between ψ_A and ψ_B. Only the exchange between same spin electrons appears in (21-49) because of the orthogonality of the spin up and spin down states. Then by a similar calculation the energy is given by

$$\varepsilon(R) = \frac{1}{1-S^2}\int d^3r_1\, d^3r_2\, d^3r_3\, \psi_A(1)\psi_A(2)\psi_B(3)H \times [\psi_A(1)\psi_B(3) - \psi_B(1)\psi_A(3)]\psi_A(2). \tag{21-50}$$

The only exchange term is between the electrons of the same spin, and as in the triplet state calculation for H_2 the exchange term gives a positive contribution to the energy.

Two electrons of opposite spin but in the same spatial state, such as the two 1s electrons in helium, are called *paired*. In the general, the overlap of the wave function of an electron on one atom with two paired electrons on another atom leads to a repulsive interaction. This explains then why the rare gases are inert. All their electrons, being in closed shells, are paired, and therefore they have only repulsive interactions with other atoms.

The chemical activity of an atom depends on its having unpaired outer electrons, since only these lead to attractive interactions. These are the *valence* electrons one learns about in elementary chemistry. Electrons in closed shells play little role in the binding of two atoms, since these electrons are closer to the nucleus than the valence electrons; at the actual separations of atoms in a molecule there is little overlap between electrons from one atom and the closed shell electrons from another. Thus the repulsive interactions from the closed shell electrons are much weaker than the attractive valence electron interactions. To a large extent, the atoms retain their identity in a molecule, since only the outermost electrons have appreciable contact with electrons from other atoms.

The *valency* of an atom (or more precisely, the valency of the state of an atom) is just the number of unpaired outer electrons; these are always s or p electrons. The unpaired d and f electrons in transition elements are too close to the nucleus to be chemically active. The valency of a configuration, in the case of the nontransition elements, is then twice the largest possible spin of the configuration.

Once two unpaired electrons from different atoms in a molecule form a singlet state and cause binding they become paired; an electron from a third atom will have only repulsive interactions with them. Chemical binding forces therefore saturate; each covalent bond in a molecule uses up a different pair of electrons, and each electron may be thought of as participating in only one bond. Thus

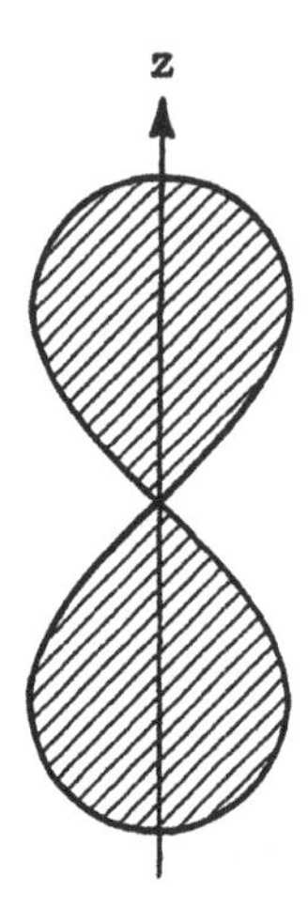

Fig. 21-7

Electron distribution in a p state with $m = 0$.

one finds that, in general, molecules formed from nontransition elements have total spin equal to zero, since each bonding pair is in a singlet state. Molecules containing transition elements often have nonzero spin due to unpaired d or f electrons.

Alkali atoms, in their ground state, have one unpaired electron which readily forms chemical bonds. Their valence is thus one. Alkalis in excited configurations have too high an energy to bind.

SPATIALLY DIRECTED ORBITALS

The alkaline earths, the second column of the periodic table, have only closed shells in their ground states. In this state they are inert. However, it costs only a few electron volts for one of the outer shell s electrons to go into a p state and increase the valency of the atom to two. This p electron can have any of three different m_l values. The p states are not spherically symmetric, like s states, but rather they tend to point along preferred directions. Thus the state with zero angular momentum along the z axis looks like Fig. 21-7, while the $m_l \pm 1$ states bulge at the equator. The p electron in bonding with an electron from another atom can have maximum overlap with its mate, which gives strongest binding, by then being in a state with

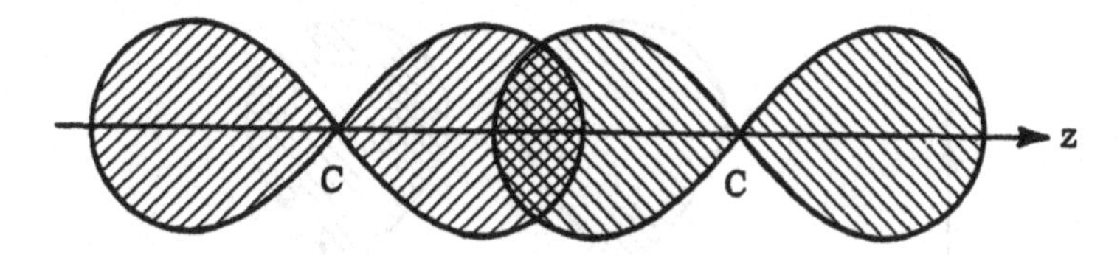

Fig. 21-8
σ bond formed by two p electrons in C_2.

$m_l = 0$ along the bonding axis, the line joining the two atoms. One denotes such states pointing along the x, y, or z axes by $|p_x\rangle$, $|p_y\rangle$, or $|p_z\rangle$. In terms of spherical harmonics, the angular parts of these states are

$$|p_z\rangle = Y_{10} = \sqrt{\frac{3}{4\pi}} \cos\theta = \sqrt{\frac{3}{4\pi}} \frac{z}{r}$$

$$|p_y\rangle = \frac{1}{\sqrt{2}} (Y_{11} - Y_{1,-1}) = \sqrt{\frac{3}{4\pi}} \frac{y}{r}$$

$$|p_x\rangle = \frac{1}{\sqrt{2}} (Y_{11} + Y_{1,-1}) = \sqrt{\frac{3}{4\pi}} \frac{x}{r}. \tag{21-51}$$

For example, $|p_y\rangle$ is the p state with the maximum amount of amplitude in the $\pm\hat{y}$ directions.

To see how these directed orbitals function in binding, let us look at the C_2 molecule. In its ground state configuration, carbon has two unpaired 2p electrons and thus has a valence of two. Suppose that the axis of the molecule is along $\hat{z}$. Then what choice of p states for each electron will have the least energy? [What we are really asking, is "what trial wave function composed of determinants of these states will give the least energy?"] Having one p electron from each atom in the state $|p_z\rangle$ gives a large overlap [Fig. 21-8], and if these two electrons form a singlet state, strong binding. Now once this bonding configuration is occupied it is energetically unprofitable to put another electron in a $|p_z\rangle$ state. This is because it would have to spin opposite to the first electron and therefore the *same* spin as the mate of the first electron on the other atom; but the exchange interaction between these two parallel spin interactions is repulsive. Thus the remaining electrons must be in states orthogonal to the $|p_z\rangle$ states. The next best possibility is if they both point perpendicular to the axis [Fig. 21-9] and along the same direction, say x, that is, are both

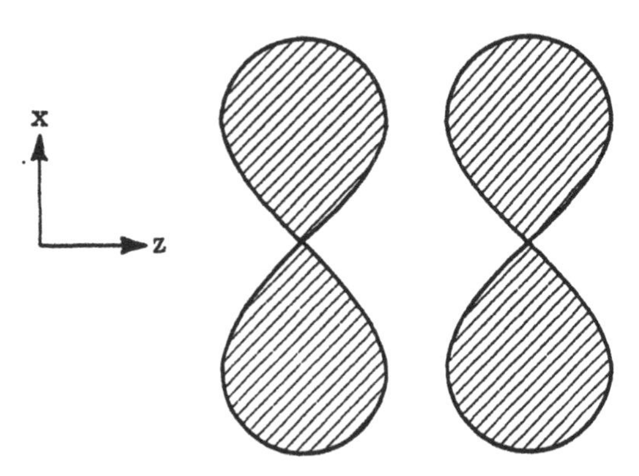

Fig. 21-9
π bond formed by two p electrons in C_2.

in the state $|p_x\rangle$, and form a spin singlet. In this state, while not having as much overlap as the two $|p_z\rangle$ electrons, they still have some overlap and contribute to the binding. Another arrangement such as one electron in $|p_x\rangle$ and the other electron on the other atom in $|p_y\rangle$ would have far less overlap than both in $|p_x\rangle$.

Thus the C_2 molecule is held together by two bonds, the first from the electrons directed along z, and the second, a weaker bond, from the electrons along x, at right angles to the bonding axis. The angular momentum of the first pair of electrons along z is zero and the bond they form is called a *σ bond* (Greek for s, as in s state). The bond in H_2 is a σ bond. Now the magnitude of the angular momentum of the state $|p_x\rangle$, or $|p_y\rangle$, along z is 1:

$$L_z^2|p_x\rangle = |p_x\rangle , \tag{21-52}$$

so the electrons in the $|p_x\rangle$ bond each have angular momentum of magnitude 1 along z. The bond they form is called a *π bond* (as in p state). π bonds are always weaker than σ bonds because the electrons in them have smaller overlap. To a first approximation the energies of the different bonds are additive, because they are reasonably localized.

Nitrogen has three 2p electrons and thus in the N_2 molecule one has one σ and two π bonds; the π bonds are at right angles to each other and to the bonding axis.

An interesting feature occurs in the O_2 molecule, because oxygen has *four* 2p electrons. The first three 2p electrons from each atom

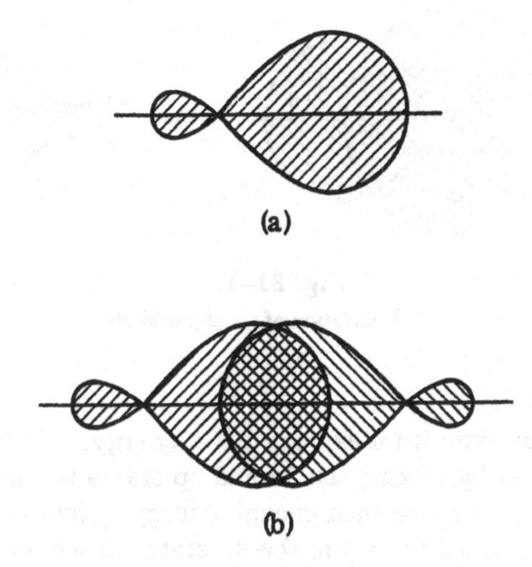

Fig. 21-10

A hybridized state (a) leads to greater electron overlap in a bond (b).

form one-σ and two π bonds as in N_2, but the problem is what does one do with the two leftover electrons, as no bonding states are left. The least harmful arrangement is the one in which these two electrons avoid each other as much as possible. This can be done by having them in antibonding orbitals (like the state ψ_- in H_2^+), and directed at right angles to each other, one in $|p_x\rangle$ say, and the other in $|p_y\rangle$; and to reduce their interaction energy, in a relative spin triplet state. Thus the total spin of O_2 is one, and the molecule is paramagnetic. This is an exception to the statement that S is zero for molecules of nontransition elements.

HYBRIDIZATION

Now one might ask, wouldn't it be better if the electron states in a σ bond protruded more in one direction [Fig. 21-10(a)], for then in the σ bond [Fig. 21-10(b)], there would be a larger overlap of the two electrons and thus stronger bonding. To construct such a state though it is necessary to mix different angular momenta, since s states by themselves have even parity, while p states by themselves have odd parity. A simple example of a molecule in which such orbitals do occur is Li_2. The 2s electron in Li is spherically symmetric about

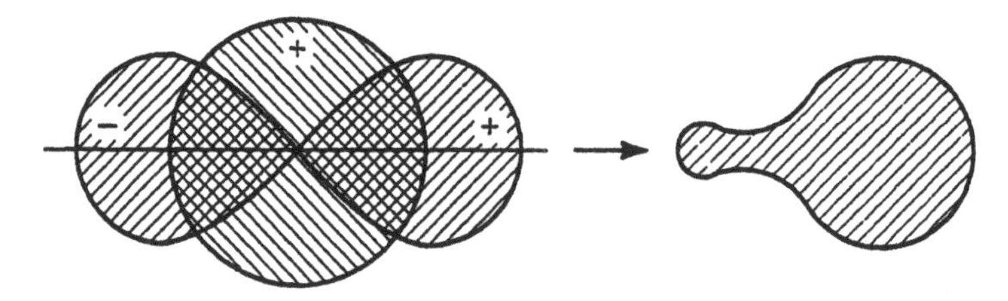

Fig. 21-11
Hybridization of s and p orbitals.

the nucleus. However it takes very little energy to excite this electron to a 2p state; by mixing the 2s and 2p states to form a lopsided orbital one can gain more than enough energy to overcome the additional energy required to mix in the 2p state. If we let $|s\rangle$ denote the 2s state, and $|p_z\rangle$ the 2p state, as in (21-51), then a trial wave function of the form

$$|s\rangle + \lambda|p_z\rangle \tag{21-53}$$

is the sum of an amplitude $|s\rangle$ which is positive in both the +z and −z directions, and an amplitude $|p_z\rangle$ which is positive in the +z direction and negative in the −z direction. Thus for $\lambda > 0$ these amplitudes add, as in Fig. 21-11, to produce a lopsided orbital. A variational calculation shows that a value of $\lambda \sim 0.3$ minimizes the binding energy of Li; the probability of the electron being p-like is only 10% of the probability of its being s-like. This phenomenon of the mixing of two angular momenta to form a lopsided orbital is called *hybridization* as we shall now see, the geometrical structure of many molecules is a result of their electrons bonding in hybridized orbitals.

The water molecule is a bent structure with an angle of 105° between the bonds as in Fig. 21-12. Because the hydrogens are off center, H_2O has a large dipole moment. The simplest explanation of this structure is to say that in the molecule the four 2p oxygen electrons are in pure p orbitals (21-51); one electron is in $|p_x\rangle$, one in $|p_y\rangle$ and the remaining two are paired in a singlet state in $|p_z\rangle$. The $|p_x\rangle$ electron then bonds with one hydrogen, and the $|p_y\rangle$ electron bonds with the other hydrogen. Because $|p_x\rangle$ and $|p_y\rangle$ are at right angles this gives a bond angle of 90°. Then with a little handwaving one can stretch this angle to 105° by arguing that the two protons, the H nuclei, really repel each other a little and distort the right angle.

One can construct a more refined picture of the water molecule by noting that the 2s and 2p levels in oxygen are close in energy, and

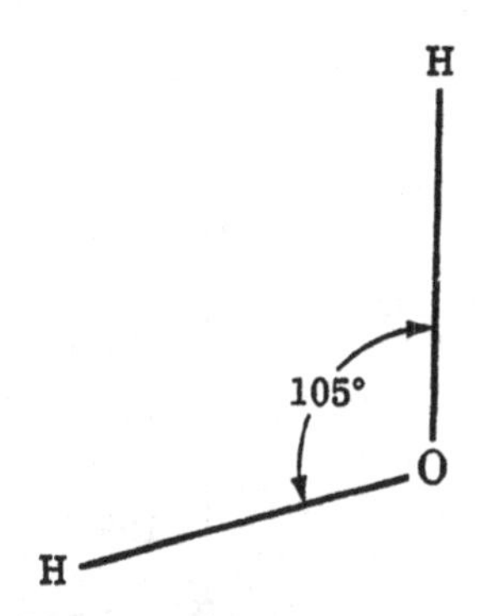

Fig. 21-12
The water molecule.

one can thus lower the energy of the molecule by hybridizing the oxygen orbitals. If the 2s and 2p states had the same energy then putting the electrons in the orthonormal linear combinations of 2s and 2p orbitals

$$|1\rangle = \frac{1}{2}(|s\rangle + |p_x\rangle + |p_y\rangle + |p_z\rangle)$$

$$|2\rangle = \frac{1}{2}(|s\rangle + |p_x\rangle - |p_y\rangle - |p_z\rangle)$$

$$|3\rangle = \frac{1}{2}(|s\rangle - |p_x\rangle + |p_y\rangle - |p_z\rangle)$$

$$|4\rangle = \frac{1}{2}(|s\rangle - |p_x\rangle - |p_y\rangle + |p_z\rangle), \tag{21-54}$$

would give the least energy in a variational calculation, since of all linear combinations of 2s and 2p orbitals, (21-54) have the maximal "directionality." We can see the shape of these orbitals in the following:

$$|p_x\rangle + |p_y\rangle + |p_z\rangle \sim \frac{x+y+z}{r} \tag{21-55}$$

has maximum positive weight along the 1, 1, 1 direction, i.e., along the main diagonal of the octant defined by the positive x, y, and z axes. Its most negative values are along the −1, −1, −1 directions and thus when we add on the spherically symmetric, positive $|s\rangle$ state we end up with a state pointing along the 1, 1, 1 direction. Similarly

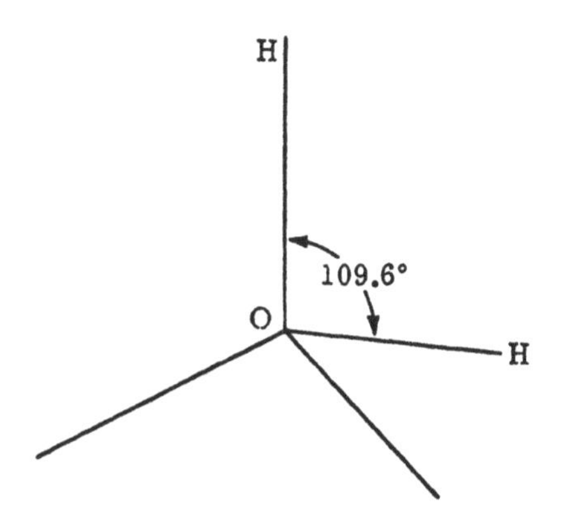

Fig. 21-13

The water molecule with fully hybridized s and p orbitals.

$|2\rangle$ points in the 1, −1, −1 direction, $|3\rangle$ in the −1, 1, −1 direction and $|4\rangle$ in the −1, −1, 1. These four orbitals thus form a tetrahedral configuration, Fig. 21-13, which by a little geometry [$\sin \theta/2 = \sqrt{2/3}$] has an angle of 109.6° between each pair of legs.

The oxygen has six n = 2 electrons – two 2s and four 2p. In this picture of H_2O, one of these electrons is in state $|1\rangle$, say, one in state $|2\rangle$, two with singlet spin in state $|3\rangle$, and two with singlet spin in $|4\rangle$. The electrons in $|1\rangle$ and $|2\rangle$ pair with the electrons from the hydrogens to bind the molecule. The bond angle in this model is thus 109.6°, closer to the observed angle, but this time too large. One would conclude that the true picture lies somewhere between the two extremes, no hybridization and complete hybridization. The reason is that the 2s electrons in the oxygen atom actually have lower energy than the 2p electrons, and thus putting the 2s electrons in the completely hybridized orbitals (21-54), to gain binding energy, requires too much excitation energy. The optimal choice of orbitals has an intermediate amount of 2s – 2p mixing; three of the four orbitals will be more p-like than s-like and one of the four more s-like. The hydrogens bind with electrons in the p-like states, and one expects a bond angle somewhere between the 90° in the nonhybridized picture and the 109.6° in the completely hybridized picture.

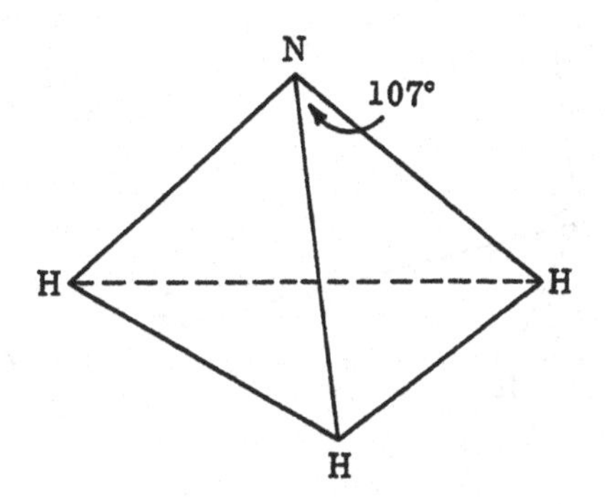

Fig. 21-14
The ammonia molecule.

The ammonia molecule, NH_3, is a similar example of intermediate hybridization. The molecule is shaped like a pyramid [Fig. 21-14] with a bond angle of 107°. If the 2p electrons in the nitrogen were in $|p_x\rangle$, $|p_y\rangle$, $|p_z\rangle$ states then this angle would be 90°. The angle is stretched by mixing in the 2s state in the bonding orbitals.

An important feature of the ammonia molecule is the splitting of the ground state due to the fact that the nitrogen can live on either side of the plane of the hydrogens. The lowest energy state is the one that has equal amplitude for the nitrogen to be on either side of the plane. The state in which the amplitude for the nitrogen to be on one side is equal in magnitude to but has the opposite sign from the amplitude for it to be on the other side, has a slightly higher energy than the first state. This "tunneling" splitting (see problem 7, Chapter 4) is utilized in the ammonia maser.

The way in which the properties of liquid water derive from the electronic structure of the water molecule is rather interesting. The molecule has a total of ten electrons; the two 1s oxygen electrons are tightly bound and are irrelevant at the moment. Very crudely we can think of the remaining eight electrons as filling up the orbitals (21-54) on the oxygen. The oxygen thus has four charged arms forming a tetrahedral structure. The two protons are held on to the ends of two of these arms by Coulomb forces. [Of course, O_2^{--}, having completely closed 2s and 2p shells is spherically symmetric; the tetragonal structure only makes sense when there are hydrogen nuclei around to break the symmetry.] The dipole moment of the molecule arises

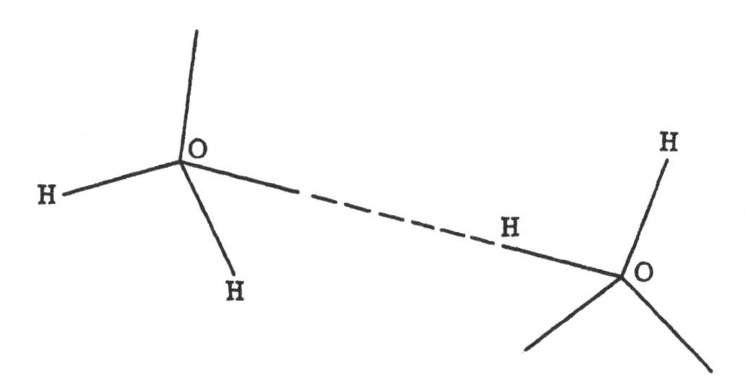

Fig. 21-15
Hydrogen bonding of two water molecules.

from there being two negatively charged arms pointed away from the two protons.

Now consider what happens when two H_2O molecules in the liquid are close together with a negatively charged arm from one molecule near a proton from the other, as in Fig. 21-15. Then the electrostatic attraction between the proton and the negatively charged arm lowers the energy of the pair of molecules, and they tend to bind together. Such a bond is called a *hydrogen bond;* the binding energy is about 0.2 eV and is thus relatively weak. In the bond the distance from the H to the O on the other molecule is about 1.8 Å, while the distance from the H to the O on its own molecule is about 1 Å. (Check that the bond energy is consistent with these spacings.)

Each water molecule is thus capable of hydrogen bonding to four other water molecules at once – one molecule at the end of each electron arm. One expects then to find large tetragonally structured clusters of water molecules in the liquid – $H_{2n}O_n$ molecules. One consequence of this is the strong temperature dependence of the viscosity of water. In cold water there are large clusters which are easily tangled together, making the water more viscous. However heating the water breaks hydrogen bonds, which reduces the size of the clusters and tends to decrease the viscosity of the water.

Many other molecules besides water can form hydrogen bonds. In fact water dissolves substances by forming hydrogen bonds with the molecules of the substance; the molecules would rather stick to the water molecules than to themselves, and so the substance dissolves.

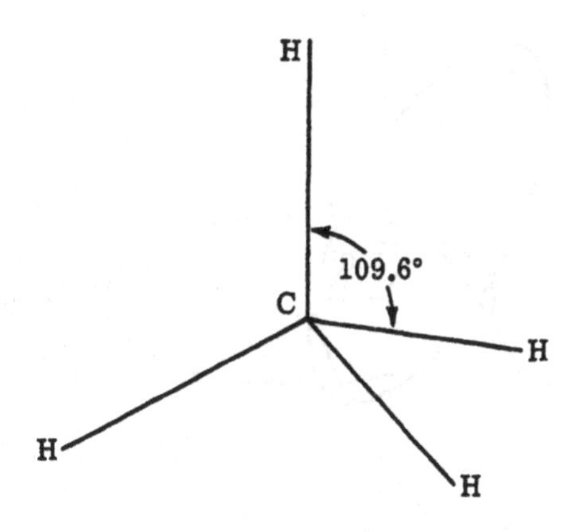

Fig. 21-16
The methane molecule.

Substances that don't form good hydrogen bonds, such as oils, don't dissolve in water.

HYDROCARBONS

As we've seen in H_2O, the geometry of the molecule is due to the particular way the electron orbitals hybridize. The vast variety of molecules containing carbon are lovely illustrations of the connection between the geometrical and electronic structure of molecules. The carbon atom in its ground state has a $(1s)^2(2s)^2(2p)^2$ electronic configuration, and in this state carbon is divalent. However the $(1s)^2(2s)(2p)^3$ state, in which carbon has a valency of four, is very close in energy to the ground state, and it is in this state that carbon forms most of its compounds.

Carbon has several different bonding structures in this state. In methane, CH_4, the 2s and 2p orbitals are completely hybridized, as in (21-54); one $n = 2$ carbon electron is in each orbital and binds to a hydrogen. The methane molecule thus has a tetragonal structure, Fig. 21-16, with a bond angle of 109.6°. Each of the bond between the hydrogens and carbon are σ bonds. The angle is exactly 109.6° from symmetry.

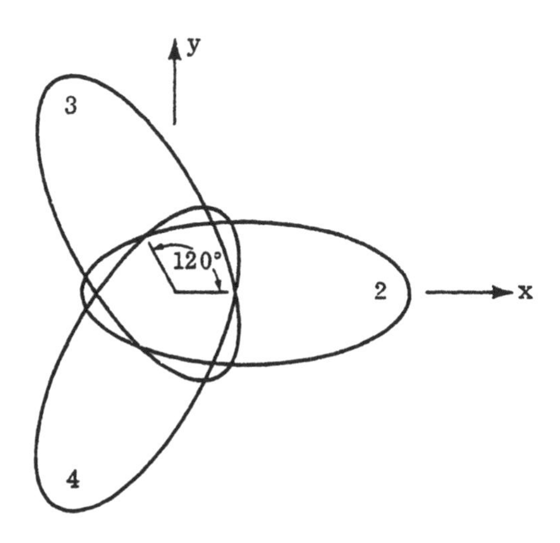

Fig. 21-17
The planar orbitals of ethylene.

In the ethylene molecule, C_2H_4, one finds a planar structure like $\begin{matrix} H & & & & H \\ & \diagdown & & \diagup & \\ & & C = C & & \\ & \diagup & & \diagdown & \\ H & & & & H \end{matrix}$, and this is explained by saying that the carbon orbitals are hybridized as follows:

$$|1\rangle = |p_z\rangle$$

$$|2\rangle = \sqrt{\frac{1}{3}}\,|s\rangle + \sqrt{\frac{2}{3}}\,|p_x\rangle$$

$$|3\rangle = \sqrt{\frac{1}{3}}\,|s\rangle - \sqrt{\frac{1}{6}}\,|p_x\rangle + \sqrt{\frac{1}{2}}\,|p_y\rangle$$

$$|4\rangle = \sqrt{\frac{1}{3}}\,|s\rangle - \sqrt{\frac{1}{6}}\,|p_x\rangle - \sqrt{\frac{1}{2}}\,|p_y\rangle . \tag{21-56}$$

The first state points along the $\pm z$ direction while the last three orbitals are at 120° intervals in the x, y plane as in Fig. 21-17. The electrons in the carbon atom on the left are in these orbitals; the electrons in $|3\rangle$ and $|4\rangle$ bind to hydrogens, while those in $|1\rangle$ and $|2\rangle$ form bonds with the other carbon, whose orbitals are mirror images of those of the first carbon:

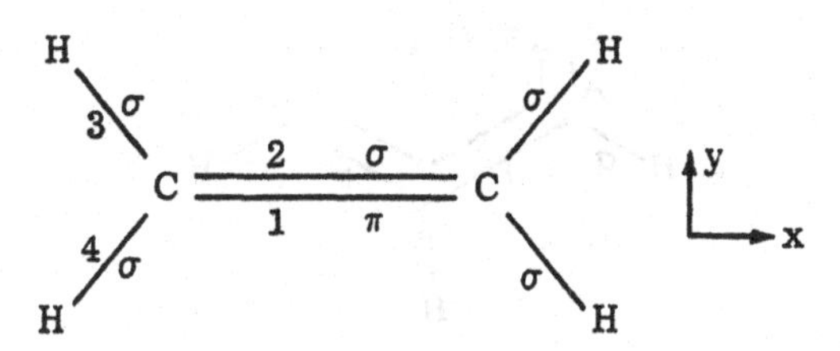

The electrons in states $|2\rangle$ form a σ bond between the carbons, while the electrons in states $|1\rangle$, being at right angles to the bonding axis, form π bonds. It is this π bond which makes the whole structure coplanar. The π bond is formed only when the normals to the planes of the CH_2 substructures are parallel; any rotation of *one* molecule about the x axis would break the π bond.

Yet another scheme occurs in acetylene, C_2H_2. This is a linear molecule arranged like H — C ≡ C — H along, say, the x axis. Here the $|s\rangle$ state is mixed only with $|p_x\rangle$ in the states

$$\frac{1}{2}\left(|s\rangle \pm |p_z\rangle\right), \tag{21-57}$$

which protrude in the $\pm x$ directions. The electrons in these states form σ bonds along the x axis:

$$\mathrm{H} \overset{\sigma}{\text{———}} \mathrm{C} \underset{\pi}{\overset{\pi}{\overset{\sigma}{\text{———}}}} \mathrm{C} \overset{\sigma}{\text{———}} \mathrm{H}$$

The remaining electrons on each carbon go into linear combinations of $|p_y\rangle$ and $|p_z\rangle$ states and form π bonds between the carbons.

In all these carbon compounds the bonds are reasonably localized in space. An interesting situation occurs in benzene, and other "aromatic" compounds. The benzene molecule is a planar ring shaped structure:

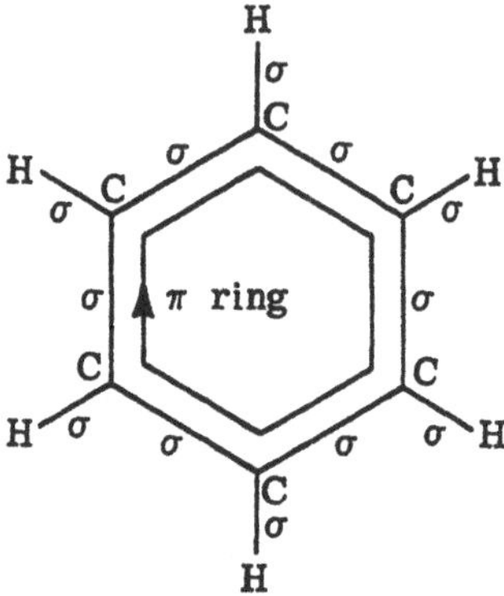

The orbitals of each carbon are hybridized as in (21-56); the electrons in the orbitals in the plane form σ bonds as in the picture. The six remaining electrons, in states at right angles to the plane form three π bonds. Between which pair of carbons do these π bonds form? There are two possibilities:

In fact, the state with the least energy is a linear combination of these two states, since by spreading out the electrons wave functions one decreases their kinetic energies. In a sense the π electrons are free to run around the benzene ring, hopping from one carbon to the next. Experimentally one finds that benzene has a very large diamagnetic susceptibility, since an electron running around the ring is a current loop surrounding a large area and thus it has a large magnetic moment.

In summary then, we find that from a few simple principles such as maximizing the electron overlap in singlet states, hybridizing orbitals and, as in carbon, increasing the valence of an atom by exciting electrons, we can get a good qualitative understanding of how molecules are put together. It should be emphasized though that none of these simple arguments enable one to predict the structure of a molecule. One must always compare detailed machine calculations to tell theoretically what the structure of a given molecule should be.

Chapter 22
RELATIVISTIC SPIN ZERO PARTICLES: KLEIN-GORDON EQUATION

It is a nontrivial problem to generalize quantum mechanics to describe relativistic particles, and as we shall see, the solution contains some rather unusual features. First we shall study spinless, or scalar, particles and then later particles with spin. The reason is that the properties of spin, an angular momentum, are closely subject to the requirements of special relativity and as a consequence the form of the wave equation for a particle depends critically on its spin. Common relativistic spinless particles are, for example, π and K mesons.

A nonrelativistic free particle has an energy-momentum relation

$$E = \frac{p^2}{2m}, \tag{22-1}$$

which is invariant under a *Galilean transformation* to a new coordinate system traveling with velocity $-\mathbf{v}$ with respect to the first. The energy E' and momentum $\mathbf{p}'$ of the particle in the new coordinate system are related to E and $\mathbf{p}$, nonrelativistically, by

$$E' = E + \mathbf{p}\cdot\mathbf{v} + \frac{mv^2}{2}$$

$$\mathbf{p}' = \mathbf{p} + m\mathbf{v}; \tag{22-2}$$

then $E' = p'^2/2m$. The transformation laws (22-2) are not restricted to a free particle. For a Galilean transformation on a general closed system, E is the total energy, $\mathbf{p}$ the total momentum and m the total mass of the system. Note that to within an additive term independent of $\mathbf{v}$, the energy-momentum relation (22-1) is the only form that is invariant under a Galilean transformation. Quantum mechanically the energy and momentum of a particle are related to the rates of change of the phase of its wave function in time and space, respectively. In order that a free particle have the energy-momentum relation (22-1), its wave function

$$\psi(\mathbf{r}t) = e^{i(\mathbf{p}\cdot\mathbf{r} - Et)/\hbar} \tag{22-3}$$

must obey the Schrödinger equation

$$i\hbar \frac{\partial\psi}{\partial t} = \frac{1}{2m}\left(\frac{\hbar}{i}\nabla\right)^2 \psi. \tag{22-4}$$

Thus the Schrödinger equation may be written down from the energy-momentum relation (22-1) by making the correspondences

$$E \to i\hbar\frac{\partial}{\partial t}, \qquad \mathbf{p} \to \frac{\hbar}{i}\nabla \tag{22-5}$$

in (22-1) and letting the result act on ψ, as in (22-4). Furthermore one can add in the coupling of a charged particle to an electromagnetic field described by a vector potential $\mathbf{A}(\mathbf{r}, t)$ and a scalar potential $\Phi(\mathbf{r}, t)$ by letting

$$E \to E - e\Phi, \qquad \mathbf{p} \to \mathbf{p} - \frac{e}{c}\mathbf{A} \tag{22-6}$$

or

$$i\hbar\frac{\partial}{\partial t} \to i\hbar\frac{\partial}{\partial t} - e\Phi, \qquad \frac{\hbar}{i}\nabla \to \frac{\hbar}{i}\nabla - \frac{e}{c}\mathbf{A}, \tag{22-7}$$

in (22-4), where e is the charge of the particle.

This connection between the energy-momentum relation and the wave equation provides us with a guide in writing down a relativistic wave equation that will give the correct relativistic energy-momentum relation

$$E = c\sqrt{p^2 + m^2c^2}. \tag{22-8}$$

The simplest guess is to say that the momentum space amplitude

$$\psi_{\mathbf{p}}(t) = \int d^3r\, e^{-i\mathbf{p}\cdot\mathbf{r}/\hbar}\psi(\mathbf{r}, t) \tag{22-9}$$

obeys the equation

$$i\hbar\frac{\partial}{\partial t}\psi_{\mathbf{p}}(t) = c\sqrt{p^2 + m^2c^2}\,\psi_{\mathbf{p}}(t); \tag{22-10}$$

the energy eigenvalues are then clearly given by (22-8). Such a relativistic Schrödinger equation is quite useful, but it has one serious limitation which we should recognize. Let us Fourier transform both sides of (22-10) back to position space. Then (22-10) becomes

$$i\hbar \frac{\partial}{\partial t}\psi(\mathbf{r}, t) = \int d^3r' \, K(\mathbf{r}-\mathbf{r}')\psi(\mathbf{r}', t) \tag{22-11}$$

where

$$K(\mathbf{r}-\mathbf{r}') = \int \frac{d^3p}{(2\pi\hbar)^3} e^{i\mathbf{p}\cdot(\mathbf{r}-\mathbf{r}')/\hbar} c\sqrt{p^2+m^2c^2}. \tag{22-12}$$

Equation (22-11) is nonlocal; this means that the value of the right side at $\mathbf{r}$ depends on the value of ψ at other points $\mathbf{r}'$. The kernel $K(\mathbf{r}-\mathbf{r}')$ is sizable as long as $\mathbf{r}'$ is within a distance $\sim\hbar/mc$ from $\mathbf{r}$. Note that $\hbar/mc$, the Compton wavelength of the particle, is the only unit of length in this problem. As a consequence of this nonlocality, the rate of change in time of ψ at the space-time point $\mathbf{r}, t$ depends on values of ψ at points $\mathbf{r}', t$ *outside* the light cone centered on $\mathbf{r}, t$. If we construct an initial wave packet localized well within a Compton wavelength of the origin, say, then the packet will be nonzero an arbitrarily short time later at points as distant as the Compton wavelength.

Thus Eq. (22-9) violates relativistic causality when used to describe particles localized to within more than a Compton wavelength, and is useful only in describing non-highly localized states. We shall not pursue this equation further here, but remark that this is not the last we shall see of problems of localization in relativistic quantum mechanics.

We can form a local wave equation by first squaring (22-8):

$$E^2 = p^2c^2 + m^2c^4, \tag{22-13}$$

and then using (22-5). This implies that the wave function ψ obeys the equation

$$\left(\frac{i\hbar}{c}\frac{\partial}{\partial t}\right)^2 \psi(\mathbf{r}t) = \left(\frac{\hbar}{i}\nabla\right)^2 \psi(\mathbf{r}t) + m^2c^2\psi(\mathbf{r}t). \tag{22-14}$$

This equation is known as the *Klein-Gordon equation.* We can also write (22-14) as

$$\left[\frac{1}{c^2}\frac{\partial^2}{\partial t^2} - \nabla^2 + \left(\frac{mc}{\hbar}\right)^2\right]\psi(\mathbf{r}, t) = 0, \tag{22-15}$$

in which form it looks like a classical wave equation with an extra $(mc/\hbar)^2$ term.

The coupling of a charged particle to the electromagnetic field can be included in a relativistically covariant manner in the Klein-Gordon equation by making the substitutions (22-7), since Φ and $\mathbf{A}$ transform relativistically as a four-vector. Thus in an electromagnetic field (22-14) becomes

$$\frac{1}{c^2}\left[i\hbar\frac{\partial}{\partial t} - e\Phi(\mathbf{r}t)\right]^2\psi(\mathbf{r},t) = \left\{\left[\frac{\hbar}{i}\nabla - \frac{e}{c}\mathbf{A}(\mathbf{r}t)\right]^2 + m^2c^2\right\}\psi(\mathbf{r},t). \qquad (22\text{-}16)$$

Before exploring the consequences of the Klein-Gordon equation let us note its behavior under a Lorentz transformation. The operator

$$\frac{1}{c^2}\frac{\partial^2}{\partial t^2} - \nabla^2$$

takes the same form in any Lorentz frame and thus for a *spinless* particle, ψ', the wave function in the new frame is given in terms of $\psi(\mathbf{r},t)$ in the original frame by

$$\psi'(\mathbf{r}',t') = \psi(\mathbf{r},t) \qquad (22\text{-}17)$$

where $\mathbf{r}'$, t' are the coordinates in the new frame of the space-time point $\mathbf{r},t$; in other words, $\psi(\mathbf{r},t)$ is a scalar.

The Klein-Gordon equation has several unusual features. First of all, unlike the nonrelativistic Schrödinger equation, it is *second order* in time. Therefore to predict the future behavior of a particle we must know both $\psi(\mathbf{r}t)$ and $\partial\psi(\mathbf{r},t)/\partial t$ at any one time. Since these are independent quantities we must basically know "twice" as much information about the particle to specify its state as nonrelativistically. Put another way, the particle has essentially an extra degree of freedom. We shall see shortly that this extra degree of freedom corresponds to specifying the *charge* of the particle, and that the Klein-Gordon equation describes both a particle and its *antiparticle* in one fell swoop.

A related consequence of the equation being second order in time is that the functions

$$\psi = e^{i(\mathbf{p}\cdot\mathbf{r} - Et)/\hbar}$$

solve the free particle equation (22-14) with *either* sign of E:

$$E = \pm c\sqrt{p^2 + m^2c^2}.$$

In other words the Klein-Gordon equation has negative energy solutions! These have the strange property that as we *increase* the magnitude of $\mathbf{p}$, the energy of the particle *decreases*. We shall see though that the negative energy eigenstates of the Klein-Gordon equation describe antiparticles, while the positive energy eigenstates describe particles.

Yet another related peculiarity is the fact that one cannot construct a *positive* probability density, like $\psi^*\psi$ nonrelativistically, that is conserved in time. For the Klein-Gordon equation $\int d^3r\, \psi^*\psi$ generally changes in time, and thus we are at a loss to interpret $\psi^*(\mathbf{r}, t)\psi(\mathbf{r}, t)$ as being the probability for finding a particle at $\mathbf{r}$.

There is one conserved density that we can construct though. Since the operator acting on ψ in (22-15) is real, ψ^* also satisfies the same equation. Thus

$$\psi^*\left[\frac{1}{c^2}\frac{\partial^2}{\partial t^2} - \nabla^2 + \left(\frac{mc}{\hbar}\right)^2\right]\psi - \psi\left[\frac{1}{c^2}\frac{\partial^2}{\partial t^2} - \nabla^2 + \left(\frac{mc}{\hbar}\right)^2\right]\psi^* = 0 \tag{22-18}$$

and by some slight rearrangement we find the continuity equation

$$\frac{\partial\rho(\mathbf{r}, t)}{\partial t} + \nabla\cdot\mathbf{j}(\mathbf{r}t) = 0, \tag{22-19}$$

where the current, as nonrelativistically, is

$$\mathbf{j}(\mathbf{r}, t) = \frac{\hbar}{2im}\,[\psi^*\nabla\psi - \psi\nabla\psi^*], \tag{22-20}$$

and the density is

$$\rho(\mathbf{r}, t) = \frac{i\hbar}{2mc^2}\left[\psi^*\frac{\partial\psi}{\partial t} - \psi\frac{\partial\psi^*}{\partial t}\right]. \tag{22-21}$$

From (22-19), the integral of this density over all space doesn't change in time. However, (22-21) needn't be positive; for example, $\rho < 0$ for a negative energy free particle eigenstate. Thus we cannot interpret $\rho(\mathbf{r})$, either, as being the particle (probability) density at $\mathbf{r}$, nor can we interpret $\mathbf{j}(\mathbf{r})$ as the particle current. The interpretation that shall emerge is that for charged particles $e\rho(\mathbf{r})$ represents the *charge density* at $\mathbf{r}$, which can have either sign, and $e\mathbf{j}(\mathbf{r})$ represents the *electric current* at $\mathbf{r}$.

One can verify that (22-19) is still satisfied in the presence of an electromagnetic field, with the current and density given by

$$\mathbf{j}(\mathbf{r}, t) = \frac{1}{2m}\left[\psi^*\left(\frac{\hbar}{i}\nabla - \frac{e}{c}\mathbf{A}\right)\psi + \psi\left(-\frac{\hbar}{i}\nabla - \frac{e}{c}\mathbf{A}\right)\psi^*\right] \tag{22-22}$$

as nonrelativistically, and

$$\rho(\mathbf{r}, t) = \frac{1}{2mc^2}\left[\psi^*\left(i\hbar\frac{\partial}{\partial t} - e\Phi\right)\psi + \psi\left(-i\hbar\frac{\partial}{\partial t} - e\Phi\right)\psi^*\right]. \tag{22-23}$$

NEGATIVE ENERGY STATES AND ANTIPARTICLES

Let us now construct a consistent interpretation of the Klein-Gordon equation. We begin by considering a free particle at rest, $\mathbf{p} = 0$. For a positive energy solution, the wave function of the particle is

$$\psi(\mathbf{r}, t) = e^{-imc^2t/\hbar} \tag{22-24}$$

since the energy of a particle at rest is mc^2. Let us make a Lorentz transformation to a frame moving with velocity $-\mathbf{v}$ with respect to the particle at rest. In that frame the particle appears to have a velocity $\mathbf{v}$, momentum $\mathbf{p} = m\mathbf{v}/\sqrt{1 - v^2/c^2}$ and energy $E_p = mc^2/\sqrt{1 - v^2/c^2}$. Thus in the new frame, denoted by primes, the wave function of the particle, according to (22-17) is

$$\psi'(\mathbf{r}', t') = e^{-imc^2t/\hbar} = e^{i(\mathbf{p}\cdot\mathbf{r}' - E_pt')/\hbar}; \tag{22-25}$$

since the quantity $\mathbf{p}\cdot\mathbf{r}' - E_pt'$ is a Lorentz scalar, it equals $-mc^2t$ in the rest frame and therefore in all frames. Equation (22-25) is the expected result that a particle with momentum $\mathbf{p}$ and energy E_p has the wave function

$$\psi'(\mathbf{r}', t') = e^{i(\mathbf{p}\cdot\mathbf{r}' - E_pt')/\hbar}$$

From (22-19) and (22-20) we see that the density ρ for this particle is

$$\rho(\mathbf{r}', t') = \frac{E_p}{mc^2}, \tag{22-26}$$

and the current $\mathbf{j}$ is

$$\mathbf{j}(\mathbf{r}', t') = \frac{\mathbf{p}}{m} = \frac{\mathbf{p}c^2}{E_p}\,\rho(\mathbf{r}', t'). \tag{22-27}$$

Thus the current is just the density times the relativistic velocity of the particle,

$$\mathbf{v} = \frac{\mathbf{p}c^2}{E_p}. \tag{22-28}$$

Notice that $\rho(\mathbf{r}, t)$ transforms as the time component of a four-vector. This is perfectly reasonable, since a unit volume in the rest frame appears smaller by a factor $\sqrt{1 - v^2/c^2}$ when seen from a moving frame. Thus a unit density in the rest frame will appear as a density $1/\sqrt{1 - v^2/c^2} = E_p/mc^2$ in a frame in which the particle is moving.

Now let us try to understand the negative energy solutions. Again we start with a particle at rest described by the wave function

$$\psi(\mathbf{r}, t) = e^{imc^2t/\hbar}; \tag{22-29}$$

this corresponds to a solution of the Klein-Gordon equation with energy $-mc^2$. The density ρ, in this state is, from (22-20),

$$\rho(\mathbf{r}, t) = -1. \tag{22-30}$$

One way to interpret a state with a negative particle density is to say that it is a state with a positive density of *antiparticles.* We shall adopt this interpretation and say that a particle at rest with energy $-mc^2$ is actually an antiparticle at rest with positive energy mc^2. This interpretation of the negative energy states leads to a consistent theoretical picture which is amply confirmed experimentally.

In a Lorentz frame traveling at velocity $-\mathbf{v}$ with respect to the antiparticle, the wave function (22-29) becomes

$$\psi'(\mathbf{r}', t') = e^{imc^2t/\hbar} = e^{-i(\mathbf{p}\cdot\mathbf{r}' - E_p t')/\hbar} \tag{22-31}$$

where $\mathbf{p} = m\mathbf{v}/\sqrt{1 - v^2/c^2}$ is the momentum of the antiparticle and $E_p = mc^2/\sqrt{1 - v^2/c^2}$ is its energy. In this new frame the antiparticle has velocity $\mathbf{v}$, momentum $\mathbf{p}$, and energy E_p. Notice that the wave function (22-31) describes the *antiparticle* as a *particle* of energy $-E_p$ and momentum $-\mathbf{p}$.

From (22-20) and (22-21) the density ρ in this frame is

$$\rho(\mathbf{r}', t') = -\frac{E_p}{mc^2} \tag{22-32}$$

and the current is

$$\mathbf{j}(\mathbf{r}', t') = -\frac{\mathbf{p}}{m} = \frac{\mathbf{p}c^2}{E_p}\rho(\mathbf{r}'t'). \tag{22-33}$$

An antiparticle moving with velocity $\mathbf{v}$ has associated with it a current moving in the opposite direction; a flow of antiparticle in one direction is equivalent to a flow of particle in the opposite direction. For a charged particle $e\rho(\mathbf{r}, t)$ is the charge density: positive for a free particle, $e > 0$, and negative for a free antiparticle, since an antiparticle has the opposite charge as the particle. Similarly $e\mathbf{j}(\mathbf{r}, t)$ is the electric current of the state ψ; for a particle, the electric current is in the direction of the particle's velocity. For an antiparticle, $e < 0$, the electric current is opposite to the velocity.

Thus the interpretation of the negative energy solutions as corresponding to antiparticles is consistent with the interpretation of the density ρ as a charge density and $\mathbf{j}$ as a current of charge (which has the opposite sign for antiparticles). Now if this interpretation is really to be correct we must show that electromagnetic fields in fact couple to antiparticles with the opposite sign of e. To show this let us take the *complex conjugate* of Eq. (22-16); the fields Φ and $\mathbf{A}$ are real, and only the explicit i's change sign. Thus we have

$$\frac{1}{c^2}\left[i\hbar\frac{\partial}{\partial t}+e\phi(\mathbf{r}t)\right]^2\psi^*(\mathbf{r}t)=\left(\left[\frac{\hbar}{i}\nabla+e\mathbf{A}(\mathbf{r}t)\right]^2+m^2c^2\right)\psi^*(\mathbf{r}t). \tag{22-34}$$

This equation says that if $\psi(\mathbf{r},t)$ is a solution to the Klein-Gordon equation with a certain sign of the charge, then $\psi^*(\mathbf{r},t)$ is a solution to the Klein-Gordon equation with the *opposite* sign of the charge and the same mass. In this sense the relativistic theory of a spin zero particle predicts the existence of its antiparticle with opposite charge and the same mass; at the same time that the theory has solutions to describe a particle it also has solutions to describe its antiparticle. This is a general feature of relativistic quantum mechanics. Now since complex conjugation reverses the signs of all frequencies (and momenta) in ψ, the complex conjugate of a negative energy solution to the original equation (22-16) is a positive energy solution to (22-34) with the opposite sign of the charge. Thus to interpret a negative energy solution we simply complex conjugate the wave function and interpret it as a positive energy solution of the opposite charge. One example of this is our interpretation of the wave function (22-31), whose complex conjugate is $e^{i(\mathbf{p}\cdot\mathbf{r}-E_pt)/\hbar}$, as describing an antiparticle of energy E_p and momentum $\mathbf{p}$. The wave function $\psi^*(\mathbf{r},t)$ is known as the *charge conjugate* wave function.

Notice that we can write the density (22-23) as

$$\begin{aligned}\rho(\mathbf{r},t)&=-\frac{1}{2mc^2}\left[\psi\left(i\hbar\frac{\partial}{\partial t}+e\Phi\right)\psi^*+\psi^*\left(-i\hbar\frac{\partial}{\partial t}+e\Phi\right)\psi\right]\\&=-\rho_c(\mathbf{r},t),\end{aligned} \tag{22-35}$$

where ρ_c is the density evaluated with the wave function ψ^* and the opposite sign of e in (22-21). Thus the quantity ρ_c is minus the density ρ. Similarly $\mathbf{j}_c$, the current evaluated with the wave function ψ^* and the opposite sign of e in (22-22) is given by

$$\mathbf{j}_c(\mathbf{r},t)=-\mathbf{j}(\mathbf{r},t). \tag{22-36}$$

We can normalize the solutions to the Klein-Gordon equation by the condition that the state represent one unit of charge:

$$\int\rho(\mathbf{r}t)\,d^3r=\pm1, \tag{22-37}$$

the upper sign is for particle solutions and the lower antiparticle solutions. The normalization (22-37) is conserved in time and is invariant under a Lorentz transformation. The complex conjugate of a normalized solution to the Klein-Gordon equation (22-16) has the opposite sign of the normalization since $\int\rho = -\int\rho_c$.

FIRST-ORDER KLEIN-GORDON EQUATION

To discuss the physics in the Klein-Gordon equation it is often convenient to transform it into two equations, each first order in time.[1] Let us define

$$\psi^0(\mathbf{r}, t) = \left[\frac{\partial}{\partial t} + \frac{ie}{\hbar}\Phi(\mathbf{r}, t)\right]\psi(\mathbf{r}, t). \tag{22-38}$$

Then from (22-16),

$$\left[\frac{\partial}{\partial t} + \frac{ie}{\hbar}\Phi(\mathbf{r}, t)\right]\psi^0(\mathbf{r}, t) = \left[c^2\left(\nabla - \frac{ie\mathbf{A}}{\hbar c}\right)^2 - \frac{m^2c^4}{\hbar^2}\right]\psi(\mathbf{r}, t). \tag{22-39}$$

The pair of first-order equations (22-38) and (22-39) for the two functions ψ and ψ^0 are equivalent to the original Klein-Gordon equation. We can make these equations more symmetric by defining the linear combinations

$$\varphi = \frac{1}{2}\left[\psi + \frac{i\hbar}{mc^2}\psi^0\right]$$

$$\chi = \frac{1}{2}\left[\psi - \frac{i\hbar}{mc^2}\psi^0\right]. \tag{22-40}$$

Then with a little juggling of (22-38) and (22-39) we see that φ and χ satisfy

$$\left(i\hbar\frac{\partial}{\partial t} - e\Phi\right)\varphi = \frac{1}{2m}\left[\frac{\hbar}{i}\nabla - \frac{e}{c}\mathbf{A}\right]^2(\varphi + \chi) + mc^2\varphi$$

$$\left(i\hbar\frac{\partial}{\partial t} - e\Phi\right)\chi = -\frac{1}{2m}\left[\frac{\hbar}{i}\nabla - \frac{e}{c}\mathbf{A}\right]^2(\varphi + \chi) - mc^2\chi. \tag{22-41}$$

We can write these equations more simply by defining a *two-component* wave function

---

[1] See the review article by H. Feshbach and F. Villars, *Rev. Mod. Phys.* **30**, 24 (1958).

$$\Psi(\mathbf{r}, t) = \begin{pmatrix} \varphi(\mathbf{r}, t) \\ \chi(\mathbf{r}, t) \end{pmatrix}. \tag{22-42}$$

Writing

$$\tau_1 = \begin{pmatrix} 0 & 1 \\ 1 & 0 \end{pmatrix}, \quad \tau_2 = \begin{pmatrix} 0 & -i \\ i & 0 \end{pmatrix}, \quad \tau_3 = \begin{pmatrix} 1 & 0 \\ 0 & -1 \end{pmatrix} \tag{22-43}$$

[these are just the Pauli spin matrices, only it is better to reserve the letter σ to refer to actual spin], equations (22-41) become

$$i\hbar \frac{\partial \Psi}{\partial t} = \left[\frac{1}{2m}\left(\frac{\hbar}{i}\nabla - \frac{e}{c}\mathbf{A}\right)^2 (\tau_3 + i\tau_2) + mc^2\tau_3 + e\Phi \right] \Psi. \tag{22-44}$$

This equation is fully equivalent to the original Klein-Gordon equation; in terms of Ψ,

$$\psi = \varphi + \chi$$

$$\psi^0 = \frac{mc^2}{i\hbar}(\varphi - \chi). \tag{22-45}$$

The two components of Ψ *do not* represent the spin of the particle – we are describing spinless particles. Rather, the internal degree of freedom represented by these components is the *charge* of the particle.

In this two-component representation, the density $\rho(\mathbf{r})$, Eq. (22-22) takes the simple form

$$\rho(\mathbf{r}t) = |\varphi|^2 - |\chi|^2 = \Psi^\dagger \tau_3 \Psi, \tag{22-46}$$

where $\Psi^\dagger$ denotes the row vector (φ^*, χ^*). The current $\mathbf{j}(\mathbf{r}, t)$ assumes the less transparent form in terms of Ψ:

$$\mathbf{j}(\mathbf{r}t) = \frac{\hbar}{2im}[\Psi^\dagger\tau_3(\tau_3 + i\tau_2)\nabla\Psi - (\nabla\Psi^\dagger)\tau_3(\tau_3 + i\tau_2)\Psi] - \frac{e\mathbf{A}}{mc}\Psi^\dagger\tau_3(\tau_3 + i\tau_2)\Psi. \tag{22-47}$$

The normalization condition is

$$\int d^3r\, \Psi^\dagger \tau_3 \Psi = \pm 1. \tag{22-48}$$

The scalar product of two such wave functions Ψ and Ψ' is defined in general as

$$\langle \Psi | \Psi' \rangle \equiv \int d^3r\, \Psi^\dagger(\mathbf{r}, t)\tau_3\Psi'(\mathbf{r}, t). \tag{22-49}$$

The wave equation (22-41) is of the form

$$i\hbar \frac{\partial \Psi}{\partial t} = H\Psi, \tag{22-50}$$

where the Hamiltonian is given by

$$H = \frac{1}{2m}\left(\mathbf{p} - \frac{e}{c}\mathbf{A}\right)^2 (\tau_3 + i\tau_2) + e\Phi + mc^2\tau_3. \tag{22-51}$$

H is seemingly not Hermitian, since $(\tau_3 + i\tau_2)^\dagger = \tau_3 - i\tau_2$. [The p as well as **A** and ϕ commute with the τ's.] It is however Hermitian in the sense that

$$\langle\Psi|(H|\Psi'\rangle) = [\langle\Psi'|(H|\Psi\rangle)]^*; \tag{22-52}$$

from (22-49) we see that this requires

$$H = \tau_3 H^\dagger \tau_3 \tag{22-53}$$

where † denotes the usual Hermitian conjugate. (22-25) is clearly satisfied since $\tau_3(\tau_3 - i\tau_2)\tau_3 = \tau_3 + i\tau_2$. As we shall see, all the good properties of Hermitian operators, e.g., real eigenvalues, etc., follow from (22-52).

Since under charge conjugation, $\varphi \to \chi^*$ and $\chi \to \varphi^*$, the two-component charge conjugate solution Ψ_c formed from ψ^* is related to Ψ by

$$\Psi_c = \tau_1 \Psi^*. \tag{22-54}$$

This is the form of the charge conjugation operation in the two-component language. Let us take the *complex conjugate* of the Hamiltonian (22-50). Then since $\tau_3 + i\tau_2$ is a real matrix, and $\mathbf{p} = (\hbar/i)\nabla$ we find $\mathbf{p}^* = -\mathbf{p}$, so

$$H^*(e) = \frac{1}{2m}\left(-\mathbf{p} - \frac{e}{c}\mathbf{A}\right)^2 (\tau_3 + i\tau_2) + e\Phi + mc^2\tau_3, \tag{22-55}$$

where we've indicated the dependence of H on the charge e. Multiplying (22-55) on both the right and left by τ_1 and using the fact that τ_1 anticommutes with both τ_2 and τ_3 we find

$$\tau_1 H^*(e)\tau_1 = -\left[\frac{1}{2m}\left(\mathbf{p} + \frac{e}{c}\mathbf{A}\right)^2 (\tau_3 + i\tau_2) - e\Phi + mc^2\tau_3\right] = -H(-e). \tag{22-56}$$

The result is minus the Hamiltonian for a particle with the opposite sign of the charge. Thus if Ψ solves the equation $i\hbar(\partial\Psi/\partial t) = H(e)\Psi$, we have

$$-i\hbar \frac{\partial \Psi^*}{\partial t} = H^*(e)\Psi^* = -\tau_1 H(-e)\tau_1 \Psi^*.$$

Multiplying through by τ_1 and using (22-54) we then see that

$$i\hbar \frac{\partial \Psi_c}{\partial t} = H(-e)\Psi_c; \tag{22-57}$$

this is the two-component statement of the fact that Ψ_c solves the Klein-Gordon equation with the opposite sign of the charge.

Let us write down the two-component solutions for free particles and antiparticles. The wave function of a free particle of momentum p, normalized to unit density, is

$$\psi_p^{(+)}(rt) = \sqrt{\frac{mc^2}{E_p}}\, e^{i(p\cdot r - E_p t)/\hbar}.$$

E_p always denotes the positive number $\sqrt{p^2c^2 + m^2c^4}$; the (+) denotes a positive energy solution. Thus from (22-38) and (22-40) the two-component wave function for this particle is

$$\Psi_p^{(+)}(r,t) = \Psi_p^{(+)} e^{i(p\cdot r - E_p t)/\hbar}, \tag{22-58}$$

where the two-component vector $\Psi_p^{(+)}$ is given by

$$\Psi_p^{(+)} = \frac{1}{2\sqrt{E_p mc^2}} \begin{pmatrix} mc^2 + E_p \\ mc^2 - E_p \end{pmatrix}. \tag{22-59}$$

Similarly for the negative energy solution corresponding to an antiparticle of momentum p,

$$\psi_p^{(-)}(r,t) = \sqrt{\frac{mc^2}{E_p}}\, e^{-i(p\cdot r - E_p t)/\hbar},$$

the two-component wave function is

$$\Psi_p^{(-)}(r,t) = \Psi_p^{(-)} e^{-i(p\cdot r - E_p t)/\hbar}, \tag{22-60}$$

where

$$\Psi_p^{(-)} = \frac{1}{2\sqrt{E_p mc^2}} \begin{pmatrix} mc^2 - E_p \\ mc^2 + E_p \end{pmatrix} = \tau_1 \Psi_p^{(+)}. \tag{22-61}$$

For a slowly moving particle

$$\Psi_{\mathbf{p}}^{(+)} \approx \begin{pmatrix} 1 \\ \\ \dfrac{-v^2}{4c^2} \end{pmatrix}, \quad \Psi_{\mathbf{p}}^{(-)} \approx \begin{pmatrix} \dfrac{-v^2}{4c^2} \\ \\ 1 \end{pmatrix}. \tag{22-62}$$

Thus in the nonrelativistic limit we expect χ to be $\sim (v/c)^2$ times φ for a particle. Dropping χ in the first of Eqs. (22-40), in this limit, we find that φ obeys the nonrelativistic Schrödinger equation, with the constant mc^2 included in the energy. Similarly dropping φ, for an antiparticle, in the lower of Eqs. (22-40) we find that χ^* becomes the nonrelativistic wave function of a particle of opposite charge.

Notice that (22-61) and (22-62) are orthogonal in the sense that

$$\Psi_{\mathbf{p}}^{(+)\dagger} \tau_3 \Psi_{\mathbf{p}}^{(-)} = 0 = \Psi_{\mathbf{p}}^{(-)\dagger} \tau_3 \Psi_{\mathbf{p}}^{(+)}. \tag{22-63}$$

In fact, any two energy eigenstates of H, belonging to different energy are orthogonal in the sense that

$$\int d^3r \, \Psi_E{}^\dagger(\mathbf{r}) \tau_3 \Psi_{E'}(\mathbf{r}) = 0 \quad \text{if } E \neq E'. \tag{22-64}$$

To prove this we let $|E\rangle$ be the state vector whose wave function is $\Psi_E(\mathbf{r})$. Then from (22-52) and (22-45) we have

$$E\langle E|E\rangle = (E\langle E|E\rangle)^* = E^*\langle E|E\rangle.$$

Then (22-52) applied to $|E\rangle$ and $|E'\rangle$ yields

$$E\langle E'|E\rangle = (E'\langle E|E'\rangle)^* = E'\langle E'|E\rangle$$

and so $\langle E'|E\rangle = 0$ if $E \neq E'$.

The free particle solutions form a complete set since any wave function, Ψ, can be expanded as a linear combination of free particle and antiparticle solutions. To see this we write Ψ in terms of its Fourier transform:

$$\Psi(\mathbf{r}t) = \int \frac{d^3p}{(2\pi\hbar)^3} \, e^{i\mathbf{p}\cdot\mathbf{r}/\hbar} \begin{pmatrix} \varphi_{\mathbf{p}}(t) \\ \\ \chi_{\mathbf{p}}(t) \end{pmatrix}. \tag{22-65}$$

However one can always expand $\begin{pmatrix} \varphi_{\mathbf{p}} \\ \chi_{\mathbf{p}} \end{pmatrix}$ as a linear combination of $\Psi_{\mathbf{p}}^{(+)}$ and $\Psi_{-\mathbf{p}}^{(-)}$, since these two vectors are linearly independent:

$$\Psi_{\mathbf{p}}(t) = \begin{pmatrix} \varphi_{\mathbf{p}} \\ \chi_{\mathbf{p}} \end{pmatrix} = u_{\mathbf{p}}(t)\Psi_{\mathbf{p}}^{(+)} + v^*_{-\mathbf{p}}(t)\Psi_{-\mathbf{p}}^{(-)}. \tag{22-66}$$

(The reason for the $-\mathbf{p}$ and * will be evident.) Thus

$$\begin{aligned}\Psi(\mathbf{r}t) &= \int \frac{d^3p}{(2\pi\hbar)^3}\left[u_{\mathbf{p}}(t)\Psi_{\mathbf{p}}^{(+)} + v^*_{-\mathbf{p}}(t)\Psi_{-\mathbf{p}}^{(-)}\right] e^{i\mathbf{p}\cdot\mathbf{r}/\hbar} \\ &= \int \frac{d^3p}{(2\pi\hbar)^3}\left[u_{\mathbf{p}}(t)\Psi_{\mathbf{p}}^{(+)} e^{i\mathbf{p}\cdot\mathbf{r}/\hbar} + v_{\mathbf{p}}^*(t)\Psi_{\mathbf{p}}^{(-)} e^{-i\mathbf{p}\cdot\mathbf{r}/\hbar}\right];\end{aligned} \tag{22-67}$$

clearly $u_{\mathbf{p}}(t)$ is the amplitude for a particle in the state Ψ to have momentum $\mathbf{p}$ and positive charge, while according to (22-54) $v_{\mathbf{p}}(t)$ is the amplitude for it to have momentum $\mathbf{p}$ and negative charge. From the orthonormality of the $\Psi_{\mathbf{p}}^{(+)}$ we have

$$\begin{aligned}u_{\mathbf{p}}(t) &= \int d^3r\, \Psi_{\mathbf{p}}^{(+)\dagger} e^{-i\mathbf{p}\cdot\mathbf{r}/\hbar}\, \tau_3\Psi(\mathbf{r}t) \\ v_{\mathbf{p}}^*(t) &= -\int d^3r\, \Psi_{\mathbf{p}}^{(-)\dagger} e^{i\mathbf{p}\cdot\mathbf{r}/\hbar}\, \tau_3\Psi(\mathbf{r}t).\end{aligned} \tag{22-68}$$

If Ψ is normalized to ± 1, then

$$\int \frac{d^3p}{(2\pi\hbar)^3}\left(|u_{\mathbf{p}}|^2 - |v_{\mathbf{p}}|^2\right) = \pm 1. \tag{22-69}$$

There is no restriction on the magnitude of either $u_{\mathbf{p}}$ or $v_{\mathbf{p}}$; only the integral (22-69) involving the *difference* of their squares is fixed. Thus one can have a state with an *arbitrarily* large amplitude for finding the particle with a certain momentum. (This is the first hint that spin zero particles must be bosons; we shall return to this point later.)

The expectation value of the kinetic energy in the state Ψ is the average kinetic energy of the particle component *plus* the average kinetic energy of the antiparticle component:

$$\int \frac{d^3p}{(2\pi\hbar)^3}\left(E_p|u_{\mathbf{p}}|^2 + E_p|v_{\mathbf{p}}|^2\right). \tag{22-70}$$

This is always a positive quantity. Similarly, the expectation value of the momentum in the state Ψ is

$$\int \frac{d^3p}{(2\pi\hbar)^3}\left(\mathbf{p}|u_{\mathbf{p}}|^2 + \mathbf{p}|v_{\mathbf{p}}|^2\right). \tag{22-71}$$

We leave it as an exercise to show that (22-70) equals

$$\int \Psi^\dagger(\mathbf{r})\tau_3 H_0 \Psi(\mathbf{r})\, d\mathbf{r} \tag{22-72}$$

where

$$H_0 = \frac{p^2}{2m}(\tau_3 + i\tau_2) + mc^2\tau_3, \tag{22-73}$$

and that (22-71) is equal to

$$\int \Psi^\dagger(\mathbf{r})\tau_3\left(\frac{\hbar}{i}\nabla\right)\Psi(\mathbf{r})\, d^3r. \tag{22-74}$$

FREE PARTICLE WAVE PACKETS

Let us consider the properties of a free particle wave packet formed of positive energy solutions:

$$\Psi^{(+)}(\mathbf{r}, t) = \int \frac{d^3p}{(2\pi\hbar)^3}\, u_{\mathbf{p}}\, e^{i(\mathbf{p}\cdot\mathbf{r} - E_p t)/\hbar}\, \Psi_{\mathbf{p}}^{(+)}. \tag{22-75}$$

If $u_{\mathbf{p}}$ is peaked about some central value $\mathbf{p}'$, then it is easy to see that the center of the wave packet moves with a group velocity

$$\mathbf{v}_g = (\nabla_{\mathbf{p}} E_p)_{\mathbf{p}=\mathbf{p}'} = \frac{\mathbf{p}'c^2}{E_{p'}}. \tag{22-76}$$

Similarly if we form a free particle wave packet of negative energy solutions

$$\Psi^{(-)}(\mathbf{r}, t) = \int \frac{d^3p}{(2\pi\hbar)^3}\, v_{\mathbf{p}}^*\, e^{-i(\mathbf{p}\cdot\mathbf{r} - E_p t)/\hbar}\, \Psi_{\mathbf{p}}^{(-)}, \tag{22-77}$$

then if $v_{\mathbf{p}}$ is peaked about the momentum $\mathbf{p}'$, this antiparticle packet also moves with the group velocity (22-76). Let's try to construct a free particle wave packet perfectly localized at the origin

$$\Psi(\mathbf{r}) = \begin{pmatrix} \varphi \\ \chi \end{pmatrix} \delta(\mathbf{r}). \tag{22-78}$$

Here φ and χ are numbers. From (22-68) we find

$$u_p = \Psi_p^{(+)\dagger}\tau_3\begin{pmatrix}\varphi \\ \chi\end{pmatrix} = \frac{E_p(\varphi+\chi)+mc^2(\varphi-\chi)}{2\sqrt{mc^2E_p}}$$

$$v_p{}^* = -\Psi_p^{(-)}\tau_3\begin{pmatrix}\varphi \\ \chi\end{pmatrix} = \frac{E_p(\varphi+\chi)-mc^2(\varphi-\chi)}{2\sqrt{mc^2E_p}}. \tag{22-79}$$

One can see that no matter how we choose φ and χ, the wave packet must always have both particle and antiparticle components. In other words, it is impossible to construct a perfectly localized wave packet of positive energy solutions alone, such as (22-75).

Suppose then that we have a wave packet of positive energy solutions and we try squeezing on it with various bits of experimental machinery, such as collimating tubes, in an attempt to make the packet very localized. Can we? Yes, but only by *creating* antiparticles as we perform operations on the wave packet!

To see this we must consider some rather strange features of the position operator in relativistic quantum mechanics. Let us multiply the wave packet (22-75) by the position operator $\mathbf{r}$. Then we can write

$$\mathbf{r}\Psi^{(+)}(\mathbf{r},t) = \int\frac{d^3p}{(2\pi\hbar)^3}u_p(t)\Psi_p^{(+)}\frac{\hbar}{i}\nabla_p e^{i\mathbf{p}\cdot\mathbf{r}/\hbar},$$

where we include the time factor in u_p. Integrating by parts we then find

$$\mathbf{r}\Psi^{(+)}(\mathbf{r},t) = \int\frac{d^3p}{(2\pi\hbar)^3}(i\hbar\nabla_p u_p(t))\Psi_p^{(+)}e^{i\mathbf{p}\cdot\mathbf{r}/\hbar} + \int\frac{d^3p}{(2\pi\hbar)^3}u_p i\hbar(\nabla_p\Psi_p^{(+)})e^{i\mathbf{p}\cdot\mathbf{r}/\hbar}. \tag{22-80}$$

But from (22-52) and (22-61)

$$\nabla_p\Psi_p^{(\pm)} = -\frac{\mathbf{p}c^2}{2E_p{}^2}\Psi_p^{(\mp)}, \tag{22-81}$$

so that

$$\mathbf{r}\Psi^{(+)}(\mathbf{r},t) = \mathbf{r}_+\Psi^{(+)}(\mathbf{r},t) + \mathbf{r}_-\Psi^{(+)}(\mathbf{r},t), \tag{22-82}$$

where

$$\mathbf{r}_+\Psi^{(+)}(\mathbf{r},t) = \int\frac{d^3p}{(2\pi\hbar)^3}[i\hbar\nabla_p u_p(t)]\Psi_p^{(+)}e^{i\mathbf{p}\cdot\mathbf{r}/\hbar} \tag{22-83}$$

and

$$r_-\Psi^+(\mathbf{r}, t) = -\int \frac{d^3p}{(2\pi\hbar)^3} u_{\mathbf{p}}(t) \frac{i\hbar \mathbf{p} c^2}{2E_p{}^2} \Psi_{\mathbf{p}}^{(-)} e^{i\mathbf{p}\cdot\mathbf{r}/\hbar}. \tag{22-84}$$

Thus multiplying a wave packet of positive energy states by the position operator mixes in a component of *negative* energy solutions, $\Psi_{\mathbf{p}}^{(-)} e^{i\mathbf{p}\cdot\mathbf{r}/\hbar}$, into the wave packet. The position operator is a sum of two parts; $\mathbf{r} = \mathbf{r}_+ + \mathbf{r}_-$. The first part, $\mathbf{r}_+$, the *even* part, acting on a positive energy component gives only a positive energy component; acting on a negative energy component it produces only a negative energy component. The second part, $\mathbf{r}_-$, called the *odd* part, has the effect of changing free particles into free antiparticles, and vice versa.

One can show that the most localized wave packet formed of positive energy states alone is an eigenstate of $\mathbf{r}_+$. From (22-83) we see that if $\mathbf{r}_+\Psi^{(+)}(\mathbf{r}) = \mathbf{r}_0\Psi^{(+)}(\mathbf{r})$, where $\mathbf{r}_0$ is the eigenvalue, the amplitudes $u_{\mathbf{p}}$ for such an eigenstate must obey

$$i\hbar\nabla_p u_{\mathbf{p}} = \mathbf{r}_0 u_{\mathbf{p}},$$

or $u_{\mathbf{p}} = e^{-i\mathbf{p}\cdot\mathbf{r}_0/\hbar}$. Thus the state

$$\Psi_{\mathbf{r}_0}^{(+)}(\mathbf{r}) = \int \frac{d^3p}{(2\pi\hbar)^3} e^{i\mathbf{p}\cdot(\mathbf{r}-\mathbf{r}_0)/\hbar} \Psi_{\mathbf{p}}^{(+)},$$

is an eigenstate of $\mathbf{r}_+$. This eigenstate, which is not normalizable, is spread out in space over distances $\sim\hbar/mc$ around $\mathbf{r}_0$. Thus the theory with the positive energy solutions alone is incapable of describing particles localized to within more than a Compton wavelength.

The presence of the odd part, $\mathbf{r}_-$, in the position operator implies that simply putting a wave packet of positive charge particles through a potential, $\Phi(\mathbf{r})$ (which multiplies $\Psi(\mathbf{r})$ by various functions of $\mathbf{r}$), has the effect of creating antiparticles, and, to conserve charge, particles as well. Thus the theory has built into it the mechanism of particle-antiparticle production by external potentials.

KLEIN'S PARADOX

We can see one example of this phenomenon very clearly by studying the behavior of a positively charged particle of momentum p striking an electrostatic barrier from the left. This example is

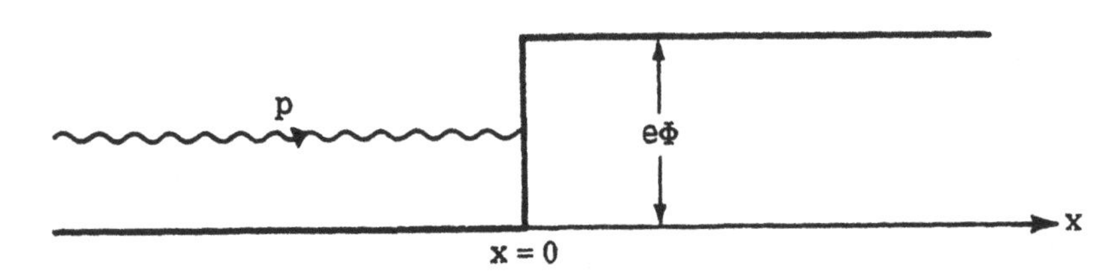

Fig. 22-1
Electrostatic potential barrier in Klein's paradox.

known, for what will be very obvious reasons, as *Klein's paradox*. Let us suppose that the barrier, Fig. 22-1, is of height $e\Phi = V$, and extends infinitely to the right. The problem can be solved in the same manner as the equivalent nonrelativistic problem by finding the stationary states of the potential. It is easiest to solve the second order Klein-Gordon equation directly. Since the particle is incident with momentum p from a region of no potential, its energy is E_p. Thus for $x < 0$ the wave function is of the form

$$\psi(x) = ae^{ipx/\hbar} + be^{-ipx/\hbar}, \tag{22-85}$$

representing an incident wave and a reflected wave. For $x > 0$, the Klein-Gordon equation is

$$(E_p - V)^2\psi(x) = -\hbar^2c^2\frac{\partial^2\psi(x)}{\partial x^2} + m^2c^4\psi(x) \tag{22-86}$$

and the solution is of the form

$$\psi(x) = de^{ikx}, \tag{22-87}$$

times the $e^{-iE_pt/\hbar}$ time dependence, where

$$\hbar^2c^2k^2 + m^2c^4 = (E_p - V)^2. \tag{22-88}$$

The boundary conditions on the wave function are the same as in the nonrelativistic case, namely, ψ and $\partial\psi/\partial x$ must be continuous at $x = 0$. Note though that ψ^0 isn't continuous at $x = 0$ since the potential has a discontinuity there. Thus the coefficients a, b, and d are related by

$$b = \frac{p - \hbar k}{p + \hbar k}a \tag{22-89}$$

$$d = \frac{2p}{p + \hbar k}\, a. \tag{22-90}$$

Now we must consider three cases. First, if $E_p > V + mc^2$, then the particle can surmount the barrier,

$$k = \frac{\sqrt{(E_p - V)^2 - m^2c^4}}{\hbar c}, \tag{22-91}$$

and the situation is the same as in the nonrelativistic case. Part of the wave is reflected, and part is transmitted.

Next suppose that the potential is stronger, but in the range

$$E_p - mc^2 < V < E_p + mc^2. \tag{22-92}$$

Then k must be imaginary and in order for the wave function to go to zero as $x \to \infty$, $k = i\kappa$, where

$$\kappa = \frac{\sqrt{m^2c^4 - (E_p - V)^2}}{\hbar c}, \tag{22-93}$$

In this case the wave is totally reflected at the barrier. Let us look though at the charge density on the right. From (22-23)

$$\rho(x) = \frac{E_p - V}{mc^2}\, |d|^2\, e^{-2\kappa x}; \tag{22-94}$$

for $E_p > V$, there is a positive, exponentially decaying charge density to the right of the barrier; but for $V > E_p$ the density is *negative!* We reflect positively charged particles from the wall and find a negative charge within the wall.

Finally if we make the potential so strong that

$$V > E_p + mc^2, \tag{22-95}$$

we expect it to be even more impossible for a particle to surmount the barrier. But look what happens: from (22-88), k is real again. This means that once again there is a particle current to the right of the barrier. The group velocity of the waves for $x > 0$ is $v_g = \partial E_p / \partial \hbar k$, and from (22-87) we find

$$\hbar^2 c^2 k = (E_p - V) \frac{\partial E_p}{\partial k}$$

or

$$v_g = \frac{\hbar c^2 k}{E_p - V}. \tag{22-96}$$

The denominator is negative; thus in order to have a wave (packet) traveling *from* the barrier *to* positive x, we must have $k < 0$.

This implies that the reflection coefficient, b/a in Eq. (22-89), is larger than one; more wave is reflected than is incident! Also the charge density on the right is

$$\rho = \frac{E_p - V}{mc^2} |d|^2, \tag{22-97}$$

again negative, and the current on the right is also negative. One way we can understand this peculiar state of affairs is to say that the incident particle induces the creation of pairs of particles and antiparticles at the barrier. The created antiparticles, having the opposite charge find $x > 0$ a region of attractive potential, and travel to the right. This explains the negative current on the right. The created particles travel to the left, and together with the incident wave, which is totally reflected, they add up to an outgoing current, on the left, larger than the incident current. The total outgoing current, on the right and left, equals the incident current, of course, since total charge is conserved.

This explanation of the solution in terms of creation of pairs doesn't violate conservation of energy. The energy of a created particle on the left is E_p, while that of an antiparticle on the right is $\sqrt{\hbar^2 c^2 k^2 + m^2 c^4} - V$, since the electrostatic potential energy has the opposite sign for a particle of the opposite charge. But from taking the positive square root of both sides of (22-88) we find

$$E_p + \sqrt{\hbar^2 c^2 k^2 + m^2 c^4} - V = 0; \tag{22-98}$$

in other words it takes zero energy to create a particle-antiparticle pair. This is because the potential V is so large that the energy of the antiparticle on the right is not merely less than mc^2, but is *negative.*

We see then that a single particle picture breaks down as soon as the potential on the right is stronger than $2mc^2$, the rest energy of a particle-antiparticle pair. In a sense the single particle theory we've developed is incapable of dealing accurately with such strong fields. After all, we should now begin to worry about the attractive Coulomb interaction between the created particles and created antiparticles. And what happens at the potential barrier when there is no incident particle? Are pairs still created, or is such a potential barrier, extending to infinity, not physically realizable? These problems, on which we shall have further comments in the next section, can only be dealt with in the framework of a many particle relativistic theory, *field theory,* which is the second quantized version of the single particle theory we've been developing here.

A wave packet $\psi(\mathbf{r}, t)$ of positive energy states is a very delicate object. Almost any potential it encounters will mix in antiparticle states. Consider the result of even the mildest possible tampering, e.g., at $t = 0$, multiplying ψ by a phase factor $e^{i\mathbf{k}\cdot\mathbf{r}/\hbar}$ where $\mathbf{k}$ is infinitesimal. Then if at $t = 0$

$$\Psi(\mathbf{r}) = \int \frac{d^3p}{(2\pi\hbar)^3} u_p \Psi_p^{(+)} e^{i\mathbf{p}\cdot\mathbf{r}/\hbar} \tag{22-99}$$

we have

$$\begin{aligned} e^{i\mathbf{k}\cdot\mathbf{r}/\hbar}\Psi(\mathbf{r}) \equiv \Psi'(\mathbf{r}) &= \int \frac{d^3p}{(2\pi\hbar)^3} u_p \Psi_p^{(+)} e^{i(\mathbf{p}+\mathbf{k})\cdot\mathbf{r}/\hbar} \\ &= \int \frac{d^3p}{(2\pi\hbar)^3} u_{p-k} \Psi_{p-k}^{(+)} e^{i\mathbf{p}\cdot\mathbf{r}/\hbar}. \end{aligned} \tag{22-100}$$

In the limit of infinitesimal $\mathbf{k}$ we can write, from (22-81)

$$\Psi_{p-k}^{(+)} \approx \Psi_p^{(+)} + \frac{\mathbf{k}\cdot\mathbf{p}c^2}{E_p^2} \Psi_p^{(-)}.$$

Thus the effect of the $e^{i\mathbf{k}\cdot\mathbf{r}/\hbar}$ factor is to introduce an amplitude for observing the particle to be in a negative energy state. Or in other words, acting with $e^{i\mathbf{k}\cdot\mathbf{r}/\hbar}$ produces an antiparticle component in the wave packet. At later times

$$\begin{aligned} \Psi'(\mathbf{r}, t) &= \int \frac{d^3p}{(2\pi\hbar)^3} u_{p-k} \Psi_p^{(+)} e^{i(\mathbf{p}\cdot\mathbf{r} - E_p t)/\hbar} \\ &\quad + \int \frac{d^3p}{(2\pi\hbar)^3} u_p \frac{\mathbf{k}\cdot\mathbf{p}c^2}{E_p^2} \Psi_p^{(-)} e^{i(\mathbf{p}\cdot\mathbf{r} + E_p t)/\hbar}. \end{aligned} \tag{22-101}$$

since the negative energy components propagate in time with phase factors $e^{iE_p t/\hbar}$. Now if u_p is peaked about some value $\mathbf{p}'$, then the positive frequency component in Ψ' has a group velocity $\sim \mathbf{p}'c/E_{p'}^2$. On the other hand the negative frequency component, introduced by the $e^{i\mathbf{k}\cdot\mathbf{r}/\hbar}$ factor, has a group velocity $-\mathbf{p}'c/E_{p'}^2$. The negative frequency component moves in the opposite direction from the positive frequency component. Put another way, if one evaluates the density or current for the state Ψ', one finds that the interference term between the positive and negative frequency parts varies in time with a frequency at least $2mc^2$. Thus multiplying the wave packet Ψ by $e^{i\mathbf{k}\cdot\mathbf{r}/\hbar}$ at time $t = 0$ introduces very rapidly oscillating motions within the wave packet. This phenomenon is known as the *zitterbewegung,* literally, trembling motion. It is excited just about whenever a wave packet of positive energy solutions travels through a potential.

Generally though the negative frequency components damp out in times $\Delta t \sim 2mc^2/\hbar$ after the wave packet leaves the region of the potential. Only when the potential is capable of transferring energy at least $2mc^2$ to the wave packet, the minimum required to conserve energy in making a pair,[2] will real, "permanent," antiparticles be created when the packet moves past the potential.

SCATTERING BY A POTENTIAL

As an illustration, let us calculate, in the Born approximation, the scattering of a particle of momentum $\mathbf{p}$ by a potential $e\Phi(\mathbf{r}, t)$. The two-component Klein-Gordon equation for this problem is

$$i\hbar \frac{\partial \Psi}{\partial t} = H_0\Psi + e\Phi\Psi \tag{22-102}$$

where H_0 is the free particle Hamiltonian. The initial state of the particle is

$$\Psi_{\mathbf{p}}^{(+)}(\mathbf{r}, t) = \Psi_{\mathbf{p}}^{(+)} e^{i(\mathbf{p}\cdot\mathbf{r} - E_p t)/\hbar};$$

let us write then

$$\Psi(\mathbf{r}t) = \Psi_{\mathbf{p}}^{(+)}(\mathbf{r}, t) + \Psi_{sc}(\mathbf{r}, t), \tag{22-103}$$

and solve for Ψ_{sc}, the scattered wave, to first order in $e\Phi$. Thus from (22-102)

$$\left(i\hbar \frac{\partial}{\partial t} - H_0\right)\Psi_{sc}(t) = e\Phi\Psi_{\mathbf{p}}^{(+)}(t),$$

---

[2] The fact the potential of Fig. 22-1 extends to infinity modifies this condition, as we have seen.

or

$$i\hbar\frac{\partial}{\partial t}\left[e^{-iH_0t/\hbar}\Psi_{sc}(t)\right] = e^{-iH_0t/\hbar}e\Phi\Psi_{p}^{(+)}(t).$$

Integrating this equation we find

$$\Psi_{sc}(t) = \frac{1}{i\hbar}\int_{-\infty}^{t} dt'\, e^{-iH_0(t-t')/\hbar}e\Phi(t')\Psi_{p}^{(+)}(t'). \tag{22-104}$$

The amplitude for observing the scattered particle in the positive energy state $\Psi_{p'}^{(+)}(\mathbf{r}, t)$ is, from (22-68),

$$u_{p'}(t) = \int d^3r\, \Psi_{p'}^{(+)\dagger}e^{-ip'\cdot\mathbf{r}/\hbar}\tau_3\Psi_{sc}(\mathbf{r}, t).$$

When we substitute (22-104) in this expression the H_0 in Ψ_{sc} acts to the left on $\Psi_{p'}^{(+)\dagger}\tau_3$ and gives simply the eigenvalue $E_{p'}$. Thus

$$u_{p'}(t) = \frac{1}{i\hbar}\int_{-\infty}^{t} dt'\, e^{-iE_{p'}(t-t')/\hbar}e^{-iE_pt'/\hbar}$$
$$\times e\Phi_{p'-p}(t')\,[\Psi_{p'}^{(+)\dagger}\tau_3\Psi_{p}^{(+)}], \tag{22-105}$$

where

$$\Phi_{p'-p}(t') = \int d^3r\, e^{i(p-p')\cdot\mathbf{r}/\hbar}\Phi(\mathbf{r}, t), \tag{22-106}$$

and from (22-59)

$$\Psi_{p'}^{(+)\dagger}\tau_3\Psi_{p}^{(+)} = \frac{E_p + E_{p'}}{2\sqrt{E_pE_{p'}}}. \tag{22-107}$$

Equation (22-105) is of exactly the same form as we find in nonrelativistic first-order time-dependent perturbation theory, with an effective matrix element

$$e\Phi_{p'-p}(t)\,\frac{(E_p + E_{p'})}{2\sqrt{E_pE_{p'}}}. \tag{22-108}$$

Therefore the scattering rate is given by the usual nonrelativistic expression with this matrix element.

Suppose that the potential has only one frequency ω:

$$\Phi_{p'-p}(t) = e^{-i\omega t}\Phi_{p'-p}. \tag{22-109}$$

Then from the golden rule we find that the rate of scattering from positive energy state p to positive energy state p' is

$$\Gamma_{p(+)\to p'(+)} = \frac{2\pi}{\hbar}\left|e\Phi_{p'-p}\right|^2 \frac{(2E_p+\hbar\omega)^2}{4E_p(E_p+\hbar\omega)}\,\delta(\hbar\omega+E_p-E_{p'}). \tag{22-110}$$

To find the scattering cross section we multiply (22-110) by the density of final states $d^3p'/(2\pi\hbar)^3 = p'^2 dE_{p'}d\Omega/[(2\pi\hbar)^3/(dE_{p'}/dp')]$ and divide by the initial flux, $v = dE_p/dp$. Thus integrating over $E_{p'}$ we arrive at

$$\frac{d\sigma}{d\Omega} = \frac{p'^2}{vv'}\,\frac{\left|e\Phi_{p'-p}\right|^2}{4\pi^2\hbar^4}\,\frac{(2E_p+\hbar\omega)^2}{4E_p(E_p+\hbar\omega)}; \tag{22-111}$$

for $\omega = 0$, a static potential, this Born approximation cross section reduces to practically the same form as the nonrelativistic result.

Now let us calculate the amplitude for the scattered wave to have a negative energy component, $\Psi_{p'}^{(-)}e^{-ip'\cdot r/\hbar}$. From (22-68) the complex conjugate of this amplitude is

$$v_{p'}^*(t) = -\int d^3r\,\Psi_{p'}^{(-)\dagger}\,e^{ip'\cdot r/\hbar}\,\tau_3\Psi_{sc}(r,t).$$

Thus as the analog of (22-105) we find

$$v_{p'}^*(t) = \frac{i}{\hbar}\int_{\infty}^{t} dt'\,e^{iE_{p'}(t-t')/\hbar}\,e^{-iE_pt'/\hbar}\,e\Phi_{-p'-p}(t') \times\left[\Psi_{p'}^{(-)\dagger}\tau_3\Psi_p^{(+)}\right], \tag{22-112}$$

where

$$\Psi_{p'}^{(-)\dagger}\tau_3\Psi_p^{(+)} = \frac{(E_p-E_{p'})}{2\sqrt{E_pE_{p'}}}. \tag{22-113}$$

Equation (22-112) is the same as the nonrelativistic expression for a transition from a state with energy E_p to one with energy $-E_{p'}$, connected by an effective matrix element

$$-e\Phi_{-p'-p}(t')\,\frac{(E_p-E_{p'})}{2\sqrt{E_pE_{p'}}}.$$

If the potential contains only one frequency ω the transition rate from the particle state p to an antiparticle state p' is, according to the golden rule:

$$\Gamma_{p(+)\to p'(-)} = \frac{2\pi}{\hbar}\left|e\Phi_{-p'-p}\right|^2 \frac{(2E_p-\omega)^2}{4E_p|E_p+\omega|}\delta(\hbar\omega+E_p+E_{p'}). \qquad (22\text{-}114)$$

Now how can we interpret such a transition physically. We can't argue that the particle simply turns into an antiparticle, since this violates charge conservation; the initial state is normalized to +1, and this normalization is a constant of the motion of the particle. Thus the final state must also have normalization +1; an antiparticle state is normalized to −1. What happens is that as the particle passes through the potential, it causes the creation of particle-antiparticle pairs; and (22-114) is the rate at which antiparticles of momentum p' are made. Notice that (22-114) vanishes unless

$$-\hbar\omega = E_p + E_{p'} \geq 2mc^2, \qquad (22\text{-}115)$$

i.e., unless the potential has sufficiently high frequency to transfer enough energy *to* the system to make a particle-antiparticle pair. (Recall that the negative frequency components of the potential cause transitions that increase the energy of the system.) Similarly the pair creation process occurs only if the potential has the spatial Fourier component $\Phi_{-p-p'}$ necessary for transferring momentum p + p' to the system.

One other subtle point: why are the conservation laws, e.g., (22-115) in the transition rate (22-114) those for creating an antiparticle with momentum p' and a particle with the *same* momentum p as the initial particle. Can't the potential create pairs in which the particle has a momentum other than p? The point is that the potential, all by itself with no incident particle, can produce pairs of all assorted momenta (consistent with energy and momentum conservation). What we have actually calculated in (22-114) is the amount of *extra* pair production due to there being present an incident particle of momentum p. In a sense (22-114) tells us the rate of *stimulated emission.* When we studied radiation we found that the rate of emission into a mode was proportional to the number of photons in the mode, plus one. Because they are bosons the same result is true for our present spinless particles. The rate at which they are created is proportional to the number already present in a given mode, plus one. The potential eΦ in the vacuum produces pairs at a certain rate. Shooting in an incident particle of momentum p increases the number of particles of momentum p by one, and thus increases the rate at which those particular pairs, the particle of which has momentum p, are produced. It is the extra production of these pairs that we have calculated in (22-114).

Since particles are not conserved in potentials that vary very rapidly in time, we no longer have simply the single particle we start out with, but rather a many-particle system. To see that pairs really are created it would be nice to calculate the probability that we observe both a particle at $\mathbf{r}$ and an antiparticle at $\mathbf{r}'$, say. Unfortunately we can't do this in the single particle theory we've developed. All we can calculate are the answers to "single particle questions" such as the probability of observing in this many-particle system a particle at $\mathbf{r}$, measured with respect to this probability in the vacuum; or the probability of observing an antiparticle at $\mathbf{r}'$ (compared with the vacuum). These questions we can answer *correctly* in terms of the wave function $\Psi(\mathbf{r}, t)$ subject, of course, to the limitation that we have no way of including interactions between all the particles and antiparticles.

BOUND STATE PROBLEMS

Let us turn now to consider bound states of spin zero relativistic particles in a static potential $\Phi(\mathbf{r})$. For a bound state of a positive particle with energy E,

$$\psi(\mathbf{r}, t) = e^{-iEt/\hbar}\psi(\mathbf{r}); \tag{22-116}$$

the charge density of the particle in the bound state is therefore, from (22-23),

$$e\rho(\mathbf{r}) = \frac{e[E - e\Phi(\mathbf{r})]}{mc^2}|\psi(\mathbf{r})|^2. \tag{22-117}$$

Thus in regions where $E > e\Phi(\mathbf{r})$, which includes the classically accessible regions, the charge density is positive; however in regions where $E < e\Phi(\mathbf{r})$, the charge density is negative. We can understand this if we think the state of the particle in the potential as a linear combination of free particle and free antiparticle states, as in (22-67). The positively charged components are found primarily in regions of smaller $e\Phi$, while the negatively charged antiparticle components are more likely to be found in regions of larger $e\Phi$, since they have the opposite charge. In this sense a charged relativistic particle has an internal electric structure which can be polarized by an electric field.

Another way of looking at the form (22-117) of the charge distribution of a bound state is to say that the potential produces particle-antiparticle pairs in the vacuum; the positively charged particles are attracted to regions of small $e\Phi$, while the antiparticles are drawn to regions of large $e\Phi$. In other words, the potential produces directly a *polarization of the vacuum.* Now this polarization actually modifies the effective potential felt by the bound particle; however we cannot take this effect into account simply in the present single particle theory.

The experimentally interesting problem of a spin zero particle bound in a Coulomb potential, such as a π^- bound to a nucleus, has the nice feature that it can be solved exactly. From (22-16) the Klein-Gordon equation for the eigenstates of a potential

$$e\Phi(r) = \frac{-Ze^2}{r} \tag{22-118}$$

is

$$\left[\left(E + \frac{Ze^2}{r}\right)^2 + \hbar^2c^2\nabla^2 - m^2c^4\right]\psi(\mathbf{r}) = 0. \tag{22-119}$$

For an eigenstate of total orbital angular momentum l, this becomes

$$\left[\left(\frac{E^2}{c^2} - m^2c^2\right) + \hbar^2\left(\frac{1}{r}\frac{\partial^2}{\partial r^2}r - \frac{l(l+1) - (Z\alpha)^2}{r^2}\right) + \frac{2Ze^2}{r}\frac{E}{c}\right]\psi(\mathbf{r}) = 0 \tag{22-120}$$

where $\alpha = e^2/\hbar c$. If we define

$$m' \equiv \frac{E}{c^2} \tag{22-121}$$

(this is effectively the relativistic mass of the bound particle),

$$l'(l'+1) \equiv l(l+1) - (Z\alpha)^2 \tag{22-122}$$

and

$$\frac{E^2}{c^2} - m^2c^2 \equiv 2m'E', \tag{22-123}$$

then (22-120) can be written

$$\left[2m'E' + \hbar^2\,\frac{1}{r}\frac{\partial^2}{\partial r^2}r - \frac{l'(l'+1)}{r^2} + \frac{2m'Ze^2}{r}\right]\psi(\mathbf{r}) = 0. \tag{22-124}$$

This equation is precisely in the form of the radial equation for the nonrelativistic Coulomb problem. The one difference is that l' is not necessarily an integer as it is in the nonrelativistic problem. This has the consequence that the orbits of the relativistic Coulomb (or Kepler) problem no longer close on themselves, but rather the orbits precess. As we shall see, the extra degeneracy of the nonrelativistic Kepler problem, which arises from the fact that the orbits close, is broken in the relativistic problem.

Now we're actually not without experience in solving the Schrödinger equation (22-124) for noninteger angular momentum. Recall that the scattering amplitude of the nonrelativistic Coulomb problem, for general complex l, takes the form [Eq. (10-30)]

$$f_l \sim e^{2i\delta_l} = \frac{\Gamma(l+1+i\gamma)}{\Gamma(l+1-i\gamma)}, \tag{22-125}$$

where

$$\gamma = \frac{-me^2}{\hbar k}, \qquad k^2 = 2mE. \tag{22-126}$$

The scattering amplitude has (Regge) poles whenever

$$l+1+i\gamma = 0, -1, -2, \ldots, \tag{22-127}$$

and these poles indicate a bound state of the system whenever l takes on a nonnegative integer value, its physical values. Thus to find the bound state energies for the present problem we merely look at the poles of the scattering amplitude at the physically allowed l', the solutions to (22-122). From (22-127) we see that these poles occur whenever

$$l'+1-\frac{im'Ze^2}{\hbar\sqrt{2m'E'}} = 0, -1, -2, \ldots \equiv -\nu. \tag{22-128}$$

Thus the eigenvalues E' are given by

$$\frac{im'Ze^2}{\hbar\sqrt{2m'E'}} = l'+\nu+1 \equiv n' \tag{22-128}$$

or

$$E' = -\frac{m'Z^2e^4}{2\hbar^2 n'^2}; \tag{22-130}$$

this is just the Bohr formula except for the primes. To find the relativistic energies we multiply both sides of (22-130) by $2m'$ and use

(22-121) and (22-123); thus

$$\left(\frac{E^2}{c^2}\right) - m^2c^2 = -\frac{E^2Z^2e^4}{c^4\hbar^2n'^2}$$

and solving:

$$E = \frac{mc^2}{[1+(Z\alpha/n')^2]^{1/2}}. \qquad (22\text{-}131)$$

It remains to solve for n' or l' in terms of l.

Solving (22-122) by adding $1/4$ onto both sides to complete the square, we find two solutions

$$l' = -\frac{1}{2} \pm \sqrt{(l+1/2)^2 - (Z\alpha)^2} \equiv l'_\pm. \qquad (22\text{-}132)$$

In order to have real l' and therefore real energy we need $(l + 1/2)^2 > (Z\alpha)^2$; for s states this is the condition that

$$Z < \frac{1}{2\alpha} = \frac{137}{2}. \qquad (22\text{-}133)$$

Let us assume that this is so. (We shall return to this point shortly.) Now we shall show that only the upper sign in (22-132) is physically allowable.

To see this we must look at the solutions to (22-124) for small distances. The general solutions are simply related to *Whittaker functions,* and like the nonrelativistic wave functions they are of the form, for angular momentum l, m:

$$\psi_{nlm}(r) \sim Y_{lm}(\Omega) r^{l'} e^{-Zr/a_0'n'} \quad \text{(times a real polynomial of order } \nu) \qquad (22\text{-}134)$$

where

$$a_0' = \frac{\hbar^2}{m'e^2}. \qquad (22\text{-}135)$$

Without the primes (22-134) is the nonrelativistic result. At large distances

$$\psi_{nlm}(r) \sim r^{n'-1} e^{-Zr/a_0'n'} \qquad (22\text{-}136)$$

while at small r,

$$\psi_{nlm}(r) \sim r^{l'}. \tag{22-137}$$

Now to see that only the upper solution l'_+ is possible we note that for $l \geq 1$, $l'_- < -1$ and consequently the solution ψ_{nlm} is not normalizable at the origin. This argument doesn't rule out the lower solution when $l = 0$. To rule this out we suppose (for any l) that both l'_+ and l'_- are possible solutions. Then the state

$$\psi_{l'_+} + i\psi_{l'_-} \tag{22-138}$$

is a possible state of the system, where $\psi_{l'_\pm}$ are *real* solutions of the form (22-134). From (22-20) we find that the radial current in the state (22-138) is

$$j_r \sim \psi_{l'_+} \frac{\partial}{\partial r} \psi_{l'_-} - \psi_{l'_-} \frac{\partial}{\partial r} \psi_{l'_+}$$

and for small r is

$$\frac{\sim r^{(l'_+)} r^{(l'_-)}}{r} = \frac{1}{r^2}. \tag{22-139}$$

This is a very divergent radial current; in particular, integrating (22-139) over the surface of a small sphere around the origin, we find

$$\int 4\pi r^2 \, d\Omega \, j_r = \text{constant},$$

independent of the radius of the sphere. This is only possible if there is a current source (or sink) at the origin. Thus, in order that one have well-behaved currents at the origin we must exclude the lower of the signs in (22-132), as the upper solution is just the continuation of the nonrelativistic ($Z\alpha \to 0$) value. Thus we have

$$l' = -\frac{1}{2} + \sqrt{(l+1/2)^2 - (Z\alpha)^2}, \tag{22-140}$$

and

$$\begin{aligned} n' = 1 + \nu + l' &= \nu - \frac{1}{2} + \sqrt{(l+1/2)^2 - (Z\alpha)^2} \\ &\equiv n - (l+1/2) + \sqrt{(l+1/2)^2 - (Z\alpha)^2}\,; \end{aligned} \tag{22-141}$$

the integer n is the same as the nonrelativistic principal quantum number; $n = l + 1, \; l + 2, \; \ldots$.

Expanding (22-131) for $Z\alpha \ll 1$ we find

$$E = mc^2 - \frac{mZ^2e^4}{2\hbar^2 n^2}\left[1 + \frac{Z^2\alpha^2}{n^2}\left(\frac{n}{l+1/2} - \frac{3}{4}\right)\right], \qquad (22\text{-}142)$$

to fourth order in α. The first term is the rest energy, the second the nonrelativistic Bohr formula, while the third is relativistic *fine structure.* Notice that relativistic effects remove the degeneracy l and always lower the energy.[3]

Let us return to the condition $Z\alpha < l + 1/2$. Essentially what happens for $Z\alpha > l + 1/2$ is that the effective centrifugal potential term $[l(l+1) - (Z\alpha)^2]/r^2$ is so attractive at small r that the particle tumbles into the origin. The increase in its kinetic energy as it does this is not sufficient to overcome the attraction. Thus in order to discuss systems with $Z\alpha > 1/2$, which certainly exist, one must include, for states with $l < Z\alpha - 1/2$, more details of the structure of the potential near the origin, taking into account, e.g., the finite size of the nucleus, and vacuum polarization effects. The condition $Z\alpha < 1/2$ can be equivalently written $Ze^2/(\hbar/mc) < mc^2/2$. This says that the potential should be weaker in magnitude than $mc^2/2$ at a distance of a Compton wavelength from the origin.

NONRELATIVISTIC LIMIT

Let us now work out the nonrelativistic limit of the Klein-Gordon theory, and calculate the lowest-order relativistic corrections to the Schrödinger equation. For a particle solution in the nonrelativistic limit, we expect $\Psi(\mathbf{r}, t)$ to vary in time as $e^{-imc^2t/\hbar}$ times a function varying slowly (compared with frequency $mc^2/\hbar$) in time. Thus to lowest order in v/c

$$i\hbar\frac{\partial\chi}{\partial t} = mc^2\chi. \qquad (22\text{-}143)$$

[3] The experimental situation on the spectra of π mesic atoms is described by D.A. Jenkins and R. Kunselman, *Phys. Rev. Letters* 17, 1148 (1966). Relativistic effects are on the order of 1% typically, but one must include vacuum polarization corrections (~½%) to get agreement with experiment. It is interesting to note that Eq. (22-11) with a Coulomb potential yields a fine structure that is, within present experimental error, the same as Eq. (22-131).

$$(2mc^2 + \cdots)\chi = -\frac{1}{2m}\left[\frac{\hbar}{i}\nabla - \frac{e}{c}\mathbf{A}\right]^2 \varphi, \tag{22-144}$$

where the dots signify terms small compared with mc^2. Hence in the nonrelativistic limit

$$\chi = -\frac{1}{4m^2c^2}\left[\frac{\hbar}{i}\nabla - \frac{e}{c}\mathbf{A}\right]^2 \varphi, \tag{22-145}$$

i.e., $\chi \sim (v/c)^2\varphi$ for a positive energy solution. This agrees with the result (22-62) for slow free particles. To lowest order in v/c we can then take $\chi = 0$ in the upper of Eqs. (22-41); the result is the nonrelativistic Schrödinger equation for φ, with the constant mc^2 included in the energy. To find the first relativistic corrections we substitute (22-145) for χ in the first of Eqs. (22-40). Thus

$$\left(i\hbar\frac{\partial}{\partial t} - e\Phi\right)\varphi = \left[mc^2 + \frac{1}{2m}\left(\frac{\hbar}{i}\nabla - \frac{e}{c}\mathbf{A}\right)^2 - \frac{1}{8m^3c^2}\left(\frac{\hbar}{i}\nabla - \frac{e}{c}\mathbf{A}\right)^4\right]\varphi. \tag{22-146}$$

Notice that the operator on the right side is just the "kinetic energy operator"

$$\sqrt{m^2c^4 + c^2\left(\frac{\hbar}{i}\nabla - \frac{e}{c}\mathbf{A}\right)^2}$$

expanded to second order in $1/mc^2$. Thus the first relativistic corrections for a spinless particle are due entirely to the relativistic modification of the kinetic energy.

For a weak uniform magnetic field, $\mathcal{H}$; Eq. (22-146) becomes, to order $(v/c)^3$:

$$i\hbar\frac{\partial\varphi}{\partial t} = -\frac{\hbar^2\nabla^2}{2m}\left[1 + \frac{\hbar^2\nabla^2}{4m^2c^2}\right]\varphi + (mc^2 + e\Phi)\varphi - \frac{e\mathcal{H}}{2mc}\cdot\mathbf{L}\left(1 + \frac{\hbar^2\nabla^2}{2m^2c^2}\right)\varphi, \tag{22-147}$$

where $\mathbf{L}$ is the angular momentum of the particle. The effective (orbital) magnetic moment operator for the particle is thus

$$\frac{e\mathbf{L}}{2mc}\left(1 - \frac{p^2}{2m^2c^2}\right); \tag{22-148}$$

relativistic effects *reduce* the magnetic moment by a term of relative order $\sim (v/c)^2$; to this order, the reduction is due simply to the relativistic increase in the particle mass.

SCALAR INTERACTIONS

Up to now we have considered only the coupling of particles to the electromagnetic field. This interaction distinguishes particles from antiparticles. There is another type of interaction which plays an important role in elementary particle physics, and that is a *scalar interaction* which treats particles and antiparticles alike. Suppose that nonrelativistically we have a potential energy $S(\mathbf{r}, t)$ that is the same for particles and antiparticles. The Schrödinger equation with this potential is

$$i\hbar \frac{\partial \psi}{\partial t} = \left[-\frac{\hbar^2 \nabla^2}{2m} + S(\mathbf{r}, t) \right] \psi(\mathbf{r}, t). \qquad (22\text{-}149)$$

We would like to be able to include such a potential in the relativistic wave equation.

One possible way, which as we shall see is a relativistic generalization of (22-149), is the following. Suppose that $S(\mathbf{r}, t)$ is a real scalar function, i.e., its value at a given space-time point is the same in all Lorentz frames. Then if we add $2mS(\mathbf{r}, t)$ onto m^2c^2 in the Klein-Gordon equation (22-14):

$$-\frac{\hbar^2}{c^2} \frac{\partial^2}{\partial t^2} \psi(\mathbf{r}, t) = [-\hbar^2 \nabla^2 + m^2c^2 + 2mS(\mathbf{r}t)]\psi(\mathbf{r}t) \qquad (22\text{-}150)$$

we find that $\psi(\mathbf{r}, t)$ is still a scalar under a Lorentz transformation. Equation (22-150) is the wave equation corresponding to a modified energy momentum relation

$$E^2 = c^2p^2 + m^2c^4 + 2mS(\mathbf{r}, t); \qquad (22\text{-}151)$$

in the nonrelativistic limit, (22-151) reduces to

$$E = mc^2 + S(\mathbf{r}t) + \frac{p^2}{2m}. \qquad (22\text{-}152)$$

Since $S(\mathbf{r}t)$ is real, both ψ and ψ^* are solutions to (22-150); thus antiparticles behave the same in the potential S as particles.

If S is independent of time then the Klein-Gordon equation for the stationary states becomes

$$\frac{E^2 - m^2c^4}{2mc^2} \psi(\mathbf{r}) = \left[-\frac{\hbar^2 \nabla^2}{2m} + S(\mathbf{r}) \right] \psi(\mathbf{r}). \qquad (22\text{-}153)$$

This equation is the relativistic generalization of the Schrödinger equation in the potential S, since in the nonrelativistic limit $E \approx mc^2$, and thus

$$\frac{E^2 - m^2c^4}{2mc^2} = \frac{(E - mc^2)(E + mc^2)}{2mc^2} \approx E - mc^2. \qquad (22\text{-}154)$$

The fact that (22-153) depends only on E^2 means that if E is an eigenvalue, so is $-E$. There is a symmetry between positive and negative energy solutions because the scalar potential S doesn't distinguish particles and antiparticles.

Solving (22-153) is equivalent to solving the nonrelativistic equation

$$E'\psi(\mathbf{r}) = \left[-\frac{\hbar^2\nabla^2}{2m} + S(\mathbf{r})\right]\psi(\mathbf{r}). \qquad (22\text{-}155)$$

The eigenvalues of this equation are related obviously to those of (22-153) by

$$E^2 = m^2c^4 - 2mc^2E'.$$

In order that E^2 be positive, only the eigenvalues of (22-155) that satisfy

$$E' > \frac{-mc^2}{2} \qquad (22\text{-}156)$$

correspond to physical solutions of the relativistic equation (22-153).

NEUTRAL PARTICLES

We have always interpreted $\rho(\mathbf{r}, t)$ as a charge density. How do neutral particles fit into the theory? There are two possible cases to consider. First, suppose that the particle is different from its antiparticle, e.g., a K^0 meson, with strangeness 1, whose antiparticle, the $\overline{K}^0$, has strangeness -1. The statement that the particle and antiparticle are different particles means that there exist interactions that distinguish them. Then we can always define a charge, though not electromagnetic, that is +1 for the particles and -1 for the antiparticles, and $\rho(\mathbf{r}, t)$ can be interpreted as a density of this charge. For K^0 mesons this charge is simply their strangeness.

Second, there is the possibility that a particle is its own antiparticle. This means that there are no interactions in nature that distinguish the particle from its antiparticle. An example is the π^0

meson. [The π^+ and π^- are antiparticles.] Then there must be a complete symmetry between positive and negative energy solutions of the Klein-Gordon equation, as in the case of the scalar coupling S. If $\psi(\mathbf{r}, t)$ satisfies the Klein-Gordon equation so must $\psi^*(\mathbf{r}, t)$, the "charge conjugate" wave function. Thus the wave function can always be chosen to be *real*, and for a real wave function the density $\rho(\mathbf{r}, t)$ is zero. This situation is analogous to the description of the photon, a neutral spin one particle, by the electromagnetic field, a real field. The physical interpretation of the solution of the Klein-Gordon equation in terms of particles is quite analogous to the interpretation of the solutions of Maxwell's equations in terms of photons.

PROBLEMS

1. Show quite generally, not assuming free particles, that under a Galilean transformation, $\mathbf{r}' = \mathbf{r} - \mathbf{v}t$, the solution $\psi'(\mathbf{r}', t)$ to the nonrelativistic Schrödinger equation in the new frame is related to the solution $\psi(\mathbf{r}, t)$ in the original frame by

$$\psi'(\mathbf{r}', t) = e^{-i(m\mathbf{v}\cdot\mathbf{r} - mv^2t/2)/\hbar}\,\psi(\mathbf{r}, t)$$

   Why, in contrast to the relativistic case for $v \ll c$, doesn't ψ transform as a scalar?
2. A π^+ meson is in a stationary state in a uniform magnetic field $\mathcal{H}$ along the z direction. Its momentum along the field is zero while its angular momentum along the field is $+\hbar$. At $t = 0$ the magnetic field is *suddenly* switched off. Calculate the distribution in angles and energy of π^- that are produced when the field is switched off. (*Hint:* In the gauge $\mathbf{A} = (\mathcal{H} \times \mathbf{r})/2$ the Klein-Gordon equation for ψ reduces to a harmonic oscillator equation.)
3. Calculate the gyromagnetic ratios, or Landé g factors, of the eigenstates of the relativistic spin zero Coulomb problem.
4. If the Coulomb potential were a scalar how would the fine structure of the bound states differ from (22-42)? Are there any restrictions on $Z\alpha$ in this case?

Chapter 23
RELATIVISTIC SPIN 1/2 PARTICLES: DIRAC EQUATION

We now want to study the behavior of particles with spin $\frac{1}{2}$ traveling at relativistic velocities. Let us think then for a few moments about the properties of spin under Lorentz transformations.[1]

LORENTZ TRANSFORMATION OF SPIN

Spin is an internal angular momentum, and as such, it must have the transformation properties of an angular momentum. What are these properties? Nonrelativistically we conceive of angular momentum as a vector, and indeed, under a simple spatial rotation it does transform as a vector. But let us consider the behavior of an orbital angular momentum.

$$\mathbf{L} = \mathbf{r} \times \mathbf{p}, \qquad L_i = \varepsilon_{ijk} r_j p_k \tag{23-1}$$

under a Lorentz transformation. If the transformation velocity is in the z direction then L_z' $(= x'p_y' - y'p_x')$, the angular momentum along z in the new frame, equals L_z, since x, y, p_x, and p_y are unchanged in the transformation. This is definitely not the transformation property of a vector; in fact, $\mathbf{L}$, as the product of two vectors, has the transformation properties of a *second-rank tensor.* So then must spin transform as a second-rank tensor.

One of the most familiar examples of a second-rank tensor is $F^{\mu\nu}$, the tensor formed of the electric and magnetic fields:

---

[1]See again the Feshbach and Villars review article, *Rev. Mod. Phys.* **30**, 24 (1958).

$$F^{\mu\nu} = \begin{pmatrix} 0 & \mathcal{E}_1 & \mathcal{E}_2 & \mathcal{E}_3 \\ -\mathcal{E}_1 & 0 & \mathcal{H}_3 & -\mathcal{H}_2 \\ -\mathcal{E}_2 & -\mathcal{H}_3 & 0 & \mathcal{H}_1 \\ -\mathcal{E}_3 & \mathcal{H}_2 & -\mathcal{H}_1 & 0 \end{pmatrix}. \tag{23-2}$$

[We shall use indices x, y, z and 1, 2, 3 interchangeably.] Under a Lorentz transformation, the fields in the new frame, denoted by primes, are given by

$$\mathcal{H}_{\parallel}' = \mathcal{H}_{\parallel}, \qquad \mathcal{E}_{\parallel}' = \mathcal{E}_{\parallel}$$

$$\mathcal{H}_{\perp}' = \frac{\mathcal{H}_{\perp} - (\mathbf{v}/c) \times \mathcal{E}}{\sqrt{1 - v^2/c^2}}, \qquad \mathcal{E}_{\perp}' = \frac{\mathcal{E}_{\perp} + (\mathbf{v}/c) \times \mathcal{H}}{\sqrt{1 - v^2/c^2}}, \tag{23-3}$$

where $\parallel$ denotes the component parallel to $\mathbf{v}$ and $\perp$ denotes the component perpendicular to $\mathbf{v}$. A pure magnetic field in one frame is a mixture of magnetic and electric fields in a new frame.

Under spatial inversion, $\mathcal{E} \to -\mathcal{E}$ but $\mathcal{H} \to \mathcal{H}$. Also $\mathbf{L} \to \mathbf{L}$ since $\mathbf{r} \to -\mathbf{r}$ and $\mathbf{p} \to -\mathbf{p}$. Thus $\mathbf{L}$, and the spin, have the same transformation properties as $\mathcal{H}$. What is the analog of $\mathcal{E}$? For the orbital angular momentum, the analog is the vector $-c\mathbf{p}t - \mathbf{r}E/c$, where E is the energy of the particle.

Since spin, $\mathbf{S} \equiv \boldsymbol{\sigma}/2$, must transform as an angular momentum we conclude that there must be another set of dynamical variables referring to the internal degrees of freedom of the particle that are analogous to $\mathcal{E}$. Let us call these variables $i\boldsymbol{\alpha}/2$. [The i/2 is simply for future convenience.] Then $\mathbf{S}$ and $i\boldsymbol{\alpha}/2$ or $\boldsymbol{\sigma}$ and $i\boldsymbol{\alpha}$ transform as $\mathcal{H}$ and $\mathcal{E}$ in (23-3):

$$\sigma_{\parallel}' = \sigma_{\parallel}, \qquad i\alpha_{\parallel}' = i\alpha_{\parallel}$$

$$\sigma_{\perp}' = \frac{\sigma_{\perp} - (\mathbf{v}/c) \times i\alpha}{\sqrt{1 - v^2/c^2}}, \qquad i\alpha_{\perp}' = \frac{i\alpha_{\perp} + (\mathbf{v}/c) \times \sigma}{\sqrt{1 - v^2/c^2}}. \tag{23-4}$$

The quantities $\boldsymbol{\sigma}$ and $\boldsymbol{\alpha}$ form a second-rank tensor $\sigma^{\mu\nu}$ analogous to $F^{\mu\nu}$:

$$\sigma^{\mu\nu} = \begin{pmatrix} 0 & i\alpha_1 & i\alpha_2 & i\alpha_3 \\ -i\alpha_1 & 0 & \sigma_3 & -\sigma_2 \\ -i\alpha_2 & -\sigma_3 & 0 & \sigma_1 \\ -i\alpha_3 & \sigma_2 & -\sigma_1 & 0 \end{pmatrix}. \tag{23-5}$$

What are the properties of these new dynamical variables $\boldsymbol{\alpha}$ (and where have they been all along)? We shall find that in the nonrelativistic limit, the expectation value of $\boldsymbol{\alpha}$ is $\sim v/c$, which explains why we've never seen them before. [Actually we have seen their effects in the spin-orbit interaction.]

We can discover the algebraic properties of $\boldsymbol{\alpha}$ by making use of the algebraic properties of the spin, and the fact that the spin generates rotations of the internal degrees of freedom. Since the spin commutes with the spatial degrees of freedom, $\mathbf{p}$, $\mathbf{r}$, etc., so must $\boldsymbol{\alpha}$. Also, since $\boldsymbol{\alpha}$ behaves as a vector under spatial rotations, it must have the commutation relation with $\mathbf{S}$:

$$\alpha_i S_j - S_j \alpha_i = i\varepsilon_{ijk}\alpha_k, \tag{23-6}$$

or

$$\alpha_i \sigma_j - \sigma_j \alpha_i = 2i\varepsilon_{ijk}\alpha_k. \tag{23-7}$$

We have not yet used the fact that $\mathbf{S}$ is a spin $1/2$. Equations (23-4) and (23-6) are valid for any spin; for spin zero, $\boldsymbol{\alpha} = 0$. The peculiar property of spin $1/2$ is that $\boldsymbol{\sigma}$ obeys the relations

$$\sigma_i \sigma_j = i\varepsilon_{ijk}\sigma_k + \delta_{ij}. \tag{23-8}$$

[*Do not* think of the *operator* $\boldsymbol{\sigma}$ as the 2×2 Pauli matrices; we shall discover that, in fact, $\boldsymbol{\sigma}$ will have to be represented by 4×4 matrices relativistically.] Now the point is that (23-8) must be true for the spin operator in all Lorentz frames. This enables us to determine the properties of $\boldsymbol{\alpha}$. For example, since $\sigma_i^2 = 1$, we have $\sigma_x'^2 = 1$. Thus taking $\mathbf{v}$ along $\hat{\mathbf{z}}$ in (23-4) we find

$$\sigma_x' = \frac{\sigma_x + iv\alpha_y/c}{\sqrt{1 - v^2/c^2}}, \tag{23-9}$$

$$\sigma_y' = \frac{\sigma_y - iv\alpha_x/c}{\sqrt{1 - v^2/c^2}} \tag{23-10}$$

and squaring σ_x':

$$\sigma_x'^2 = 1 = \frac{\sigma_x^2 + i(v/c)(\sigma_x\alpha_y + \alpha_y\sigma_x) - (v/c)^2\alpha_y^2}{1 - v^2/c^2}.$$

Since this is true for *all* v, the coefficient of v/c on top must vanish; $\sigma_x\alpha_y + \alpha_y\sigma_x = 0$, and also $\alpha_y^2 = 1$. Thus we have in general that if $i \neq j$,

$$\sigma_i \alpha_j = -\alpha_j \sigma_i \tag{23-11}$$

while from (23-7), if i = j,

$$[\sigma_i, \alpha_i] = 0. \tag{23-12}$$

Next, multiplying (23-10) on the right by σ_z', equal to σ_z, we find

$$\sigma_y' \sigma_z' = i\sigma_x' = \frac{i\sigma_x - iv\alpha_x\sigma_z/c}{\sqrt{1 - v^2/c^2}}.$$

Comparing with (23-9) we conclude that

$$\alpha_x \sigma_z = -i\alpha_y. \tag{23-13}$$

Multiplying (23-9) on the right by σ_z' implies similarly

$$\alpha_y \sigma_z = i\alpha_x. \tag{23-14}$$

Thus in general

$$\alpha_i \sigma_j = i\varepsilon_{ijk}\alpha_k, \quad i \neq j. \tag{23-15}$$

Lastly, let us multiply σ_x' by σ_y':

$$\sigma_x' \sigma_y' = i\sigma_z' = \frac{(\sigma_x + iv\alpha_y/c)(\sigma_y - iv\alpha_x/c)}{1 - v^2/c^2}. \tag{23-16}$$

Since $\sigma_z' = \sigma_z$ we find on comparing coefficients of v^2/c^2 in numerator and denominator that

$$\alpha_y \alpha_x = -i\sigma_z. \tag{23-17}$$

Similarly, from the product $\sigma_y' \sigma_x'$ we find

$$\alpha_x \alpha_y = i\sigma_z. \tag{23-18}$$

Adding (23-17) and (23-18) we then see that the different components of α anticommute:

$$\alpha_i \alpha_j + \alpha_j \alpha_i = 0, \quad i \neq j; \tag{23-19}$$

and subtracting (23-17) from (23-18) we find

$$\alpha_i \alpha_j - \alpha_j \alpha_i = 2i\varepsilon_{ijk}\sigma_k. \tag{23-20}$$

We may then summarize the algebraic relations of the dynamic variable α associated with a spin $1/2$ as follows:

$$\alpha_i\alpha_j + \alpha_j\alpha_i = 2\delta_{ij}$$

$$\alpha_i\alpha_j - \alpha_j\alpha_i = 2i\varepsilon_{ijk}\sigma_k$$

$$\alpha_i\sigma_j - \sigma_j\alpha_i = 2i\varepsilon_{ijk}\alpha_k$$

$$\alpha_i\sigma_j + \sigma_j\alpha_i = 0, \quad i \neq j. \tag{23-21}$$

$\boldsymbol{\alpha}$ obeys exactly the same relations as $\boldsymbol{\sigma}$. How do we know that $\boldsymbol{\alpha}$ isn't equal to $\boldsymbol{\sigma}$? The answer is that under the parity operation we have $\boldsymbol{\sigma} \rightarrow \boldsymbol{\sigma}$, since an angular momentum is unchanged by spatial inversion; the space-space components of a second-rank tensor do not change sign under parity. However, the time-space components, such as the electric field or $i\boldsymbol{\alpha}$ do change sign: $\boldsymbol{\alpha} \rightarrow -\boldsymbol{\alpha}$. Thus $\boldsymbol{\alpha}$ and $\boldsymbol{\sigma}$ behave differently under parity.

Let β be the operator that carries out the parity operation in the spin space. Two inversions bring one back to the original coordinate system; however, since the spin representation of rotations is double valued we have a choice of letting the square of the parity operation include a rotation by 2π about some axis or not. Then we can have either $\beta^2 = +1$ or -1. In the first case the eigenvalues of β are ± 1 and in the second $\pm i$. The only important feature is that the ratio of the two eigenvalues in each case is -1; this shall become the statement that the relative intrinsic parity of a particle and an antiparticle is -1. For convenience we choose

$$\beta^2 = 1, \quad \text{or} \quad \beta^{-1} = \beta. \tag{23-22}$$

Since $\boldsymbol{\sigma}$ doesn't change sign under parity,

$$\beta^{-1}\boldsymbol{\sigma}\beta = \boldsymbol{\sigma}, \quad \text{or} \quad \beta\boldsymbol{\sigma} = \boldsymbol{\sigma}\beta, \tag{23-23}$$

while $\beta^{-1}\boldsymbol{\alpha}\beta = -\boldsymbol{\alpha}$, so that

$$\beta\boldsymbol{\alpha} = -\boldsymbol{\alpha}\beta. \tag{23-24}$$

Let us try to construct an explicit matrix representation of the operators $\boldsymbol{\alpha}$, β, and $\boldsymbol{\sigma}$, analogous to the nonrelativistic representation of the spin by the 2×2 Pauli matrices. It is easy to see that such matrices must have even dimension; consider the determinant of the matrix representing $\beta^{-1}\alpha_i\beta$: from (23-22) and (23-24) we have on the one hand

$$\det \beta^{-1}\alpha_i\beta = \det(-\alpha_i) = (-1)^N \det \alpha_i \tag{23-25}$$

where N is the dimensionality of the representation. On the other hand

$$\det \beta^{-1}\alpha_i\beta = \det \beta^{-1}\beta\alpha_i = \det \alpha_i, \tag{23-26}$$

thus comparing (23-25) and (23-26) we see that N must be even [det α_i can't vanish since $\alpha_i^2 = 1$]:

$$N = 2, 4, \ldots .$$

However N = 2 is impossible; one can write any 2 × 2 matrix as a linear combination of the three 2 × 2 spin matrices and the unit matrix. Thus from (23-23) β would have to commute with all 2 × 2 matrices, so it would be impossible to construct $\boldsymbol{\alpha}$ to satisfy (23-24).

The smallest dimension that the matrices representing $\boldsymbol{\alpha}$, β, and $\boldsymbol{\sigma}$ can have is four. That means relativistically a spin $\frac{1}{2}$ particle must have no fewer than *four* internal states — twice as many as nonrelativistically. This doubling of the number of internal states is similar to the phenomenon that occurred in the Klein-Gordon equation for a spin zero particle, and as there, the doubling means that the theory will encompass both particles and antiparticles at once.

One explicit 4×4 representation for the internal operators is the following

$$\sigma_x = \begin{pmatrix} 0 & 1 & 0 & 0 \\ 1 & 0 & 0 & 0 \\ 0 & 0 & 0 & 1 \\ 0 & 0 & 1 & 0 \end{pmatrix} \qquad \sigma_y = \begin{pmatrix} 0 & -i & 0 & 0 \\ i & 0 & 0 & 0 \\ 0 & 0 & 0 & -i \\ 0 & 0 & i & 0 \end{pmatrix}$$

$$\sigma_z = \begin{pmatrix} 1 & 0 & 0 & 0 \\ 0 & -1 & 0 & 0 \\ 0 & 0 & 1 & 0 \\ 0 & 0 & 0 & -1 \end{pmatrix} \qquad \beta = \begin{pmatrix} 1 & 0 & 0 & 0 \\ 0 & 1 & 0 & 0 \\ 0 & 0 & -1 & 0 \\ 0 & 0 & 0 & -1 \end{pmatrix}$$

$$\alpha_x = \begin{pmatrix} 0 & 0 & 0 & 1 \\ 0 & 0 & 1 & 0 \\ 0 & 1 & 0 & 0 \\ 1 & 0 & 0 & 0 \end{pmatrix} \qquad \alpha_y = \begin{pmatrix} 0 & 0 & 0 & -i \\ 0 & 0 & i & 0 \\ 0 & -i & 0 & 0 \\ i & 0 & 0 & 0 \end{pmatrix}$$

$$\alpha_z = \begin{pmatrix} 0 & 0 & 1 & 0 \\ 0 & 0 & 0 & -1 \\ 1 & 0 & 0 & 0 \\ 0 & -1 & 0 & 0 \end{pmatrix} \tag{23-27}$$

It is left as an exercise to check that these matrices have all the algebraic properties of $\boldsymbol{\alpha}$, β, and $\boldsymbol{\sigma}$. Notice that the matrices (23-27) are all Hermitian. Though this representation is by no means the only such representation of $\boldsymbol{\alpha}$, β, and $\boldsymbol{\sigma}$, we shall, for convenience, call it the "standard representation." An example of a representation with $\beta^2 = -1$ is (27-22) with β multiplied by i.

One can write the matrices (23-27) in a shorthand form as

$$\boldsymbol{\sigma} = \begin{pmatrix} \boldsymbol{\tau} & 0 \\ 0 & \boldsymbol{\tau} \end{pmatrix}, \quad \boldsymbol{\alpha} = \begin{pmatrix} 0 & \boldsymbol{\tau} \\ \boldsymbol{\tau} & 0 \end{pmatrix}, \quad \beta = \begin{pmatrix} 1 & 0 \\ 0 & -1 \end{pmatrix}, \tag{23-28}$$

where the elements stand for 2×2 matrices; 1 is the 2×2 unit matrix, while the τ's in the matrices are the 2×2 Pauli matrices. It is easy to check the commutation properties of the $\boldsymbol{\sigma}$, $\boldsymbol{\alpha}$, and β from those of the τ's.

Notice that the trace of each of the matrices (23-27) is zero. This is a general property of matrices that obey anticommutation relations. For example,

$$\operatorname{tr}\beta\alpha_i\beta = \operatorname{tr}(-\alpha_i) = -\operatorname{tr}\alpha_i$$

$$\operatorname{tr}\beta\alpha_i\beta = \operatorname{tr}\beta\beta\alpha_i = \operatorname{tr}\alpha_i;$$

thus $\operatorname{tr}\alpha_i = 0$.

The tensor $\sigma^{\mu\nu}$, Eq. (23-5), has a simple physical interpretation. Exactly as the space-space components, the spin operators, generate in the spin degrees of freedom a rotation of the coordinate system, the $i\alpha$'s along the border of $\sigma^{\mu\nu}$ generate a rotation of the space axes with the time axis — a Lorentz transformation. In other words, in the same way that the operators $\boldsymbol{\sigma}'$, $\boldsymbol{\alpha}'$, β' in a spatially rotated frame are given by

$$\boldsymbol{\sigma}' = R_\phi \boldsymbol{\sigma} R_\phi^{-1}, \quad \boldsymbol{\alpha}' = R_\phi \boldsymbol{\alpha} R_\phi^{-1}, \quad \beta' = R_\phi \beta R_\phi^{-1} = \beta \tag{23-29}$$

where

$$R_\phi = e^{-i\boldsymbol{\sigma}\cdot\boldsymbol{\phi}/2}, \tag{23-30}$$

and $\boldsymbol{\phi}$ is the vector that denotes the rotation, the operators $\boldsymbol{\sigma}'$, $\boldsymbol{\alpha}'$, β' in a Lorentz transformed frame are given by expressions of the form

$$\boldsymbol{\sigma}' = \mathcal{L}_V \boldsymbol{\sigma} \mathcal{L}_V^{-1}, \quad \boldsymbol{\alpha}' = \mathcal{L}_V \boldsymbol{\alpha} \mathcal{L}_V^{-1}, \quad \beta' = \mathcal{L}_V \beta \mathcal{L}_V^{-1}. \tag{23-31}$$

The Lorentz transformation operator is given by

$$\mathcal{L}_V = e^{-i(i\boldsymbol{\alpha})\cdot\boldsymbol{\omega}/2} = e^{\boldsymbol{\alpha}\cdot\boldsymbol{\omega}/2}, \quad \mathcal{L}_V^{-1} = e^{-\boldsymbol{\alpha}\cdot\boldsymbol{\omega}/2}, \tag{23-32}$$

where ω is a vector in the direction of the velocity $\mathbf{v}$ of the primed frame with respect to the unprimed frame, and of magnitude given by

$$\tanh \omega = \frac{v}{c}. \tag{23-33}$$

We can check that $\mathcal{L}$ generates a Lorentz transformation by comparing (23-31) with (23-4). For example,

$$\sigma_{\parallel}' = \mathcal{L}\sigma_{\parallel}\mathcal{L}^{-1} = \mathcal{L}\mathcal{L}^{-1}\sigma_{\parallel} = \sigma_{\parallel} \tag{23-34}$$

as in (23-4). Also

$$\boldsymbol{\sigma}_{\perp}' = e^{\boldsymbol{\alpha}\cdot\boldsymbol{\omega}/2}\sigma_{\perp}e^{-\boldsymbol{\alpha}\cdot\boldsymbol{\omega}/2} = e^{\boldsymbol{\alpha}\cdot\boldsymbol{\omega}}\sigma_{\perp}, \tag{23-35}$$

since $\boldsymbol{\sigma}_{\perp}$ anticommutes with $\alpha_{\parallel}$ and therefore bringing the $\boldsymbol{\sigma}_{\perp}$ through to the right changes the sign of the exponent. Now we can evaluate $e^{\boldsymbol{\alpha}\cdot\boldsymbol{\omega}}\sigma_{\perp}$ by using the relation

$$e^{\boldsymbol{\alpha}\cdot\boldsymbol{\omega}} = \cosh \omega + \boldsymbol{\alpha}\cdot\hat{\omega} \sinh \omega. \tag{23-36}$$

This expression is the analog of $e^{i\boldsymbol{\sigma}\cdot\boldsymbol{\phi}} = \cos \phi + i\boldsymbol{\sigma}\cdot\hat{\boldsymbol{\phi}} \sin \phi$ and is proven by expanding the exponential in a power series and using the fact that $(\boldsymbol{\alpha}\cdot\hat{\boldsymbol{\omega}})^2 = 1$. Thus from (23-4)

$$\boldsymbol{\sigma}_{\perp}' = \cosh \omega\,[1 + \boldsymbol{\alpha}\cdot\hat{\boldsymbol{\omega}} \tanh \omega]\boldsymbol{\sigma}_{\perp} = \frac{\boldsymbol{\sigma}_{\perp} + (\boldsymbol{\alpha}\cdot\mathbf{v}/c)\boldsymbol{\sigma}_{\perp}}{\sqrt{1 - v^2/c^2}},$$

since from (23-33), $\cosh \omega = (1 - v^2/c^2)^{-1/2}$. Using (23-15) and a little vector algebra we have

$$(\boldsymbol{\alpha}\cdot\mathbf{v}/c)\boldsymbol{\sigma}_{\perp} = -i\frac{\mathbf{v}}{c} \times \boldsymbol{\alpha}, \tag{23-37}$$

and therefore (23-32) transforms $\boldsymbol{\sigma}$ correctly. By a similar calculation one can check that $\mathcal{L}$ induces the correct transformation of $\boldsymbol{\alpha}_{\perp}$.

We should mention that by methods similar to those that we've used to find the properties of a relativistic spin $1/2$, and its associated dynamical variables $\boldsymbol{\alpha}$ and β, one can find the relativistic properties of higher spins.

The operator β in a new Lorentz frame is given by

$$\beta' = \mathcal{L}_v\beta\mathcal{L}_v^{-1} = e^{\boldsymbol{\alpha}\cdot\boldsymbol{\omega}}\,\beta \tag{23-38}$$

since β anticommutes with $\boldsymbol{\alpha}$. Using (23-36) we then have

$$\beta' = \frac{\beta - (\mathbf{v}/c)\cdot\beta\boldsymbol{\alpha}}{\sqrt{1 - v^2/c^2}}, \tag{23-39}$$

again since $\alpha\beta = -\beta\alpha$. It looks from this equation as if β transforms as the time component of a four-vector of which $\beta\boldsymbol{\alpha}$ is the space part. In fact, under a Lorentz transformation

$$\beta'\boldsymbol{\alpha}'_\perp = \mathcal{L}_V \beta\boldsymbol{\alpha}_\perp \mathcal{L}_V^{-1} = \beta\boldsymbol{\alpha}_\perp \tag{23-40}$$

since both β and $\boldsymbol{\alpha}_\perp$ anticommute with $\mathcal{L}_V$ and therefore $\beta\alpha_\perp$ commutes with $\mathcal{L}_V$. Furthermore one can check that

$$\beta'\alpha'_\parallel = \frac{\beta\alpha_\parallel - (v/c)\beta}{\sqrt{1 - v^2/c^2}}; \tag{23-41}$$

thus β and $\beta\boldsymbol{\alpha}$ do indeed transform as the time and space components of a four-vector.

This four-vector is often called γ^μ (μ = 0, 1, 2, 3); one defines

$$\gamma^0 = \beta, \qquad \gamma^i = \beta\alpha_i. \tag{23-42}$$

In the notation of (23-28), the standard representation of γ, the space part of γ^μ, is[2]

$$\boldsymbol{\gamma} = \begin{pmatrix} 0 & \boldsymbol{\tau} \\ -\boldsymbol{\tau} & 0 \end{pmatrix}. \tag{23-43}$$

$\boldsymbol{\gamma}$ is anti-Hermitian while γ^0 is Hermitian. The γ^μ have many interesting properties. For one,

$$(\gamma^0)^2 = 1, \qquad (\gamma^i)^2 = -1, \qquad i = 1, 2, 3, \tag{23-44}$$

since $(\gamma^i)^2 = \beta\alpha_i\beta\alpha_i = -\beta^2\alpha_i{}^2 = -1$. Also if $\mu \neq \nu$ then γ^μ anticommutes with γ^ν. Clearly γ^0 anticommutes with $\boldsymbol{\gamma}$ since β anticommutes with $\boldsymbol{\alpha}$. Furthermore if $i \neq j$ then

$$\gamma^i\gamma^j = \beta\alpha_i\beta\alpha_j = -\beta^2\alpha_i\alpha_j = \beta^2\alpha_j\alpha_i = -\gamma^j\gamma^i.$$

We can express these properties of the γ^μ by writing

$$\gamma^\mu\gamma^\nu + \gamma^\nu\gamma^\mu = 2g^{\mu\nu}, \tag{23-45}$$

where the *metric tensor* $g^{\mu\nu}$ is

---

[2] *Warning:* There exist several different definitions of the γ matrices; this can be a source of confusion in reading the literature. Ours is the Feynman definition and is that used by Bjorken and Drell in their book *Relativistic Quantum Mechanics* [McGraw-Hill, New York (1964)].

$$g^{\mu\nu} = \begin{pmatrix} 1 & 0 & 0 & 0 \\ 0 & -1 & 0 & 0 \\ 0 & 0 & -1 & 0 \\ 0 & 0 & 0 & -1 \end{pmatrix}. \tag{23-46}$$

We leave it as a simple exercise to verify that

$$\sigma^{\mu\nu} = \frac{i}{2}[\gamma^\mu, \gamma^\nu]. \tag{23-47}$$

This relation is an example of another property of the γ^μ: every 4×4 matrix can be written as a unique linear combination of products of the γ^μ. One can verify that the 16 matrices

$$1, \gamma^\mu, \sigma^{\mu\nu}, \gamma_5\gamma^\mu, \gamma_5, \tag{23-48}$$

where

$$\gamma_5 \equiv \gamma^0\gamma^1\gamma^2\gamma^3, \tag{23-49}$$

are linearly independent and complete in the sense that any 4×4 matrix can be written as a linear combination of them. Notice that the traces of the matrices (23-48), with the exception of 1, all vanish.

The operator γ_5 is interesting. It anticommutes with the γ^μ, since each γ^μ anticommutes with three of the γ's and commutes with itself in (23-49). Thus it commutes with $\alpha_i = \gamma^0\gamma^i$ and hence is unchanged under a Lorentz transformation (23-32). It appears to be a scalar. However

$$\beta\gamma_5\beta = -\gamma_5, \tag{23-50}$$

so that γ_5 changes sign under the parity operation. Thus γ_5 is a *pseudo-scalar.* Similarly $\gamma_5\gamma^\mu$ is a pseudo-vector or *axial vector.* (this is a four-vector whose space part doesn't change sign under parity, and whose time component does). Thus the operators (23-48) transform respectively as a scalar, vector, second-rank tensor, axial vector, and pseudo-scalar; this is one indication that they are linearly independent.

DIRAC EQUATION

Now let us formulate the wave equation obeyed by a spin $1/2$ particle. Since $\gamma^\mu = (\beta, \beta\boldsymbol{\alpha})$ is a four-vector and so is $(E/c, \mathbf{p})$, the scalar product of these two vectors

$$\frac{\beta E}{c} - \beta \boldsymbol{\alpha} \cdot \mathbf{p} \tag{23-51}$$

must be a scalar; it must have the same value for a particle in all Lorentz frames. To see what this quantity is let us square (23-51):

$$\begin{aligned}\left(\frac{\beta E}{c} - \beta \boldsymbol{\alpha} \cdot \mathbf{p}\right)^2 &= \beta^2\left(\frac{E}{c}\right)^2 + (\beta\boldsymbol{\alpha} \cdot \mathbf{p})^2 - (\beta\beta\boldsymbol{\alpha} + \beta\boldsymbol{\alpha}\beta) \cdot \frac{\mathbf{p}E}{c} \\ &= \left(\frac{E}{c}\right)^2 - p^2 = m^2c^2,\end{aligned} \tag{23-52}$$

since $\beta^2 = 1$, $\beta\boldsymbol{\alpha} + \boldsymbol{\alpha}\beta = 0$ and $(\beta\alpha_i)^2 = -1$. Thus the scalar (23-51) can equal either $\pm mc$ for a particle. The sign depends on the sign we've chosen for β; we could equally well have called the parity operator $-\beta$ (or $\pm i\beta$) and have changed nothing. Let us assume that the sign of β is chosen so that the scalar (23-51) equals $+mc$. Thus

$$\beta E - \beta c \boldsymbol{\alpha} \cdot \mathbf{p} = mc^2. \tag{23-53}$$

The wave function of a particle, $\psi(\mathbf{r}, t)$ is a four-component spinor, because the operators $\boldsymbol{\sigma}$, $\boldsymbol{\alpha}$, and β are 4×4 matrices. Since (23-53) is a scalar, we see, on making the correspondence

$$E \rightarrow i\hbar \frac{\partial}{\partial t}, \qquad \mathbf{p} \rightarrow i\hbar\nabla, \tag{23-54}$$

in (23-53) that the wave function of a free particle obeys

$$\beta i\hbar \frac{\partial \psi}{\partial t} = \beta c \boldsymbol{\alpha} \cdot \frac{\hbar}{i} \nabla\psi + mc^2\psi, \tag{23-55}$$

or multiplying through by β:

$$i\hbar \frac{\partial \psi}{\partial t} = \left[c\boldsymbol{\alpha} \cdot \frac{\hbar}{i} \nabla + \beta mc^2\right] \psi. \tag{23-56}$$

This is the *Dirac equation* for a relativistic spin $1/2$ particle.

Dirac didn't go to any of the trouble we did to write down his equation. He simply applied the calculation (23-52) in reverse to take the "square root" of the energy-momentum relation, and deduced directly that $\boldsymbol{\alpha}$ and β had to have their commutation properties and be 4×4 matrices.

From (23-56) we see that the Hamiltonian of a relativistic spin $1/2$ particle is

$$H = c\boldsymbol{\alpha} \cdot \mathbf{p} + \beta mc^2. \tag{23-57}$$

In the presence of an electromagnetic field the Dirac equation becomes, by the usual coupling,

$$\left(i\hbar\frac{\partial}{\partial t} - e\Phi\right)\psi = \left[c\boldsymbol{\alpha}\cdot\left(\frac{\hbar}{i}\nabla - \frac{e}{c}\mathbf{A}\right) + \beta mc^2\right]\psi. \tag{23-58}$$

Notice that the vector potential is directly coupled to the internal degrees of freedom. The value of the gyromagnetic ratio of the electron, $g = 2$, falls out of the Dirac equation (23-58).

To see this, let us look at the nonrelativistic limit of the Dirac equation. To do that we write α and β as 4×4 matrices in the representation (23-28) and write

$$\psi = \begin{pmatrix}\varphi\\ \chi\end{pmatrix} \tag{23-59}$$

where φ and χ are each two-component spinors. Then

$$i\hbar\frac{\partial}{\partial t}\begin{pmatrix}\varphi\\ \chi\end{pmatrix} = c\left(\frac{\hbar}{i}\nabla - \frac{e}{c}\mathbf{A}\right)\cdot\begin{pmatrix}0 & \tau\\ \tau & 0\end{pmatrix}\begin{pmatrix}\varphi\\ \chi\end{pmatrix} + \begin{pmatrix}1 & 0\\ 0 & -1\end{pmatrix} mc^2\begin{pmatrix}\varphi\\ \chi\end{pmatrix} + e\Phi\begin{pmatrix}\varphi\\ \chi\end{pmatrix}$$

or

$$i\hbar\frac{\partial\varphi}{\partial t} = c\left(\frac{\hbar}{i}\nabla - \frac{e}{c}\mathbf{A}\right)\cdot\tau\chi + (e\Phi + mc^2)\varphi$$

$$i\hbar\frac{\partial\chi}{\partial t} = c\left(\frac{\hbar}{i}\nabla - \frac{e}{c}\mathbf{A}\right)\cdot\tau\varphi + (e\Phi - mc^2)\chi. \tag{23-60}$$

[Compare these equations with the corresponding equations for a spin zero particle.] It should be emphasized that although we shall apply (23-60) to derive the nonrelativistic limit, (23-60) is often a quite useful way of writing the Dirac equation for relativistic problems.

In the nonrelativistic limit, the wave function of the particle behaves as $e^{-imc^2t/\hbar}$ times a function with frequencies small compared with $mc^2/\hbar$. Thus

$$i\hbar\frac{\partial\chi}{\partial t} = mc^2\chi + \cdots \tag{23-61}$$

and we have from the second of Eqs. (23-60) that to lowest order

$$mc^2\chi = c\left(\frac{\hbar}{i}\nabla - \frac{e}{c}\mathbf{A}\right)\cdot\tau\varphi - mc^2\chi$$

or

$$\chi = \frac{1}{2mc}\left(\frac{\hbar}{i}\nabla - \frac{e}{c}\mathbf{A}\right)\cdot\tau\varphi. \tag{23-62}$$

χ is thus $\sim v/c$ smaller than φ in the nonrelativistic limit. Plugging this back into the first of Eqs. (23-60) we then find

$$i\hbar\frac{\partial\varphi}{\partial t} = \frac{1}{2m}\left[\left(\frac{\hbar}{i}\nabla - \frac{e}{c}\mathbf{A}\right)\cdot\tau\right]^2\varphi + (e\Phi + mc^2)\varphi. \tag{23-63}$$

Now from the properties of the Pauli matrices

$$(\mathbf{a}\cdot\tau)(\mathbf{b}\cdot\tau) = (\mathbf{a}\cdot\mathbf{b}) + i\tau\cdot(\mathbf{a}\times\mathbf{b});$$

thus

$$\left[\left(\frac{\hbar}{i}\nabla - \frac{e}{c}\mathbf{A}\right)\cdot\tau\right]^2\varphi = \left(\frac{\hbar}{i}\nabla - \frac{e}{c}\mathbf{A}\right)^2\varphi - \frac{e\hbar}{c}\tau\cdot(\nabla\times\mathbf{A} + \mathbf{A}\times\nabla)\varphi.$$

Now $(\nabla\times\mathbf{A} + \mathbf{A}\times\nabla)\varphi$ is simply $\mathcal{H}\varphi$, where $\mathcal{H} = \nabla\times\mathbf{A}$ is the magnetic field. Thus (23-63) becomes

$$i\hbar\frac{\partial\varphi}{\partial t} = \frac{1}{2m}\left(\frac{\hbar}{i}\nabla - \frac{e}{c}\mathbf{A}\right)^2\varphi - \frac{e\hbar}{2mc}\sigma\cdot\mathcal{H}\varphi + (e\Phi + mc^2)\varphi, \tag{23-64}$$

where we denote the Pauli matrices by σ here. This is exactly the nonrelativistic equation, the *Pauli equation*, obeyed by a charged spin $1/2$ particle in an electromagnetic field. But notice that the magnetic moment associated with the spin of the particle is *predicted* to be $e\hbar/2mc$, or $g = 2$. We don't have to put it in; it is there. This is one of the great triumphs of the Dirac equation.[3]

The scalar product $\psi^\dagger\psi$ of the wave function with itself obeys a continuity equation. In our standard representation α and β are Hermitian and therefore $\psi^\dagger$ obeys the adjoint equation to (23-58):

$$\left(-i\hbar\frac{\partial}{\partial t} - e\Phi\right)\psi^\dagger = c\left(-\frac{\hbar}{i}\nabla - \frac{e}{c}\mathbf{A}\right)\psi^\dagger\cdot\alpha + mc^2\psi^\dagger\beta \tag{23-65}$$

The α and β are on the right since $\psi^\dagger$ is a row spinor. Multiplying

[3] The Dirac equation doesn't include the influence of vacuum fluctuations, e.g., of the electromagnetic field, on the magnetic moment. For electrons these produce the small $\alpha/2\pi$ correction to g. Nucleons, on the other hand, have strong interactions with the vacuum fluctuations of the pion field and other fields of strongly interacting particles, which produce very sizeable corrections to g.

(23-58) on the left by $\psi^\dagger$, (23-65) on the right by ψ and subtracting we then find

$$\frac{\partial(\psi^\dagger\psi)}{\partial t} + \nabla\cdot(\psi^\dagger c\alpha\psi) = 0. \qquad (23\text{-}66)$$

Thus $\psi^\dagger\psi$ is a positive conserved density, which can be interpreted as a probability density. Then

$$\mathbf{j} = c\psi^\dagger\alpha\psi \qquad (23\text{-}67)$$

is the probability current. The operator $c\alpha$ plays the role of a velocity operator; it is the derivative of the Hamiltonian (23-57) with respect to $\mathbf{p}$. We shall discuss the expression in detail later.

Let us consider the behavior of the Dirac equation under a Lorentz transformation. In a new coordinate system the equation is [cf. (23-55)]

$$\beta' i\hbar \frac{\partial}{\partial t'}\hat{\psi}(\mathbf{r}', t') = \beta' c\alpha' \cdot \frac{\hbar}{i}\nabla'\hat{\psi}(\mathbf{r}', t') + mc^2\hat{\psi}(\mathbf{r}', t'), \qquad (23\text{-}68)$$

where $\hat{\psi}(\mathbf{r}', t')$ is the wave function in the new frame. Now since

$$\beta' i\hbar\frac{\partial}{\partial t'} - \beta' c\alpha' \cdot \frac{\hbar}{i}\nabla' = \beta i\hbar\frac{\partial}{\partial t} - \beta c\alpha\cdot\frac{\hbar}{i}\nabla,$$

(this quantity is a Lorentz scalar) we see that $\hat{\psi}(\mathbf{r}', t')$ satisfies the same equation as $\psi(\mathbf{r}, t)$ and therefore

$$\hat{\psi}(\mathbf{r}', t') = \psi(\mathbf{r}, t) \qquad (23\text{-}69)$$

if $\mathbf{r}'t'$ and $\mathbf{r}t$ are the same space-time point.

It turns out to be most convenient not to use the equation (23-68) in the new frame, but rather to choose a new basis in the spin space such that β' and α' are represented by the *same* matrices as β and α in the original frame. To accomplish this transformation we use (23-31) to write (23-68) as

$$\mathcal{L}_V\beta\mathcal{L}_V^{-1} i\hbar\frac{\partial}{\partial t'}\hat{\psi}(\mathbf{r}', t') = \mathcal{L}_V\beta c\alpha\mathcal{L}_V^{-1}\cdot\frac{\hbar}{i}\nabla'\hat{\psi}(\mathbf{r}', t') + mc^2\hat{\psi}(\mathbf{r}', t'),$$

and multiplying through by $\mathcal{L}_V^{-1}$, we find

$$\beta i\hbar\frac{\partial}{\partial t'}\psi'(\mathbf{r}'t') = c\beta\alpha\cdot\frac{\hbar}{i}\nabla'\psi'(\mathbf{r}'t') + mc^2\psi'(\mathbf{r}'t'), \qquad (23\text{-}70)$$

where

$$\psi'(\mathbf{r}'t') = \mathcal{L}_V^{-1}\psi(\mathbf{r}'t') = \mathcal{L}_V^{-1}\psi(\mathbf{r}, t). \tag{23-71}$$

Equation (23-70) is the form of the Dirac equation that has the same matrices α and β in *all* Lorentz frames. Then the wave function in the new frame is related to the wave function in the old frame by (23-71).

FREE PARTICLE SOLUTIONS

Now let us construct free particle solutions to the Dirac equation. To begin with we construct solutions for a particle at rest, and then construct solutions for moving particles by Lorentz transformation. For a particle at rest,

$$\psi(\mathbf{r}, t) = e^{-iEt/\hbar}u, \tag{23-72}$$

where u is a spinor independent of space and time. Putting (23-72) in the Dirac equation (23-72) we find

$$Eu = \beta mc^2 u. \tag{23-73}$$

Thus, since the eigenvalues of β are ± 1, we have two possibilities. If u is an eigenstate of β with eigenvalue $+1$, then $E = mc^2$; if the eigenvalue is -1, then $E = -mc^2$. Again, as in the Klein-Gordon equation, we find negative energy solutions, and again these solutions will be associated with positive energy antiparticles, although, as we shall see, in a somewhat different way than for spin zero particles.

In the basis (23-27) we can then write four linearly independent solutions to the Dirac equation:

$$u_{0\uparrow}^{(+)} = \begin{pmatrix} 1 \\ 0 \\ 0 \\ 0 \end{pmatrix}, \quad u_{0\downarrow}^{(+)} = \begin{pmatrix} 0 \\ 1 \\ 0 \\ 0 \end{pmatrix}$$

$$u_{0\downarrow}^{(-)} = \begin{pmatrix} 0 \\ 0 \\ 1 \\ 0 \end{pmatrix}, \quad u_{0\uparrow}^{(-)} = \begin{pmatrix} 0 \\ 0 \\ 0 \\ 1 \end{pmatrix}. \tag{23-74}$$

The upper index b = ($\pm$) denotes the eigenvalue of β, the 0 denotes that $\mathbf{p} = 0$, and the arrow denotes the value of the spin that will eventually be associated physically with these states. The spinors $u_{0\uparrow}^{(+)}$ and $u_{0\downarrow}^{(-)}$ are eigenstates of σ_z with eigenvalue $+1$, while

$u_{0\downarrow}^{(+)}$ and $u_{0\uparrow}^{(-)}$ are eigenstates of σ_z with eigenvalue -1. Thus while $u_{0\uparrow}^{(-)}$ is the spinor of a negative energy particle with spin up, it will be associated with a positive energy antiparticle with spin down.

The states with $\beta = +1$ vary in time as $e^{-imc^2t/\hbar}$, while those with $\beta = -1$ vary as $e^{imc^2t/\hbar}$. Since β is the parity operator for the spin degrees of freedom we see that the positive and negative energy states have *opposite* intrinsic parity (the parity of the wave function in the rest frame). The assignment $\beta = 1$ to the positive energy solutions is arbitrary and is due to our having taken $+mc^2$ in (23-53); had we taken $-mc^2$, the positive energy states would have $\beta = -1$. Had we taken $\beta^2 = -1$, the relative parities would still be -1.

Now to construct the solutions for a particle of momentum p we start with a particle at rest and make a Lorentz transformation to a frame moving with velocity $\mathbf{v} = -\mathbf{p}c^2/E_p$ where $E_p = +\sqrt{p^2c^2 + m^2c^4}$. Then from (23-71) the wave function in the new frame is given by

$$\psi'(\mathbf{r}', t') = e^{-\boldsymbol{\alpha}\cdot\boldsymbol{\omega}/2}\, e^{\mp imc^2t/\hbar}\, u_{0,\sigma}^{(\pm)}. \tag{23-75}$$

Now the quantity $E_p't' - \mathbf{p}'\cdot\mathbf{r}'$ is a Lorentz scalar; since in the rest frame it equals mc^2t we have

$$e^{\mp imc^2t/\hbar} = e^{\pm i(\mathbf{p}'\cdot\mathbf{r}' - E_p't')/\hbar}.$$

Dropping the primes then, (23-75) becomes

$$\psi(\mathbf{r}, t) = e^{\pm i(\mathbf{p}\cdot\mathbf{r} - E_pt)/\hbar}\, e^{-\boldsymbol{\alpha}\cdot\boldsymbol{\omega}/2}\, u_{0,\sigma}^{(\pm)}. \tag{23-76}$$

Finally we must evaluate the new spinors

$$u_{\mathbf{p},\sigma}^{(\pm)} = e^{-\boldsymbol{\alpha}\cdot\boldsymbol{\omega}/2}\, u_{0,\sigma}^{(\pm)} = \left[\cosh\frac{\omega}{2} - \boldsymbol{\alpha}\cdot\hat{\mathbf{v}}\sinh\frac{\omega}{2}\right] u_{0,\sigma}^{(\pm)}. \tag{23-77}$$

Since $\mathbf{v} = -\mathbf{p}c^2/E_p$, we have

$$\cosh\frac{\omega}{2} = \sqrt{\frac{E_p + mc^2}{2mc^2}}, \qquad \hat{\mathbf{v}}\tanh\frac{\omega}{2} = \frac{-\mathbf{p}c}{E_p + mc^2}. \tag{23-78}$$

Thus

$$u_{\mathbf{p}\sigma}^{(\pm)} = \sqrt{\frac{E_p + mc^2}{2mc^2}}\left(1 + \frac{c\mathbf{p}\cdot\boldsymbol{\alpha}}{E_p + mc^2}\right) u_{0,\sigma}^{(\pm)}. \tag{23-79}$$

and in the standard representation (23-27) we have

$$u_{p\uparrow}^{(+)} = \sqrt{\frac{E_p + mc^2}{2mc^2}} \begin{pmatrix} 1 \\ 0 \\ \dfrac{p_z c}{E_p + mc^2} \\ \dfrac{(p_x + ip_y)c}{E_p + mc^2} \end{pmatrix}$$

$$u_{p\downarrow}^{(+)} = \sqrt{\frac{E_p + mc^2}{2mc^2}} \begin{pmatrix} 0 \\ 1 \\ \dfrac{(p_x - ip_y)c}{E_p + mc^2} \\ \dfrac{-p_z c}{E_p + mc^2} \end{pmatrix}$$

$$u_{p\downarrow}^{(-)} = \sqrt{\frac{E_p + mc^2}{2mc^2}} \begin{pmatrix} \dfrac{p_z c}{E_p + mc^2} \\ \dfrac{(p_x + ip_y)c}{E_p + mc^2} \\ 1 \\ 0 \end{pmatrix}$$

$$u_{p\uparrow}^{(-)} = \sqrt{\frac{E_p + mc^2}{2mc^2}} \begin{pmatrix} \dfrac{(p_x - ip_y)c}{E_p + mc^2} \\ \dfrac{-p_z c}{E_p + mc^2} \\ 0 \\ 1 \end{pmatrix} \tag{23-80}$$

The arrow refers to the spin associated with the state in the rest frame [this is minus the σ_3 eigenvalue for (-) spinors]. A particle in a σ_z eigenstate in its rest frame appears to be in a σ_z eigenstate to an observer moving with respect to the particle *only* if the observer is moving along the z direction.

The positive energy solutions

$$u_{p\sigma}^{(+)}\, e^{i(\mathbf{p}\cdot\mathbf{r} - E_p t)/\hbar} \tag{23-81}$$

represent particles with momentum **p**, energy E_p and spin orientation σ. The negative energy solutions

$$u_{p\sigma}^{(-)}\, e^{-i(\mathbf{p}\cdot\mathbf{r} - E_p t)/\hbar} \tag{23-82}$$

are eigenstates of the momentum, energy, and spin operators with eigenvalues $-\mathbf{p}$, $-E_p$, and $-\sigma$; they will be associated physically with anti- particles with momentum **p**, energy E_p, and spin σ.

The spinors (23-74) are normalized to one and orthogonal to each other. The spinors (23-80) are no longer normalized to one; instead they obey the orthogonality relations:

$$u_{bp,\sigma}^{(b)\dagger}\, u_{b'p,\sigma}^{(b')} = \frac{E_p}{mc^2}\, \delta_{bb'}\delta_{\sigma\sigma'}, \tag{23-83}$$

where b takes on the values $\pm$ (the β eigenvalue in the rest frame). Thus we see that the scalar product (23-83) has the Lorentz transformation properties of the time component of a four-vector. This is the correct transformation law for a density (recall that $\psi^\dagger\psi$ is a probability density). Mathematically the normalization is not preserved under a Lorentz transformation because the operator $\mathcal{L}$ is not unitary; $\mathcal{L}^\dagger \neq \mathcal{L}^{-1}$ generally, and therefore $\mathcal{L}$ changes the length of a vector. In the representation (23-27), $\boldsymbol{\alpha}$ is Hermitian and therefore

$$\mathcal{L}^\dagger = \mathcal{L}. \tag{23-84}$$

Thus if $u_p = \mathcal{L}^{-1}u_0$, then $u_p^\dagger = u_0^\dagger(\mathcal{L}^{-1})^\dagger = u_0^\dagger\mathcal{L}^{-1}$ and

$$u_p^\dagger u_p = u_0^\dagger \mathcal{L}^{-2} u_0.$$

It is possible though to define a norm that is invariant under a Lorentz transformation, by noticing that since β anticommutes with $\boldsymbol{\alpha}$ we have

$$(\mathcal{L}^{-1})^\dagger\beta = \mathcal{L}^{-1}\beta = \beta\mathcal{L}. \tag{23-85}$$

Thus if the spinor u transforms as $u' = \mathcal{L}^{-1}u$, the spinor

$$\bar{u} = u^\dagger\beta \tag{23-86}$$

is given in the new frame by

$$\overline{u'} = u'^\dagger\beta = (\mathcal{L}^{-1}u)^\dagger\beta = u^\dagger(\mathcal{L}^{-1})^\dagger\beta = u^\dagger\beta\mathcal{L} = \bar{u}\mathcal{L}. \tag{23-87}$$

Consequently the product $\bar{u}_1 u_2$ of any two spinors is a Lorentz invariant:

$$\bar{u}_1' u_2' = (\bar{u}_1 \mathscr{L})(\mathscr{L}^{-1} u_2) = \bar{u}_1 u_2.$$

In particular, since $\bar{u}^{(b)}_{p\sigma} u^{(b')}_{p\sigma'}$ is given in the rest frame by $b\delta_{bb'}\delta_{\sigma\sigma'}$ we have that for all p:

$$\bar{u}^{(b)}_{p\sigma} u^{(b')}_{p\sigma'} = b\delta_{bb'}\delta_{\sigma\sigma'}, \quad b = \pm\ . \tag{23-88}$$

For example, $\bar{u}^{(\pm)}_{p\sigma} u^{(\pm)}_{p\sigma} = \pm 1.$

The spinors $u^{(\pm)}_{p\sigma}$ obey the completeness relation that says that the unit 4×4 matrix can be written as the sum of the outer products of the four spinors (23-80):

$$\sum_{b,\sigma} b u^{(b)}_{p\sigma} \bar{u}^{(b)}_{p\sigma} = 1. \tag{23-89}$$

To prove this relation we notice that from (23-74),

$$\sum_{b\sigma} b u^{(b)}_{0\sigma} \bar{u}^{(b)}_{0\sigma} = \sum_{b\sigma} u^{(b)}_{0\sigma} u^{(b)\dagger}_{0\sigma} = 1.$$

Multiplying this equation on the right by $\mathscr{L} = e^{\boldsymbol{\alpha}\cdot\omega/2}$ and on the left by $\mathscr{L}^{-1}$, we find (23-89).

We note that the spinors $u^{(\pm)}_{p\sigma}$ obey

$$(\beta E_p - c\beta\boldsymbol{\alpha}\cdot\mathbf{p}) u^{(\pm)}_{p\sigma} = \pm mc^2 u^{(\pm)}_{p\sigma} \tag{23-90}$$

and

$$u^{(\pm)\dagger}_{p\sigma}(\beta E_p - c\boldsymbol{\alpha}\cdot\mathbf{p}\beta) = \pm mc^2 u^{(\pm)\dagger}_{p\sigma}\ . \tag{23-91}$$

Multiplying (23-91) on the right by β, we have

$$\bar{u}^{(\pm)}_{p\sigma}(\beta E_p - c\beta\boldsymbol{\alpha}\cdot\mathbf{p}) = \pm mc^2 \bar{u}^{(\pm)}_{p\sigma}. \tag{23-92}$$

CURRENTS

In general, if $\psi(\mathbf{r}, t)$ is a solution of the Dirac equation, then under a Lorentz transformation in which

$$\psi(\mathbf{r}, t) \rightarrow \psi'(\mathbf{r}', t') = \mathscr{L}^{-1}\psi(\mathbf{r}, t),$$

the spinor

$$\bar{\psi}(\mathbf{r}, t) = \psi^\dagger(\mathbf{r}, t)\beta \tag{23-93}$$

transforms as

$$\bar{\psi}(\mathbf{r}, t) \to \bar{\psi}'(\mathbf{r}', t') = [\mathcal{L}^{-1}\psi(\mathbf{r}, t)]^\dagger\beta = \bar{\psi}(\mathbf{r}, t)\mathcal{L}. \tag{23-94}$$

Thus the product $\bar{\psi}(\mathbf{r}, t)\psi(\mathbf{r}, t)$ transforms as a Lorentz scalar. Since under $\gamma^\mu \to \mathcal{L}\gamma^\mu\mathcal{L}^{-1}$ the operators γ^μ transform as the components of a four-vector, the product

$$\bar{\psi}(\mathbf{r}t)\gamma^\mu\psi(\mathbf{r}t) = (\rho(\mathbf{r}, t), c^{-1}\mathbf{j}(\mathbf{r}, t))$$

transforms as a four-vector, the particle current time c^{-1}, under a Lorentz transformation. Similarly the quantities

$$\bar{\psi}\sigma^{\mu\nu}\psi, \qquad \bar{\psi}\gamma_5\gamma^\mu\psi, \qquad \bar{\psi}\gamma_5\psi.$$

transform respectively as a second-rank tensor, axial vector, and pseudo-scalar.

The *positive* density

$$\rho(\mathbf{r}, t) = \psi^\dagger(\mathbf{r}, t)\psi(\mathbf{r}, t) \tag{23-95}$$

and the current

$$\mathbf{j}(\mathbf{r}, t) = c\psi^\dagger(\mathbf{r}, t)\alpha\psi(\mathbf{r}, t) \tag{23-96}$$

obey the continuity equation $\partial\rho/\partial t + \nabla \cdot \mathbf{j} = 0$. Thus for a state ψ, the quantity $\int \rho(\mathbf{r}t)\, d^3r$ is a constant of the motion. We may therefore interpret the density $\rho(\mathbf{r}t)$ as a probability density, as in the nonrelativistic theory. This is an important difference between the relativistic spin $1/2$ theory and spin 0 theory. For spin 0 the conserved density had to be interpreted as a charge density which could be either positive or negative; it could not be interpreted simply as a probability density.

An important consequence of this difference is that for spin 0 it is impossible for a particle to make a transition from a state normalized to +1 to a state normalized to −1, since the normalization remains constant for all time. Thus we were able to associate the negative energy states with particles and negatively normalized states with antiparticles; the conservation of charge prevents a particle from making a transition from a positive energy state to a negative energy state. However, for spin $1/2$ particles both positive and negative energy states have positive normalization, and there is nothing in the theory, so far, to prevent a particle in a positive energy state from making a transition to a negative energy state, radiating away several high energy photons. We shall return to this difficulty later.

The position operator in the Dirac theory has the same strange features that it did in the Klein-Gordon theory. Acting on a wave packet composed of positive energy free particle states, it mixes in negative energy free particle states, i.e., if

$$\psi^{(+)}(\mathbf{r}) = \sum_{\sigma} \int \frac{d^3p}{(2\pi\hbar)^3} a_{\mathbf{p}\sigma} u_{\mathbf{p}\sigma}^{(+)} e^{i\mathbf{p}\cdot\mathbf{r}/\hbar} \tag{23-97}$$

then

$$\begin{aligned} \mathbf{r}\psi^{(+)}(\mathbf{r}) &= \sum_{\sigma} \int \frac{d^3p}{(2\pi\hbar)^3} a_{\mathbf{p}\sigma} u_{\mathbf{p}\sigma}^{(+)} \frac{\hbar\nabla_{\mathbf{p}}}{i} e^{i\mathbf{p}\cdot\mathbf{r}/\hbar} \\ &= \sum_{\sigma} \int \frac{d^3p}{(2\pi\hbar)^3} (i\hbar\nabla_{\mathbf{p}} a_{\mathbf{p}\sigma}) u_{\mathbf{p}\sigma}^{(+)} e^{i\mathbf{p}\cdot\mathbf{r}/\hbar} \\ &\quad + \sum_{\sigma} \int \frac{d^3p}{(2\pi\hbar)^3} a_{\mathbf{p}\sigma} (i\hbar\nabla_{\mathbf{p}} u_{\mathbf{p}\sigma}^{(+)}) e^{i\mathbf{p}\cdot\mathbf{r}/\hbar} . \end{aligned} \tag{23-98}$$

The first term has only positive energy components. However, the $i\hbar\nabla_{\mathbf{p}} u_{\mathbf{p}\sigma}^{(+)}$ term is a linear combination of both positive and negative energy components, as one can see by acting with $\nabla_{\mathbf{p}}$ on the spinors (23-80). As in the Klein-Gordon theory, we can write the position operator as

$$\mathbf{r} = \mathbf{r}_{(+)} + \mathbf{r}_{(-)}. \tag{23-99}$$

The even part $\mathbf{r}_{(+)}$ acting on a wave packet of positive energy free particle states produces only positive energy free particle states, and acting on a negative energy wave packet it again yields a negative energy wave packet. The odd part $\mathbf{r}_{(-)}$ turns a positive energy wave packet into a negative energy wave packet, and vice versa.

The eigenstates of the even part $\mathbf{r}_{(+)}$ are spread out over a distance on the order of the Compton wavelength of the particle. As in the Klein-Gordon theory, one requires both positive and negative energy free particle solutions to produce a more localized wave packet. Similarly whenever a positive energy wave packet travels through a potential, negative energy components are mixed in and the zitterbewegung is excited.

Closely related to these properties of the position operator is the following peculiarity of the Dirac theory. As we see from the expression (23-96) for the current, the operator $c\boldsymbol{\alpha}$ plays the role of a velocity operator. This is further confirmed by the fact that

$$-i\hbar[\mathbf{r}, H] = c\boldsymbol{\alpha}, \tag{23-100}$$

so that in the Heisenberg representation $d\mathbf{r}/dt = c\boldsymbol{\alpha}$. Now if we take the square of, say, the z component of this velocity operator, we find

$$(c\alpha_z)^2 = c^2\alpha_z^{\,2} = c^2. \tag{23-101}$$

Thus the eigenvalues of each component of the velocity operator are $\pm c$; a particle in an eigenstate of the velocity operator travels at the speed of light! The point is that the velocity operator is not simply related to the momentum operator relativistically. The eigenstates of any component of $\boldsymbol{\alpha}$ are linear combinations of positive and negative energy free particle states, and are not realized in any physical situation. For any arbitrary state, the expectation value of $c\boldsymbol{\alpha}$ has a magnitude somewhere between 0 and c.

The physical content of the expression (23-96) for the current can be made more transparent by using the fact that ψ obeys the Dirac equation, and $\psi^\dagger$ the adjoint equation, to rewrite (23-96) as

$$j_i = c\psi^\dagger\alpha_i\psi = \frac{1}{2mc}\,\psi^\dagger\alpha_i\left\{\beta\left(i\hbar\frac{\partial}{\partial t} - e\Phi\right) - c\beta\boldsymbol{\alpha}\cdot\left(\frac{\hbar\nabla}{i} - \frac{e\mathbf{A}}{c}\right)\psi\right\}$$
$$+\frac{1}{2mc}\left\{\left(-i\hbar\frac{\partial}{\partial t} - e\Phi\right)\psi^\dagger\beta + \left(-\frac{\hbar\nabla}{i} - \frac{e\mathbf{A}}{c}\right)\psi^\dagger\cdot\beta\boldsymbol{\alpha}\right\}\alpha_i\psi. \tag{23-102}$$

The terms involving Φ cancel. The $\partial/\partial t$ terms reduce to

$$-\frac{\hbar}{2mc}\frac{\partial}{\partial t}(\bar{\psi}i\alpha_i\psi). \tag{23-103}$$

In the remaining terms we write

$$\alpha_i\alpha_j = \frac{1}{2}[\alpha_i, \alpha_j] + \frac{1}{2}\{\alpha_i, \alpha_j\} = i\epsilon_{ijk}\sigma_k + \delta_{ij}. \tag{23-104}$$

Thus (23-102) reduces, after a little bookkeeping, to

$$\mathbf{j}(\mathbf{r}t) = \frac{1}{2m}\left[\bar{\psi}\left(\frac{\hbar}{i}\nabla - \frac{e}{c}\mathbf{A}\right)\psi + \left(-\frac{\hbar}{i}\nabla - \frac{e}{c}\mathbf{A}\right)\bar{\psi}\cdot\psi\right]$$
$$+\frac{\hbar}{2m}\left[\nabla\times(\bar{\psi}\boldsymbol{\sigma}\psi) - \frac{1}{c}\frac{\partial}{\partial t}(\bar{\psi}i\boldsymbol{\alpha}\psi)\right]. \tag{23-105}$$

By a similar calculation we can rewrite the density as

$$\rho(\mathbf{r}t) = \psi^\dagger(\mathbf{r}t)\psi(\mathbf{r}t) \tag{23-106}$$

$$= \frac{1}{2mc^2}\left[\bar{\psi}\left(i\hbar\frac{\partial}{\partial t} - e\Phi\right)\psi + \left(-i\hbar\frac{\partial}{\partial t} - e\Phi\right)\bar{\psi}\cdot\psi\right] + \frac{\hbar}{2mc}\nabla\cdot[\bar{\psi}i\alpha\psi].$$

These equations for the current and density, known as the *Gordon decomposition,* have a simple physical interpretation. They are in the form

$$\mathbf{j} = \mathbf{j}_{conv} + \mathbf{j}_{int}, \qquad \rho = \rho_{conv} + \rho_{int} \tag{23-107}$$

where the "convective" parts of the current and density:

$$\mathbf{j}_{conv} = \frac{\hbar}{2im}(\bar{\psi}\nabla\psi - \nabla\bar{\psi}\cdot\psi) - \frac{e\mathbf{A}}{mc}\bar{\psi}\psi \tag{23-108}$$

$$\rho_{conv} = \frac{i\hbar}{2mc^2}\left(\bar{\psi}\frac{\partial\psi}{\partial t} - \frac{\partial\bar{\psi}}{\partial t}\psi\right) - \frac{e\Phi}{mc^2}\bar{\psi}\psi \tag{23-109}$$

resemble the expressions for the current and density for a spinless particle (with $\bar{\psi}$, rather than $\psi^\dagger$), and are determined by the rate of change of ψ in space and time. The "internal" current and density are of the form

$$\mathbf{j}_{int} = c\nabla\times\mathbf{M} + \frac{\partial\mathbf{P}}{\partial t}, \qquad \rho_{int} = -\nabla\cdot\mathbf{P} \tag{23-110}$$

where

$$\mathbf{M} = \frac{\hbar}{2mc}\bar{\psi}\sigma\psi, \qquad \mathbf{P} = \frac{\hbar}{2mc}\bar{\psi}(-i\alpha)\psi, \tag{23-111}$$

and are closely related to the internal state of the particle.

For a particle of charge e, e**M** is essentially the magnetic moment density of the particle arising from its spin. A spinning particle at rest has only a magnetic moment. But a magnetic moment seen from a moving frame appears to be a combination of a magnetic moment and an electric moment. e**P** is this electric polarization belonging to the spin. Notice that **M** and **P** transform as the components of a second rank tensor. The quantity $-e(\mathbf{M}\cdot\mathcal{H} + \mathbf{P}\cdot\mathcal{E})$, where $\mathcal{H}$ and $\mathcal{E}$ are the magnetic and electric fields, is a Lorentz scalar. From this point of view, we see that the extra internal dynamic variables $-i\alpha$ which relativity forced us to introduce, represent essentially a polarizability of the internal structure of the particle.

Equations (23-110) are the usual forms for the current and density associated with a magnetic moment and an electric polarization.

Notice that the internal and convective parts of ρ are separately conserved, since from (23-110),

$$\frac{\partial \rho_{\text{int}}}{\partial t} + \nabla \cdot \mathbf{j}_{\text{int}} = 0, \tag{23-112}$$

and subtracting from the continuity equation for ρ and $\mathbf{j}$, we have

$$\frac{\partial \rho_{\text{conv}}}{\partial t} + \nabla \cdot \mathbf{j}_{\text{conv}} = 0. \tag{23-113}$$

For a plane wave, $u_{p\sigma}^{(\pm)} e^{\pm i(\mathbf{p}\cdot\mathbf{r} - E_p t)/\hbar}$, **M** and **P** are independent of space and therefore

$$\mathbf{j}_{\text{int}} = 0, \qquad \rho_{\text{int}} = 0; \tag{23-114}$$

also

$$\rho = \rho_{\text{conv}} = \frac{E_p}{mc^2} > 0,$$

$$\mathbf{j} = \mathbf{j}_{\text{conv}} = \frac{\mathbf{p}}{m} = \frac{\mathbf{p}c^2}{E_p}\rho \tag{23-115}$$

for both positive and negative energy. The total current associated with the free particle positive energy wave packet (23-97) is

$$\mathbf{J} = \int \mathbf{j}(\mathbf{r})\, d^3r = \int \mathbf{j}_{\text{conv}}(\mathbf{r})\, d^3r = \sum_{\sigma} \int \frac{dp}{(2\pi\hbar)^3} |a_{p\sigma}|^2 \frac{\mathbf{p}}{m}. \tag{23-116}$$

For ψ normalized to one,

$$\sum_{\sigma} \int \frac{d^3p}{(2\pi\hbar)^3} |a_{p\sigma}|^2 \frac{E_p}{mc^2} = 1,$$

so that **J** can be written as

$$\mathbf{J} = \langle c\boldsymbol{\alpha} \rangle = \sum_{\sigma} \int \frac{d^3p}{(2\pi\hbar)^3} |a'_{p\sigma}|^2 \frac{\mathbf{p}c^2}{E_p}, \tag{23-117}$$

where

$$a'_{p\sigma} = \sqrt{\frac{E_p}{mc^2}}\, a_{p\sigma}$$

and

$$\sum_{\sigma} \int \frac{d^3p}{(2\pi\hbar)^3} |a'_{p\sigma}|^2 = 1.$$

Thus, despite the weird properties of $c\boldsymbol{\alpha}$, looked upon as a velocity operator, the total current (23-117) is just the average over the wave packet of pc^2/E_p, the usual relativistic group velocity.

NONRELATIVISTIC LIMIT

Let us now derive the correction terms to the nonrelativisitic Pauli equation (23-64). To do this we solve for χ in terms of φ to higher order in v/c than (23-62). From the second of Eqs. (23-60) we can write exactly

$$\chi = \frac{1}{2mc}\left(\frac{\hbar}{i}\nabla - \frac{e}{c}\mathbf{A}\right)\cdot\boldsymbol{\tau}\varphi - \frac{1}{2mc^2}\left(i\hbar\frac{\partial}{\partial t} - mc^2 - e\Phi\right)\chi. \qquad (23\text{-}118)$$

Neglecting the χ on the right gives Eq. (23-62); the first correction to (23-62) is found by iterating (23-118) once:

$$\chi = \frac{1}{2mc}\left(\frac{\hbar}{i}\nabla - \frac{e}{c}\mathbf{A}\right)\cdot\boldsymbol{\tau}\varphi - \frac{(i\hbar\frac{\partial}{\partial t} - mc^2 - e\Phi)}{4m^2c^3}\left(\frac{\hbar}{i}\nabla - \frac{e}{c}\mathbf{A}\right)\cdot\boldsymbol{\tau}\varphi. \qquad (23\text{-}119)$$

Substituting this expression for χ in the first of Eqs. (23-60) we find the first relativistic correction term to the Pauli equation for φ:

$$-\frac{1}{4m^2c^2}\left(\frac{\hbar}{i}\nabla - \frac{e}{c}\mathbf{A}\right)\cdot\boldsymbol{\sigma}\left[i\hbar\frac{\partial}{\partial t} - mc^2 - e\Phi\right]\left(\frac{\hbar}{i}\nabla - \frac{e}{c}\mathbf{A}\right)\cdot\boldsymbol{\sigma}\varphi, \qquad (23\text{-}120)$$

where $\boldsymbol{\sigma}$ now stands for the 2 × 2 Pauli matrices. The term (23-120) is $\sim v^2/c^2$ smaller than the kinetic energy $p^2/2m$.

Now (23-120) can be written as

$$-\frac{1}{4m^2c^2}\left[\left(\frac{\hbar}{i}\nabla - \frac{e}{c}\mathbf{A}\right)\cdot\boldsymbol{\sigma}\right]^2\left(i\hbar\frac{\partial}{\partial t} - mc^2 - e\Phi\right)\varphi$$
$$- \frac{ie\hbar}{4m^2c^2}\left[\left(\frac{\hbar}{i}\nabla - \frac{e}{c}\mathbf{A}\right)\cdot\boldsymbol{\sigma}\right][\boldsymbol{\varepsilon}\cdot\boldsymbol{\sigma}]\varphi \qquad (23\text{-}121)$$

where

$$\boldsymbol{\varepsilon} = -\nabla\Phi - \frac{1}{c}\frac{\partial \mathbf{A}}{\partial t}$$

is the electric field. To lowest order in v/c

$$\left(i\hbar\frac{\partial}{\partial t} - mc^2 - e\Phi\right)\varphi = \frac{p^2}{2m}\varphi. \tag{23-122}$$

Thus using $(\mathbf{a}\cdot\boldsymbol{\sigma})(\mathbf{b}\cdot\boldsymbol{\sigma}) = \mathbf{a}\cdot\mathbf{b} + i\boldsymbol{\sigma}\cdot(\mathbf{a}\times\mathbf{b})$, (23-121) becomes

$$-\left[\frac{p^4}{8m^3c^2} + \frac{e\hbar}{4m^2c^2}\boldsymbol{\sigma}\cdot(\boldsymbol{\varepsilon}\times\mathbf{p}) + \frac{ie\hbar}{4m^2c^2}\mathbf{p}\cdot\boldsymbol{\varepsilon}\right]\varphi. \tag{23-123}$$

The first term is the first relativistic correction to the kinetic energy; the second term is just the spin-orbit coupling. The last term however presents us with a difficulty — it isn't Hermitian.

A non-Hermitian term in the wave equation means that the normalization integral $\int\varphi^\dagger\varphi\, d^3r$ can change in time. The reason for this is simple. The original Dirac wave function obeys the normalization condition

$$\int d^3r\, \psi^\dagger\psi = \int d^3r[\varphi^\dagger\varphi + \chi^\dagger\chi] = 1; \tag{23-124}$$

but to lowest order, $\chi = \hbar\nabla\cdot\boldsymbol{\tau}\varphi/2imc$, so that $\chi^\dagger\chi = \varphi^\dagger(p^2/4m^2c^2)\varphi$. Thus the integral of $\varphi^\dagger\varphi$ remains constant only to within terms of order $(v/c)^2$; it is the integral $\int d^3r\, \varphi^\dagger(1 + p^2/4m^2c^2)\varphi$ or equivalently [to order $(v/c)^2$] the integral

$$\int d^3r\left[\left(1 + \frac{p^2}{8m^2c^2}\right)\varphi\right]^\dagger\left[\left(1 + \frac{p^2}{8m^2c^2}\right)\varphi\right] \tag{23-125}$$

that remains constant, and equal to 1, including $(v/c)^2$ terms.

This suggests that the correct nonrelativistic limit of the Dirac wave function, the limit whose normalization remains constant in time, is

$$\Psi(\mathbf{r}, t) \equiv \left(1 + \frac{p^2}{8m^2c^2}\right)\varphi(\mathbf{r}, t). \tag{23-126}$$

From (23-125) we have

$$\int d^3r\, \Psi^\dagger\Psi = 1; \tag{23-127}$$

thus we expect the equation for Ψ to have no non-Hermitian terms.

To construct the equation for Ψ we take the equation for φ, to order $(v/c)^2$,

$$\left[i\hbar\frac{\partial}{\partial t} - mc^2 - \frac{(p - eA/c)^2}{2m} + \frac{p^4}{8m^3c^2}\right]\varphi$$

$$= -\left[\frac{e\hbar}{2mc}\,\sigma\cdot\mathcal{H} + \frac{e\hbar}{4m^2c^2}\,\sigma\cdot(\varepsilon\times p)\,\varphi\right] + \left[e\Phi - \frac{ie\hbar}{4m^2c^2}\,p\cdot\varepsilon\right]\varphi, \qquad (23\text{-}128)$$

multiply on the left by $1 + p^2/8m^2c^2$ and try to bring the $1 + p^2/8m^2c^2$ through to the φ to form Ψ. To terms $\sim c^{-2}$, this factor commutes with all terms in (23-128) except the $e\Phi$. Now

$$\frac{1}{8m^2c^2}\,[p^2, e\Phi] - \frac{ie\hbar}{4m^2c^2}\,p\cdot\varepsilon = \frac{\hbar^2}{8m^2c^2}\,(\nabla^2 e\Phi)$$

to lowest order in v/c. Thus we find that Ψ obeys the equation

$$i\hbar\frac{\partial\Psi}{\partial t} = \left[mc^2 + \frac{1}{2m}\left(p - \frac{eA}{c}\right)^2 - \frac{p^4}{8m^3c^2}\right]\Psi$$

$$-\left[\frac{e\hbar}{2mc}\,\sigma\cdot\mathcal{H} + \frac{e\hbar}{4m^2c^2}\,\sigma\cdot(\varepsilon\times p)\right]\Psi + \left[e\Phi + \frac{\hbar^2}{8m^2c^2}\,(\nabla^2 e\Phi)\right]\Psi. \qquad (23\text{-}129)$$

This equation is the correct nonrelativistic limit of the Dirac equation; all the terms in it are Hermitian. The first term in [ ] on the right is the expansion of the kinetic energy operator $[c^2(p - eA/c)^2 + m^2c^4]^{1/2}$ to order $(v/c)^2$. The terms involving σ explicitly are the Pauli magnetic moment term and the spin orbit coupling. The correction to the potential term is known as the *Darwin term.*

From Poisson's equation,

$$\nabla^2 e\Phi(r) = -4\pi e Q(r), \qquad (23\text{-}130)$$

where $Q(r)$ is the charge density producing the electrostatic potential Φ. Thus the Darwin term is essentially a contact interaction between the particle and the charge $Q(r)$. For a Coulomb potential, $-Ze^2/r$, the Darwin term is

$$\frac{\pi\hbar^2}{2m^2c^2}\,Ze^2\delta(r);$$

this term tends to raise the energy of s-states since they are nonvanishing at the origin.

There exists a simple explanation of the Darwin term in terms of the zitterbewegung of the electron. Really, one argues, a relativistic particle is "spread out" over a distance $\sim \hbar/mc$, and in fact, it samples, at any time, the potential averaged over a region $\sim \hbar/mc$ about its position. Thus the potential term $e\Phi(\mathbf{r})$ in the Schrödinger equation is effectively replaced by $\overline{e\Phi(\mathbf{r}+\delta\mathbf{r})}$, where one averages over values of $\delta\mathbf{r} \sim \hbar/mc$ about $\mathbf{r}$. Expanding one has

$$e\Phi(\mathbf{r}+\delta\mathbf{r}) \approx e\Phi(\mathbf{r}) + \delta\mathbf{r}\cdot\nabla e\Phi(\mathbf{r}) + \frac{1}{2}(\delta\mathbf{r}\cdot\nabla)^2 e\Phi(\mathbf{r}),$$

and if one averages over $\delta\mathbf{r}$, and assumes spherically symmetric deviations, then

$$\overline{e\Phi(\mathbf{r}+\delta\mathbf{r})} \approx e\Phi(\mathbf{r}) + \frac{1}{6}\overline{(\delta r)^2}\,\nabla^2 e\Phi(\mathbf{r}). \tag{23-131}$$

If we write $\overline{(\delta r)^2} \approx (\hbar/mc)^2$ we find a correction term to $e\Phi$ of exactly the same form, order of magnitude, and sign as the Darwin term. The disquieting feature of this argument, however, is that the Darwin term is *absent* to this order in the spin zero case even though a spin zero particle is equally "spread out." The form of the Darwin term depends critically on the spin of the particle, and a satisfactory "physical" explanation would have to take this into account.

To see the origin of the spin-orbit term let us write the spin dependent terms in (23-129) as

$$-\frac{e\hbar}{2mc}\,\boldsymbol{\sigma}\cdot\left[\mathcal{H}+\frac{\boldsymbol{\varepsilon}\times\mathbf{p}}{2mc}\right] = -\frac{e\hbar}{2mc}\,\boldsymbol{\sigma}\cdot\left[\mathcal{H}+\frac{\boldsymbol{\varepsilon}\times\mathbf{p}}{mc}\right] + \frac{e\hbar}{4m^2c^2}\,\boldsymbol{\sigma}\cdot(\boldsymbol{\varepsilon}\times\mathbf{p}). \tag{23-132}$$

$\mathcal{H} + \boldsymbol{\varepsilon}\times\mathbf{p}/mc$ is, to order v/c, the magnetic field seen by the particle in its rest frame. The first term on the right is the interaction of the magnetic moment of the particle with this field.

The second term on the right side of (23-132) can be explained by a semiclassical argument starting from the fact that a spin on a particle traveling on a curved orbit appears to precess; this is called the *Thomas precession.* The rate of precession $\delta\boldsymbol{\Omega}$ of the spin on a particle with velocity $\mathbf{v}$ ($\ll c$) and acceleration $\dot{\mathbf{v}}$ is, according to Thomas,

$$\delta\boldsymbol{\Omega} = \dot{\mathbf{v}}\times\frac{\mathbf{v}}{2c^2}; \tag{23-133}$$

we shall derive this result below. This precession increases the classical rotational kinetic energy by an amount

$$\delta E = \mathbf{S} \cdot \delta \boldsymbol{\Omega} \tag{23-134}$$

where $\mathbf{S}$ is the spin angular momentum. [Equation (23-134) is analogous to the statement that increasing the linear velocity of a particle with momentum $\mathbf{p}$ increases its translational kinetic energy by $\delta(mv^2/2) = m\mathbf{v} \cdot \delta\mathbf{v} = \mathbf{p} \cdot \delta\mathbf{v}$.] Now to lowest order in v/c, the acceleration $\dot{\mathbf{v}}$ is given by Newton's equation

$$m\dot{\mathbf{v}} = e\boldsymbol{\varepsilon}. \tag{23-135}$$

Thus

$$\delta E = \frac{\mathbf{S} \cdot (e\boldsymbol{\varepsilon} \times \mathbf{v})}{2mc^2} = \frac{e\hbar}{4m^2c^2}\, \boldsymbol{\sigma} \cdot (\boldsymbol{\varepsilon} \times \mathbf{p}); \tag{23-136}$$

this is just the second term in (23-132).

We can derive (23-133) as follows. If $|0\rangle$ is the internal state of a particle with velocity $\mathbf{v}$, as seen in its own rest frame, then according to (23-71), the state as seen in the lab frame (moving with velocity $-\mathbf{v}$ with respect to the particle) is

$$|\mathbf{v}\rangle = \mathcal{L}_{-\mathbf{v}}^{-1}|0\rangle = \mathcal{L}_{\mathbf{v}}|0\rangle. \tag{23-137}$$

Suppose that the particle accelerates to velocity $\mathbf{v} + \delta\mathbf{v}$ while its internal state remains $|0\rangle$ in its rest frame. Then its new state as seen in the lab is

$$|\mathbf{v}+\delta\mathbf{v}\rangle = \mathcal{L}_{\mathbf{v}+\delta\mathbf{v}}|0\rangle = \mathcal{L}_{\mathbf{v}+\delta\mathbf{v}}\mathcal{L}_{\mathbf{v}}^{-1}|\mathbf{v}\rangle. \tag{23-138}$$

Now we leave it as an exercise to verify that for $\delta\mathbf{v} \perp \mathbf{v}$ and $v \ll c$:

$$\mathcal{L}_{\mathbf{v}+\delta\mathbf{v}}\mathcal{L}_{\mathbf{v}}^{-1} \approx e^{-i\delta\boldsymbol{\theta}\cdot\boldsymbol{\sigma}/2}\, e^{\boldsymbol{\alpha}\cdot(\delta\mathbf{v}/2c)} \tag{23-139}$$

where

$$\delta\boldsymbol{\theta} = \delta\mathbf{v} \times \frac{\mathbf{v}}{2c^2}. \tag{23-140}$$

(23-139) is the product of an infinitesimal Lorentz transformation by velocity $\delta\mathbf{v}$ and an infinitesimal rotation by angle $|\delta\theta|$ about $\delta\theta$. Thus the state $|\mathbf{v} + \delta\mathbf{v}\rangle$ appears to be the old state $|\mathbf{v}\rangle$ slightly Lorentz transformed and slightly rotated by $\delta\boldsymbol{\theta}$. If the particle is constantly accelerating then the internal state appears in the lab to be precessing with angular velocity

$$\delta\Omega = \frac{\delta\theta}{\delta t} = \frac{\delta \mathbf{v}}{\delta t} \times \frac{\mathbf{v}}{2c^2} = \dot{\mathbf{v}} \times \frac{\mathbf{v}}{2c^2};$$

this is the Thomas precession.

The Thomas contribution to the spin-orbit term is present even when the forces acting on the particle are not electromagnetic. Thus a neutral spin $\frac{1}{2}$ particle in a potential V will experience a spin-orbit interaction

$$\frac{\hbar}{4m^2c^2}\,\sigma \cdot (\mathbf{F} \times \mathbf{p}), \tag{23-141}$$

where $\mathbf{F} = -\nabla V$. In a sense then the presence of a spin-orbit splitting is a test of whether the internal degree of freedom of the particle is in fact an angular momentum, governed by relativistic kinematics, or is instead a nonkinematic degree of freedom such as isotopic spin.

SECOND-ORDER DIRAC EQUATION

It is useful for some problems to have a form of the Dirac equation that is second order in time. To derive this, let us first write the Dirac equation as

$$\beta\left(i\hbar\frac{\partial}{\partial t} - H\right)\psi(\mathbf{r}, t) = 0 \tag{23-142}$$

where

$$H = c\alpha \cdot \left(\frac{\hbar}{i}\nabla - \frac{e}{c}\mathbf{A}\right) + \beta mc^2 + e\Phi. \tag{23-143}$$

Then let us define the operator P to be

$$P = \frac{\beta(i\hbar\partial/\partial t - H) + 2mc^2}{2mc^2}, \tag{23-144}$$

and multiply (23-142) on the left by P. We find after some algebra that

$$\left[\frac{1}{c^2}\left(i\hbar\frac{\partial}{\partial t} - e\Phi\right)^2 - \left(\frac{\hbar}{i}\nabla - \frac{e}{c}\mathbf{A}\right)^2 - m^2c^2 + \frac{e\hbar}{c}(\sigma \cdot \mathcal{H} - i\alpha \cdot \mathcal{E})\right]\psi = 0. \tag{23-145}$$

In deriving (23-143) we have used the relations

$$\left[\left(\frac{\hbar}{i}\nabla-\frac{e}{c}\mathbf{A}\right)\cdot\boldsymbol{\alpha}\right]^2=\left(\frac{\hbar}{i}\nabla-\frac{e}{c}\mathbf{A}\right)^2-\frac{e\hbar}{c}\boldsymbol{\sigma}\cdot\boldsymbol{\mathcal{H}} \tag{23-146}$$

and

$$\left[\frac{h}{i}\nabla-\frac{e}{c}\mathbf{A},\ i\hbar\frac{\partial}{\partial t}-e\Phi\right]=-i\hbar e\boldsymbol{\mathcal{E}}. \tag{23-147}$$

Equation (23-145) is the Klein-Gordon equation with an additional $\boldsymbol{\sigma}\cdot\boldsymbol{\mathcal{H}} - i\boldsymbol{\alpha}\cdot\boldsymbol{\mathcal{E}}$ term representing the direct coupling of the electromagnetic fields to the magnetic (and electric) moment of the particle.

Now while every solution to the Dirac equation is a solution of (23-145), the converse is not always true. However if ψ is a solution to (23-145), then

$$\varphi = P\psi \tag{23-148}$$

solves the Dirac equation. To see this we write the second-order equation as

$$P\beta\left(i\hbar\frac{\partial}{\partial t}-H\right)\psi=\beta\left(i\hbar\frac{\partial}{\partial t}-H\right)P\psi=0. \tag{23-149}$$

Thus (23-145) is equivalent to the equation

$$i\hbar\frac{\partial}{\partial t}(P\psi)=H(P\psi), \tag{23-150}$$

so that $\varphi = P\psi$ solves the Dirac equation. If ψ is already a solution of the Dirac equation then $P\psi = \psi$. Thus P acts as a projection operator, reducing solutions of the second-order equation to solutions of the first-order Dirac equation.

DIRAC HYDROGEN ATOM

One enormous success of the Dirac equation was that it correctly described the fine structure in the spectrum of the hydrogen atom. To see this let us solve for the energy levels of a relativistic electron in a Coulomb potential

$$e\Phi=-\frac{Ze^2}{r}. \tag{23-151}$$

Rather than solving the Dirac equation directly, a problem discussed in all standard books, let us find the energy levels from the second order equation (23-145).[4] For a stationary state of energy E in the potential (23-151), Eq. (23-145) becomes

$$\left[\frac{1}{c^2}\left(E+\frac{Ze^2}{r}\right)^2-\left(\frac{\hbar\nabla}{i}\right)^2-m^2c^2+\frac{i\hbar Ze^2}{r^2c}\,\alpha_r\right]\psi=0 \tag{23-152}$$

where

$$\alpha_r = \boldsymbol{\alpha}\cdot\hat{\mathbf{r}}. \tag{23-153}$$

If we write

$$\left(\frac{\hbar\nabla}{i}\right)^2 = -\frac{\hbar^2}{r^2}\frac{\partial^2}{\partial r^2}\,r^2+\frac{L^2}{r^2}, \tag{23-154}$$

Equation (23-152) becomes

$$\left[\frac{E^2-m^2c^4}{c^2}+\frac{2EZe^2}{rc^2}+\frac{\hbar^2}{r^2}\frac{\partial^2}{\partial r^2}\,r^2 - \frac{L^2-(Ze^2/c)^2-i\hbar(Ze^2/c)\alpha_r}{r^2}\right]\psi=0. \tag{23-155}$$

This equation is almost in the same form as the Klein-Gordon equation for a Coulomb potential, and by a few tricks we shall be able to reduce it to that equation.

To do this, let us define an operator

$$K=\beta\left(1+\boldsymbol{\sigma}\cdot\frac{\mathbf{L}}{\hbar}\right). \tag{23-156}$$

K has the following properties, which we leave as an exercise to show:

$$[K,\boldsymbol{\alpha}\cdot\mathbf{p}]=0,\qquad [K,\boldsymbol{\alpha}\cdot\mathbf{r}]=0,\qquad [K,r^2]=0 \tag{23-157}$$

$$[K,\mathbf{J}]=0,\qquad \mathbf{J}=\mathbf{L}+\frac{\hbar\boldsymbol{\sigma}}{2}. \tag{23-158}$$

Consequently K commutes with the Hamiltonian for the relativistic hydrogen atom

---

[4] This method is given in a paper by P.C. Martin and R.J. Glauber, *Phys. Rev.* **109**, 1307 (1958); the explicit calculations of the wave functions can be found there.

$$H = c\alpha \cdot p + \beta mc^2 - \frac{Ze^2}{r}. \tag{23-159}$$

This means that K is a constant of the motion, and since it commutes with the total angular momentum **J**, we may label the energy levels of the hydrogen atom by the eigenvalues of K, J^2, and J_z. In fact, K is a constant of the motion for any spherically symmetric, spin independent potential. Physically K measures the degree to which the spin and orbital angular momentum of the particle are aligned.

We can find the eigenvalues, k, of K by noting first that

$$K^2 = 1 + \left(\sigma \cdot \frac{L}{\hbar}\right)^2 + 2\sigma \cdot \frac{L}{\hbar} = 1 + \frac{L^2}{\hbar^2} + \sigma \cdot \frac{L}{\hbar} = \frac{J^2}{\hbar^2} + \frac{1}{4}. \tag{23-160}$$

Thus since the eigenvalues of J^2 are $\hbar^2 j(j+1)$, the eigenvalues of K^2 must be

$$k^2 = j(j+1) + \frac{1}{4} = \left(j + \frac{1}{2}\right)^2 . \tag{23-161}$$

Furthermore from the fact that

$$\{K, \gamma_5\} = 0 \tag{23-162}$$

we see that if k is an eigenvalue of K:

$$K|k\rangle = k|k\rangle,$$

then

$$K\gamma_5|k\rangle = -\gamma_5 K|k\rangle = -k\gamma_5|k\rangle,$$

so that $-k$ is also an eigenvalue of K. Thus the eigenvalues of K must be

$$k = \pm 1, \pm 2, \pm 3, \ldots, \tag{23-163}$$

since $j = 1/2, 3/2, \ldots$. Zero is not an eigenvalue of K. Also any eigenstate of k is also an eigenstate of J^2 with eigenvalue

$$j = |k| - \frac{1}{2}. \tag{23-164}$$

Now to find the energy eigenvalues from Eq. (23-155) let us define an operator

$$\Lambda = -\beta K - i\frac{Ze^2}{\hbar c}\,\alpha_r. \tag{23-165}$$

This operator has the properties

$$[\Lambda . K] = 0, \qquad [\Lambda, J] = 0 \tag{23-166}$$

and

$$\Lambda^2 = K^2 - \left(\frac{Ze^2}{\hbar c}\right)^2. \tag{23-167}$$

The reason Λ is useful is that

$$\hbar^2\Lambda(\Lambda+1) = L^2 - \left(\frac{Ze^2}{c}\right)^2 - i\hbar\left(\frac{Ze^2}{c}\right)\alpha_r \tag{23-168}$$

is just the operator coefficient of $1/r^2$ in Eq. (23-155). Thus that equation can be written as

$$\left[\frac{E^2 - m^2c^4}{\hbar^2c^2} + \frac{2EZe^2}{r\hbar^2c^2} + \frac{1}{r^2}\frac{\partial^2}{\partial r^2}r^2 - \frac{\Lambda(\Lambda+1)}{r^2}\right]\psi(\mathbf{r}) = 0. \tag{23-169}$$

In this form the equation is remarkably like the spin zero case where we had the number $l'(l' + 1)$ in place of the operator $\Lambda(\Lambda + 1)$ above the r^2. This suggests that we can solve (23-169) as follows.

If $\psi(\mathbf{r})$ is an eigenstate of Λ, then $\Lambda(\Lambda + 1)$ in Eq. (23-169) becomes a number, which we call $l'(l' + 1)$; thus the energy eigenvalues are given in terms of l', as in the spin zero case, by

$$E = \frac{mc^2}{[1 + (Ze^2/\hbar cn')^2]^{1/2}} \tag{23-170}$$

where

$$n' = l' + 1 + \nu, \qquad \nu = 0, 1, 2, \ldots . \tag{23-171}$$

Furthermore, since Λ, K, J^2, and J_z all commute, we can construct the solutions to (23-169) to be eigenstates of K, J^2, and J_z as well as Λ.

There is only one minor hitch in this method, namely the fact that Λ fails to commute with H, the Dirac Hamiltonian. This implies that the solutions ψ of (23-165) are not directly the eigenfunctions of H; rather we must operate on ψ with [cf. (23-144)]

$$P = \frac{\beta(E + Ze^2/r) - \beta c\boldsymbol{\alpha}\cdot\mathbf{p} + mc^2}{2mc^2} \tag{23-172}$$

to form the eigenfunctions of H. The eigenvalues E of (23-169) are the eigenvalues of H, since from (23-150), the second order equation

(23-169) is equivalent to

$$H(P\psi) = E(P\psi). \tag{23-173}$$

Since P and Λ don't commute, $P\psi$ will generally be a linear combination of different Λ eigenfunctions.

To find the energy levels then we need to know the possible eigenvalues of Λ. Consider an eigenstate of Λ and K, with eigenvalue k. From Eq. (23-167) we see that Λ acting on this state gives the number λ^2, where

$$\lambda = +\sqrt{k^2 - \left(\frac{Ze^2}{\hbar c}\right)^2}. \tag{23-174}$$

Thus the possible eigenvalues of Λ are $\pm\lambda$. When $\Lambda(\Lambda + 1)$ acts on a Λ, K eigenstate it becomes $\pm\lambda(\pm\lambda + 1)$. Writing

$$\pm\lambda(\pm\lambda + 1) = l'(l' + 1),$$

we find two possible l' values for each eigenvalue of Λ:

$$l' = \lambda, \quad -\lambda - 1 \qquad \text{for } \Lambda = \lambda$$

$$l' = -\lambda, \quad \lambda - 1 \qquad \text{for } \Lambda = -\lambda.$$

For each eigenvalue of Λ the two l' solutions add up to -1. The lesser of the solutions, $-\lambda$ and $-\lambda - 1$, may be excluded by the same argument we used in the spin zero case to exclude the lesser of the two solutions for l' in terms of l.

Thus we have two cases to consider:

$$\Lambda = \lambda, \qquad l' = \lambda$$

or

$$\Lambda = -\lambda, \qquad l' = \lambda - 1.$$

The possible energy eigenvalues are given in terms of l' by (23-170). It is convenient to write

$$n' = n - |k| + \lambda = n - j - \frac{1}{2} + \sqrt{\left(j + \frac{1}{2}\right)^2 - \left(\frac{Ze^2}{\hbar c}\right)^2}. \tag{23-175}$$

The n, which is basically the nonrelativistic principal quantum number for the hydrogen levels takes on the values

$$n = |k|, |k| + 1, |k| + 2, \ldots \quad \text{for } \Lambda = -\lambda \tag{23-176a}$$

$$n = |k|+1, |k|+2, |k|+3, \ldots \quad \text{for } \Lambda = \lambda. \tag{23-176b}$$

For each value of $j = |k| - 1/2$ we find therefore two sequences of energy levels:

	. . .	. . .
	$n = |k| + 2$	$|k| + 2$
E ↑	$|k| + 1$	$|k| + 1$
	$|k|$	

The left sequence corresponds to $\Lambda = -\lambda$ and the right sequence to $\Lambda = +\lambda$. The energies of these levels are given by

$$E = mc^2 \left\{1 + \frac{(Ze^2/\hbar c)^2}{\left[n - j - (1/2) - \sqrt{(j+1/2)^2 - (Ze^2/\hbar c)^2}\right]^2}\right\}^{-1/2}; \tag{23-177}$$

for Z = 1, these are the possible energy levels of the hydrogen atom with total angular momentum j.

It is interesting to note that the formula (23-177) for the energy levels of the spin $1/2$ particle is the same as we found for a spin zero particle, only with l for spin zero replaced by j for spin $1/2$. In order that the energy be real we are restricted to

$$j + \frac{1}{2} < \frac{Ze^2}{\hbar c}.$$

For $j = 1/2$ this requires $Z < 137$.

While the solutions to the Dirac equation are not Λ eigenstates they are still K eigenstates, since $[K, H] = 0$. We would like to learn then how to classify the Dirac eigenfunctions, $P\psi$, by eigenvalues of K. We can accomplish this simply by peeking at the nonrelativistic limit. Then $\Lambda \to -\beta K$ and for a positive energy nonrelativistic state, $\beta = +1$. Thus in this limit, $\Lambda = -K$. We expect then that the solutions of the second-order equation with one sign of Λ will correspond to solutions of the first-order equation with the opposite sign of K. Thus we expect that the eigenfunctions of H labeled by the sequence (23-176a) have $k > 0$, while those labeled by the sequence (23-176b) have $k < 0$.

The total orbital angular momentum **L** is not a constant of the motion and neither is L^2. However it is convenient to label the energy levels by the l value that they would have in the nonrelativistic limit. To find this l value we note that K obeys the relation

$$K(K-\beta) = \frac{L^2}{\hbar^2}. \tag{23-178}$$

In the nonrelativistic limit $\beta \approx 1$, and therefore if we *define* a nonnegative integer l by

$$k(k-1) = l(l+1), \tag{23-179}$$

then l becomes the total orbital angular momentum quantum number in the nonrelativistic limit. Solving (23-179) for l in terms of k we have

$$l = k - 1 = j - \frac{1}{2} \text{ for } k > 0$$

$$l = |k| = j + \frac{1}{2} \text{ for } k < 0. \tag{23-180}$$

As we mentioned K measures the alignment of the spin and the orbital angular momentum. For $k > 0$, they are essentially parallel and so $j = l + 1/2$, while for $k < 0$ they are essentially antiparallel, and $j = l - 1/2$.

From a detailed calculation of the wave functions one sees that the upper two components of the wave function, the large components, are eigenstates of total orbital angular momentum with eigenvalue l, while the lower two components, the small components, are eigenstates of total orbital angular momentum with eigenvalue $l + 1$ for $k > 0$ and $l - 1$ for $k < 0$.

The complete level scheme of the hydrogen atom for n = 1, 2, and 3 then looks as follows:

n = 3:	k = 1 $S_{1/2}$	k = 2 $P_{3/2}$; k = −1 $P_{1/2}$	k = 3 $D_{5/2}$; k = −2 $D_{3/2}$
n = 2:	k = 1 $S_{1/2}$	k = 2 $P_{3/2}$; k = −1 $P_{1/2}$	
n = 1:	k = 1 $S_{1/2}$		

The complete degeneracy of a given n in the nonrelativistic case is lifted by relativistic effects. The $1S_{1/2}$, $2P_{3/2}$, $3D_{5/2}$, $4F_{7/2}$, etc.

levels are nondegenerate. There still remains a degeneracy between the $2S_{1/2}$ and $2P_{1/2}$, the $3S_{1/2}$ and $3P_{1/2}$, the $3P_{3/2}$ and $3D_{3/2}$, etc., levels. We shall come back to this degeneracy shortly.

For small $Ze^2/\hbar c$ we may expand (23-177); then [cf. Eq. (22-142)]

$$E = mc^2 - \frac{mZ^2e^2}{2\hbar^2 n^2}\left[1 + \frac{Z^2\alpha^2}{n^2}\left(\frac{n}{j+1/2} - \frac{3}{4}\right)\right], \tag{23-181}$$

where $\alpha = e^2/\hbar c$. The $nP_{3/2} - nP_{1/2}$ splitting is $+(Z\alpha)^2/2n^3$ Rydbergs; similarly the $nD_{5/2} - nD_{3/2}$ splitting is $+(Z\alpha)^2/6n^3$ Rydbergs. In hydrogen, $Z = 1$, the $2P_{3/2}$ level lies about 10^4 Mc above the $2P_{1/2}$. For each nl the level with the smaller j lies below that with the larger j.

We noticed above that all the levels, with the exception of the $1S_{1/2}$, $2P_{3/2}$, $3D_{5/2}$, etc., are two-fold degenerate. The two states of a degenerate pair are eigenstates of K with opposite eigenvalues, e.g., the $2P_{3/2}$ has $k = 2$ while the $2D_{3/2}$ has $k = -2$. There is a simple mathematical reason for this structure of the levels. To see this consider the operator[5]

$$B = \gamma_5\left(\Lambda + \frac{KH}{mc^2}\right). \tag{23-182}$$

It is a straightforward exercise to verify that B is Hermitian, and that

$$[B, H] = 0, \qquad \{B, K\} = 0. \tag{23-183}$$

Thus B is also a constant of the motion. However since it fails to commute with K, the energy eigenstates of the hydrogen atom cannot be both K and B eigenstates simultaneously.

Suppose that $|k, E\rangle$ is an eigenstate of K and H:

$$K|k, E\rangle = k|k, E\rangle$$

$$H|k, E\rangle = E|k, E\rangle. \tag{23-184}$$

Then the state $B|k, E\rangle$ is again an eigenstate of K and H, but with eigenvalues $-k$ and E. To see this we use that fact that B commutes with H but anticommutes with K. Thus

$$HB|k, E\rangle = BH|k, E\rangle = EB|k, E\rangle$$

$$KB|k, E\rangle = -BK|k, E\rangle = -kB|k, E\rangle. \tag{23-185}$$

[5] The discovery of this constant of the motion is due to M.H. Johnson and B.A. Lippmann, *Phys. Rev.* 78, 329(A) (1950).

This means that if the state $B|k, E\rangle$ does not vanish, then it is a different state with the same energy, but with the opposite eigenvalue of K. This explains the two-fold degeneracy of the hydrogen levels, e.g., B acting on the $3P_{3/2}$ produces the $3D_{3/2}$ level, and vice versa. The reason that the $1S_{1/2}$, $2P_{3/2}$, ..., levels are nondegenerate is that B acting on their wave functions gives zero. The point is that these levels are in fact Λ eigenstates, with eigenvalue $-\lambda$, and have energy $mc^2/[1 + (Ze^2/\hbar c\lambda)^2]^{1/2}$. Thus for these levels

$$\Lambda + \frac{K\mathcal{H}}{mc^2} \to -\lambda + |k|E = 0.$$

In summary then, the operators B and K and J are the constants of the motion of the Dirac hydrogen atom, and their algebraic properties explain all the degeneracy. Each level has, of course, a $(2j + 1)$-fold degeneracy from rotational invariance.

HYPERFINE STRUCTURE

The actual level scheme in hydrogen is somewhat modified from the results of the Dirac equation. On the one hand the two-fold degeneracy is removed by the interaction of the electron with the vacuum fluctuations of the electromagnetic radiation field. The shift of the levels due to this effect is called the *Lamb shift*. On the other hand there is a hyperfine splitting of each level into two, due to the interaction of the electron with the magnetic moment of the proton. Let us discuss the *hyperfine* structure first.

We can find the hyperfine splitting of an s-state by a simple nonrelativistic first-order perturbation theory calculation. The interaction of the electron spin with the magnetic moment of the proton is

$$H' = \frac{|e|\hbar}{2mc}\,\sigma \cdot \mathcal{H}(r) = \mu_B \sigma \cdot \mathcal{H}(r) \tag{23-186}$$

where $\mathcal{H}$ is the magnetic field due to the magnetic moment of the proton. This magnetic moment is

$$\mathfrak{M}_p = \frac{|e|\hbar g_p}{4Mc}\,\sigma_p = \frac{1}{2}\, g_p \mu_p \sigma_p \tag{23-187}$$

where g_p is the gyromagnetic ratio of the proton, M is the proton mass and $\hbar\sigma_p/2$ is the spin of the proton. The magnetic field from this moment is, assuming the proton to be localized at the origin:

$$\mathcal{H}(\mathbf{r}) = -\nabla\times(\sigma_p\times\nabla)\frac{\mu_p}{2r}g_p. \tag{23-188}$$

Thus

$$\begin{aligned} H' &= -g_p\mu_B\mu_p\,\sigma\cdot(\nabla\times(\sigma_p\times\nabla))\frac{1}{2r} \\ &= -g_p\mu_B\mu_p[\sigma\cdot\sigma_p\nabla^2-(\sigma\cdot\nabla)(\sigma_p\cdot\nabla)]\frac{1}{2r} \end{aligned} \tag{23-189}$$

The first-order shift of a level is

$$\langle H'\rangle = -g_p\mu_B\mu_p\int d^3r\,|\psi(\mathbf{r})|^2[\langle\sigma\cdot\sigma_p\rangle\nabla^2-\langle(\sigma\cdot\nabla)(\sigma_p\cdot\nabla)\rangle]\frac{1}{2r} \tag{23-190}$$

where the brackets on the right denote the expectation value in the spin state of the proton and electron; $\psi(\mathbf{r})$ is the nonrelativistic wave function of the level. Let us consider only s-states. It is clear from the spherical symmetry of an s-state that the term $\langle(\sigma\cdot\nabla)(\sigma_p\cdot\nabla)\rangle$ reduces to $(1/3)\langle\sigma\cdot\sigma_p\rangle\nabla^2$. Thus

$$\begin{aligned} \langle H'\rangle &= -\frac{1}{3}\mu_B\mu_p g_p\int d^3r\,|\psi(\mathbf{r})|^2\langle\sigma\cdot\sigma_p\rangle\nabla^2\frac{1}{r} \\ &= \frac{4\pi}{3}\mu_B\mu_p g_p\langle\sigma\cdot\sigma_p\rangle|\psi(0)|^2. \end{aligned} \tag{23-191}$$

Now for a hydrogen atom s-state

$$|\psi(0)|^2 = \frac{1}{\pi(na_0)^3}, \tag{23-192}$$

where n is the principal quantum number. Thus (23-191) becomes

$$\langle H'\rangle = \frac{2}{3}\left(\frac{e^2}{2a_0}\right)g_p\frac{m}{M}\frac{\alpha^2}{n^3}\langle\sigma\cdot\sigma_p\rangle. \tag{23-193}$$

For a relative triplet spin state $\langle\sigma\cdot\sigma_p\rangle = 1$ while for a singlet $\langle\sigma\cdot\sigma_p\rangle = -3$. Consequently the singlet state lies lower than the triplet.

The total splitting of the ground state is thus

$$\Delta E = \frac{8}{3}\left(\frac{e^2}{2a_0}\right)g_p\frac{m}{M}\alpha^2; \tag{23-194}$$

this corresponds to radiation of 21 cm wavelength. It is the transition from the triplet to the singlet ground state that radio astronomers listen to. (23-194) agrees, to this order in α, with the result of a relativistic calculation.

The singlet ground state is lowered three times as much as the triplet ground state is raised. Now the triplet state is three-fold degenerate. This means that the "center of gravity" of the ground state multiplet, i.e., the average energy of the four states, is, to first order, unchanged by the hyperfine interaction.

The hyperfine splitting of the $2S_{1/2}$ level is $1/8$ that of the ground state. It turns out that the splitting of the $2P_{1/2}$ level is $1/24$ that of the ground state.

THE LAMB SHIFT

The coupling

$$\hat{H}_{\text{int}} = -\frac{e}{c}\int d^3r\, \mathbf{j}(\mathbf{r}) \cdot \mathbf{A}(\mathbf{r}t) \tag{23-195}$$

of the electron to the quantum mechanical radiation field causes a shift in the energy levels of the electron in a hydrogen atom. This has the effect of removing the degeneracy due to the constant of the motion B. We can get a first understanding of this effect, and at the same time a glimpse at some of the fundamental difficulties in quantum electrodynamics, by doing a *nonrelativistic* second-order perturbation theory calculation.[6] Consider an electron in the state $|n\rangle$ with energy ε_n. Through the interaction (23-195) the electron is capable of spontaneously emitting a photon, going to some state $|n'\rangle$. This produces a second-order shift in the energy of $|n\rangle$ given by

$$\Delta E_n = \sum_{n'} \sum_{\mathbf{k}\lambda} \frac{|\langle n', \mathbf{k}\lambda|\hat{H}_{\text{int}}|n, 0\rangle|^2}{\varepsilon_n - \varepsilon_{n'} - ck} \tag{23-196}$$

where $|n, 0\rangle$ is the initial state with the electron in $|n\rangle$ and no photons present, and $|n', \mathbf{k}\lambda\rangle$ is the intermediate state with the electron in $|n'\rangle$ and one photon of momentum $\mathbf{k}$ and polarization λ present. The energy of this intermediate state is $\varepsilon_{n'} + ck$.

From Eq. (13-70) we have

---

[6]See also the simple calculation by T. Welton, *Phys. Rev.* 74, 1157 (1948), based directly on the idea that the electron interacts with the fluctuating zero-point electromagnetic field.

$$\langle n', k\lambda|\hat{H}_{int}|n, 0\rangle = -\frac{e}{c}\sqrt{\frac{2\pi\hbar^2 c^2}{\omega_k V}}\,\langle n'|j_k\cdot\lambda^*|n\rangle \tag{23-197}$$

where j_k is the kth Fourier component of the current $j(r)$; thus

$$\Delta E_n = \int \frac{d^3k}{(2\pi\hbar)^3}\frac{2\pi\hbar^2 e^2}{ck}\sum_{n'}\frac{\sum_\lambda |\langle n'|j_k\cdot\lambda^*|n\rangle|^2}{\varepsilon_n - \varepsilon_{n'} - ck}$$
$$= \int \frac{k^2 dk}{4\pi^2\hbar}\frac{e^2}{ck}\sum_{n'}\frac{\int d\Omega \sum_\lambda |\langle n'|j_k\cdot\lambda^*|n\rangle|^2}{\varepsilon_n - \varepsilon_{n'} - ck}. \tag{23-198}$$

In the dipole approximation we may replace j_k by $j_0 = p/m$, where p is the electron momentum operator. Then doing the angular integral and summing over polarizations we find

$$\int d\Omega \sum_\lambda |\langle n'|p\cdot\lambda^*|n\rangle|^2 = 4\pi\,\frac{2}{3}\,|\langle n'|p|n\rangle|^2. \tag{23-199}$$

The factor $^2/_3$ comes from the fact that there are only two independent polarizations for each k. Thus we have

$$\Delta E_n = \frac{2e^2}{3\pi\hbar c^3 m^2}\int_0^\infty \omega d\omega \sum_{n'}\frac{|\langle n'|p|n\rangle|^2}{\varepsilon_n - \varepsilon_{n'} - \omega} \tag{23-200}$$

where $\omega = ck$. Notice that the ω integral is divergent! This means that the interaction with the radiation field produces, according to this calculation, an *infinite* shift downward in the energy of the electron.

This result was a great theoretical difficulty for many years, and it was not really resolved until 1947 by Bethe, Schwinger, and Weisskopf.[7]

The point is that if one does a similar calculation for a free electron one again finds an infinite result. In the dipole approximation (23-200) gives, for a free electron in a momentum eigenstate $|p\rangle$:

$$\Delta E_p = -\frac{2e^2}{3\pi\hbar c^3 m^2}\sum_q |\langle q|p|p\rangle|^2 \int_0^\infty d\omega$$
$$= -\frac{2e^2}{3\pi\hbar c^3 m^2}\langle p|p^2|p\rangle \int_0^\infty d\omega \tag{23-201}$$
$$= -\frac{2e^2}{3\pi\hbar c^3 m^2}\int_0^\infty d\omega\; p^2,$$

[7] See H. Bethe, *Phys. Rev.* 72, 339 (1947).

since the operator p has only diagonal elements in the momentum basis. But notice that (23-201), the electromagnetic self-energy of a free electron of momentum p, is proportional to p^2. This means that we can interpret (23-201) as representing a shift in the mass of the electron. In other words, if we say that m_0 is the mass and $p^2/2m_0$ is the kinetic energy of a free electron of momentum p *neglecting* electromagnetic interactions, then the energy including (23-201) is

$$\left[\frac{1}{m_0} - \frac{4e^2}{3\pi\hbar c^3 m^2}\int_0^\infty d\omega\right]\frac{p^2}{2}. \tag{23-202}$$

But since we always make observations on free electrons with the interaction (23-195) always present, (23-202) must be just the expression we normally write down for the energy of a free electron, namely $p^2/2m$, where m is the observed mass. Thus we should make the identification

$$\frac{1}{m} = \frac{1}{m_0} - \frac{4e^2}{3\pi\hbar c^3 m^2}\int_0^\infty d\omega. \tag{23-203}$$

The electromagnetic self-energy can thus be interpreted as giving a shift in the mass of the electron from its "bare" value m_0 to its observed value m. This shift is called the *renormalization* of the mass.

We can then argue that the reason (23-200) is infinite is that it includes an infinite energy change that is already counted when we use the observed mass in the Hamiltonian rather than the bare mass. In other words, we should really start out with the Hamiltonian for the hydrogen atom in the presence of the radiation field given by

$$H = \frac{p^2}{2m_0} - \frac{e^2}{r} + H_{int}. \tag{23-204}$$

Then using (23-203) we can rewrite H as

$$H = \frac{p^2}{2m} - \frac{e^2}{r} + \left[H_{int} + \frac{2p^2e^2}{3\pi\hbar c^3 m^2}\int_0^\infty d\omega\right]. \tag{23-205}$$

Thus if we write the observed free particle mass in the kinetic energy (which we always do) we should not count that part of H_{int} that produces the mass shift, i.e., we should regard

$$H_{int} + \frac{2p^2e^2}{3\pi\hbar c^3 m^2}\int_0^\infty d\omega \tag{23-206}$$

as the effective interaction of an electron of renormalized mass m with the radiation field.

Returning then to the calculation of the level shift, we see that to first order in $e^2/\hbar c$ we must add the expectation value of the second term in (23-206) onto (23-200) in order to avoid counting the electromagnetic interaction twice, once in m and once in (23-200). Thus more correctly the shift of the level $|n\rangle$ is given by

$$\Delta E'_n = \frac{2e^2}{3\pi\hbar c^3 m^2} \int_0^\infty \omega d\omega \left(\sum_{n'} \frac{|\langle n'|p|n\rangle|^2}{\varepsilon_n - \varepsilon_{n'} - \omega} + \frac{\langle n|p^2|n\rangle}{\omega} \right).$$

By completeness

$$\langle n|p^2|n\rangle = \sum_{n'} |\langle n'|p|n\rangle|^2$$

so that

$$\Delta E'_n = \frac{2e^2}{3\pi\hbar c^3 m^2} \sum_{n'} |\langle n'|p|n\rangle|^2 \int_0^\infty d\omega \, \frac{(\varepsilon_{n'} - \varepsilon_n)}{\varepsilon_{n'} - \varepsilon_n + \omega}. \tag{23-207}$$

The ω integral is still divergent; but now only logarithmically. This divergence is not present in a more sophisticated relativistic calculation; we can imagine that such a calculation would yield a result like (23-207) only with an integrand that falls off more rapidly at high frequencies, $\omega \gtrsim mc^2$. We can simulate the results of such a calculation by the simple trick of cutting off the ω integral at $\omega = mc^2$. Then

$$\Delta E'_n = \frac{2e^2}{3\pi\hbar c^3 m^2} \sum_{n'} |\langle n'|p|n\rangle|^2 (\varepsilon_{n'} - \varepsilon_n) \ln \left| \frac{mc^2}{\varepsilon_{n'} - \varepsilon_n} \right|, \tag{23-208}$$

where we have neglected $\varepsilon_{n'} - \varepsilon_n$ compared with mc^2. [The absolute value occurs because the ω integral is really a principal value integral.]

At this point Eq. (23-208) must be evaluated numerically. However, the result can be cast into a more transparent form by defining an average excitation energy by

$$\ln|\varepsilon_m - \varepsilon_n|_{av} = \frac{\sum_{n'} |\langle n'|p|n\rangle|^2 (\varepsilon_{n'} - \varepsilon_n) \ln|\varepsilon_{n'} - \varepsilon_n|}{\sum_{n'} |\langle n'|p|n\rangle|^2 (\varepsilon_{n'} - \varepsilon_n)}. \tag{23-209}$$

Then

$$\Delta E'_n = \frac{2e^2}{3\pi\hbar c^3 m^2}\left(ln\,\frac{mc^2}{|\varepsilon_m - \varepsilon_n|_{av}}\right)\sum_{n'}|\langle n'|p|n\rangle|^2(\varepsilon_{n'} - \varepsilon_n); \tag{23-210}$$

Now letting H_0 be the usual hydrogen atom Hamiltonian we have

$$\begin{aligned}\sum_{n'}|\langle n'|p|n\rangle|^2(\varepsilon_{n'} - \varepsilon_n) &= \langle n|p(H_0 - \varepsilon_n)\cdot p|n\rangle \\ &= -\frac{1}{2}\langle n|[p,\cdot[p,H_0]]|n\rangle .\end{aligned} \tag{23-211}$$

The commutators give the second derivative of H_0 with respect to $\mathbf{r}$. Thus (23-211) is

$$\frac{\hbar^2}{2}\langle n|\nabla^2\left(\frac{-e^2}{r}\right)|n\rangle = -\frac{e^2\hbar^2}{2}\int d^3r|\psi(r)|^2\nabla^2\frac{1}{r} = 2\pi e^2\hbar^2|\psi(0)|^2. \tag{23-212}$$

For an s-state we have, using Eq. (23-192),

$$\Delta E'_n = \frac{1}{3\pi}\,\frac{e^2}{2a_0}\left(\frac{2\alpha}{n}\right)^3 ln\,\frac{mc^2}{|\varepsilon_m - \varepsilon_n|_{av}}\,. \tag{23-213}$$

Bethe has calculated that for the 2S level, $|\varepsilon_m - \varepsilon_n|_{av} = 17.8$ Ry. Thus the logarithm equals 7.63 and

$$\Delta E'_n = +1040 \text{ megacycles.} \tag{23-214}$$

The $2P_{1/2}$ levels turns out to be shifted downward by a few megacycles. The result (23-214) is in remarkable agreement with the observed $2S_{1/2} - 2P_{1/2}$ shift of 1057 Mc. A full quantum electrodynamic calculation, including a similar subtraction of two infinite quantities to take into account mass renormalization, gives striking agreement with experiment.[8]

The experiment of Lamb and Retherford[9] to observe the relative shift of the $2S_{1/2} - 2P_{1/2}$ levels made use of the fact that the $2S_{1/2}$ level has a very long lifetime. The reason is that the $2S_{1/2} \rightarrow 1S_{1/2}$ transition is forbidden by parity conservation. It can only occur if two photons are emitted; consequently the lifetime is $\sim 1/7$ sec. On the other hand, the $2P_{1/2}$ level decays to the ground state in $\sim 10^{-9}$ sec.

Now one can shorten the lifetime of the $2S_{1/2}$ level substantially by subjecting it to an external electric or magnetic field. This has

[8]See G.W. Erickson and D.R. Yennie, *Ann. Phys.* (N.Y.) 35, 271, 447 (1965).

[9]*Phys. Rev.* 72, 241 (1947).

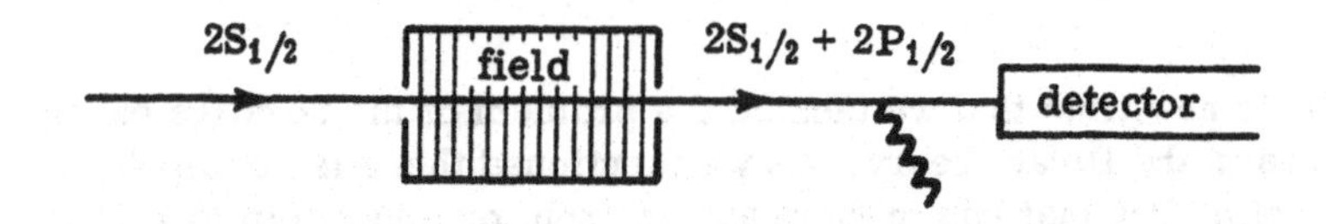

Fig. 23-1

Very schematic representation of the measurement of the Lamb shift.

the effect of mixing in a $2P_{1/2}$ component into the $2S_{1/2}$ state; the state can then decay to the ground state through its $2P_{1/2}$ component. Put another way, parity and angular momentum are no longer conserved in an external field, so that the $2S_{1/2} \rightarrow 1S_{1/2}$ transition is no longer forbidden. The amount of mixing of the $2P_{1/2}$ level into the $2S_{1/2}$ level depends on their energy difference [this difference occurs in energy denominators in perturbation theory] and on the strength of the external field.

In the experiment, they constructed a beam of excited hydrogen atoms; after a short distance of travel the only excited state left was the $2S_{1/2}$. They then passed this beam through a field, Fig. 23-1, which shortened the lifetime of the $2S_{1/2}$ state. Their detector was sensitive only to excited atoms, and by measuring the depletion of the $2S_{1/2}$ atoms caused by the field they were able to deduce the $2S_{1/2} - 2P_{1/2}$ level separation.

Taking into account both the Lamb shift and the hyperfine splitting one has the following level scheme for n = 2:

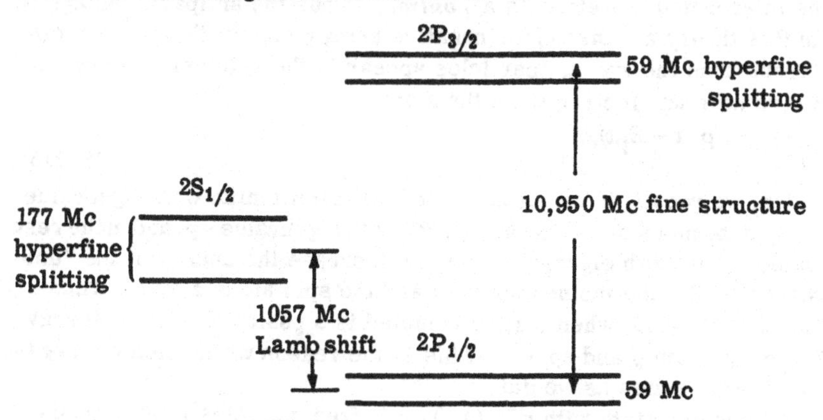

The 21 cm hyperfine splitting of the ground state corresponds to 1420 Mc.

DIRAC HOLE THEORY

It is now time that we tackled the problem of the negative energy states in the Dirac theory. As we mentioned there is no simple conservation law that can prevent an electron, or other spin $1/2$ particle in a positive energy state from making a radiative transition to a negative energy state. Dirac proposed an ingenious way out of this problem: since spin $1/2$ particles obey the exclusion principle, all one must do to insure stability is to say that the negative energy states are completely filled. Then a particle can't make a transition from a positive to a negative energy state for this would put two particles into the same (negative energy) state.

The vacuum state in this picture consists of an infinite sea of particles in negative energy states. The particle and charge density at every point are infinite, but this need not be a problem if we realize that one always only measures deviations from the vacuum. In the absence of a potential, the charge density of the negative energy sea is uniform, and one can argue that this charge density can produce no forces, since by isotropy, the forces have no special direction to point. [One can also argue that from Maxwell's equations $\nabla \cdot \boldsymbol{\varepsilon} = 4\pi\rho$; in order that this make sense, ρ must be thought of as the charge density measured with respect to the uniform infinite sea of charge.]

Now this theory gives us a special bonus. Suppose that we *remove* a negative energy electron from the vacuum. What is left behind is a *hole* in the negative energy sea. Measured with respect to the vacuum the hole appears to have positive charge and positive energy; since it is an absence of negative charge and negative energy, it can be interpreted therefore as a *positron*. Thus the antiparticles appear in this theory as *unoccupied* negative energy states; this is very different from the way antiparticles appear in the spin zero theory. If we remove an electron from the state

$$u_{p\sigma}^{(-)} e^{-i(\mathbf{p}\cdot\mathbf{r} - E_p t)/\hbar}, \tag{23-215}$$

which, we recall, is an *eigenstate* of the Hamiltonian with eigenvalue $-E_p$, of the momentum operator, $\hbar\nabla/i$, with eigenvalue $-\mathbf{p}$, and in the rest frame, of σ_z with eigenvalue $-\sigma$, we increase the energy of the "universe" by E_p, the momentum by $\mathbf{p}$ and the spin along z by σ. Thus the state (23-215) when it is unoccupied is a positron state of energy E_p, momentum $\mathbf{p}$ and spin σ. This is the reason we labeled the negative energy states as we did.

Consider a state with $n_{p\sigma}^{(+)}$ (= 0 or 1) free particles in each positive energy state, and $n_{p\sigma}^{(-)}$ (= 0 or 1) free particles in each negative energy state (23-215). The total energy of this system is

$$E_{tot} = \sum_{p\sigma} [E_p n_{p\sigma}^{(+)} + (-E_p) n_{p\sigma}^{(-)}]. \tag{23-216}$$

Let us write

$$\bar{n}_{p\sigma} = 1 - n_{p\sigma}^{(-)}; \tag{23-217}$$

$\bar{n}_{p\sigma}$ is the number of antiparticles since it equals 1 when the negative energy state p, σ is unoccupied. Then we can write (23-216) as

$$E_{tot} = \sum_{p\sigma}(E_p n_{p\sigma}^{(+)} + E_p \bar{n}_{p\sigma}) + \sum_{p\sigma} (-E_p). \tag{23-218}$$

The last term, $\sum_{p\sigma} (-E_p)$ is the energy of the vacuum. The energy measured with respect to the vacuum is thus

$$E'_{tot} = \sum_{p\sigma} E_p (n_{p\sigma}^{(+)} + \bar{n}_{p\sigma}); \tag{23-219}$$

this energy, a sum of $+E_p$ for each actual particle and antiparticle, is the energy one measures in an experiment and is nonnegative. Similarly we can write the total momentum of the state as

$$\mathbf{P}_{tot} = \sum_{p\sigma} [\mathbf{p} n_{p\sigma}^{(+)} + (-\mathbf{p}) n_{p\sigma}^{(-)}] = \sum_{p\sigma} \mathbf{p}(n_{p\sigma}^{(+)} + \bar{n}_{p\sigma}] + \sum_{p\sigma} (-\mathbf{p}). \tag{23-220}$$

The last term is the momentum of the vacuum, while $\sum_{p\sigma} \mathbf{p}(n_{p\sigma}^{(+)} + \bar{n}_{p\sigma})$ is the total momentum of the "physical" particles and antiparticles. Also the total electromagnetic current four vector, $J^\mu = ej^\mu$, is

$$\langle J^\mu \rangle = \sum_{p\sigma} (n_{p\sigma}^{(+)} ej_{p\sigma}^\mu + n_{p\sigma}^{(-)} ej_{p\sigma}^\mu),$$

where $j_{p\sigma}^\mu$ is the expectation value of the probability current four-vector $(c\rho, \mathbf{j})$ of a particle in the state $p\sigma$; it is the same for positive and negative energy states. In terms of $\bar{n}_{p\sigma}$,

$$\langle J^\mu \rangle = \sum_{p\sigma} [(ej_p{}^\mu) n_{p\sigma}^{(+)} + (-ej_p{}^\mu) \bar{n}_{p\sigma}] + \sum_{p\sigma} ej_p{}^\mu . \tag{23-222}$$

The first term is the experimentally observed current. We see from the $-e$ that appears in the first term that the antiparticles have a charge opposite to that of the particles. They have the same mass as the particles.

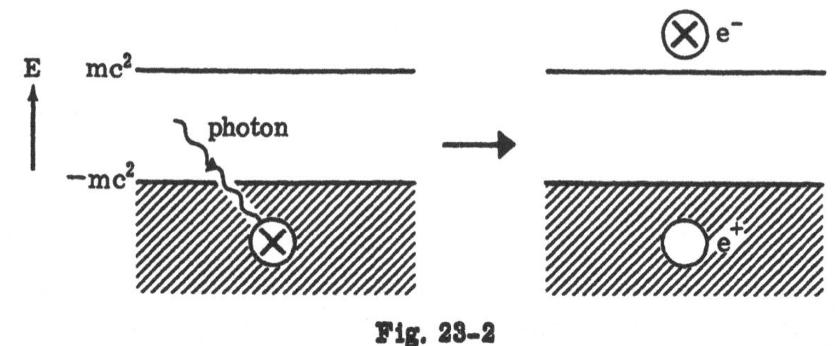

Fig. 23-2
Pair production by a photon in the hole theory.

This treatment of the negative energy states is called the Dirac *hole theory*. By introducing the hole theory to make the positive energy states stable against radiative collapse, one automatically has a theory that describes positive energy antiparticles as well. A very nice feature of this theory is the simple way it describes pair production. Suppose that a photon of energy > $2mc^2$ traveling through the vacuum is absorbed by a negative energy electron, Fig. 23-2, and the electron is excited to a positive energy state; what remains is a hole in the negative energy sea, i.e., a positron, and a positive energy electron. Thus pair production is simply the excitation of a particle from a negative to a positive energy state.

We can construct a simple one-to-one correspondence between the negative energy states of a particle of charge e and the positive energy states of a particle of the opposite charge. To do this, let us take the complex conjugate of the Dirac equation (23-58) for charge e. Then in the standard representation:

$$\left(-i\hbar\frac{\partial}{\partial t}-e\Phi\right)\psi^* = \left[c\boldsymbol{\alpha}^*\cdot\left(-\frac{\hbar}{i}\nabla-\frac{e}{c}\mathbf{A}\right)+\beta mc^2\right]\psi^*. \tag{23-223}$$

Now $\alpha_x^* = \alpha_x$, $\alpha_z^* = \alpha_z$, but $\alpha_y^* = -\alpha_y$. Thus from the anticommutation relations of the α's we can write

$$\boldsymbol{\alpha}^* = -\alpha_y\boldsymbol{\alpha}\alpha_y. \tag{23-224}$$

Hence (23-223) becomes

$$\left(i\hbar\frac{\partial}{\partial t}+e\Phi\right)\psi^* = \left[-c\alpha_y\boldsymbol{\alpha}\alpha_y\left(\frac{\hbar}{i}\nabla+\frac{e}{c}\mathbf{A}\right)-\beta mc^2\right]\psi^*;$$

multiplying through on the left by $\beta\alpha_y = \gamma_y$ we have then

$$\left(i\hbar\frac{\partial}{\partial t}+e\Phi\right)(\beta\alpha_y\psi^*)=\left[c\alpha\cdot\left(\frac{\hbar}{i}\nabla+\frac{e}{c}A\right)+\beta mc^2\right](\beta\alpha_y\psi^*). \tag{23-225}$$

Thus the *charge conjugate* wave function

$$\psi_c(\mathbf{r},t)=i\beta\alpha_y\psi^*(\mathbf{r},t)=i\gamma_y\psi^* \tag{23-226}$$

obeys the Dirac equation with the opposite sign of e, and hence describes a particle of charge −e and mass m.

Explicitly the charge conjugation operation is

$$\psi_c(\mathbf{r},t)=\begin{pmatrix}0&0&0&1\\0&0&-1&0\\0&-1&0&0\\1&0&0&0\end{pmatrix}\psi^*(\mathbf{r},t). \tag{23-227}$$

The complex conjugation clearly changes the sign of all momenta and energies, so that if ψ is a negative energy solution, ψ_c is a positive energy solution of the charge conjugate equation. The matrix in (21-228) simply inverts the spinor part of the wave function, changing the sign of the second and third components. This changes an eigenstate of σ_z into one with the opposite eigenvalue. Thus, for example, if

$$\psi(\mathbf{r},t)=u_{p\sigma}^{(-)}\,e^{i(\mathbf{p}\cdot\mathbf{r}-E_pt)/\hbar}$$

then

$$\psi_c(\mathbf{r},t)=u_{p\sigma}^{(+)}\,e^{-i(\mathbf{p}\cdot\mathbf{r}-E_pt)/\hbar} \tag{23-228}$$

The operation of charge conjugation gives us a way of writing down the wave function of an antiparticle; if a particular state of an antiparticle corresponds to a negative energy state $\psi(\mathbf{r},t)$ being unoccupied, then $\psi_c(\mathbf{r},t)$ is the wave function of the antiparticle. The probability current four-vector in a state ψ is

$$j^\mu=(c\rho,\mathbf{j})=c\bar{\psi}\gamma^\mu\psi.$$

The probability current four-vector for the charge conjugate state is

$$j_c^{\,\mu}=c\bar{\psi}_c\gamma^\mu\psi_c. \tag{23-229}$$

Now $\psi_c=i\gamma_y\psi^*$ so that

$$\bar{\psi}_c=(\psi^*)^\dagger(i\gamma_y)\gamma^0=-\bar{\psi}^*i\gamma_y \tag{23-230}$$

Thus

$$j_c{}^{\mu} = -c\bar{\psi}^* i\gamma_y \gamma^{\mu} i\gamma_y \psi^* .$$

But from (23-224) we have

$$\gamma_y \gamma^{\mu} \gamma_y = \gamma^{\mu *} \tag{23-231}$$

so that

$$j_c{}^{\mu} = c\psi^* (\gamma^{\mu})^* \psi^* = (j^{\mu})^* = j^{\mu} , \tag{23-232}$$

since j^{μ} is real. Thus the probability current of the charge conjugate state equals that of the original state. It is left as an exercise to show that the expectation value of the spin in the charge conjugate state ψ_c also equals that in ψ, i.e.,

$$\langle \sigma \rangle_c = \psi_c{}^{\dagger} \sigma \psi_c = \psi^{\dagger} \sigma \psi = \langle \sigma \rangle . \tag{23-233}$$

One very unsatisfactory feature of the Dirac hole theory is the very unsymmetrical way it treats particles and antiparticles. We could, instead of describing positrons as holes in a sea of negative electrons, have equally well begun with the Dirac equation for positrons; then the vacuum would consist of an infinite sea of negative energy positrons, and electrons would appear as holes in this positron sea. Thus we must conclude that the negative energy seas can't have any physical reality. Rather, the hole theory is a mathematical model that enables one to do correct "bookkeeping" within the framework of the single particle Dirac theory. Field theory, the second quantized Dirac theory, allows one to treat both particles and antiparticles on an equal footing.

It is important to realize that the definition of the vacuum in the hole theory depends upon the potentials that are present. If we have a time independent potential $\Phi(\mathbf{r})$, then to define the vacuum we must solve for the energy eigenstates, ψ_E, of the potential, and fill up those eigenstates with negative energy. Since the states ψ_E do not, in general, have a probability density that is uniform in space, the negative energy sea has a varying charge density. Compared with the free particle vacuum, the vacuum in the presence of Φ has positive charge wherever the electrostatic potential is very negative, and negative charge where it is very positive. Thus applying a potential to the vacuum has the effect of polarizing the vacuum, as in the spin zero theory. The vacuum polarization has the effect of modifying the effective potential felt by a particle. We are not really able to take this into account in the present theory.

Since the plane wave solutions form a complete set we can expand the eigenstates of $\Phi(\mathbf{r})$ in terms of them, viz.,

$$\psi_E(\mathbf{r}) = \sum_\sigma \int \frac{d^3p}{(2\pi\hbar)^3}\left[a_{p\sigma}u_{p\sigma}^{(+)}\sqrt{\frac{mc^2}{E_p}}\, e^{i\mathbf{p}\cdot\mathbf{r}/\hbar} + b_{p\sigma}u_{p\sigma}^{(-)}\sqrt{\frac{mc^2}{E_p}}\, e^{-i\mathbf{p}\cdot\mathbf{r}/\hbar}\right]. \tag{23-234}$$

The factor $\sqrt{mc^2/E_p}$ ensures that the plane wave states have unit normalization. Then if ψ_E is normalized to 1, the expansion coefficients obey

$$\sum_\sigma \int \frac{d^3p}{(2\pi\hbar)^3}\,(|a_{p\sigma}|^2 + |b_{p\sigma}|^2) = 1. \tag{23-235}$$

The question is, given a particle in the eigenstate ψ_E of the potential Φ, how can we interpret physically the meaning of $a_{p\sigma}$ and $b_{p\sigma}$. The point to realize is that in the vacuum with Φ present there is an amplitude for each state $u_{p\sigma}^{(+)}e^{i\mathbf{p}\cdot\mathbf{r}/\hbar}$ to be occupied, and an amplitude for each state $u_{p\sigma}^{(-)}e^{-i\mathbf{p}\cdot\mathbf{r}/\hbar}$ to be unoccupied. These are particles and antiparticles of the vacuum polarization. The coefficient $a_{p\sigma}$ is, in a sense, the "additional" amplitude for the positive energy free particle state $\mathbf{p}$, σ to be occupied when the particle is in the state ψ_E. In other words, $a_{p\sigma}$ is the amplitude for *removing* a particle of momentum $\mathbf{p}$ and spin σ and returning to the vacuum. On the other hand, $b_{p\sigma}$ is the additional amplitude for the free particle negative energy state $p\sigma$ to be occupied when the particle is in the state ψ_E. It is thus the amplitude for removing a particle from the state $u_{p\sigma}^{(-)}e^{-i\mathbf{p}\cdot\mathbf{r}/\hbar}$ and returning to the vacuum, i.e., $b_{p\sigma}$ is the amplitude for *adding* an antiparticle of momentum $\mathbf{p}$ and spin σ, and returning to the vacuum. The same interpretation applies equally to the u_p and v_p, Eq. (22-67), in the Klein-Gordon theory.

It is possible to define a vacuum only when the potentials are static or vary slowly in time, and are weak. If the potential can produce real pairs, or if it is so strong that the gap (= $2mc^2$ for free particles) between the negative and positive energy states is closed, then it is not at all clear what one means by a vacuum (cf. the Klein paradox).

The connection between spin and statistics is already apparent in the Klein-Gordon equation for spin zero particles, and the Dirac equation for spin $1/2$ particles. Recall that when we expanded an arbitrary state of a spin zero particle in positive and negative energy plane waves, we had a normalization condition, Eq. (22-69):

$$\int \frac{d^3p}{(2\pi\hbar)^3}\,(|u_p|^2 - |v_p|^2) = \pm 1. \tag{23-236}$$

$|u_p|^2$ is the density of positive charge in the state p, while $|v_p|^2$ is the density of negative charge in this state. There is no limitation in the theory on the magnitude of $|u_p|^2$ or $|v_p|^2$; Eq. (23-236) is the only constraint. This means that it is possible to have an arbitrarily large charge density of either sign in a momentum state p. From a particle point of view this means that one can have an arbitrary number of particles of either charge in any momentum state. Thus spin zero particles are not subject to the exclusion principle and must be bosons.

On the other hand, Eq. (23-235) places a limitation on the magnitude of both $|a_{p\sigma}|^2$ and $|b_{p\sigma}|^2$, because it has a plus sign, rather than a minus sign as in (23-236). The state $p\sigma$ for positive and negative energy can have only limited occupation, indicating that spin $^1/_2$ particles must be fermions. Of course, the hole theory was based on the assumption that the particles were fermions. It is not possible, as for spin $^1/_2$ bosons, to interpret the negative energy states in the Dirac theory in such a way that particles would be stable against radiative collapse into states with more and more negative energy. One can only have a positive energy, as (23-216), if spin $^1/_2$ particles are fermions.

Once one fills the negative energy states, the Dirac theory becomes a many particle theory. We cannot take into account, in the Dirac equation, interactions between these particles. Thus the Dirac equation gives valid results only when these interactions can be neglected. In the hydrogen atom, the modification of the Coulomb potential by vacuum polarization accounts for about 2.5% of the Lamb shift. Also one must worry about the fact that these particles obey the exclusion principle. Thus in calculating the motion of a given particle, say in a scattering problem, one must make sure that it never spends time in a state occupied by another particle, such as a particle of the negative energy sea. This problem was elegantly resolved by Feynman[10] who showed how one can correctly take the exclusion principle into account in the framework of the single particle Dirac equation.

PROBLEMS

1. Suppose that an electron of momentum p incident from the left strikes a (one-dimensional) potential barrier

$$e\phi(x) = V, \quad x > 0$$

$$e\phi(x) = 0, \quad x < 0.$$

[10] *Phys. Rev.* 76, 749 (1949).

(a) Calculate the reflection from the barrier for $V < 2mc^2$, and for $V > 2mc^2$ [Klein paradox]. Interpret the results for $V > 2mc^2$ in terms of the Dirac hole theory.

(b) Consider $0 < V \ll 2mc^2$, and no incident particle. Describe, as a function of x, the charge density due to the polarization of the vacuum by the potential.

2. Consider the relativistic hydrogen atom.

   (a) Show that $[K, \boldsymbol{\alpha}\cdot\mathbf{p}] = 0$, $[K, \boldsymbol{\alpha}\cdot\mathbf{r}] = 0$, $[K, H] = 0$ where $K = \beta(\boldsymbol{\sigma}\cdot\mathbf{L} + 1)$.

   (b) Let $B = \gamma_5(\Lambda + KH/mc^2)$, where $\Lambda = -\beta K - i(Ze^2/\hbar c)\alpha_r$. Show that $[B, H] = 0$, i.e., that B is also a constant of the motion, but that $BK + KB = 0$. What is $[B, \Lambda]$?

   (c) Verify that B is Hermitian.

3. The hyperfine interaction splits the ground state of the hydrogen atom. Calculate the lifetime of the upper level of the ground state doublet. This will require knowing the relativistic ground state wave function.

4. Estimate the part of the fine structure of the hydrogen atom that is due to the spin-orbit interaction.

INDEX

Printed in the United States
by Baker & Taylor Publisher Services